Apfel, Weigand (Eds.)

Bioorganometallic Chemistry

Also of Interest

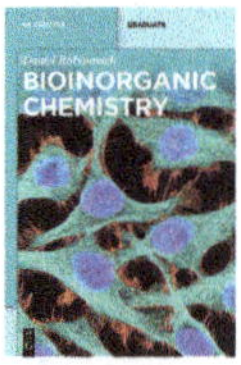

Bioinorganic Chemistry.
Rabinovich, 2020
ISBN 978-3-11-049204-0,
e-ISBN 978-3-11-049443-3

Organic Chemistry.
Fundamentals and Concepts
McIntosh, 2018
ISBN 978-3-11-056512-6,
e-ISBN 978-3-11-056514-0

Protein Chemistry.
Backman, 2020
ISBN 978-3-11-056616-1,
e-ISBN 978-3-11-056618-5

Rare Earth Chemistry.
Poettgen, Juestel, Strassert, 2020
ISBN 978-3-11-065360-1,
e-ISBN 978-3-11-065492-9

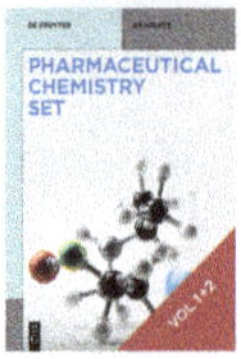

Pharmaceutical Chemistry.
Vol 1: Drug Design and Action
Vol 2: Drugs and Their Biological Targets
Campos Rosa, Camacho Quesada, 2017
Set: ISBN 978-3-11-056680-2

Bioorganometallic Chemistry

Edited by
Ulf Peter Apfel und Wolfgang Weigand

DE GRUYTER

Editors
Prof. Dr. Ulf-Peter Apfel
Ruhr-Universität Bochum
Fakultät für Chemie u. Biochemie
Anorganische Chemie
Universitätsstr. 150
44801 Bochum
Germany
ulf.apfel@rub.de
and
Fraunhofer UMSICHT
Osterfelder Strasse 3
46047 Oberhausen

Prof. Dr. Wolfgang Weigand
Friedrich-Schiller-Universität Jena
Institut für Anorganische und
Analytische Chemie
Humboldtstr. 8
07743 Jena
Germany
wolfgang.weigand@uni-jena.de

ISBN 978-3-11-049650-5
e-ISBN (PDF) 978-3-11-049657-4
e-ISBN (EPUB) 978-3-11-049397-9

Library of Congress Control Number: 2019951948

Bibliographic information published by the Deutsche Nationalbibliothek
The Deutsche Nationalbibliothek lists this publication in the Deutsche Nationalbibliografie; detailed bibliographic data are available on the Internet at http://dnb.dnb.de.

Cover image: Molekuul / Science Photo Library
Typesetting: Integra Software Services Pvt. Ltd.
Printing and binding: CPI books GmbH, Leck

www.degruyter.com

Contents

List of contributing authors —— VII

Wolfgang Weigand and Ulf-Peter Apfel
1 Introduction —— 1

Part I: **Reduction and Oxidation Catalysts**

Ulf-Peter Apfel, Wolfgang Weigand, Marius Horch, Ingo Zebger, Oliver Lenz and Takashi Fujishiro
2 Hydrogen development —— 13

Holger Dobbek
3 CO_2 Reduction —— 137

Andrew Jasniewski, Caleb Hiller, Yilin Hu and Markus Ribbe
4 The study of nitrogen reduction by nitrogenase —— 159

Xenia Engelmann, Teresa Corona and Kallol Ray
5 Oxidation of methane: methane monooxygenases —— 207

Part II: **Organometallic Enzyme Reactions**

Bernhard Kräutler
6 Organometallic B_{12}-derivatives in life processes —— 243

Paul A. Lindahl
7 Acetyl-coenzyme A synthase: a beautiful metalloenzyme —— 285

Part III: **Medical Applications**

Daniel Siegmund and Nils Metzler-Nolte
8 Medicinal organometallic chemistry —— 319

Part IV: **Spectroscopy Methods**

Leland B. Gee, Hongxin Wang and Stephen P. Cramer
9 Nuclear resonance vibrational spectroscopy —— 353

Maurice van Gastel
10 EPR spectroscopy — 395

Serena DeBeer
11 Introduction to X-ray spectroscopy – including X-ray absorption, X-ray emission and resonant inelastic X-ray scattering — 407

Index — 433

List of contributing authors

Teresa Corona
Department of Chemistry
Humboldt-Universität zu Berlin
Brook-Taylor-Strasse 2
12489 Berlin
Germany

Serena DeBeer
Max Planck Institute for Chemical Energy Conversion
Stiftstrasse 34-3
D-45470 Mülheim an der Ruhr
Germany
and
Cornell University
Department of Chemistry and Chemical Biology
Ithaca
NY 14853
USA

Holger Dobbek
Department of Biology
Structural Biology/Biochemistry
Humboldt-Universität zu Berlin
Germany

Xenia Engelmann
Department of Chemistry
Humboldt-Universität zu Berlin
Brook-Taylor-Strasse 2
12489 Berlin
Germany

Takashi Fujishiro
Department of Biochemistry and Molecular Biology
Saitama University
255 Shimo-Okubo
Sakura-ku
Saitama City
Saitama 338-8570
Japan

Leland B. Gee
Department of Chemistry
Stanford University
Stanford
CA 94305
USA

Dr. Marius Horch
Department of Physics
Freie Universität Berlin
Arnimallee 14
14195 Berlin
Germany

Andrew Jasniewski
Department of Molecular Biology and Biochemistry
University of California
Irvine
CA
USA

Oliver Lenz
TU Berlin Institute of Chemistry
Strasse des 17 .Juni 135
10623 Berlin
Germany

Ingo Zebger
TU Berlin Institute of Chemistry
Straße des 17. Juni 135
10623 Berlin
Germany

Caleb Hiller
Department of Molecular Biology and Biochemistry
University of California
Irvine
CA
USA

https://doi.org/10.1515/9783110496574-203

Yilin Hu
Department of Molecular Biology and Biochemistry
University of California
Irvine
CA
USA

Stephen P. Kramer
Department of Chemistry
University of California
Davis
CA 95616
USA

Bernhard Kräutler
Institute of Organic Chemistry and Centre of Molecular Biosciences
University of Innsbruck
A-6020 Innsbruck
Austria

Paul A. Lindahl
Department of Chemistry
Texas A&M University
College Station
TX 77843-3255
USA

Prof. Dr. Nils Metzler-Nolte
Inorganic Chemistry I – Bioinorganic Chemistry
Faculty of Chemistry and Biochemistry
Ruhr University Bochum
44797 Bochum
Germany

Kallol Ray
Department of Chemistry
Humboldt-Universität zu Berlin
Brook-Taylor-Strasse 2
12489 Berlin
Germany

Markus Ribbe
Department of Molecular Biology and Biochemistry
and
Department of Chemistry
University of California
Irvine
CA
USA

Dr. Daniel Siegmund
Fraunhofer UMSICHT
Osterfelder Strasse 3
46047 Oberhausen
Germany

Maurice van Gastel
Max Planck Institute for Chemical Energy Conversion
Stiftstrasse 34-3
D-45470 Mülheim an der Ruhr
Germany

Hongxin Wang
Department of Chemistry
University of California
Davis
California 95616
USA

Wolfgang Weigand and Ulf-Peter Apfel

1 Introduction

Two decades ago, bioorganometallic chemistry was still a relatively new offshoot and just a "side branch" of the well-established bioinorganic and organometallic chemistry, respectively. The bioorganometallic chemistry, at first associated mainly with a few examples of active centers of enzymes containing a metal carbon bond as vitamin B_{12} (Chapter 6), the three hydrogenases (Chapter 2) or acetyl-CoA synthase (Chapter 7), has developed almost explosively within the last 20 years. In this respect, it should be recalled that the first important review articles have been published by Jaouen et al. [1] and Beck et al. [2]. Nowadays there exists an enormous wealth of research in this field, and numerous important review articles and books [3, 4] have been published in the last few years targeting selected aspects of bioorganometallic chemistry. Meanwhile, this field has become its own discipline, which is taught within the scope of the regular curricula for master's degree programs in "chemistry" and "biochemistry."

This fact led us believe that it would be a useful and important undertaking to publish now a textbook for master and PhD students who are interested to gain a first insight into that subject. We hope that this book will be an inspiring introduction for researchers into this fascinating topic and will even further accelerate the development of this research field.

What is bioorganometallic chemistry?

In living systems, transition metal complexes provide the active centers of several enzymes, which catalyze important reactions. These enzymes are defined as metalloenzymes, the biorelevant transition metals V, Mo, W, Mn, Fe, Co, Ni, Cu, Zn occur in organisms only in trace amounts with mostly catalytic functions: These active centers are involved, for example, in H^+/H_2 conversion by hydrogenases, in nitrogen fixation by the nitrogenases, in interconversion of CO_2 and CO by the CO dehydrogenases and in methylation of CO by the acetyl CoA synthase. The understanding of the functioning of these metalloenzymes and the design of active bioinspired molecules are great challenges for chemists and they require a multidisciplinary approach at the interface of chemistry, biology and physics. These metalloenzymes are characterized by the fact that they exhibit a metal–carbon bond or are forming a metal–carbon bond during the activation of the substrate–organometallic structures results.

Historically, for the first time a metal–carbon bond was detected in **vitamin B_{12}**, its coenzyme and methylcobalamin, which became apparent when the organometallic nature of coenzyme B_{12} was discovered by Dorothy Hodgkin by X-ray diffraction

Wolfgang Weigand, Friedrich-Schiller-Universität Jena, Institut für Anorganische und Analytische Chemie, Humboldtstr. 8, 07743 Jena, Germany
Ulf-Peter Apfel, Ruhr-Universität Bochum, Fakultät für Chemie u. Biochemie, Anorganische Chemie, Universitätsstr. 150, 44801 Bochum, Germany

https://doi.org/10.1515/9783110496574-001

methods in 1965 [5]. In various organisms, **organometallic B_{12}-cofactors** help to drive metabolism and to control gene expression, and are attached to B_{12}-binding macromolecules, either proteins [6] or RNA [7]. These organometallic compounds, one of the axial sites is coordinated to a carbon atom, have a unique capacity for biological catalysis and for controlling life processes. This natural organometallic system will be discussed in Chapter 6.

Along this line, the important formation of molecular hydrogen and its bioconsumption will be discussed. Chapter 2 focuses on the rapid development over the last two decades of enzymes responsible for hydrogen evolution reactions and hydrogen oxidation: Hydrogenases are metalloenzymes that reversibly catalyze the conversion of molecular hydrogen (H_2) to protons and electrons (eq. (1.1)):

$$2H^+ + 2e^- \rightleftharpoons H_2 \tag{1.1}$$

This class of enzymes was described for the first time in 1931 [8]. Based on the metallic active sites, three distinct classes of hydrogenases are known: **[FeFe]-hydrogenase** (Section 2.1), **[NiFe]-hydrogenase** (Section 2.2) and **[Fe]-hydrogenase** (Section 2.3). Three types of hydrogenases have different protein structures and metallic active sites, although they commonly have the ability to heterolytically cleave H_2 to a hydride and a proton, which is as the result of convergent evolution. [NiFe]- and [FeFe]-hydrogenases can both catalyze the heterolytic cleavage of H_2 to a proton and a hydride and further extraction of two electrons from a hydride.

Among the various enzymes containing organometallic cofactors, **[FeFe]-hydrogenases** (Section 2.1) comprise 2-aza-propane-1,3-ditholato, CO and CN^- ligands [9, 10]. The organometallic complexes $Fe_2(CO)_6(\mu\text{-}SR)_2$ are known for 90 years at a time when nobody had remotely considered that such a type of complex was synthesized by Nature [11, 12] (Figure 1.1).

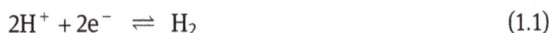

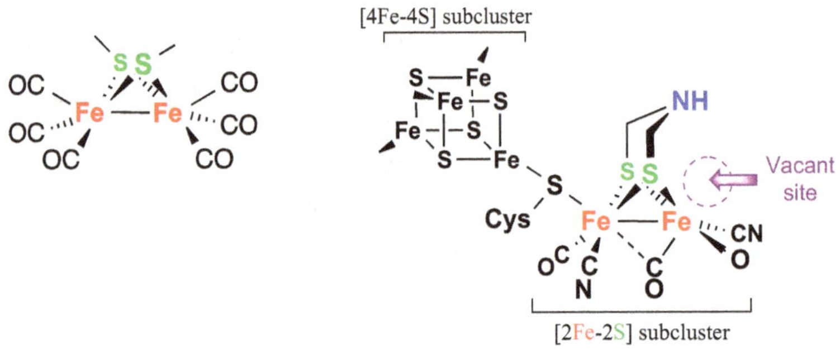

Figure 1.1: Organometallic complex $Fe_2(CO)_6(\mu\text{-}SMe)_2$ [11] (left); the active H-cluster site of a [FeFe]-hydrogenase [9, 10] (right).

These enzymes are found in fermentative bacteria as well as cyanobacteria and green algae. The most commonly investigated [FeFe]-hydrogenases are from *Clostridium pasteurianum* (*Cp*), *Desulfovibrio desulfuricans* (*Dd*) and *Chlamydomonas reinhardtii* (*Cr*). These hydrogenases can be regarded as "natural power stations," as they efficiently allow for the reversible interchange of protons to hydrogen (eq. (1.1) [13, 14]. The [FeFe]-hydrogenase from *Clostridium pasteurianum*, for example, shows very high activity with a turnover frequency of 3,400 μmol (H_2) min/mg (enzyme) under mild conditions (−0.413 V vs standard hydrogen electrode, pH 7) [9]. These enzymes solely operate under strictly anaerobic conditions, and a complete loss of activity is observed when they are exposed to air through a decay mechanism, which is still not fully understood.

[NiFe]-hydrogenases (Section 2.2) form a versatile class of ancient metalloenzymes that catalyze the reversible cleavage of H_2 into protons and electrons under ambient conditions [13]. As a consequence, these biomolecules represent ideal prototype catalysts and valuable targets for sustainable energy conversion approaches and other (bio)technological applications. A rational utilization of these enzymes, however, requires a thorough understanding of their catalytic mechanism and the underlying structural peculiarities. In this chapter, the current state of knowledge on [NiFe]-hydrogenases will be presented, together with open questions and challenges for future research. After a general introduction to the structure, mechanism and physiological function of these enzymes, we will highlight unique features of the technologically important subclass of O_2-tolerant hydrogenases, which sustain catalysis under aerobic conditions. The evolution and biosynthesis of [NiFe] hydrogenases will be discussed as well, together with strategies for the design of bioengineered and synthetic $H_2/2H^+$ cycling catalysts. We close our treatise with a summary of envisaged applications and future perspectives.

Finally, **[Fe]-hydrogenase** (Section 2.3) contains a mononuclear Fe as its active site, and directly uses a hydride for catalytic reduction of the substrate methenyl-tetrahydromethanopterin to methylene-tetrahydromethanopterin after heterolytic cleavage of H_2 at its Fe center requiring methenyl-tetrahydromethanopterin (Figure 1.1) [15, 16]. In other words, [Fe]-hydrogenase cannot heterolytically cleave H_2 without methenyl-tetrahydromethanopterin, although the other two classes of hydrogenases can with their dinuclear metallic active sites. Thus, [Fe]-hydrogenase utilizes a unique strategy for hydrogenase catalysis by using its unprecedented structural and functional features, which are distinct from those of other two types of hydrogenases.

Likewise in the conversion of CO_2, bioorganometallic complexes play a key role in its reduction. In Chapter 3, the focus is on the reduction of CO_2 in organisms. Microorganisms evolved different ways to overcome the stability of CO_2, converting the molecule into more reduced carbon compounds. **Carbon monoxide dehydrogenases** (CODHs) are part of the microbial repertoire to catalyze the two-electron conversion between CO and CO_2. It is suggested that hydroxide attacks a CO that is coordinated at a Ni atom to give a metallocarboxylic acid. By analogy, according to the Hieber base reaction, which was reported in 1932 [17], hydroxide attacks at the iron coordinated-CO ligand achieving a carbonyl metallate hydride (eq. (1.2)). This reaction can be seen as the metal-catalyzed water gas shift reaction (eq. (1.3), Scheme 1.1).

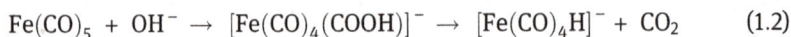

$$Fe(CO)_5 + OH^- \rightarrow [Fe(CO)_4(COOH)]^- \rightarrow [Fe(CO)_4H]^- + CO_2 \quad (1.2)$$

$$CO + H_2O \rightarrow CO_2 + H_2 \quad (1.3)$$

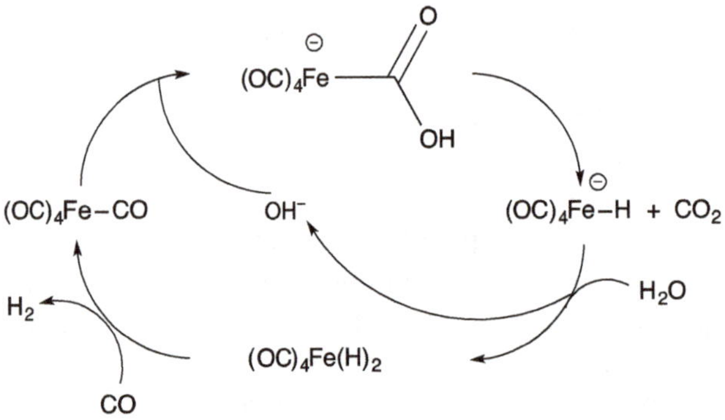

Scheme 1.1: Metal-catalyzed water gas shift reaction.

Two principal types of CODHs have been found, which employ different overall architectures, cofactors and metals in their active sites. A CODH-containing Cu and Mo in the active site allows aerobic bacteria to oxidize CO while living in an oxygen-containing environment. In contrast, a Ni- and Fe-containing enzyme is found in bacteria and archaea that live in the absence of dioxygen, under strictly anaerobic conditions, where it catalyzes not only the oxidation of CO to CO_2, but also the efficient reduction of CO_2 without observable overpotential. The CODH is a wonderful example for the close analogy between biochemistry and organometallic chemistry.

The current state of knowledge on the structure and function of these enzymes, as well as what we can learn from related synthetic catalysts, is detailed in Chapter 3.

Chapter 4 deals with the **nitrogenase** [18], which is a highly complex biological system that can cleave the strong triple bond of dinitrogen under ambient temperature and pressure. This compares to energy-intensive industrial methods (Haber–Bosch process) that require high temperatures, pressures, equipment and hydrogen gas [19]. The Mo-nitrogenase from *A. vinelandii* NifDK is the best characterized nitrogenase system. The two-component NifDK/NifH system uses unique metallocofactors to facilitate this reactivity that are not found elsewhere in Nature, and these cofactors have captivated research groups over many decades. Advances in spectroscopic characterization and crystallography have provided a robust understanding of the proteins, cofactors and some of their interactions, but there are still questions that remain to be answered. The cluster appears as two partial cubane units, $[Fe_4S_3]$ and $[MoFe_3S_3]$, with the fascinating organometallic μ_6-insterstitial carbide as a shared vertex, and additionally is bridged by three μ_2-sulfide ligands. The Mo-nitrogenase mechanism has been studied extensively, but there is still no definitive proposal due to the complexity of targeting the M-cluster-bound intermediates during catalysis. V- and Fe-nitrogenases are also proposed to function similar to Mo-nitrogenase, but further detailed kinetic and mechanistic studies are required to probe these suggestions. The future of nitrogenase research is bright and will continue to capture the attention of investigators for years to come (Figure 1.2).

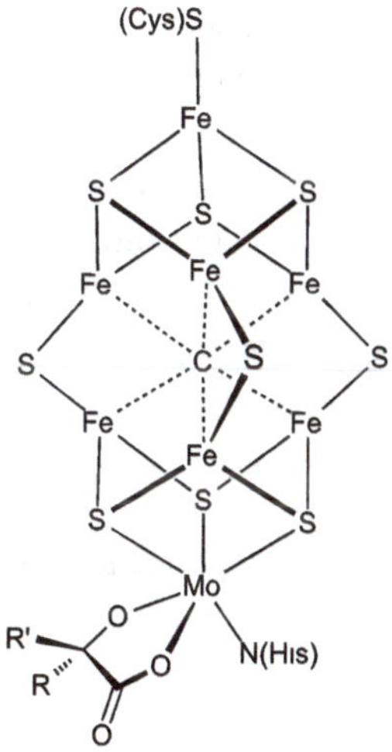

Figure 1.2: M-cluster cofactor from *A. vinelandii* NifDK.

In Chapters 2–4 the reduction of small molecules H_2, CO_2 and N_2 by metalloenzymes has been discussed, while Chapter 5 takes up on the biological oxidation of methane. A class of bacteria known as *methanotrophs* employs enzymes called **methane monooxygenases (MMOs)**, which utilize CH_4 as a sole carbon source under ambient conditions. These aerobic proteobacterias are able to hydroxylate the nonpolar and strong C–H bonds in CH_4 using O_2, protons and electrons under mild conditions to form CH_3OH as the first step of CH_4 metabolism. However, their sophisticated enzyme machinery cannot be used and scaled up for industrial purpose, which is too expensive. There are two forms of MMOs, which have been evolved to perform this reaction: one is membrane-bound and is a copper-containing particulate MMO (*p*MMO), and the other is water-soluble cytoplasmic and is an iron-containing soluble methane monooxygenase (sMMO). In both cases, the MMOs utilize transition metal centers to activate dioxygen in order to attack the strong C–H bonds of CH_4, but the structures, active sites, cofactor requirements and mechanisms of CH_4 activations are completely different in the two cases. Notably, compared to *s*MMO, *p*MMO is more efficient in oxidizing CH_4 and is the most efficient CH_4 oxidizer known to date. It is capable of activating CH_4 at a rate of one CH_4 molecule per second per enzyme [20], and is preferably used in biology for methane. The pMMO hydroxylase consists of the PmoB, PmoA and PmoC subunits. Unfortunately, pMMO being a membrane-bound protein, it is extremely difficult to isolate and to purify it from the plasma membrane for biochemical and biophysical studies and, thus, studies on the pMMO are quite rare and a lot of issues are still under debate [21]. For example, even the nature of the active site in pMMO is not unambiguous; it has been controversially discussed to contain a mono-, di- and trinuclear copper centers, as well as a dinuclear iron center [22, 23]. In contrast, the oxidative chemistry of sMMO has been investigated extensively for over 30 years and is better understood; this will be discussed in detail in this chapter.

Chapters 6 and 7 deal with C–C bond formation reactions in organisms. We have already mentioned organometallic B_{12}-derivatives in life processes (Chapter 6). The **acetyl-coenzyme A synthase** (Chapter 7) catalyzes the synthesis of the acetyl group of acetyl-CoA from CO and a methyl group from a methylated corrinoid iron–sulfur protein, both of which ultimately originate from CO_2. Both carbon moieties bind to a particular nickel in the enzyme (called Ni_p) that is the heart of the so-called A-cluster active site. The thiol coenzyme A (CoA) is a carrier molecule of nucleotide origin that is regenerated when the acetyl group is transferred for use in downstream metabolic processes. The synthesis of the acetyl group of acetyl-CoA shows a great similarity to the well-known organometallic insertion reaction, where CO inserts into metal–alkyl bonds to give metal acyls (eq. (1.4)). This type of reaction is also proposed for the catalytic Monsanto acetic acid process that converts methanol and CO to acetic acid by use of an Rh catalyst $[RhI_2(CO)_2]^-$ [24].

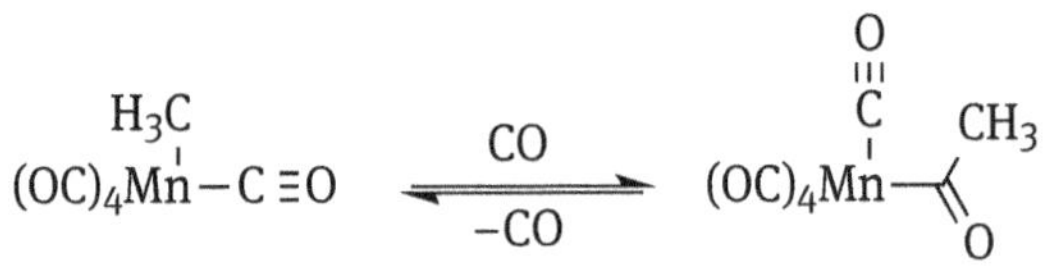

$$\underset{\text{(OC)}_4\overset{|}{\text{Mn}}-\text{C}\equiv\text{O}}{\text{H}_3\text{C}} \underset{\text{−CO}}{\overset{\text{CO}}{\rightleftharpoons}} (\text{OC})_4\text{Mn}\overset{\overset{\text{O}}{\overset{|||}{\underset{|}{\text{C}}}\,\text{CH}_3}}{\diagdown_{\text{O}}} \tag{1.4}$$

Chapter 8 takes up the essential question for potential benefits of **organometallic complexes as drug candidates**. First of all, it is evident that organometallic compounds provide a huge structural and stereochemical diversity. In particular, their unique three-dimensional shape (e.g., square-planar, octahedral or even higher coordinated) can be hardly achieved with solely organic molecules. Further, the exchange of ligands facilitates their synthetic accessibility and allows for a simplified rational design of structurally demanding complexes. One of the most important aspects that sets metal complexes apart from organic molecules is the availability of a broader range of modes of action. These may be depending on well-controllable oxidative and reductive processes, the catalytic activity of metal complexes or structural alteration by the aforementioned ligand exchange reactions. Another important point to consider is the possibility to use radioactive metal isotopes as diagnostic tools or radiotherapeutic drugs. Despite the huge diversity of metals, the lead structures and compounds presented later in this chapter can almost entirely be classified and understood by one of or by a combination of several of these key concepts [25].

Finally, in Chapters 9, 10 and 11 spectroscopic methods important for the characterization of biological complexes are discussed. Chapter 9 introduces **nuclear resonance vibrational spectroscopy** and its applications, with emphasis on biological metallocofactors. The history of the method and its relationship to Mössbauer spectroscopy [26, 27] are discussed, followed by a theoretical treatment that includes both qualitative fundamentals and a necessary mathematical treatment. The physical observables available from the spectroscopy are covered. A cursory overview of the experimental components, procedures, data analysis and interpretation is included. Finally, multiple applications are shown that focus on increasingly complex electron transfer proteins. The chapter concludes with an outlook on the future of the burgeoning technique. Chapter 10 will give an introduction to electron paramagnetic resonance (EPR) spectroscopy which is called by the author as "one of the most mystique-surrounded techniques". Chapter 11 focuses on **X-ray-based spectroscopies** (X-ray absorption, X-ray emission, resonant inelastic X-ray scattering), which have played a major role in describing the geometric and electronic structure of countless metalloprotein active sites. A major advantage of X-ray spectroscopy is the element selectivity, which allows for the changes that occur at a protein active site to be probed in an element selective way. In addition, X-ray spectroscopic methods can be applied to samples in any form (solutions, lyophilized powders or single crystals). Hence, these approaches are well suited for studying reactive intermediates and are particularly useful for proteins that are not readily crystallized.

This book shall help to understand the strong link between organometallic chemistry as well as its application to homogenous catalysis and enzymatic reactions mediated by transition metal complexes. Organisms use organometallic complexes "packaged" in proteins to activate simple molecules, such as H_2, CO_2, CO, N_2, CH_4, under mild and physiologically compatible conditions. Nature utilizes readily available metals like iron, nickel or manganese in the catalytic active sites of enzymes. Inter alia proteins have the role of a protective and controlling shell in these catalytic processes. An excellent example is the special structural arrangement of the [$2Fe_H$] cofactor in the [FeFe] hydrogenases, which is called the "rotated state" (Figure 1.3)

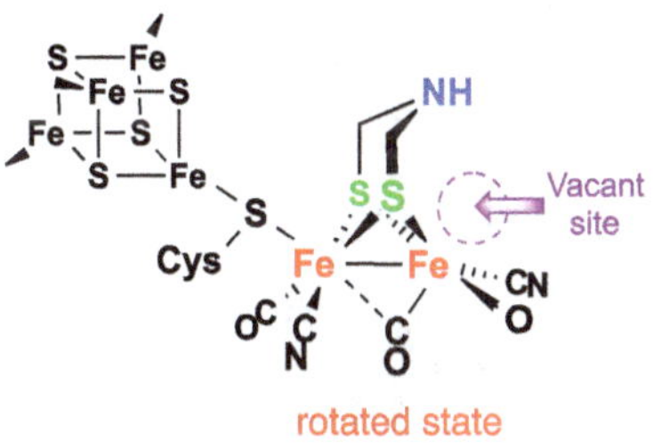

Figure 1.3: "Rotated state" in the active H-cluster site of a [FeFe]-hydrogenase.

with respect to the pseudo-symmetric arrangement of the early enzyme mimetics and forces the cofactor to display a vacant coordination site on Fe_d and a characteristic bridging μ_2-CO ligand (Section 3.1).

References

[1] Jaouen, G., Vessieres, A., and Butler, IS. Bioorganometallic chemistry: a future direction for transition metal organometallic chemistry? Acc Chem Res 1993, 26, 361–369.
[2] Severin, K., Bergs, R., and Beck, W. Bioorganometallic chemistry-transition metal complexes with α–amino acids and peptides. Angew Chem Int Ed 1998, 37, 1634–1654.
[3] Jaouen, G. (Ed.). Bioorganometallics, Biomoilecules, Labeling, Medicine, Weinheim, Wiley-VCH, 2006.
[4] Jaouen, G., and Salmain, M. (Eds.). Biorgonametallic Chemistry, Applications in Drug Discovery, Biocatalysis, and Imaging, Weinheim, Wiley-VCH, 2015.
[5] Hodgkin, DC. X-ray analysis of complicated molecules. Science 1965, 150, 979–988.
[6] Gruber, K., Puffer, B., and Kräutler, B. Vitamin B_{12}-derivatives – enzyme cofactors and ligands of proteins and nucleic acids. Chem Soc Rev 2011, 40, 4346–63.
[7] Winkler, WC., and Breaker, RR. Regulation of bacterial gene expression by riboswitches. Ann Rev Microbiol 2005, 59, 487–517.
[8] Stephenson, M., and Stickland, LH. Hydrogenase: a bacterial enzyme activating molecular hydrogen. Biochem J (London) 1931, 25, 205–214.
[9] Peters, JW., Lanzilotta, WN., Lemon, BJ., and Seefeldt, LC. X-Ray crystal structure of the Fe-only hydrogenase (CpI) from clostridium pasteurianum to 1.8 angstrom resolution. Science 1998, 282, 1853–1858.
[10] Nicolet, Y., Piras, C., Legrand, P., Hatchikian, CE., and Fontecilla-Camps JC. Desulfovibrio desulfuricans iron hydrogenase: The structure shows unusual coordination to an active site Fe binuclear center. Structure 1999, 7, 13–23.
[11] Reihlen, H., Gruhl, A., and Hessling, G. Über den photochemischen und oxydativen abbau von carbonylen. Liebigs Ann Chem 1929, 472, 268–287.
[12] Li, Y., and Rauchfuss, TB. Synthesis of diiron(I) dithiolato carbonyl complexes. Chem Rev 2016, 116, 7043–7077.
[13] Lubitz, W., Ogata, H., Rüdiger, O., and Reijerse, E. Hydrogenases. Chem Rev 2014, 114, 4081–4148.
[14] Stripp, ST., and Happe, T. How algae produce hydrogen-news from the photosynthetic hydrogenase. Dalton Trans 2009, 45, 9960–9969.
[15] Shima, S., and Thauer, RK. A third type of hydrogenase catalyzing H_2 activation. Chem Rec 2007, 7, 37–46.
[16] Shima, S., and Ermler, U. Structure and function of [Fe]-hydrogenase and its iron-guanylylpyridinol (FeGP) cofactor. Eur J Inorg Chem 2011, 2011, 963–972.
[17] Hieber, W., and Leutert, F. Über metallcarbonyle. XII. Die basenreaktion des eisenpentacarbonyls und die bildung des eisencarbonylwasserstoffs. Z Anorg Allg Chem 1932, 204, 145–164.
[18] Ribbe, MW., Hu, Y., Hodgson, KO., and Hedman, B. Biosynthesis of nitrogenase metalloclusters. Chem Rev 2014, 114, 4063–4080.
[19] Dybkjaer, I. Ammonia, Catalysis and Manufacture, Heidelberg, Springer, 1995.
[20] Semrau, JD., DiSpirito, AA., and Yoon, S. Methanotrophs and copper. FEMS Microbiol Rev 2010, 34, 496–531.

[21] Chan, SI., and Yu, SSF. Controlled oxidation of hydrocarbons by the membrane-bound methane monooxygenase: The case for a tricopper cluster. Acc Chem Res 2008, 41, 969–979.
[22] Culpepper, MA., and Rosenzweig, AC. Architecture and active site of particulate methane monooxygenase. Crit Rev Biochem Mol Biol 2012, 47, 483–492.
[23] Cao, L., Caldararu, O., Rosenzweig, AC., and Ryde, U. Quantum refinement does not support dinuclear copper sites in crystal structures of particulate methane monooxygenase. Angew Chem Int Ed 2018, 57, 162 –166.
[24] Crabtree, RH. The Organometallic Chemistry of the Transition Metals, sixth edition, Hoboken, New Jersey, John Wiley & Sons, Inc., 2014.
[25] Gianferrara, T., Bratsos, I., and Alessio, EA. Categorization of metal anticancer compounds based on their mode of action. Dalton Trans 2009, 7588–7598.
[26] Mössbauer, R. Kernresonanzfluoreszenz von gammastrahlung in Ir191. Zeitschrift für Physik 1958, 151, 124–143.
[27] Mössbauer, R. Kernresonanzabsorption von gammastrahlung in Ir191. Naturwissenschaften 1958, 45, 538–539.

Part I: Reduction and Oxidation Catalysts

Ulf-Peter Apfel, Wolfgang Weigand, Dr. Marius Horch, Ingo Zebger, Oliver Lenz and Takashi Fujishiro

2 Hydrogen development

2.1 [FeFe]-hydrogenases

Ulf-Peter Apfel and Wolfgang Weigand

Among the various enzymes containing organometallic cofactors, [FeFe]-hydrogenases are among this class of enzymes comprising metal–carbon bonds [1]. These enzymes are found in fermentative bacteria as well as cyanobacteria and green algae [2]. The most commonly investigated [FeFe]-hydrogenases are from *Clostridium pasteurianum* (*Cp*), *Desulfovibrio desulfuricans* (*Dd*) and *Chlamydomonas reinhardtii* (*Cr*). These hydrogenases can be regarded as "natural power stations," as they efficiently allow for the reversible interchange of protons to hydrogen (eq. (2.1)) [3, 4]. The [FeFe]-hydrogenase from *Clostridium pasteurianum*, for example, shows very high activity with a turnover frequency of 3,400 μmol (H_2) min^{-1} mg^{-1} (enzyme) under mild conditions (−0.413 V vs standard hydrogen electrode, pH 7) [5]. These enzymes solely operate under strictly anaerobic conditions and a complete loss of activity is observed when they are exposed to air through a decay mechanism, which is still not fully understood [6–10].

$$2H^+ + 2e^- \rightleftharpoons H_2 \tag{2.1}$$

2.1.1 The active site

Understanding the structural features of the hydrogenase enzyme class can reveal how this enzyme can perform H_2 interchange at optimal efficiency. The crystal structure of the [FeFe]-hydrogenase was first determined from the isolated enzyme from the organisms *D. desulfuricans* as well as *C. pasteurianum* [5, 11]. The active

Ulf-Peter Apfel, Inorganic Chemistry I, Ruhr-university Bochum & Fraunhofer UMSICHT, Bochum, Germany, ulf.apfel@rub.de
Wolfgang Weigand, Friedrich-Schiller-Universität Jena, Institut für Anorganische und Analytische Chemie, Humboldtstr. 8, 07743 Jena, Germany
Dr. Marius Horch, Department of Physics, Freie Universität Berlin, Arnimallee 14, 14195 Berlin, Germany, marius.horch@fu-berlin.de
Ingo Zebger, TU Berlin Institute of Chemistry, Straße des 17. Juni 135, 10623, Berlin, Germany, ingo.zebger@tu-berlin.de
Oliver Lenz, TU Berlin Institute of Chemistry, Faculty II Straße des 17. Juni 135, 10623, Berlin, Germany, oliver.lenz@tu-berlin.de
Takashi Fujishiro, Department of Biochemistry and Molecular Biology, Saitama University, Japan, tfujishiro@mail.saitama-u.ac.jp

https://doi.org/10.1515/9783110496574-002

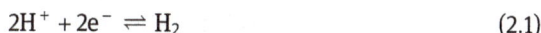

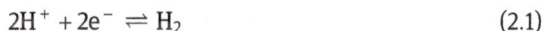

site structures, while obtained from different organisms, reveal the overall folding core active site cluster domain with remarkable similarities.

The active cofactor of the enzymes catalytic core is referred to as the hydrogen-forming or H-cluster. This cluster consists of a [$2Fe_H$] subcluster and a [4Fe–4S]-cluster (Figure 2.1). Both clusters are covalently linked to the protein backbone via a cysteine residue as well as further hydrogen bonding interactions. Remarkably, this cysteine residue is the only covalent attachment of the [$2Fe_H$] moiety to the entire protein scaffold. The H-cluster is a common structural scheme within the known classes of [FeFe]-hydrogenases and the major difference is in the number of [4Fe–4S] clusters that direct the electron transfer toward/from the [$2Fe_H$] subcluster. *C. pasteurianum* contains three [4Fe–4S] clusters [5], while *D. desulfuricans* possesses only two [4Fe–4S] cluster in its electron transport path [11]. The protein scaffold can in many ways be regarded as an "oversized" natural organic ligand, although it is worth mentioning that the protein environment not only serves as an anchor for the H-cluster but also allows for a controlled shuttling of protons and hydrogen to and from the active site as well as controlling the geometry of the [2Fe]-cluster [12, 13].

With respect to H^+/H_2 shuttling, the enzyme contains a distinct hydrophobic gas channel that reaches from the molecular surface to the active site. The stability as well as activity of the enzyme depends upon mutations within this gas channel. The gas channel is notably unselective toward H_2 or the inhibitors CO and O_2; hence, the enzymes operate under anaerobic conditions. Molecular dynamic simulations helped to establish and assign a well-paved proton channel from the structures of *Dd* and *CpI* [14, 15]. It is assumed that the protons travel to the active site from the enzyme's molecular surface via a Grotthuss type mechanism following a conserved proton channel established by glutamic acid (Glu279, Glu282), serine (Ser319, Ser320), water (H_2O-29) and cysteine (Cys299). Variation of the conserved amino acids within this proton-exchange path results in a significant decrease of the enzymatic activity [16].

The [$2Fe_H$] cluster is the active catalytic center where the reversible proton reduction or H_2 oxidation occurs. The [$2Fe_H$]-center is best described as a butterfly-shaped [2Fe–2S] cluster. The proximal iron (Fe_P) is coordinated by one CN^- as well as one terminal CO and linked to the [4Fe–4S] cluster via a cysteine amino acid. Thus, the [$2Fe_H$] and [4Fe–4S] clusters are positioned approximately 4 Å apart, allowing for fast electron transfer between both clusters.

The distal iron (Fe_d) is likewise coordinated to a CN^- and CO ligand, but contrary to Fe_p, it offers a square pyramidal environment with one vacant coordination site. The iron ions Fe_p and Fe_d share a (quasi)bridging carbonyl ligand, μCO.

The CO ligands of the [$2Fe_H$] cluster are located in hydrophobic pockets and allow for regulation of the electronic properties of the various intermediates of this enzyme during catalysis, which is a common feature of typical organometallic catalysts that utilize strong field ligands. The cyanides have been proposed to provide additional anchors to the [$2Fe_H$] cluster within the enzyme binding pocket by hydrogen bonding. The protein environment is important not only because it directly

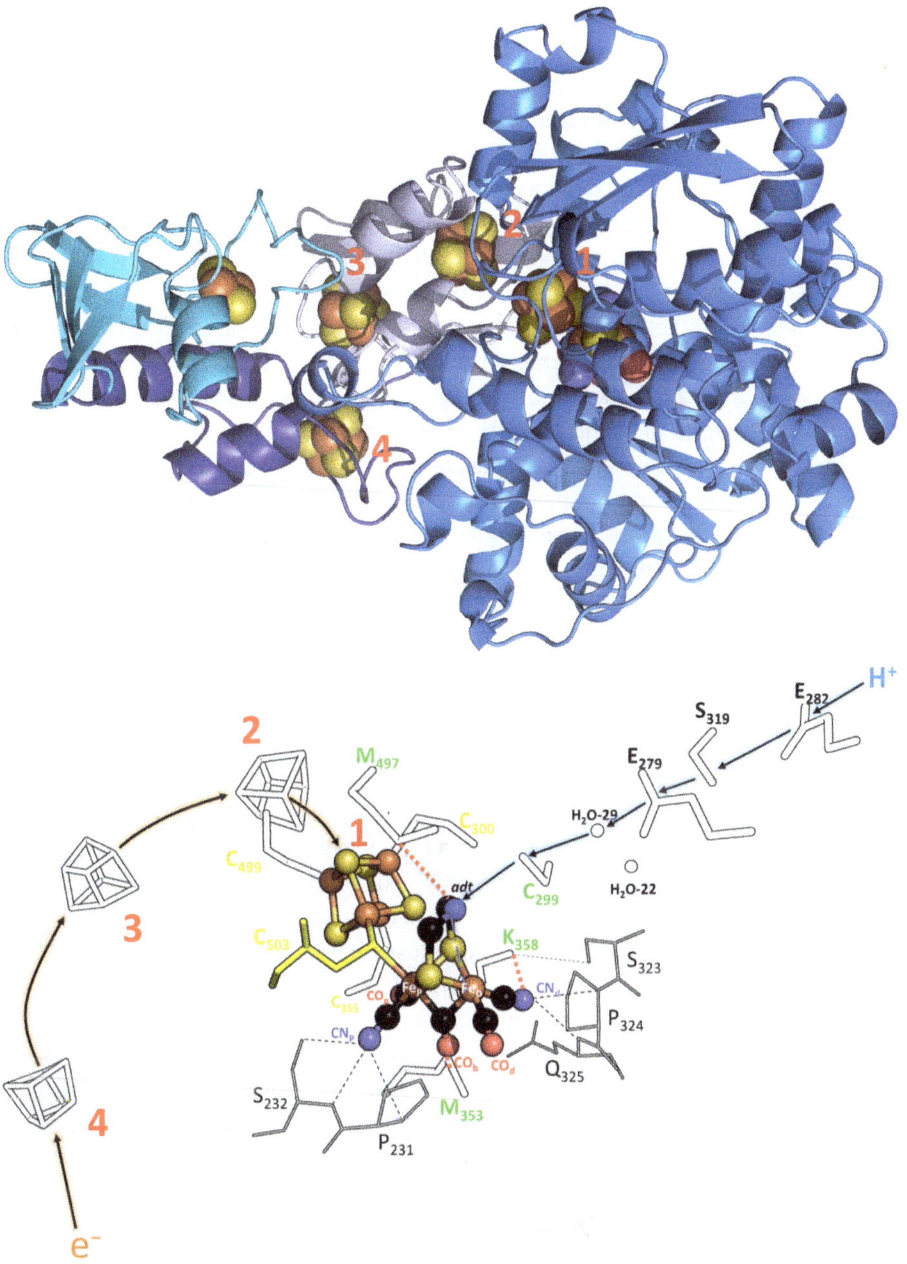

Figure 2.1: (Top) Crystal structure of the [FeFe]-hydrogenase from *Clostridium pasteurianum* (pdb-code: 3C8Y). (Bottom) Zoom in into the H-cluster including the electron (orange arrows) as well as proton (blue arrows) transfer pathways. The figure also shows the linking of the [2FeH] cluster to the protein environment via the cyanides CN_p and CN_d as well as to the [4Fe–4S] cluster via cysteine C_{503}.

controls electronic exchange but it also controls the [$2Fe_H$] clusters shape, that is, into its active specific geometry.

Moreover, both iron centers are further bridged by a CO-ligand as well as a unique dithiolate bridge. The dithiolate bridge is critical to the function and reactivity of the enzyme and thus not surprisingly located at the end of the proton channel [16]. While there has been much debate on the nature of this dithiol bridge especially the atom located at the bridge head position, it is now generally well accepted that the bridge is composed of a 2-aza-propane-1,3-ditholate bridge (adt, $-SCH_2NHCH_2S-$) [17, 18]. The [$2Fe_H$] cluster can thus be best described as a "frustrated Lewis pair", efficiently combining a Lewis acid (Fe_d^I) and base (NH) in close proximity. This special assembly can be rationalized as the origin for the unique performance of this enzyme and allows pulling apart molecular hydrogen by heterolytic cleavage as well as its formation from an iron bound hydride and an amine centered proton. The special structural arrangement of the [$2Fe_H$] cofactor is called the "rotated state" with respect to the pseudo-symmetric arrangement of the early enzyme mimetics and forces the cofactor to display a vacant coordination site on Fe_d and a characteristic μCO ligand [19]. This vacant coordination site is responsible for the enzymes' activity and blocking this site with a strongly bound ligand such as CO leads to a reversible inhibition of the enzyme [20].

2.1.2 How is it assembled?

The exact structural composition of the H-cluster actually raised more questions than it answered. This initial turmoil stems from two general facts:

1) The CN^- and CO ligands are both highly toxic in their "free form" and their formation and transportation within living materials was unknown.
2) Structurally comparable [2Fe–2S] complexes have been known since the early 20th century [21] and an analog $[Fe_2(CO)_4(CN)_2\{(SCH_2)_2NH\}]^{2-}$ complex was reported briefly after the discovery of the hydrogenases' active site structure [22]. Synthesis of such complexes, however, required elaborate synthetic protocols utilizing synthetic synthons and require the exclusion of air and moisture. These brought into question as to the origin and formation of this unprecedented cluster, which lead to extensive debate about its natural origin.

Biological formation of the [$2Fe_H$] active site is indeed more complicated than the synthesis of synthetic active site mimics in a laboratory and requires a sophisticated enzymatic machinery. The in vivo assembly, the synthesis in living organisms, of the [FeFe]-hydrogenase is accomplished in two main steps: the expression of the structural genes, which will not be discussed herein, and a subsequent insertion of the different iron–sulfur clusters.

The [4Fe–4S] cluster assembly is accomplished by iron–sulfur cluster biogenesis from a labile pool of Fe^{2+} and S^{2-}, which is comparable to the artificial self-assembly of [4Fe–4S] clusters. In contrast, the in vivo synthesis of the [$2Fe_H$] cofactor requires three maturase enzymes (Latin: maturare = ripening) to afford a functional enzyme that allows for hydrogen generation and oxidation, namely, HydE, HydF and HydG [23, 24]. Only after complete synthesis by these three maturase enzymes, the organometallic cofactor is transferred to its final protein environment (HydA) [25]. A proposed mechanism based on the current state-of-the-art that shows all steps of the cofactor formation is presented in Figure 2.2 and the fundamental properties of each enzyme implemented in this scheme is highlighted in Table 2.1 [23–25]. The exact mechanism of formation is however, up to now, not fully understood.

The maturase enzymes were first identified in *Chlamydomonas reinhardtii*, a single-cell green alga, but seem to be ubiquitous in all [FeFe]-hydrogenase-expressing organisms [26]. HydE and HydG were shown to be critical for the assembly of the [$2Fe_H$] cofactor [25]. Both maturase enzymes show high sequence homology and consist of an iron–sulfur cluster-binding pocket built up by three cysteine residues located near the C-terminal. Both maturases show a GTPase domain, a realm that allows for the hydrolysis of guanosine triphosphate, at their N-terminal. HydE and HydG both belong to the radical S-adenosyl methionine (SAM) family, a class of enzymes that contains highly reactive methyl groups and these enzymes are commonly observed in transmethylation processes [27, 28]. It is thus not surprising that an additional [4Fe–4S] cluster close to the N-terminal of the protein possesses a vacant coordination site and enables binding of S-adenosyl methionine (AdoMet), a substrate of SAM enzymes.

While the substrate of HydE is yet unknown, tyrosine was shown to be the substrate for HydG [29]. Here, tyrosine is converted to p-cresol, CN^- and CO [30–33]. This process is facilitated by the [4Fe–4S] cluster located at the N-terminal of HydG. The formation of CN^- and CO from tyrosine was unequivocally confirmed by ^{13}C-isotope labeling [34]. The tyrosine degradation mechanism by HydG was further supported by the spectroscopic observation of an oxido-benzyl radical when HydG reacts with tyrosine [35]. These experimental findings supported the formation of both CO and CN^- ligands from tyrosine by HydG but the pathway with respect towards the assembly of [2Fe–2S] cluster with toxic ligands is still not well understood. It is believed that the CO and CN^- ligands are directly installed on an iron site upon formation of the cluster. This assumption is supported by rapid freeze quench Fourier transform infrared spectroscopic experiments that showed typical IR signatures of a $Fe(CO)_2CN$ moiety obtained after the turnover of two tyrosine molecules [36]. These chemical transformations appear to occur at the N-terminal of HydG, facilitated by a [4Fe–4S] cluster. It is worth noting that at no time during the generation of CN^- and CO, these highly toxic molecules exist in their "free form." The immediate binding of both ligands upon

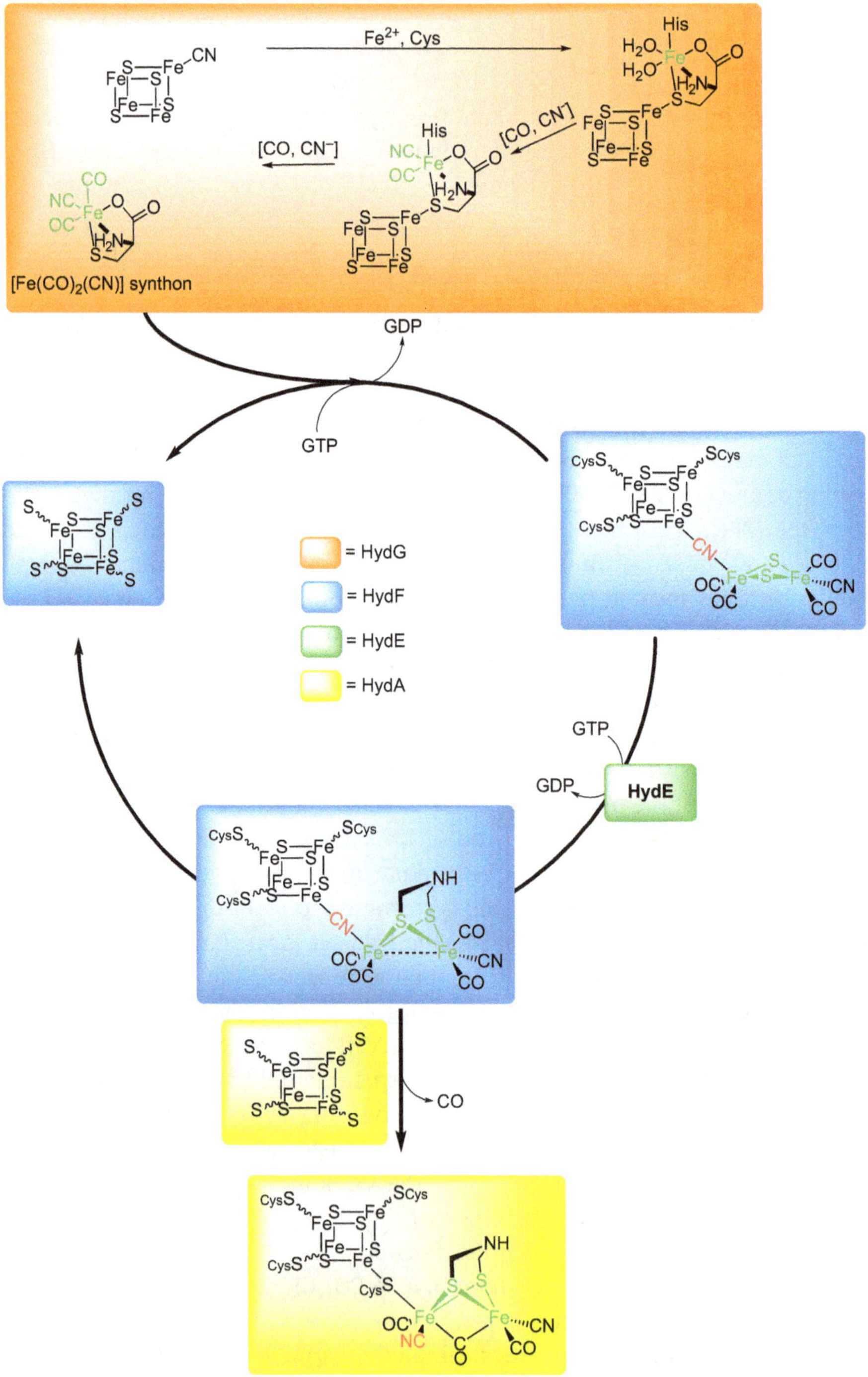

Figure 2.2: Proposed natural H-cluster maturation pathway [3].

Table 2.1: Maturation enzymes and their (tentative) purposes.

	Function	Substrate
HydG	– Generation of CN^- and CO – Assembly of a $Fe_2S_2(CO)_4(CN)_2$ fragment	Tyrosine
HydF	– Transfer enzyme – Construction platform	–
HydE	– Presumably formation of the adt-bridgehead	Most likely a $Fe_2S_2(CO)_4(CN)_2$ cluster
HydA	– H_2 generation and consumption	H^+/H_2

formation as well as their high binding affinity to iron avoids any poisoning effects.

In this process, HydG was found as the source for the [$2Fe_H$] cofactor of the [FeFe]-hydrogenase. HydG is also thought to be responsible for the formation of the adt-bridgehead. Like the formation of the CN^- and CO ligands, the formation of the adt-bridgehead is also suggested to proceed via tyrosine degradation (Figure 2.3) [37]. It is assumed that during this process a glycine radical is generated that reacts with the formed [2Fe–2S] cluster to yield the final [$2Fe_H$] cofactor. However, the participation of HydG in this process is not experimentally confirmed. Due to the high sequence homology of HydE and HydG, it is also plausible that HydE conducts the formation of the adt-bridgehead. The iron sulfur cluster binding site at the C-terminal of HydE most likely takes over the [2Fe–2S] cluster generated by HydG and further converts this cluster to its final form. Taking the uniqueness of the special [2Fe–2S] intermediate into account, it is understandable that, until now, no substrate for HydE was experimentally confirmed. In the absence of HydE the [FeFe]-hydrogenase is virtually inactive.

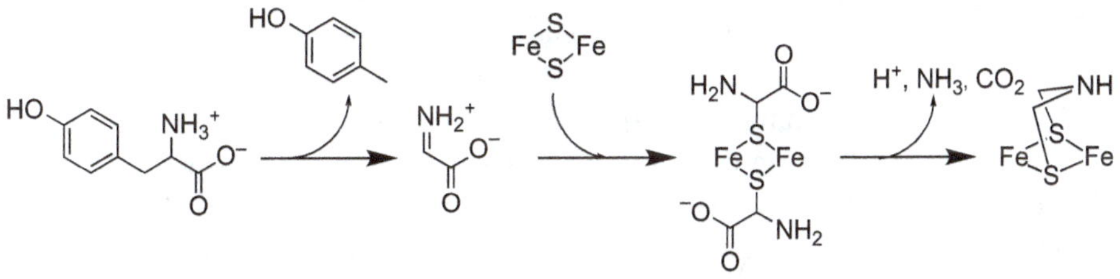

Figure 2.3: Hypothetic tyrosine degradation pathway leading to the iron bound adt-bridge.

The transfer of the [2Fe–2S] intermediate from HydE to HydG and reverse is most likely conducted by HydF and is presumably driven by GTPase activity [38]. Notably, HydF is not elementary for the final maturation of HydA. HydF acts solely as transfer-protein that transports the cofactor to the different protein assembly positions.

Although the expression of the HydA, E, F and G gene codes all appear critical toward generating a functional [FeFe]-hydrogenases in the various hydrogenase containing organisms, *Escherichia coli* can likewise serve as an amenable host for the expression of such gene codes and affords fully assembled active [FeFe]-hydrogenase enzymes [29]. Even cell-free maturation was shown to provide a successful assembly of a functional enzyme. This occurs through individual expression of key genes followed by subsequent mixing to allow for a cell-free maturation of the components in a test tube. This method provides formation of functional [FeFe]-hydrogenase enzymes through a simplified and faster work-up procedure.

A major breakthrough was achieved that resulted in shortening the natural biosynthetic pathway by application of synthetic diiron disulfide cluster to apo-HydF [17]. The apo-HydF from *Thermotoga maritima* lacking the [$2Fe_H$] cofactor was shown to bind to an artificial synthetic $[Fe_2(adt)(CO)_4(CN)_2]^{2-}$ complex. The cluster binds to the enzyme via a bridging cyanide ligand to the [4Fe–4S] cluster of HydF with the N-atom being connected to the [4Fe–4S] cluster and the C atom coordinating the [$2Fe_H$] moiety. However, binding of this artificial mimic seems to be different from its natural counterpart. The IR bands of the artificially generated complex are broader than in the natural form, which suggests higher flexibility. The artificial cofactor form can be oxidized and this observation is in direct contrast to the natural form that is redox inactive. Not only can $[Fe_2(adt)(CO)_4(CN)_2]^{2-}$ be bound to HydF, other alternative synthetic mimics like odt ($-SCH_2\mathbf{O}CH_2S-$) and pdt ($-SCH_2\mathbf{CH_2}CH_2S-$) dithiolate linkers (Figure 2.4) can be attached to HydF [17].

These artificially bound cofactors can be transferred from HydF to HydA as was shown for the insertion into the apo-hydrogenase from the algae *Chlamydomonas reinhardtii* bypassing the convoluted expression procedure to give active HydA. While *E. coli* was notable because it allowed for the gene expression of HydA, it did not yield a functional protein in the absence of HydE, F as well as G and is incompetent to facilitate the synthesis of the [$2Fe_H$] cofactor. Adding a HydF-bound mimic to apo-HydA, which contains a preformed [4Fe–4S] cluster obtained from iron salts and a sulfide source yields enzymes with wild-type-like activity [17]. This method offers the possibility to generate increased amounts of HydA from *E. coli*, thereby yielding fully functional enzymes in large scale thus allowing to further evaluate of their properties. Furthermore, synthetic cofactors containing the pdt, odt and adt bridges can be introduced into the enzymatic environment. This allowed molecular control of the bridgehead and for definitive assignment as to the atom identity, a topic that was previously strongly

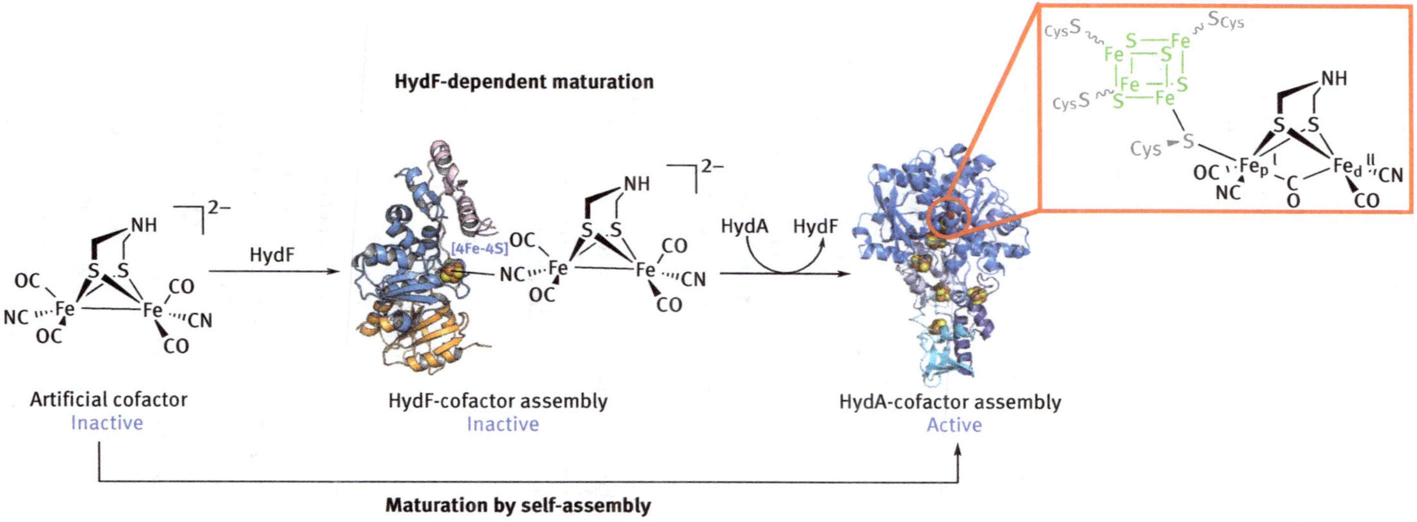

Figure 2.4: Artificial maturation processes of the [FeFe]-hydrogenase utilizing synthetic [$2Fe_H$] cofactor mimics.

debated. While synthetic cofactors containing the adt, pdt and odt bridging ligands are essentially inactive for the hydrogen conversion, only the adt containing mimic revealed full enzymatic activity upon maturation into HydA [17, 39]. Thus, it was unambiguously determined that the nitrogen-containing bridgehead must be present in the natural enzyme to afford an active hydrogenase and agrees well with the concept of the NH group serving as a Lewis base during the heterolytic cleavage of H_2 as well as its formation.

This synthetic method for generating active enzymes with synthetic cofactors was simplified when the HydF maturase factor was found to be unnecessary for the artificial assembly of HydA [18]. Even in the absence of HydF, structural mimics comprising a $[Fe_2(CO)_4(CN)_2]$ fragment can readily diffuse into the cofactor binding niche in a self-assembly process. The method is not specific to the enzymatic system used and similar maturase experiments were successfully performed with the bacterial [FeFe]-hydrogenases from *Clostridium pasteurianum* and *Megasphaera elsdenii* [18, 40]. The simplicity of this method generates on demand large amounts of the highly pure functional enzyme. This allows access to direct alteration of enzymatic properties further fueling in-depth investigation of the enzymes' working mechanism through targeted manipulations of the cofactor. For example, the structural homologs adt, pdt, odt and sdt ($-SCH_2\mathbf{S}CH_2S-$) synthetic cofactors were synthesized and all revealed structural similarity to the native enzyme. As such, differences in reactivity stem from a single atom of the bridgehead and are not due to altered structural or electronic parameters.

This artificial maturation procedure also allowed researchers to obtain insight into the maturation process of the [FeFe]-hydrogenase enzyme of *C. pasteurianum* (Figure 2.5) [41]. Researchers showed that the initial step of the maturation process involved the cofactor entering to a position close to the accessory [4Fe–4S] cluster but still maintains a symmetric $[Fe_2(CO)_4(CN)_2]$ moiety. The initial binding step was followed by oxidation of the accessory cluster to the $[Fe_4S_4]^{2+}$ state, allowing for rapid binding to cysteine (Cys503). The cofactor cannot bind to the accessory cluster in the reduced state and only after binding to Cys503 can fast rearrangement and CO loss afford the rotated state as is seen in the crystal. A comparable mechanism was also proposed for *D. desulfuricans*-based spectroelectrochemical IR measurements [42]. Thus, the maturation mechanism can be best described as associative mechanism in terms of classical organometallic chemistry.

The classical biochemical approaches allowed only for a complete labeling of the H-cluster with 57-Fe and for the spin density distribution by ENDOR and Mössbauer spectroscopy to be determined [43]. The artificial maturation mechanism, however, allowed for a selective enrichment of either the $[2Fe_H]$ cluster or the [4Fe–4S] cluster with 57-Fe as a spectroscopic handle [44–47]. This method gave the chance to synthetically label specific atoms and information concerning the enzymatic states could be obtained by various spectroscopic techniques. The key spectroscopic techniques included nuclear magnetic resonance (NMR), electron paramagnetic resonance (EPR),

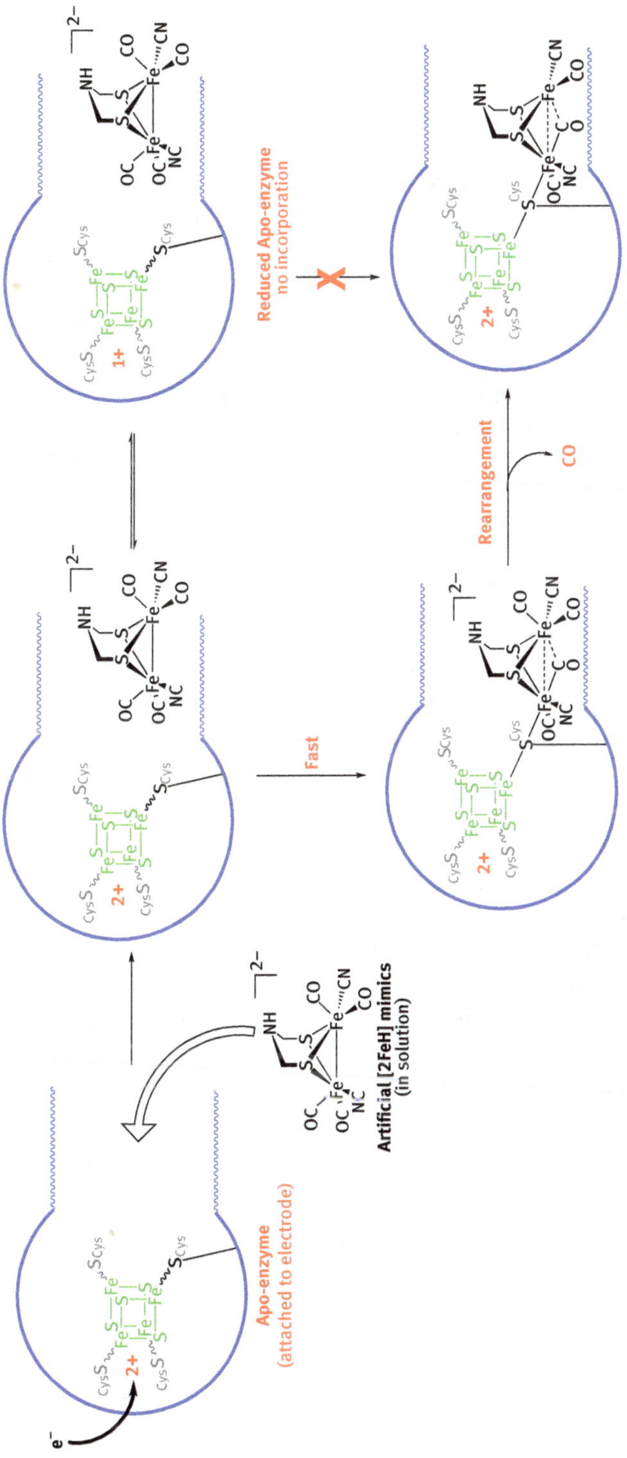

Figure 2.5: Proposed maturation process for the assembly [FeFe]-hydrogenase enzyme utilizing synthetic [$2Fe_H$] cofactor mimics [41].

nuclear resonance vibrational (NRVS) and infrared (IR) spectroscopic methods that can be selectively used to probe the enzymes properties. For example, ^{13}CN labeling in the [2FeH] cluster in CrHydA1-PDT and CrHydA1-ADT highlighted that significant spin density is located on this ligand and electron density is shifted from the [$2Fe_H$] cofactor to the [4Fe–4S] cluster back and forth [48].

Here, numerous chemically modified cofactors used to reconstitute the apo protein and these artificial cofactors demonstrate alterations such as substitution of the bridgehead amine with N–CH_3 or S, exchanging methyl groups on the CH_2 of the bridge or exchange of CN groups by CO (Figure 2.6) [49–54]. While adt or the isotope-labeled adt mimics reveal enzyme like activity, except for the N–CH_3 and monocyanide-modified cofactor mimics, all other variations of the natural cofactor were virtually inactive. These results clearly demonstrate that even small changes on the [$2Fe_H$] clusters bridgehead or the anchoring ligands can lead to catalytic breakdown [49].

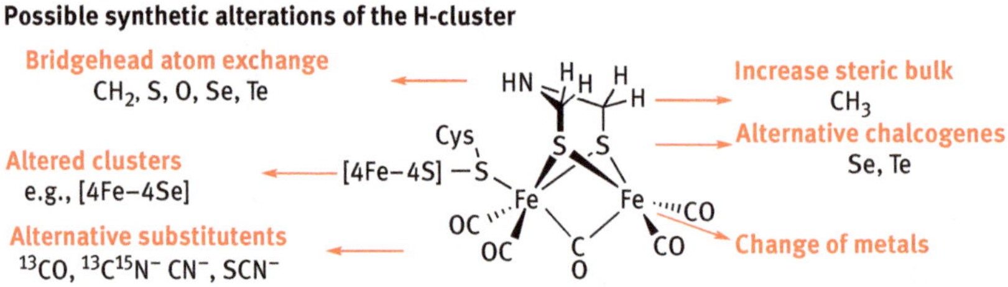

Figure 2.6: Potential manipulations of the [FeFe]-hydrogenases' cofactor.

While alterations of the bridgehead rendered the enzyme inactive, other parts of the enzyme were altered by S/Se exchange and afforded functional enzymes (Figure 2.7). At first, a selective reconstitution of a [4Fe–4Se] cluster was performed, followed by enzyme maturation with an adt mimic [55]. The successful incorporation of selenium into the H-cluster was confirmed by X-ray crystallography. The Se-enriched enzymes revealed wildtype-like enzymatic activity. These results show that the redox potentials within the electron transport chain are not as fine-tuned as was previously expected. As a result of this research, one could envision even more possible alterations to the H-cluster. Similar S/Se exchange experiments were performed with the [$2Fe_H$] cluster and the incorporation of the [$Fe_2Se_2(CO)_4(CN)_2$] assembly was confirmed by X-ray crystallography [56]. It is known from synthetic mimics that the S/Se exchange leads to higher electron density at the iron centers and adSe (–**Se**CH_2NHCH_2**Se–**) mimics reveal full enzymatic activity. Notably, the selenium-modified enzymes biased the enzyme toward proton reduction over hydrogen oxidation. While the artificial maturation of the [FeFe]-hydrogenase is still in an early stage of development, the method already shows that it is useful to perform rapid screenings of mutant libraries. With such libraries, it might be possible to generate hydrogenases with improved

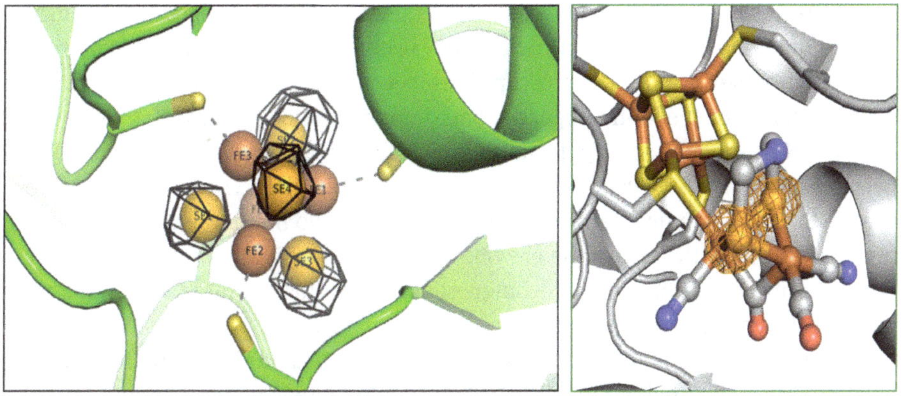

Figure 2.7: Crystallographic structures of selenium-modified enzyme variants. (Left) [4Fe–4Se] cluster in apo-HydA from C. reinhardtii [55]. (right) CpI-ADSe crystal structure of the H-cluster comprising a $[2Fe(Se)_H]$-active site mimic [56].

properties such as high oxygen tolerance, increased activity and reduced temperature sensitivity.

2.1.3 How does it work?

Even though the structural composition as well as its natural formation is well understood, the mechanism of the catalytically reversible hydrogen evolution is still not fully elucidated. While the exact mechanism is still under debate, some aspects are known to play key roles such as the mixed valence redox states, formation of a rotated state and the bridgehead ligand found in the $[2Fe_H$ cluster] [57]. The role of bridging versus terminal hydride ligands within the H_2 evolution pathway [45, 58–60] and the interplay between proton and electron transfer steps within the H-cluster [20, 47, 61] are generally accepted as fundamental steps for the enzymes unprecedented catalytic performance. The $[2Fe_H]$ molecular structure has been unambiguously determined crystallographically [39]; it is questionable whether assigning transient catalytic intermediates to static structures is valid, in particular due to the time and conditions required to obtain suitable crystals. As such, crystal structures should be regarded as a thermodynamic dead end of a stable configuration of the enzyme. In addition and in contrast to single crystal X-ray diffraction of small molecules, the resolution of enzymatic crystal structures is insufficient to reveal fine details expected between the different states [62]. Alternative possibilities to gain more reliable insight into the working mechanism of such systems can be found in Mössbauer, NRVS, EPR and IR spectroscopic techniques, which allow for the investigation of specific short-lived intermediates during the catalytic process. While each technique has its strengths and weaknesses, they allow for the selective real-time investigation of the chemistry occurring at the H-cluster without perturbation of the enzyme environment. FTIR spectroscopy has been used to investigate the catalytic conversions in the [FeFe]-hydrogenase based on the unique spectroscopic features of CO and CN-ligands in the [$2Fe_H$] cluster, which show well-resolved intense bands between 2,150 and 1,800 cm^{-1} [50]. The IR bands from the CO and CN-ligands are unique to each catalytic intermediate and these bands do not happen to overlap with other signals from the rest of the enzyme. In combination with quantum mechanics/ molecular mechanics modeling techniques, significant insight into the working principles of the enzymes was generated.

An overview of the different catalytic relevant states, including their characteristic properties, is provided in Table 2.2 [61]. While there is no doubt about the existence of the various enzymatic states, the plausible connection between those intermediates and their importance to the catalytic cycle is still under debate.

Table 2.2: Spectroscopic data on relevant intermediates of the [FeFe]-hydrogenase [48, 61, 63].

Redox species	[4Fe–4S] cluster	Formal charge of Fe_p/Fe_d	IR frequency (cm^{-1})		EPR	Mössbauer
			CN	CO		
H_{ox}	2+	I/II	2088, 2070	1964, 1940, 1802		
$H_{red'}$	1+	I/II	2084, 2066	1962, 1933, 1792		
H_{red}	2+	I/I	2070, 2033	1915, 1961, 1891		
H_{sred}	1+	I/I	2068, 2026	1918, 1953, 1882		
H_{hyd}	1+	II/II	2088,2076	1978, 1960, 1860		
$H_{ox}H$	2+	I/II	2092, 2074	1970, 1946, 1812	X	X
$H_{red'}H$	1+	I/II	2086,2068	1966, 1938, 1800	X	X
H_{ox}–CO	2+	I/II	2091, 2081	2012, 1968, 1962, 1808		
$H_{red'}$–CO	1+	I/II	–	2002, 1967, 1951, 1792		
$H_{ox}H$–CO	2+	I/II[a]	2094, 2086	2006, 1972, 1966, 1816	X	X

[a]A stronger electron delocalization was observed in the $H_{ox}H$–CO state, suggesting the real charge of Fe_p/Fe_d to be 1.5.

It is a general consensus that both the [4Fe–4S] and [$2Fe_H$] clusters are redox active; however, it remains that assignment of oxidation state is generally difficult due to the multiple electron configurations that can be considered. While the [4Fe–4S] cluster cycles between a 1+ and 2+ overall charge with average metal oxidation states of $Fe^{+2.25}$ and $Fe^{+2.5}$, [61] the [$2Fe_H$] subcluster formally can be found in the reduced ([Fe^IFe^I]), oxidized ([Fe^IFe^{II}]) and "super-oxidized" ([$Fe^{II}Fe^{II}$]) states. These clusters can exist in multiple protonation states further complicated by proton-coupled electron transfer (PCET) processes making assignment of the catalytic mechanism even more difficult [47, 61]. The elucidation of catalytic pathways in organometallic molecular catalysts is a delicate undertaking, unraveling enzymatic mechanisms is significantly more challenging due to the multitude of processes occurring along with less spectroscopic methods in the chemists tool box. These difficulties arise from the following problems:

1) Typical techniques, like NMR, FTIR and mass spectrometry that usually give insight into the mechanism of small molecule catalysts, often cannot resolve the chemistry occurring at specific parts of the cofactor in a protein environment. Here the target signals are usually overlapped by the signals of the protein host.
2) While potential enzymatic intermediates can be isolated, it is always questionable if such species are the real targets of the catalytic pathway or are secondary products of a side pathway.
3) Alterations of specific structural moieties in the active site or the second coordination sphere that are commonly performed to stabilize specific intermediates when using synthetic organometallic mimics are not as straightforward to perform within the enzyme. Thus, "stopping" the catalytic cycle at a specific intermediate is even more complicated.
4) A direct comparison of the chemistry of small molecular catalysts with enzymatic systems is not feasible due to the absence of the protein environment. [FeFe]-hydrogenases are a good example for this. Mimics of the type $[Fe_2(CO)_4(CN)_2(L)]^{2-}$ are essentially inactive under the conditions of enzymatic H_2 formation. Although usually having a [Fe^IFe^I] core and showing structural aspects of the native cofactor, due to the absence of a rotated state moiety, the iron sites are usually not basic enough to form hydrides from protons and require further reduction to either [Fe^IFe^0] or [Fe^0Fe^0] state to allow for H_2 formation. This task is usually achieved by significantly lowering the applied electrochemical potential [64].

Enzymatic mechanisms are usually based on constricted information as compared to their small molecular counterparts. The currently state-of-the-art proposed catalytic mechanism of the [FeFe]-hydrogenase is shown in Figure 2.8. The resting state of the enzyme is known as the H_{ox}. It is notable to state that while enzyme inhibition by CO was observed, this inhibition is solely present in the oxidized states of the [$2Fe_H$] cluster, H_{ox} and $H_{ox}H$. Under reductive conditions, decarbonylation of the [$2Fe_H$] cluster was observed and leads to reactivation of the enzyme [20]. The H_{ox} state consists of a

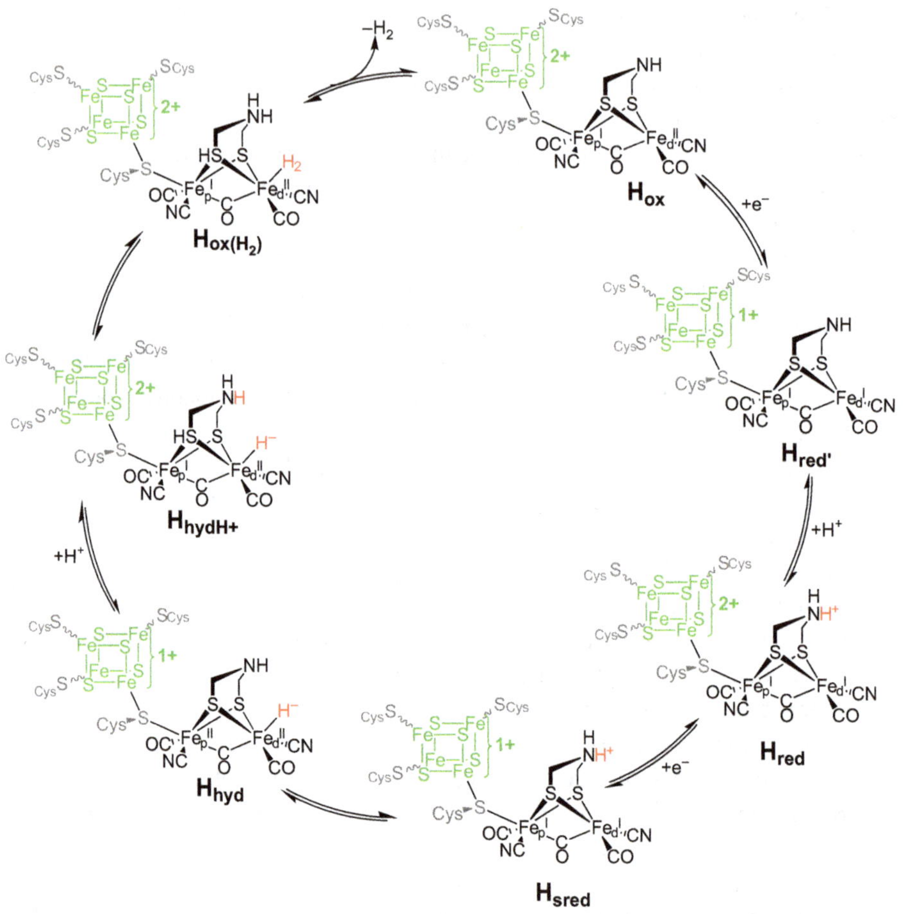

Figure 2.8: Tentative H_2 formation mechanism (after Ref. 65).

$[4Fe–4S]^{2+}$ and a $[Fe^IFe^{II}]$ cluster that possesses a bridging carbonyl and a vacant coordination site at Fe_D. In this scenario, single electron reduction subsequently leads to the $H_{red'}$ state, which is constituted by a reduced $[4Fe–4S]^+$ and an unaltered $[Fe^IFe^{II}]$ cluster [63]. Following this reduction, a PCET step is thought to afford H_{red}. This protonation most likely occurs at the amine in the adt-bridge. At the same time, an electron transfer from the $[4Fe–4S]^+$ site to the $[2Fe_H]$ cluster yields the $[Fe^IFe^I]$ cluster [58, 65]. In the following step, the $[4Fe–4S]^{2+}$ site is again reduced forming the H_{sred} state [66]. Internal proton transfer from the adt-bridge to the $[Fe^IFe^I]$ center results in the formation of a $[Fe^{II}Fe^{II}]$ intermediate bearing a terminal hydride, which was verified by IR spectroscopy and NRVS under strongly reducing conditions for blocking the following proton transfer or by applying an increased H_2 pressure to the enzyme at pH 5 [45, 59, 60, 67]. A second protonation event then finally completes the $2e^-/2H^+$ cycle forming

H_2 [68]. Based on the activity of the enzyme, the cycle is presumed to operate in the same way for the reverse process, that is, hydrogen oxidation.

As always in science, there are alternative explanations. Contrary to the mechanism depicted in Figure 2.8, an alternative mechanism suggested is more complex and differentiates between a slow and a fast catalytic process for the H_2 formation. Likewise with the previous mechanism, H_{ox} is assumed to be the resting state of the enzyme. However, single-electron reduction of the [4Fe–4S] cluster to afford $H_{red'}$ herein is assumed to be associated with an additional protonation of a cysteine (Figure 2.9), which is anchoring the accessory [4Fe–4S] cluster of the H-cluster [61, 69]. This observation is based on spectroelectrochemical redox titrations and density function theory (DFT) calculations, suggesting that the initial protonation does not take place at the adt bridge. Notably, a second proton channel to Cys499 was subsequently found in the enzymatic structure of CpI (Figure 2.9), further supporting this route. It is assumed that this "regulatory" proton pathway serves to bias the location of electrons at the [4Fe–4S] cluster and thus keeps the redox potentials of the H-cluster at a constant level. From this $H_{red'}$ state, the catalytic pathway can enter either a fast or slow cycle. In the fast cycle, PCET is assumed to occur affording the H_{hyd} state [46]. Here, the $[4Fe–4S]^+$ cluster is still in a reduced state while being protonated at Cys499. The $[2Fe_H]$ cluster now possesses

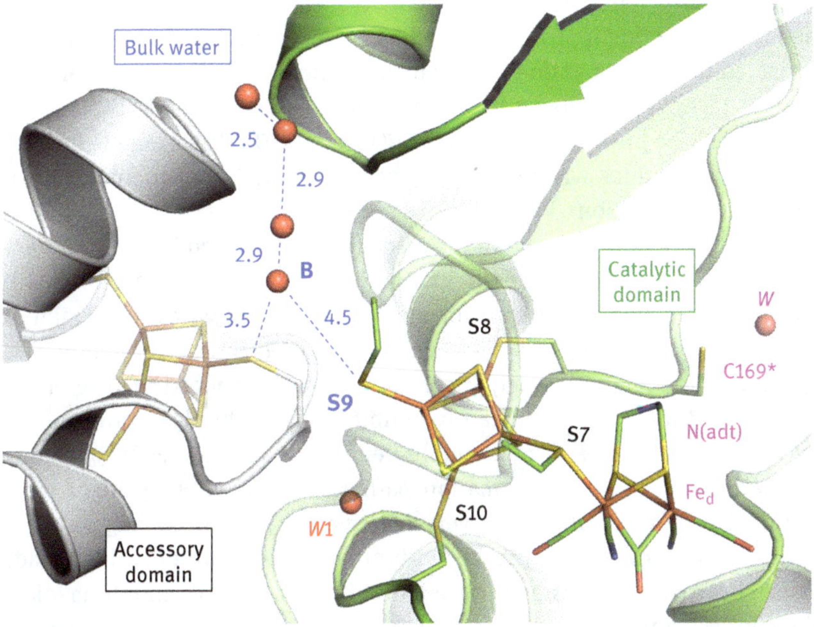

Figure 2.9: Putative proton path toward the [4Fe–4S] cluster in Clostridium pasteurianum (pdb 4XDC). The cysteines are labeled S7–S10 and the four oxygen atoms found in the putative proton pathway are highlighted in red. The important distances are given in Å.

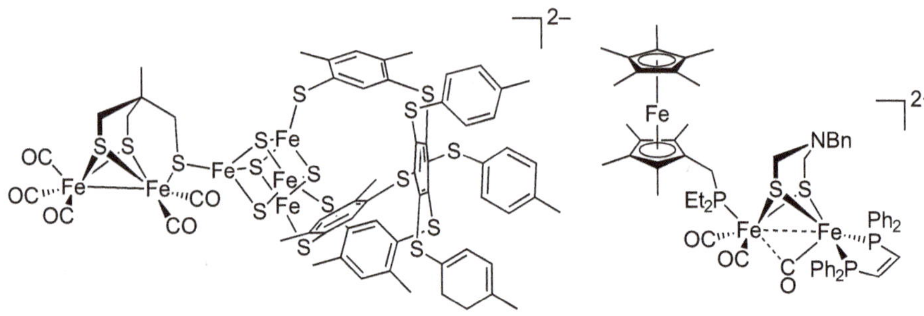

Figure 2.10: Synthetic H-cluster mimics.

a terminal hydride bound to Fe_D in the $[Fe^{II}Fe^{II}]$ moiety, which was recently supported by NRVS spectroscopy [45]. Comparable to the mechanism provided in Figure 2.8, it is believed that the second protonation takes place at the amine of the adt bridge and is immediately transferred to the distal iron site (Fe_D).

Contrary to the fast formation of H_{hyd}, deprotonation of the [4Fe–4S]-cluster in the $H_{red'}$ state can occur concomitantly with a protonation of the amine in the adt bridge. This protonation/deprotonation leads to a change in the electron distribution. The H-cluster now is built up by a $[4Fe–4S]^{2+}$ and a $[Fe^IFe^I]$ cluster (H_{red}). Further single-electron reduction leads to the H_{sred} state. Notably, the $[2Fe_H]$ cluster is assumed to bear a bridging hydride, which was recently shown via IR spectroscopy, isotope editing and quantum chemical calculations [58]. These experiments, however, were recently questioned by low temperature FTIR and NRVS data that contrarily revealed the presence of a bridging CO ligand [70]. This example furthermore shows the difficulties when comparing spectroscopic informations obtained at different reaction conditions.

The H_{red} and H_{sred} states shown in the mechanism presented in Figure 2.8 are part of the main route, and their participation in the overall mechanism is still questioned. It was previously shown for molecular [FeFe]-hydrogenase mimics that the formation of hydrogen is significantly hindered by the formation of a bridging hydride, especially in comparison to terminal hydrides [71]. The bridging hydride should therefore significantly inhibit the catalytic H_2 turnover of the [FeFe]-hydrogenase enzymes. It is therefore believed that H_{red} and H_{sred} belong to an alternative but slower H_2 formation pathway. Notably, H_{sred} might react with additional protons to afford the very reactive H_{hyd} again. As suggested in that alternative mechanism, H_{hyd} reacts with an additional proton, which leads to H_2 release. However, it is assumed herein that $H_{ox}H$ is formed, an intermediate that still comprises a protonated [4Fe–4S] cluster. Deprotonation finally leads to the reformation of the catalytic resting state H_{ox}. Whether H_{ox} and/or $H_{ox}H$ constitutes for fast hydrogen turnover remains to be explained.

2.1.4 Chemical mimics

Due to the high heating value of hydrogen (142 MJ kg^{-1} s), it is a constant endeavor to synthesize molecular systems showing comparable activity as is reported for the [FeFe]-hydrogenase enzymes. Mimicking structural aspects of the enzymes therefore was suggested to generate possible functional mimetics [72]. Furthermore, such mimics can be used to understand the basic underlying chemistry of the natural enzymes.

Notably, the [$2Fe_H$] cluster chemistry, although unwitting, was investigated in the early twentieth century by Reihlen [73], Hieber [21, 74, 75] and later on by Dahl and Wei [76], Huttner [77, 78], King [79] and Seyferth [80, 81]. They developed a rich structural chemistry comprising numerous [$Fe_2SR_2(CO)_6$] complexes with numerous modifications of the dithiolato-linker as well as CO-ligand substitutions. However, only after the discovery of the crystallographic structure of the [FeFe]-hydrogenase, such complexes were seen in an entirely new scientific context. Subsequently, new complexes were synthesized in light of establishing potential catalysts for H_2 formation and accessing specific intermediates of the natural enzyme.

Noteworthy, only two models that contain the entire H-cluster were reported so far (Figure 2.10) and researchers mainly focused on specific properties of the synthetic [$2Fe_H$] mimics [82]. Incorporating redox cofactors, such as [Fe_4S_4] or synthetic moieties such as ferrocene, appear to be a key factor for facilitating the rapid combination/separation of charge states [83–88].

Investigation limited to models in the enzymatic resting state (H_{ox}) with its bridging carbonyl ligand [89], the mixed valent Fe^IFe^{II} core and the vacant coordination site was key to understand the underlying chemistry of the enzyme. The stability of early H_{ox} mimics was, however, low but was improved significantly when CO ligands were replaced with monophosphines, diphosphines and/or carbenes (Figure 2.11) [19, 90–92]. The steric shielding provided by these synthetic ligands stabilize the vacant site either through the dithiolate linker or the CO substitutes that are necessary to force the diiron site into accepting a rotating state. In addition, due to the implemented phosphine and carbene ligands, the electron density on iron is increased and further stabilizes the mixed valent [Fe^IFe^{II}] species.

While the H_{ox} species are accessible, hydride formation and subsequently the generation of H_2 are facilitated from the enzymatic $H_{red'}$ intermediate, a [Fe^IFe^I] state. Only a few [Fe^IFe^I] state mimics comprising a rotated state structure were reported (Figure 2.12). Likewise, as the H_{ox} mimics, the steric shielding of the vacant site is suggested to play a crucial role. The complex $Fe_2(CO)_4(\kappa^2$-dmpe){μ-$(SCH_2)_2$-*N*-Bn}] (Bn = benzyl) and [$Fe_2(CO)_4(\kappa^2$-dppv){μ-$(SCH_2)_2$-CEt_2}] adopted the rotated reduced state, an asymmetrical disubstituted center, a bulky dithiolate bridge and intramolecular remote agostic interactions that are all key factors to stabilize such mimics [93, 94]. It was later shown that while the asymmetric substitution pattern and the sterically demanding dithiolate linker are indeed mandatory to afford

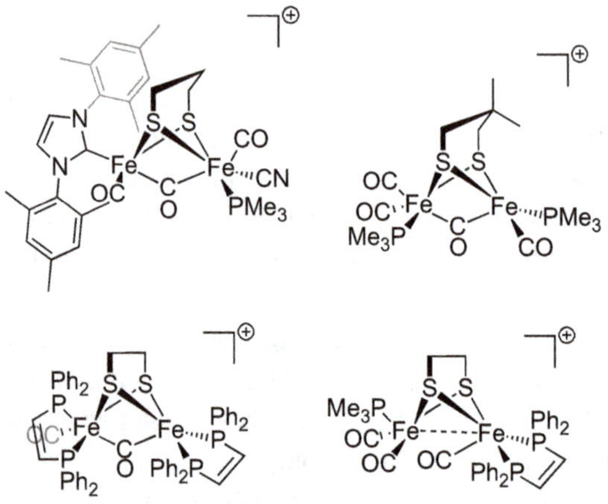

Figure 2.11: Synthetic H_{ox} mimics comprising a rotated state.

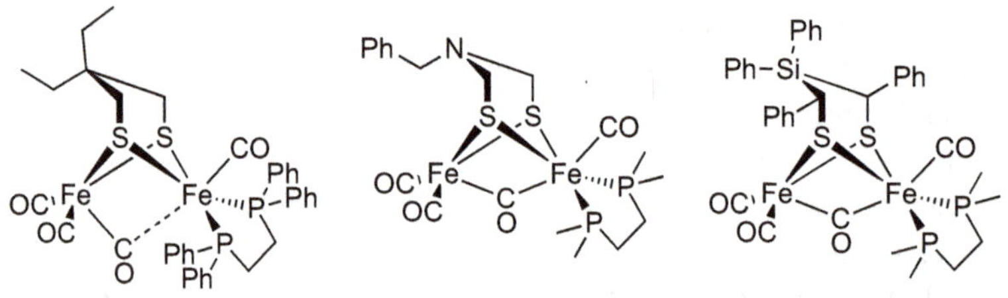

Figure 2.12: Synthetic $H_{red'}$ mimics comprising a rotated state.

structural mimics, agostic interactions are not required [95]. However, such $H_{red'}$ mimics are quite instable and will rapidly form the unrotated species in solution.

Comparable to the understanding of the structural parameters to form a rotated state structure, the protonation behavior of [FeFe]-hydrogenase mimics is a second key feature in understanding the enzymatic mechanism. Since the functioning at the molecular level is closely related to the chemistry of the [2FeH] cluster, the understanding also helps to allow for spectroscopic comparison with the enzymatic system. The formation of iron-hydrides as well as protonation of the nitrogen atom in the dithiolato bridge has been the focus of much research.

Metal–metal protonation of [2FeH] mimics is thermodynamically favored; the adt linker is kinetically preferred [96, 97]. Due to the absence of directed proton channels in model complexes, protonation at both the adt-linker and the iron site can occur. The protonation herein strongly depends on the pK_a of the applied proton source.

The protonation of the diiron site requires high electron density on the [2Fe2S] core. Therefore, protonation of the symmetric hexacarbonyl complexes [$Fe_2(CO)_6$(μ-dithiolate)] (dithiolate ≠ adt) cannot be achieved with common acids due to the high kinetic barriers that must be overcome [64, 98]. Only in the presence of the super acid $[SiEt_3][B(C_6F_5)_4]/HCl$, a direct protonation was observed [99]. Reduction to an $[Fe^0Fe^I]$ or an $[Fe^0Fe^0]$ is commonly required to achieve protonation of the hexacarbonyl diiron subsite [100–103]. Likewise, high electron density at the diiron site can be achieved by substitution of the CO ligands by phosphines, carbenes or cyanides. Hydrides subsequently generated with such substituted molecules are considerably more stable as compared to their unsubstituted counterparts [91, 96, 104–113].

Metal–hydride vibrations produce hard to detect signals by IR spectroscopy especially due to the weak intensity of such M–H bands; the CO signals provide valuable information on the electron structure of the diiron site. For $[2Fe_H]$ mimics comprising bridging hydrides, shifts of the CO bands of ~50 cm^{-1} were reported [106, 114]. In addition to IR techniques, ^1H NMR spectroscopy provides significant insight into the diiron subsite chemistry. These bridging hydride signals are commonly observed between –8 and –20 ppm [115]. Notably, the bridging hydride mode in [2Fe–2S] complexes was also shown by crystallography. These mimics show similar structural and electronic properties of the H_{red} and H_{sred} state. While structurally fully characterized and the bridging hydride mode is thermodynamically favored, these complexes were generally shown to possesses low activity toward H_2 formation and slow H_2 release [91, 111, 116, 117].

It is also possible to obtain terminal iron hydrides. In contrast to their bridging congeners, the terminal iron hydride mimics can be reduced at anodically shifted potentials relative to complexes containing bridging hydrides and allow for H_2 formation at moderate and enzyme like conditions. Mimics comprising a terminal hydride are, however, rare and were only observed when the diiron site revealed an unsymmetric substitution pattern or bearing sterically demanding phosphine ligands [91, 112, 115, 117–124]. Generally, proton resonances for such species are usually found between –3 and –5 ppm in the ^1H NMR spectrum. Nevertheless, terminal hydrides are only stable at low temperatures and they show rapid geometry rearrangement reactions to afford the thermodynamically favored bridging hydrides upon warming the sample (Figure 2.13) [121]. Isomerization can be significantly slowed down by utilizing sterically crowding ligands, and the

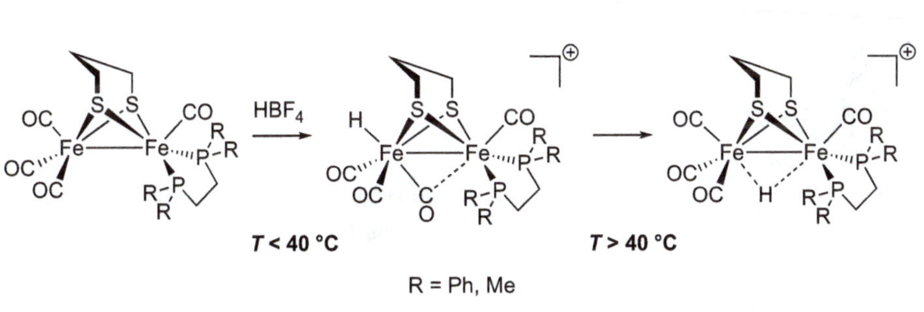

Figure 2.13: Hydride formation and equilibrium between terminal and bridging hydrides.

terminal hydride could be stabilized in $[Fe_2H(CO)_2(\kappa^2\text{-dppv})_2(\mu\text{-pdt})]^+$ (dppv = $Ph_2P(CH)_2PPh_2$) up to a temperature of 20 °C [117].

Following this method of generating terminal hydrides, it was also possible obtain single crystals of $[Fe_2H(xdt)(CO)_2(PMe_3)_4]^+$ (xdt = adt, pdt) (Figure 2.14) and to obtain in depth structural information on such hydridic species [125].

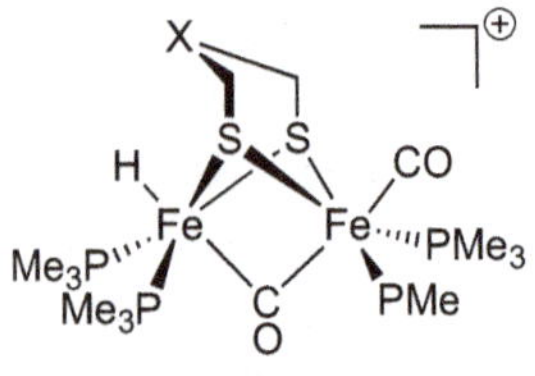

Figure 2.14: Crystallographically verified [2FeH] mimics containing a terminal hydride.

While thermodynamically more stable, protonation reactions on $[2Fe_H]$ mimics in the presence of the adt-bridge are usually kinetically controlled. This behavior is comparable to the chemistry observed in the enzyme where the proton is expected to be subsequently shuffled to the distal iron center in a fast isomerization process.

Indeed, the formation of an ammonium species can be observed showing a distinct ^1H NMR signal at ~5.9 ppm when $[Fe_2(CO)_6(adt)]$ reacts with acids [116].

The basicity of the adt bridge is much lower than expected for secondary amines. Protonation can only be afforded by using moderately or very strong acids [96, 116]. Theoretical calculations revealed that the low basicity of the adt-nitrogen stems from an orbital interaction between the adt-N-lone pair and the antibonding σ* (C–S) orbital (Figure 2.15) [126]. The interaction subsequently leads to a decrease of the electron density on the adt-nitrogen and thus lowering its basicity. In addition, an increase of the C–S bond distance was suggested and confirmed in the molecular structure of $[Fe_2(CO)_6(adt)]$ [116, 127].

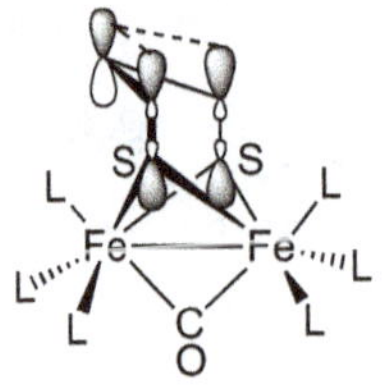

Figure 2.15: Orbital interactions in the adt-bridge that leads to an increase of the adt-nitrogens' basicity. L = ligands.

The migration of a proton from the adt-nitrogen and with it the rate of the formation of a hydridic species depends on the basicity of the diiron center. In bis-disphosphine complexes comprising either NH or $NHCH_2Ph$ groups in the dithiolate bridgehead, the proton migration from the ligand to the metal was found to be very rapid [117, 120, 127, 128]. In the case of isopropyl substituents at the dithiolate bridge or when applying dppe substituted complexes, this tautomerization was not observed [129]. The importance of the structural features of the mimics is clear, but it is worth mentioning that the proton migration can also be affected by the anion of the acid [97, 127] as well as the solvent used [130]. This observation shows that second sphere effects play a major role for the reactivity of [2FeH] mimics and by extension highlights the importance of the protein environment of the natural enzyme.

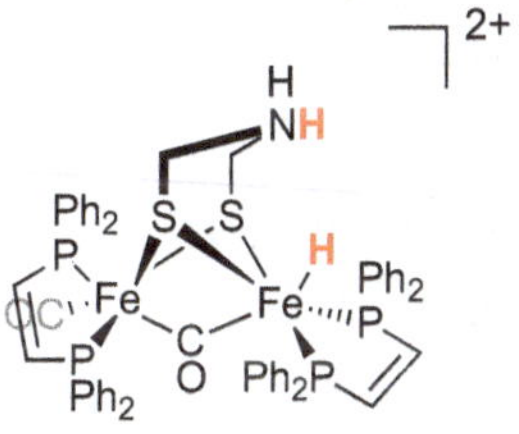

Figure 2.16: Double protonated [2FeH] mimic.

In the presence of an excess of protons, the proton migration can be followed by a second protonation at the adt-nitrogen [117, 127, 130–133]. While this reaction usually leads to the fast release of hydrogen, a doubly protonated species could be spectroscopically observed and was structurally analyzed showing the structure $[Fe_2(CO)_2(dppe)_2(H)(adtH)]^{2+}$ (Figure 2.16) [127]. This species was postulated to represent a key intermediate in the fast enzymatic H_2 formation mechanism and shows that both a hydride and a proton can be positioned in close proximity to each other without generating H_2. This type of dihydrogen bonding has also been described for various other organometallic species where the metal and basic ligand play an important role providing a stabile interaction [134]. In terms of fundamental chemistry, this example clearly shows that the special arrangement in the [2FeH] cluster acts as a frustrated Lewis pair. And indeed, under electrocatalytic conditions, hydrogen formation utilizing CF_3CO_2H as proton source, this mimic revealed a turn over frequency of 58,000 s^{-1}, rendering this complex one of the fastest reported. It is suggested that hydrogen release forms the double protonated species, which leads to a rotated oxidized form that again can enter the electrocatalytic reduction cycle.

References

[1] Holm, R. H., Kennepohl, P., and Solomon, E. I. Structural and functional aspects of metal sites in biology. Chem Rev 1996, 96(7), 2239–2314. https://doi.org/10.1021/cr9500390.

[2] Vignais, P. M., and Billoud, B.. Occurrence, classification, and biological function of hydrogenases: an overview. Chem Rev 2007, 107(10), 4206–4272. https://doi.org/10.1021/cr050196r.
[3] Lubitz, W., Ogata, H., Rüdiger, O., and Reijerse, E. Hydrogenases. Chem Rev 2014, 114(8), 4081–4148. https://doi.org/10.1021/cr4005814.
[4] Stripp, S. T., and Happe, T. How algae produce hydrogen—news from the photosynthetic hydrogenase. Dalton Trans 2009, No. 45, 9960. https://doi.org/10.1039/b916246a.
[5] Peters, J. W., Lanzilotta, W. N., Lemon, B. J., and Seefeldt, L. C. X-ray crystal structure of the fe-only hydrogenase (CpI) from clostridium pasteurianum to 1.8 angstrom resolution. Science 1998, 282(5395), 1853–1858.
[6] Stripp, S. T. Molecular background of oxygen sensitivity in [FeFe] hydrogenases. PNAS 2009, 106(41), 17331–17336.
[7] Stripp, S. T., Goldet, G., Brandmayr, C., Sanganas, O., Vincent, K. A., Haumann, M., Armstrong, F. A., and Happe, T. How oxygen attacks [FeFe] hydrogenases from photosynthetic organisms. Proc Natl Acad Sci 2009, 106(41), 17331–17336. https://doi.org/10.1073/pnas.0905343106.
[8] Koo, J., Shiigi, S., Rohovie, M., Mehta, K., and Swartz, J. R. Characterization of [FeFe] hydrogenase O 2 sensitivity using a new, physiological approach. J Biol Chem 2016, jbc.M116.737122. https://doi.org/10.1074/jbc.M116.737122.
[9] Rodríguez-Maciá, P., Birrell, J. A., Lubitz, W., and Rüdiger, O. Electrochemical investigations on the inactivation of the [FeFe] hydrogenase from desulfovibrio desulfuricans by O 2 or light under hydrogen-producing conditions. Chem Plus Chem 2016. https://doi.org/10.1002/cplu.201600508.
[10] Swanson, K. D., Ratzloff, M. W., Mulder, D. W., Artz, J. H., Ghose, S., Hoffman, A., White, S., Zadvornyy, O. A., Broderick, J. B., Bothner, B.; et al. [FeFe]-hydrogenase oxygen inactivation is initiated at the H cluster 2Fe subcluster. J Am Chem Soc 2015, 137(5), 1809–1816. https://doi.org/10.1021/ja510169s.
[11] Nicolet, Y., Piras, C., Legrand, P., Hatchikian, C. E., and Fontecilla-Camps, J. C. Desulfovibrio desulfuricans iron hydrogenase: the structure shows unusual coordination to an active site fe binuclear center. Structure 1999, 7(1), 13–23. https://doi.org/10.1016/S0969-2126(99)80005-7.
[12] Lampret, O., Adamska-Venkatesh, A., Konegger, H., Wittkamp, F., Apfel, U.-P., Reijerse, E. J., Lubitz, W., Rüdiger, O., Happe, T., and Winkler, M. Interplay between CN – ligands and the secondary coordination sphere of the H-cluster in [FeFe]-hydrogenases. J Am Chem Soc 2017, 139(50), 18222–18230. https://doi.org/10.1021/jacs.7b08735.
[13] Duan, J., Mebs, S., Senger, M., Laun, K., Wittkamp, F., Heberle, J., Happe, T., Hofmann, E., Apfel, U.-P., Winkler, M.; et al. The geometry of the catalytic active site in [FeFe]-hydrogenases is determined by hydrogen bonding and proton transfer. Chem Rxiv 2019. https://doi.org/10.26434/chemrxiv.7756214.v1.
[14] Cornish, A. J., Ginovska, B., Thelen, A., Da Silva, J. C. S., Soares, T. A., Raugei, S., Dupuis, M., Shaw, W. J., and Hegg, E. L. Single-amino acid modifications reveal additional controls on the proton pathway of [FeFe]-hydrogenase. Biochemistry 2016, 55(22), 3165–3173. https://doi.org/10.1021/acs.biochem.5b01044.
[15] Ginovska-Pangovska, B., Ho, M.-H., Linehan, J. C., Cheng, Y., Dupuis, M., Raugei, S., and Shaw, W. J. Molecular dynamics study of the proposed proton transport pathways in [FeFe]-hydrogenase. Biochim Biophys Acta BBA – Bioenerg 2014, 1837(1), 131–138. https://doi.org/10.1016/j.bbabio.2013.08.004.
[16] Duan, J., Senger, M., Esselborn, J., Engelbrecht, V., Wittkamp, F., Apfel, U.-P., Hofmann, E., Stripp, S. T., Happe, T., and Winkler, M.. Crystallographic and spectroscopic assignment of

the proton transfer pathway in [FeFe]-hydrogenases. Nat Commun 2018, 9 (1). https://doi.org/10.1038/s41467-018-07140-x.
[17] Berggren, G., Adamska, A., Lambertz, C., Simmons, T. R., Esselborn, J., Atta, M., Gambarelli, S., Mouesca, J.-M., Reijerse, E., Lubitz, W.; et al. Biomimetic assembly and activation of [FeFe]-hydrogenases. Nature 2013, 499(7456), 66–69. https://doi.org/10.1038/nature12239.
[18] Esselborn, J., Lambertz, C., Adamska-Venkatesh, A., Simmons, T., Berggren, G., Noth, J., Siebel, J., Hemschemeier, A., Artero, V., Reijerse, E.; et al. Spontaneous activation of [FeFe]-hydrogenases by an inorganic [2Fe] active site mimic. Nat Chem Biol 2013, 9(10), 607–609. https://doi.org/10.1038/nchembio.1311.
[19] Liu, T., Darensbourg, M. Y., and Mixed-Valent, A., Fe(II)Fe(I), diiron complex reproduces the unique rotated state of the [FeFe]hydrogenase active site. J Am Chem Soc 2007, 129(22), 7008–7009. https://doi.org/10.1021/ja071851a.
[20] Laun, K., Mebs, S., Duan, J., Wittkamp, F., Apfel, U.-P., Happe, T., Winkler, M., Haumann, M., and Stripp, S. Spectroscopical investigations on the redox chemistry of [FeFe]-hydrogenases in the presence of carbon monoxide. Molecules 2018, 23 (7), 1669. https://doi.org/10.3390/molecules23071669.
[21] Hieber, W., and Gruber, J. Zur kenntnis der eisencarbonylchalkogenide. Z Für Anorg Allg Chem 1958, 296 (1–6), 91–103.
[22] Li, H., and Rauchfuss, T. B. Iron carbonyl sulfides, formaldehyde, and amines condense to give the proposed azadithiolate cofactor of the Fe-only hydrogenases. J Am Chem Soc 2002, 124(5), 726–727. https://doi.org/10.1021/ja016964n. Guodong Rao, Lizhi Tao, Daniel L. M. Suess, R. David Britt, Nature Chemistry doi: 10.1038/s41557-018-0026-7.
[23] Posewitz, M. C., King, P. W., Smolinski, S. L., Zhang, L., Seibert, M., and Ghirardi, M. L. Discovery of two novel radical S -adenosylmethionine proteins required for the assembly of an active [Fe] hydrogenase. J Biol Chem 2004, 279(24), 25711–25720. https://doi.org/10.1074/jbc.M403206200.
[24] Posewitz, M. C., King, P. W., Smolinski, S. L., Smith, R. D., Ginley, A. R., Ghirardi, M. L., and Seibert, M. Identification of genes required for hydrogenase activity in chlamydomonas reinhardtii : figure 1. Biochem Soc Trans 2005, 33(1), 102–104. https://doi.org/10.1042/BST0330102.
[25] McGlynn, S. E., Shepard, E. M., Winslow, M. A., Naumov, A. V., Duschene, K. S., Posewitz, M. C., Broderick, W. E., Broderick, J. B., and Peters, J. W. HydF as a scaffold protein in [FeFe] hydrogenase H-cluster biosynthesis. FEBS Lett 2008, 582(15), 2183–2187. https://doi.org/10.1016/j.febslet.2008.04.063.
[26] McGlynn, S. E., Ruebush, S. S., Naumov, A., Nagy, L. E., Dubini, A., King, P. W., Broderick, J. B., Posewitz, M. C., and Peters, J. W. In vitro activation of [FeFe] hydrogenase: new insights into hydrogenase maturation. JBIC J Biol Inorg Chem 2007, 12(4), 443–447. https://doi.org/10.1007/s00775-007-0224-z.
[27] Rubach, J. K., Brazzolotto, X., Gaillard, J., and Fontecave, M. Biochemical characterization of the HydE and hydg iron-only hydrogenase maturation enzymes from thermatoga maritima. FEBS Lett 2005, 579(22), 5055–5060. https://doi.org/10.1016/j.febslet.2005.07.092.
[28] Sofia, H. J. Radical SAM, a novel protein superfamily linking unresolved steps in familiar biosynthetic pathways with radical mechanisms: functional characterization using new analysis and information visualization methods. Nucleic Acids Res 2001, 29(5), 1097–1106. https://doi.org/10.1093/nar/29.5.1097.
[29] Kuchenreuther, J. M., Grady-Smith, C. S., Bingham, A. S., George, S. J., Cramer, S. P., and Swartz, J. R. High-yield expression of heterologous [FeFe] hydrogenases in escherichia coli. PLoS ONE 2010, 5 (11), e15491. https://doi.org/10.1371/journal.pone.0015491.

[30] Shepard, E. M., Duffus, B. R., George, S. J., McGlynn, S. E., Challand, M. R., Swanson, K. D., Roach, P. L., Cramer, S. P., Peters, J. W., and Broderick, J. B. [FeFe]-hydrogenase maturation: hydg-catalyzed synthesis of carbon monoxide. J Am Chem Soc 2010, 132(27), 9247–9249. https://doi.org/10.1021/ja1012273.
[31] Driesener, R. C., Challand, M. R., McGlynn, S. E., Shepard, E. M., Boyd, E. S., Broderick, J. B., Peters, J. W., and Roach, P. L. [FeFe]-hydrogenase cyanide ligands derived from S-adenosylmethionine-dependent cleavage of tyrosine. Angew Chem Int Ed 2010, 49(9), 1687–1690. https://doi.org/10.1002/anie.200907047.
[32] Peters, J. W., Szilagyi, R. K., Naumov, A., and Douglas, T. A radical solution for the biosynthesis of the H-cluster of hydrogenase. FEBS Lett 2006, 580(2), 363–367. https://doi.org/10.1016/j.febslet.2005.12.040.
[33] Driesener, R. C., Duffus, B. R., Shepard, E. M., Bruzas, I. R., Duschene, K. S., Coleman, N. J.-R., Marrison, A. P. G., Salvadori, E., Kay, C. W. M., Peters, J. W.; et al. Biochemical and kinetic characterization of radical S -adenosyl- l -methionine enzyme HydG. Biochemistry 2013, 52 (48), 8696–8707. https://doi.org/10.1021/bi401143s.
[34] Kuchenreuther, J. M., George, S. J., Grady-Smith, C. S., Cramer, S. P., and Swartz, J. R. Cell-free H-cluster synthesis and [FeFe] hydrogenase activation: all five CO and CN– ligands derive from tyrosine. PLoS ONE 2011, 6 (5), e20346. https://doi.org/10.1371/journal.pone.0020346.
[35] Kuchenreuther, J. M., Myers, W. K., Stich, T. A., George, S. J., NejatyJahromy, Y., Swartz, J. R., and Britt, R. D. A radical intermediate in tyrosine scission to the CO and CN- ligands of FeFe hydrogenase. Science 2013, 342(6157), 472–475. https://doi.org/10.1126/science.1241859.
[36] Kuchenreuther, J. M., Myers, W. K., Suess, D. L. M., Stich, T. A., Pelmenschikov, V., Shiigi, S. A., Cramer, S. P., Swartz, J. R., Britt, R. D., and George, S. J. The HydG enzyme generates an Fe(CO)2(CN) synthon in assembly of the FeFe hydrogenase H-cluster. Science 2014, 343 (6169), 424–427. https://doi.org/10.1126/science.1246572.
[37] Pilet, E., Nicolet, Y., Mathevon, C., Douki, T., Fontecilla-Camps, J. C., and Fontecave, M. The role of the maturase HydG in [FeFe]-hydrogenase active site synthesis and assembly. FEBS Lett 2009, 583(3), 506–511. https://doi.org/10.1016/j.febslet.2009.01.004.
[38] Vallese, F., Berto, P., Ruzzene, M., Cendron, L., Sarno, S., De Rosa, E., Giacometti, G. M., and Costantini, P. Biochemical analysis of the interactions between the proteins involved in the [FeFe]-hydrogenase maturation process. J Biol Chem 2012, 287(43), 36544–36555. https://doi.org/10.1074/jbc.M112.388900.
[39] Esselborn, J., Muraki, N., Klein, K., Engelbrecht, V., Metzler-Nolte, N., Apfel, U.-P., Hofmann, E., Kurisu, G., and Happe, T. A structural view of synthetic cofactor integration into [FeFe]-hydrogenases. Chem Sci 2016, 7(2), 959–968. https://doi.org/10.1039/C5SC03397G.
[40] Caserta, G., Adamska-Venkatesh, A., Pecqueur, L., Atta, M., Artero, V., Roy, S., Reijerse, E., Lubitz, W., and Fontecave, M. Chemical assembly of multiple metal cofactors: the heterologously expressed multidomain [FeFe]-hydrogenase from megasphaera elsdenii. Biochim Biophys Acta BBA – Bioenerg 2016, 1857(11), 1734–1740. https://doi.org/10.1016/j.bbabio.2016.07.002.
[41] Megarity, C. F., Esselborn, J., Hexter, S. V., Wittkamp, F., Apfel, U.-P., Happe, T., and Armstrong, F. A. Electrochemical investigations of the mechanism of assembly of the active-site H-cluster of [FeFe]-hydrogenases. J Am Chem Soc 2016, 138(46), 15227–15233. https://doi.org/10.1021/jacs.6b09366.
[42] Rodríguez-Maciá, P., Reijerse, E., Lubitz, W., Birrell, J. A., and Rüdiger, O. Spectroscopic evidence of reversible disassembly of the [FeFe] hydrogenase active site. J Phys Chem Lett 2017, 8(16), 3834–3839. https://doi.org/10.1021/acs.jpclett.7b01608.
[43] Adamska-Venkatesh, A., Simmons, T. R., Siebel, J. F., Artero, V., Fontecave, M., Reijerse, E., and Lubitz, W.. Artificially maturated [FeFe] hydrogenase from chlamydomonas reinhardtii: a

HYSCORE and ENDOR study of a non-natural H-cluster. Phys Chem Chem Phys 2015, 17(7), 5421–5430. https://doi.org/10.1039/C4CP05426A.
[44] Gilbert-Wilson, R., Siebel, J. F., Adamska-Venkatesh, A., Pham, C. C., Reijerse, E., Wang, H., Cramer, S. P., Lubitz, W., and Rauchfuss, T. B. Spectroscopic investigations of [FeFe] hydrogenase maturated with [57 Fe 2 (Adt)(CN) 2 (CO) 4] 2–. J Am Chem Soc 2015, 137(28), 8998–9005. https://doi.org/10.1021/jacs.5b03270.
[45] Reijerse, E. J., Pham, C. C., Pelmenschikov, V., Gilbert-Wilson, R., Adamska-Venkatesh, A., Siebel, J. F., Gee, L. B., Yoda, Y., Tamasaku, K., Lubitz, W.; et al. Direct observation of an iron-bound terminal hydride in [FeFe]-hydrogenase by nuclear resonance vibrational spectroscopy. J Am Chem Soc 2017, 139(12), 4306–4309. https://doi.org/10.1021/jacs.7b00686.
[46] Mebs, S., Kositzki, R., Duan, J., Senger, M., Wittkamp, F., Apfel, U.-P., Happe, T., Stripp, S. T., Winkler, M., and Haumann, M.. Hydrogen and oxygen trapping at the H-cluster of [FeFe]-hydrogenase revealed by site-selective spectroscopy and QM/MM calculations. Biochim Biophys Acta BBA – Bioenerg 2017. https://doi.org/10.1016/j.bbabio.2017.09.003.
[47] Mebs, S., Duan, J., Wittkamp, F., Stripp, S. T., Happe, T., Apfel, U.-P., Winkler, M., and Haumann, M. Differential protonation at the catalytic six-iron cofactor of [FeFe]-hydrogenases revealed by 57 Fe nuclear resonance x-ray scattering and quantum mechanics/molecular mechanics analyses. Inorg Chem 2019, 58(6), 4000–4013. https://doi.org/10.1021/acs.inorgchem.9b00100.
[48] Lubitz, W., Reijerse, E., and van Gastel, M. [NiFe] and [FeFe] hydrogenases studied by advanced magnetic resonance techniques. Chem Rev 2007, 107(10), 4331–4365. https://doi.org/10.1021/cr050186q.
[49] Siebel, J. F., Adamska-Venkatesh, A., Weber, K., Rumpel, S., Reijerse, E., and Lubitz, W.. Hybrid [FeFe]-hydrogenases with modified active sites show remarkable residual enzymatic activity. Biochemistry 2015, 54(7), 1474–1483. https://doi.org/10.1021/bi501391d.
[50] Wittkamp, F., Senger, M., Stripp, S. T., and Apfel, U.-P. [FeFe]-hydrogenases: recent developments and future perspectives. Chem Commun 2018. https://doi.org/10.1039/C8CC01275J.
[51] Singleton, M. L., Bhuvanesh, N., Reibenspies, J. H., and Darensbourg, M. Y. Synthetic support of de novo design: sterically bulky [FeFe]-hydrogenase models. Angew Chem Int Ed 2008, 47(49), 9492–9495. https://doi.org/10.1002/anie.200803939.
[52] Apfel, U.-P., Pétillon, F. Y., Schollhammer, P., Talarmin, J., and Weigand, W. [FeFe] Hydrogenase Models: An Overview. In Bioinspired Catalysis; Weigand, W., Schollhammer, P., Eds.; Wiley-VCH Verlag GmbH & Co. KGaA: Weinheim, Germany, 2014; pp 79–104.
[53] Almazahreh, L. R., Imhof, W., Talarmin, J., Schollhammer, P., Görls, H., El-khateeb, M., and Weigand, W. Ligand effects on the electrochemical behavior of [Fe 2 (CO) 5 (L){μ-(SCH 2) 2 (Ph)P□O}] (L = PPh 3, P(OEt) 3) hydrogenase model complexes. Dalton Trans 2015, 44(16), 7177–7189. https://doi.org/10.1039/C5DT00064E.
[54] Harb, M. K.., Apfel, U.-P., Sakamoto, T., El-khateeb, M., and Weigand, W.. Diiron dichalcogenolato (se and te) complexes: models for the active site of [FeFe] hydrogenase. Eur J Inorg Chem 2011, 2011(7), 986–993. https://doi.org/10.1002/ejic.201001112.
[55] Noth, J., Esselborn, J., Güldenhaupt, J., Brünje, A., Sawyer, A., Apfel, U.-P., Gerwert, K., Hofmann, E., Winkler, M., and Happe, T. [FeFe]-hydrogenase with chalcogenide substitutions at the H-cluster maintains Full H 2 evolution activity. Angew Chem Int Ed 2016, 55(29), 8396–8400. https://doi.org/10.1002/anie.201511896.
[56] Kertess, L., Wittkamp, F., Sommer, C., Esselborn, J., Rüdiger, O., Reijerse, E. J., Hofmann, E., Lubitz, W., Winkler, M., Happe, T.; et al. Chalcogenide substitution in the [2Fe] cluster of [FeFe]-hydrogenases conserves high enzymatic activity. Dalton Trans 2017, 46(48), 16947–16958. https://doi.org/10.1039/C7DT03785F.

[57] Haumann, M., and Stripp, S. T. The molecular proceedings of biological hydrogen turnover. Acc Chem Res 2018. https://doi.org/10.1021/acs.accounts.8b00109.
[58] Mebs, S., Senger, M., Duan, J., Wittkamp, F., Apfel, U.-P., Happe, T., Winkler, M., Stripp, S. T., and Haumann, M. Bridging hydride at reduced H-cluster species in [FeFe]-hydrogenases revealed by infrared spectroscopy, isotope editing, and quantum chemistry. J Am Chem Soc 2017, 139(35), 12157–12160. https://doi.org/10.1021/jacs.7b07548.
[59] Rumpel, S., Sommer, C., Reijerse, E., Farès, C., and Lubitz, W. Direct detection of the terminal hydride intermediate in [FeFe] hydrogenase by NMR spectroscopy. J Am Chem Soc 2018, 140 (11), 3863–3866. https://doi.org/10.1021/jacs.8b00459.
[60] Mulder, D. W., Guo, Y., Ratzloff, M. W., and King, P. W. Identification of a catalytic iron-hydride at the H-cluster of [FeFe]-hydrogenase. J Am Chem Soc 2016. https://doi.org/10.1021/jacs.6b11409.
[61] Senger, M., Mebs, S., Duan, J., Shulenina, O., Laun, K., Kertess, L., Wittkamp, F., Apfel, U.-P., Happe, T., Winkler, M.; et al. Protonation/reduction dynamics at the [4Fe–4S] cluster of the hydrogen-forming cofactor in [FeFe]-hydrogenases. Phys Chem Chem Phys 2018. https://doi.org/10.1039/C7CP04757F.
[62] Nicolet, Y., de Lacey, A. L., Vernède, X., Fernandez, V. M., Hatchikian, E. C., and Fontecilla-Camps, J. C. Crystallographic and FTIR spectroscopic evidence of changes in Fe coordination upon reduction of the active site of the Fe-only hydrogenase from desulfovibrio d esulfuricans. J Am Chem Soc 2001, 123(8), 1596–1601. https://doi.org/10.1021/ja0020963.
[63] Adamska-Venkatesh, A., Krawietz, D., Siebel, J., Weber, K., Happe, T., Reijerse, E., and Lubitz, W. New redox states observed in [FeFe] hydrogenases reveal redox coupling within the H-cluster. J Am Chem Soc 2014, 136(32), 11339–11346. https://doi.org/10.1021/ja503390c.
[64] Felton, G. A. N., Mebi, C. A., Petro, B. J., Vannucci, A. K., Evans, D. H., Glass, R. S., and Lichtenberger, D. L. Review of electrochemical studies of complexes containing the Fe2S2 core characteristic of [FeFe]-hydrogenases including catalysis by these complexes of the reduction of acids to form dihydrogen. J Organomet Chem 2009, 694(17), 2681–2699. https://doi.org/10.1016/j.jorganchem.2009.03.017.
[65] Sommer, C., Adamska-Venkatesh, A., Pawlak, K., Birrell, J. A., Rüdiger, O., Reijerse, E. J., and Lubitz, W. Proton coupled electronic rearrangement within the H-cluster as an essential step in the catalytic cycle of [FeFe] hydrogenases. J Am Chem Soc 2017, 139(4), 1440–1443. https://doi.org/10.1021/jacs.6b12636.
[66] Adamska, A., Silakov, A., Lambertz, C., Rüdiger, O., Happe, T., Reijerse, E., and Lubitz, W.. Identification and characterization of the "super-reduced" state of the H-cluster in [FeFe] hydrogenase: a new building block for the catalytic cycle? Angew Chem Int Ed 2012, 51(46), 11458–11462. https://doi.org/10.1002/anie.201204800.
[67] Winkler, M., Senger, M., Duan, J., Esselborn, J., Wittkamp, F., Hofmann, E., Apfel, U.-P., Stripp, S. T., and Happe, T. Accumulating the hydride state in the catalytic cycle of [FeFe]-hydrogenases. Nat Commun 2017, 8, 16115. https://doi.org/10.1038/ncomms16115.
[68] Pelmenschikov, V., Birrell, J. A., Pham, C. C., Mishra, N., Wang, H., Sommer, C., Reijerse, E., Richers, C. P., Tamasaku, K., Yoda, Y.; et al. Reaction coordinate leading to H 2 production in [FeFe]-hydrogenase identified by nuclear resonance vibrational spectroscopy and density functional theory. J Am Chem Soc 2017, 139(46), 16894–16902. https://doi.org/10.1021/jacs.7b09751.
[69] Senger, M., Laun, K., Wittkamp, F., Duan, J., Haumann, M., Happe, T., Winkler, M., Apfel, U.-P., and Stripp, S. T. Proton-coupled reduction of the catalytic [4Fe-4S] cluster in [FeFe]-hydrogenases. Angew Chem Int Ed 2017, 56(52), 16503–16506. https://doi.org/10.1002/anie.201709910.

[70] DeBeer et al. JACS. https://doi.org/10.1021/jacs.9b09745.
[71] Filippi, G., Arrigoni, F., Bertini, L., De Gioia, L., and Zampella, G. DFT dissection of the reduction step in H 2 catalytic production by [FeFe]-hydrogenase-inspired models: can the bridging hydride become more reactive than the terminal isomer? Inorg Chem 2015, 54(19), 9529–9542. https://doi.org/10.1021/acs.inorgchem.5b01495.
[72] Tard, C., and Pickett, C. J. Structural and functional analogues of the active sites of the [Fe]-, [NiFe]-, and [FeFe]-hydrogenases †. Chem Rev 2009, 109(6), 2245–2274. https://doi.org/10.1021/cr800542q.
[73] Reihlen, H., Friedolsheim, A. V., and Oswald, W. Über stickoxyd- und kohlenoxydverbindungen des scheinbar einwertigen eisens und nickels. zugleich erwiderung an die herren W. Manchot und W. Hieber. Justus Liebigs Ann Chem 1928, 465(1), 72–96. https://doi.org/10.1002/jlac.19284650106.
[74] Hieber, W., and Spacu, P.. Über Metallcarbonyle. XXVI. Einwirkung Organischer Schwefelverbindungen Auf Die Carbonyle von Eisen Und Kobalt. Z Für Anorg Allg Chem 1937, 233(4), 353–364. https://doi.org/10.1002/zaac.19372330402.
[75] Hieber, W., and Scharfenberg, C. Einwirkung Organischer Schwefelverbindungen Auf Die Carbonyle Des Eisens (XXXI. Mitteil. Über Metallcarbonyle). Berichte Dtsch Chem Ges B Ser 1940, 73(9), 1012–1021. https://doi.org/10.1002/cber.19400730914.
[76] Wei, C. H., and Dahl, L. F. The molecular structure of a tricyclic complex, [SFe(CO) 3] 2. Inorg Chem 1965, 4(1), 1–11. https://doi.org/10.1021/ic50023a001.
[77] Winter, A., Zsolnai, L., and Huttner, G. Deprotonierung und substitutionsreaktionen dreikerniger eisencluster Fe3(CO)9(H)(SR). Chem Ber 1982, 115(4), 1286–1304. https://doi.org/10.1002/cber.19821150405.
[78] Winter, A., Zsolnai, L., and Hüttner, G. Zweikernige und dreikernige carbonyleisenkomplexe mit 1.2-und 1.3-dithiolatobrückenliganden / dinuclear and trinuclear carbonyliron complexes containing 1,2-and 1,3-dithiolato bridging ligands. Z Für Naturforschung B 1982, 37(11), 1430–1436. https://doi.org/10.1515/znb-1982-1113.
[79] King, R. B. Organosulfur derivatives of the metal carbonyls. IV. the reactions between certain organic sulfur compounds and iron carbonyls. J Am Chem Soc 1963, 85(11), 1584–1587. https://doi.org/10.1021/ja00894a009.
[80] Seyferth, D., Womack, G. B., Gallagher, M. K., Cowie, M., Hames, B. W., Fackler, J. P., and Mazany, A. M. Novel anionic rearrangements in hexacarbonyldiiron complexes of chelating organosulfur ligands. Organometallics 1987, 6(2), 283–294. https://doi.org/10.1021/om00145a009.
[81] Seyferth, D., and Henderson, R. S. Novel bridging sulfide anion complexes of the hexacarbonyldiiron unit: a new route to alkylthio complexes of iron. J Am Chem Soc 1979, 101 (2), 508–509. https://doi.org/10.1021/ja00496a053.
[82] Tard, C., Llu, X., Ibrahim, S. K., Bruschi, M., Gioia, L. D., Davies, S. C., Yang, X., Wang, L.-S., Sawers, G., and Pickett, C. J. Synthesis of the H-cluster framework of iron-only hydrogenase. Nature 2005, 433(7026), 610–613. https://doi.org/10.1038/nature03298.
[83] Camara, J. M., and Rauchfuss, T. B. Combining acid–base, redox and substrate binding functionalities to give a complete model for the [FeFe]-hydrogenase. Nat Chem 2011, 4(1), 26–30. https://doi.org/10.1038/nchem.1180.
[84] Schilter, D., Gray, D. L., Fuller, A. L., and Rauchfuss, T. B. Synthetic models for nickel–iron hydrogenase featuring redox-active ligands. Aust J Chem 2017, 70 (5), 505. https://doi.org/10.1071/CH16614.
[85] de Hatten, X., Bothe, E., Merz, K., Huc, I., and Metzler-Nolte, N. A ferrocene-peptide conjugate as a hydrogenase model system. Eur J Inorg Chem 2008, 2008(29), 4530–4537. https://doi.org/10.1002/ejic.200800566.

[86] Ghosh, S., Hogarth, G., Hollingsworth, N., Holt, K. B., Kabir, S. E., and Sanchez, B. E. Hydrogenase biomimetics: Fe 2 (CO) 4 (μ-Dppf)(μ-Pdt) (Dppf = 1,1′-bis(Diphenylphosphino) ferrocene) both a proton-reduction and hydrogen oxidation catalyst. Chem Commun 2014, 50 (8), 945–947. https://doi.org/10.1039/C3CC46456C.

[87] Kaur-Ghumaan, S., Sreenithya, A., and Sunoj, R. B. Synthesis, characterization and DFT studies of 1, 1′-bis(Diphenylphosphino)ferrocene substituted diiron complexes: bioinspired [FeFe] hydrogenase model complexes. J Chem Sci 2015, 127(3), 557–563. https://doi.org/10.1007/s12039-015-0809-y.

[88] Häßner, M., Fiedler, J., and Ringenberg, M. R. (Spectro)Electrochemical and electrocatalytic investigation of 1,1′-dithiolatoferrocene–hexacarbonyldiiron. Inorg Chem 2019, 58(3), 1742–1745. https://doi.org/10.1021/acs.inorgchem.8b02971.

[89] Georgakaki, I. Fundamental properties of small molecule models of Fe-only hydrogenase: computations relative to the definition of an entatic state in the active site. Coord Chem Rev 2003, 238–239, 255–266. https://doi.org/10.1016/S0010-8545(02)00326-0.

[90] van der Vlugt, J. I., Rauchfuss, T. B., and Wilson, S. R. Electron-rich diferrous-phosphane-thiolates relevant to Fe-only hydrogenase: is cyanide "nature's trimethylphosphane"? Chem – Eur J 2006, 12(1), 90–98. https://doi.org/10.1002/chem.200500752.

[91] van der Vlugt, J. I., Rauchfuss, T. B., Whaley, C. M., and Wilson, S. R. Characterization of a diferrous terminal hydride mechanistically relevant to the Fe-only hydrogenases. J Am Chem Soc 2005, 127(46), 16012–16013. https://doi.org/10.1021/ja055475a.

[92] Boyke, C. A., Rauchfuss, T. B., Wilson, S. R., Rohmer, M.-M., and Bénard, M. [Fe 2 (SR) 2 (μ-CO) (CNMe) 6] 2+ and analogues: a new class of diiron dithiolates as structural models for the H o x A ir state of the Fe-only hydrogenase. J Am Chem Soc 2004, 126(46), 15151–15160. https://doi.org/10.1021/ja049050k.

[93] Wang, W., Rauchfuss, T. B., Moore, C. E., Rheingold, A. L., De Gioia, L., and Zampella, G. Crystallographic characterization of a fully rotated, basic diiron dithiolate: model for the H red state? Chem – Eur J 2013, 19(46), 15476–15479. https://doi.org/10.1002/chem.201303351.

[94] Munery, S., Capon, J.-F., De Gioia, L., Elleouet, C., Greco, C., Pétillon, F. Y., Schollhammer, P., Talarmin, J., and Zampella, G. New Fe I -Fe I complex featuring a rotated conformation related to the [2 Fe] H subsite of [Fe-Fe] hydrogenase. Chem – Eur J 2013, 19(46), 15458–15461. https://doi.org/10.1002/chem.201303316.

[95] Goy, R., Bertini, L., Elleouet, C., Görls, H., Zampella, G., Talarmin, J., De Gioia, L., Schollhammer, P., Apfel, U.-P., and Weigand, W. A sterically stabilized Fe I –Fe I semi-rotated conformation of [FeFe] hydrogenase subsite model. Dalton Trans 2015, 44(4), 1690–1699. https://doi.org/10.1039/C4DT03223C.

[96] Stanley, J. L., Heiden, Z. M., Rauchfuss, T. B., Wilson, S. R., De Gioia, L., and Zampella, G. Desymmetrized diiron azadithiolato carbonyls: a step toward modeling the iron-only hydrogenases. Organometallics 2008, 27(1), 119–125. https://doi.org/10.1021/om7009599.

[97] Eilers, G., Schwartz, L., Stein, M., Zampella, G., de Gioia, L., Ott, S., and Lomoth, R.. Ligand versus metal protonation of an iron hydrogenase active site mimic. Chem – Eur J 2007, 13 (25), 7075–7084. https://doi.org/10.1002/chem.200700019.

[98] Rauchfuss, T. B. Diiron azadithiolates as models for the [FeFe]-hydrogenase active site and paradigm for the role of the second coordination sphere. Acc Chem Res 2015, 48 (7), 2107–2116. https://doi.org/10.1021/acs.accounts.5b00177.

[99] Matthews, S. L., and Heinekey, D. M. A carbonyl-rich bridging hydride complex relevant to the Fe–Fe hydrogenase active site. Inorg Chem 2010, 49(21), 9746–9748. https://doi.org/10.1021/ic1017328.

[100] Cheah, M. H., Tard, C., Borg, S. J., Liu, X., Ibrahim, S. K., Pickett, C. J., and Best, S. P. Modeling [Fe–Fe] hydrogenase: evidence for bridging carbonyl and distal iron coordination vacancy in an electrocatalytically competent proton reduction by an iron thiolate assembly that operates through Fe(0)–Fe(II) levels. J Am Chem Soc 2007, 129(36), 11085–11092. https://doi.org/10.1021/ja071331f.
[101] Borg, S. J., Behrsing, T., Best, S. P., Razavet, M., Liu, X., and Pickett, C. J. Electron transfer at a dithiolate-bridged diiron assembly: electrocatalytic hydrogen evolution. J Am Chem Soc 2004, 126(51), 16988–16999. https://doi.org/10.1021/ja045281f.
[102] Felton, G. A. N., Vannucci, A. K., Chen, J., Lockett, L. T., Okumura, N., Petro, B. J., Zakai, U. I., Evans, D. H., Glass, R. S., and Lichtenberger, D. L. Hydrogen generation from weak acids: electrochemical and computational studies of a diiron hydrogenase mimic. J Am Chem Soc 2007, 129(41), 12521–12530. https://doi.org/10.1021/ja073886g.
[103] Aguirre de Carcer, I., DiPasquale, A., Rheingold, A. L., and Heinekey, D. M.. Active-site models for iron hydrogenases: reduction chemistry of dinuclear iron complexes. Inorg Chem 2006, 45(20), 8000–8002. https://doi.org/10.1021/ic0610381.
[104] Schwartz, L., Eilers, G., Eriksson, L., Gogoll, A., Lomoth, R., and Ott, S. Iron hydrogenase active site mimic holding a proton and a hydride. Chem Commun 2006, No. 5, 520–522. https://doi.org/10.1039/b514280f.
[105] Zhao, X., Chiang, C.-Y., Miller, M. L., Rampersad, M. V., and Darensbourg, M. Y. Activation of alkenes and H 2 by [Fe]-H 2 ase model complexes. J Am Chem Soc 2003, 125(2), 518–524. https://doi.org/10.1021/ja0280168.
[106] Zhao, X., Georgakaki, I. P., Miller, M. L., Yarbrough, J. C., and Darensbourg, M. Y. H/D Exchange reactions in dinuclear iron thiolates as activity assay models of Fe–H 2 ase. J Am Chem Soc 2001, 123(39), 9710–9711. https://doi.org/10.1021/ja0167046.
[107] Zhao, X., Georgakaki, I. P., Miller, M. L., Mejia-Rodriguez, R., Chiang, C.-Y., and Darensbourg, M. Y. Catalysis of H/D scrambling and other H/D exchange processes by [Fe]-hydrogenase model complexes. Inorg Chem 2002, 41(15), 3917–3928. https://doi.org/10.1021/ic020237r.
[108] Justice, A. K., Zampella, G., Gioia, L. D., and Rauchfuss, T. B. Lewis vs. brønsted-basicities of diiron dithiolates: spectroscopic detection of the "rotated structure" and remarkable effects of ethane- vs. propanedithiolate. Chem Commun 2007, No. 20, 2019–2021. https://doi.org/10.1039/B700754J.
[109] Justice, A. K., Zampella, G., De Gioia, L., Rauchfuss, T. B., van der Vlugt, J. I., and Wilson, S. R. Chelate control of diiron(I) dithiolates relevant to the [Fe–Fe]- hydrogenase active site. Inorg Chem 2007, 46(5), 1655–1664. https://doi.org/10.1021/ic0618706.
[110] Adam, F. I., Hogarth, G., and Richards, I. Models of the iron-only hydrogenase: reactions of [Fe2(CO)6(μ-Pdt)] with small bite-angle diphosphines yielding bridge and chelate diphosphine complexes [Fe2(CO)4(Diphosphine)(μ-Pdt)]. J Organomet Chem 2007, 692(18), 3957–3968. https://doi.org/10.1016/j.jorganchem.2007.05.050.
[111] Thomas, C. M., Rüdiger, O., Liu, T., Carson, C. E., Hall, M. B., and Darensbourg, M. Y. Synthesis of carboxylic acid-modified [FeFe]-hydrogenase model complexes amenable to surface immobilization. Organometallics 2007, 26(16), 3976–3984. https://doi.org/10.1021/om7003354.
[112] Barton, B. E., Zampella, G., Justice, A. K., De Gioia, L., Rauchfuss, T. B., and Wilson, S. R. Isomerization of the hydride complexes [HFe 2 (SR) 2 (PR 3)x(CO) 6−x] + (x = 2, 3, 4) relevant to the active site models for the [FeFe]-hydrogenases. Dalton Trans 2010, 39(12), 3011–3019. https://doi.org/10.1039/B910147K.
[113] Xu, F., Tard, C., Wang, X., Ibrahim, S. K., Hughes, D. L., Zhong, W., Zeng, X., Luo, Q., Liu, X., and Pickett, C. J. Controlling carbon monoxide binding at Di-iron units related to the iron-only

hydrogenase sub-site. Chem Commun 2008, No. 5, 606–608. https://doi.org/10.1039/B712805C.

[114] Gloaguen, F., Lawrence, J. D., and Rauchfuss, T. B. Biomimetic hydrogen evolution catalyzed by an iron carbonyl thiolate. J Am Chem Soc 2001, 123(38), 9476–9477. https://doi.org/10.1021/ja016516f.

[115] Orain, P.-Y., Capon, J.-F., Gloaguen, F., Pétillon, F. Y., Schollhammer, P., Talarmin, J., Zampella, G., De Gioia, L., and Roisnel, T. Investigation on the protonation of a trisubstituted [Fe 2 (CO) 3 (PPh 3)(κ 2 -Phen)(μ-Pdt)] complex: rotated versus unrotated intermediate pathways. Inorg Chem 2010, 49(11), 5003–5008. https://doi.org/10.1021/ic100108h.

[116] Wang, F., Wang, M., Liu, X., Jin, K., Dong, W., and Sun, L.. Protonation, electrochemical properties and molecular structures of halogen-functionalized diiron azadithiolate complexes related to the active site of iron-only hydrogenases. Dalton Trans 2007, No. 34, 3812. https://doi.org/10.1039/b706178a.

[117] Barton, B. E., and Rauchfuss, T. B. Terminal hydride in [FeFe]-hydrogenase model has lower potential for H 2 production than the isomeric bridging hydride. Inorg Chem 2008, 47(7), 2261–2263. https://doi.org/10.1021/ic800030y.

[118] Adam, F. I., Hogarth, G., Kabir, S. E., and Richards, I. Models of the iron-only hydrogenase: synthesis and protonation of bridge and chelate complexes [Fe2(CO)4{Ph2P(CH2)NPPh2}(μ-Pdt)] (N=2–4) – evidence for a terminal hydride intermediate. Comptes Rendus Chim 2008, 11(8), 890–905. https://doi.org/10.1016/j.crci.2008.03.003.

[119] Ghosh, S., Hogarth, G., Hollingsworth, N., Holt, K. B., Richards, I., Richmond, M. G., Sanchez, B. E., and Unwin, D. Models of the iron-only hydrogenase: a comparison of chelate and bridge isomers of Fe2(CO)4{Ph2PN(R)PPh2}(μ-Pdt) as proton-reduction catalysts. Dalton Trans 2013, 42 (19), 6775. https://doi.org/10.1039/c3dt50147g.

[120] Barton, B. E., Olsen, M. T., and Rauchfuss, T. B. Aza- and oxadithiolates are probable proton relays in functional models for the [FeFe]-hydrogenases. J Am Chem Soc 2008, 130(50), 16834–16835. https://doi.org/10.1021/ja8057666.

[121] Ezzaher, S., Capon, J.-F., Gloaguen, F., Pétillon, F. Y., Schollhammer, P., Talarmin, J., Pichon, R., and Kervarec, N. Evidence for the formation of terminal hydrides by protonation of an asymmetric iron hydrogenase active site mimic. Inorg Chem 2007, 46(9), 3426–3428. https://doi.org/10.1021/ic0703124.

[122] Orain, P.-Y., Capon, J.-F., Kervarec, N., Gloaguen, F., Pétillon, F., Pichon, R., Schollhammer, P., and Talarmin, J. Use of 1,10-phenanthroline in diiron dithiolate derivatives related to the [Fe–Fe] hydrogenase active site. Dalton Trans 2007, No. 34, 3754–3756. https://doi.org/10.1039/b709287c.

[123] Yang, D., Li, Y., Wang, B., Zhao, X., Su, L., Chen, S., Tong, P., Luo, Y., and Synthesis, Qu, J. and Electrocatalytic property of diiron hydride complexes derived from a thiolate-bridged diiron complex. Inorg Chem 2015, 54(21), 10243–10249. https://doi.org/10.1021/acs.inorgchem.5b01508.

[124] Justice, A. K., De Gioia, L., Nilges, M. J., Rauchfuss, T. B., Wilson, S. R., and Zampella, G.. Redox and structural properties of mixed-valence models for the active site of the [FeFe]-hydrogenase: progress and challenges. Inorg Chem 2008, 47(16), 7405–7414. https://doi.org/10.1021/ic8007552.

[125] Zaffaroni, R., Rauchfuss, T. B., Gray, D. L., De Gioia, L., and Zampella, G. Terminal vs bridging hydrides of diiron dithiolates: protonation of Fe 2 (Dithiolate)(CO) 2 (PMe 3) 4. J Am Chem Soc 2012, 134(46), 19260–19269. https://doi.org/10.1021/ja3094394.

[126] Lawrence, J. D., Li, H., Rauchfuss, T. B., Bénard, M., and Rohmer, M.-M. Diiron azadithiolates as models for the iron-only hydrogenase active site: synthesis, structure, and stereoelectronics. Angew Chem Int Ed 2001, 40(9), 1768–1771.

[127] Carroll, M. E., Barton, B. E., Rauchfuss, T. B., and Carroll, P. J. Synthetic models for the active site of the [FeFe]-hydrogenase: catalytic proton reduction and the structure of the doubly protonated intermediate. J Am Chem Soc 2012, 134(45), 18843–18852. https://doi.org/10.1021/ja309216v.
[128] Olsen, M. T., Barton, B. E., and Rauchfuss, T. B. Hydrogen activation by biomimetic diiron dithiolates. Inorg Chem 2009, 48(16), 7507–7509. https://doi.org/10.1021/ic900850u.
[129] Ezzaher, S., Orain, P.-Y., Capon, J.-F., Gloaguen, F., Pétillon, F. Y., Roisnel, T., Schollhammer, P., and Talarmin, J. First insights into the protonation of dissymetrically disubstituted Di-iron azadithiolate models of the [FeFe]H2ases active site. Chem Commun 2008, No. 22, 2547. https://doi.org/10.1039/b801373j.
[130] Ezzaher, S., Capon, J.-F., Gloaguen, F., Pétillon, F. Y., Schollhammer, P., Talarmin, J., and Kervarec, N. Influence of a pendant amine in the second coordination sphere on proton transfer at a dissymmetrically disubstituted diiron system related to the [2Fe] H subsite of [FeFe]H 2 ase. Inorg Chem 2009, 48(1), 2–4. https://doi.org/10.1021/ic801369u.
[131] Wang, N., Wang, M., Liu, J., Jin, K., Chen, L., and Sun, L.. Preparation, facile deprotonation, and rapid H/D exchange of the μ-hydride diiron model complexes of the [FeFe]-hydrogenase containing a pendant amine in a chelating diphosphine ligand. Inorg Chem 2009, 48(24), 11551–11558. https://doi.org/10.1021/ic901154m.
[132] Wang, Y., Wang, M., Sun, L., and Ahlquist, M. S. G. Pendant amine bases speed up proton transfers to metals by splitting the barriers. Chem Commun 2012, 48 (37), 4450. https://doi.org/10.1039/c2cc00044j.
[133] Lounissi, S., Zampella, G., Capon, J.-F., De Gioia, L., Matoussi, F., Mahfoudhi, S., Pétillon, F. Y., Schollhammer, P., and Talarmin, J.. Electrochemical and theoretical investigations of the role of the appended base on the reduction of protons by [Fe 2 (CO) 4 (κ 2 -PNP R)(μ-S(CH 2) 3 S] (PNP R ={Ph 2 PCH 2 } 2 NR, R=Me, Ph). Chem – Eur J 2012, 18(35), 11123–11138.
[134] Gordon, J. C., and Kubas, G. J. Perspectives on how nature employs the principles of organometallic chemistry in dihydrogen activation in hydrogenases †. Organometallics 2010, 29(21), 4682–4701. https://doi.org/10.1021/om100436c.

2.2 [NiFe] hydrogenases

Marius Horch, Oliver Lenz, Ingo Zebger

2.2.1 Introduction and scope

As stated in the previous chapter, H_2-converting enzymes can be grouped into three major classes of hydrogenases based on the metal content of their catalytic sites [1–3]. In this chapter, we will focus on [NiFe] hydrogenases, which represent the most widespread, diverse and best studied of these classes [1–4]. Historically, the huge body of experimental data available for these enzymes can be explained by the fact that [NiFe] hydrogenases were the first of these classes to be identified, isolated and characterized in detail. In particular, 35 years after the first purification of a [NiFe] hydrogenase [5], initial crystallographic insights into the structure of such an enzyme were reported in 1995 [6, 7] – earlier than for the other two classes [8–10], but still more than 60 years after the existence of hydrogenases had been postulated [11] and almost a century after the first observation of biological H_2 oxidation [12]. Although [NiFe] hydrogenases have been extensively explored by biochemical and biophysical techniques before the availability of crystal structure data [13], these long time spans clearly illustrate the considerable challenges that were and continue to be associated with the characterization of these enzymes. As a consequence, the plethora of insights from decades of [NiFe] hydrogenase research is opposed by numerous open questions, and several structural and functional aspects of these enzymes are still matters of debate in this highly active research field.

In the following sections, we will give an introduction to [NiFe] hydrogenases, laying emphasis on enzymatic structure–function relationships. After highlighting the biological and technological relevance of base metal-catalyzed H_2 oxidation, the current understanding of the corresponding enzymatic mechanism will be reviewed. In this respect, we will specifically address possible catalytic functions of the biologically unprecedented CN^- and CO ligands, the latter of which represent a common organometallic motif in all known types of hydrogenases (see also Sections 2.1 and2.3) [14]. This introduction to [NiFe] hydrogenase function will be complemented by an overview of synthetic analogues, and finally, existing and envisaged applications of both types of catalysts will be presented. Notably, the technological exploitation of H_2-cycling catalysts is so far limited, since most [NiFe] hydrogenases and functional synthetic analogues are inactivated by O_2. Thus, a second central section of this chapter will deal with the toxic effects of O_2 on hydrogenase, counterstrategies developed in a subclass of [NiFe] hydrogenases and the possible transferability of these principles to synthetic catalysts. To further assess relations between biological and chemical H_2 conversion, we will also briefly discuss the evolution and prebiotic analogues of [NiFe] hydrogenases as well as selected aspects of (bio)synthesis. Besides summarizing the *status quo* in [NiFe] hydrogenase research, this chapter explicitly aims to highlight open questions and challenges that cumulate in perspectives for future studies and applications. In this respect, the reader is strongly encouraged to critically analyze the outlined problems and hypotheses.

2.2.2 Learning from nature: base metal catalysts for H_2 oxidation

The major interest in hydrogenases and their synthetic analogues arises from the fact that H_2 represents an ideally clean fuel, which releases a huge amount of Gibbs free energy but no greenhouse gases upon combustion:

$$H_2 + 1/2\ O_2 \rightarrow H_2O, \quad \Delta G^\circ = -237\,\text{kJ}\,\text{mol}^{-1} \tag{2.2}$$

In light of anthropogenic global warming and the depletion of fossil fuels [15], H_2 is thus considered as a renewable source of energy [16], and major efforts have been made to explore future strategies for its cost-effective and sustainable exploitation. To reach this goal, however, several challenges have to be met in order to (1) realize safe and effective H_2 storage, (2) exploit renewable sources of H_2 and (3) create sustainable approaches for H_2 activation and energy conversion [17]. Here, we will focus on the latter point, since the applicability of H_2 as a fuel stands or falls by the accessibility of the free energy "stored" in the dihydrogen bond.

The conversion of H_2 into two protons and electrons is often quoted as the "simplest chemical reaction." Yet, from a kinetic point of view, H_2 splitting is far from trivial, since the dihydrogen bond is remarkably strong and completely nonpolar [18]. As a consequence, H_2 is a poor acid and essentially inert [18, 19], requiring activation by suitable catalysts. Platinum electrodes have been classically used in this context; however, Pt is expensive, fluctuating in price and resource-limited [20]. In addition, irreversible poisoning by typical trace impurities like H_2S or CO represents a major problem in practical applications, ruling out Pt-based electrodes as workhorse catalysts for future applications [20].

While the concept of a hydrogen-based economy dates back to the 1970s [21], H_2 as a fuel has found little attention before the turn of the millennium.[1] In contrast, microorganisms appear to have used H_2 as a source of energy since the earliest days of life on earth, and [NiFe] hydrogenases are most frequently involved in its utilization [2, 3, 23, 24]. Remarkably, many of these enzymes catalyze H_2 cleavage at rates comparable to Pt electrodes by using an active site that contains the base metals Ni and Fe only (see Section 2.2.3.3) [25]. In contrast to Pt, both elements are earth-abundant and cheap, with Fe representing the most common element on the Earth. In addition, [NiFe] hydrogenases are not resource-limited as they can be produced, often in large amounts, by the cellular machinery of the homo- or heterologous host organism. Some of these enzymes are even insensitive toward both O_2 and CO (see Section 2.2.4) [1, 26], which represents a major advantage over all currently known synthetic molecular catalysts for H_2 cycling. Besides their potential for several other applications (see Section 2.2.6), [NiFe] hydrogenases thus represent ideal prototype catalysts for sustainable H_2 oxidation [16].

As stated earlier, H_2 combustion is highly exergonic; however, as written in eq. (2.2), this "Knallgas" reaction between H_2 and O_2 proceeds in an uncontrolled manner, which is inappropriate for both biological and industrial utilization. Moreover, energy is released as heat, which is unsuited as a secondary energy carrier, again both from a cellular and technological point of view. To cope with these problems, anode and cathode reactions need to be separated to allow a safe reaction and conversion of (redox) chemical energy to a universally accessible form:

$$H_2 \rightarrow 2H^+ + 2e^- \tag{2.3}$$

$$1/2\ O_2 + 2H^+ + 2e^- \rightarrow H_2O \tag{2.4}$$

1 Remarkably, H_2 has been prophesized as a future fuel by J. B. S. Haldane and Jules Verne as early as 1924 and 1874, respectively.

Within the cell, reaction (2.3) is catalyzed by a [NiFe] hydrogenase, while the reduction of the electron acceptor, for example, O_2 (reaction (2.4)), involves a terminal oxidase (Figure 2.17A) [26]. The two enzymes are conductively connected *via* an electron transport chain of redox proteins exhibiting increasing reduction potentials, and the free energy released at the individual steps is transformed to a proton gradient across the cytoplasmic membrane. Finally, the generated chemiosmotic potential is used to drive the synthesis of adenosine triphosphate (ATP), a universal cellular energy carrier that is utilized to drive endergonic reactions. In a similar sense, a [NiFe] hydrogenase may be coupled to an oxidase within a biofuel cell (Figure 2.17B) [27, 28]. Within such a device, both enzymes are generally immobilized on biocompatible electrodes, and, similar to the cellular arrangement, anode and cathode reactions are spatially separated but connected through a conductive element. Within the biofuel cell, however, chemical energy is directly transformed to electrical energy, which, analogous to cellular ATP, represents a universally deployable form. Notably, ordinary fuel cells need to be designed in a way that excludes direct reaction of the educts (here H_2 and O_2) at either of the two electrodes. In contrast, enzymes are highly selective and specific with respect to their substrate and the catalyzed reaction, respectively. As a consequence, short circuits and explosive reactions can be omitted in H_2 biofuel cells without requiring a separation of the two half cells *per se*.

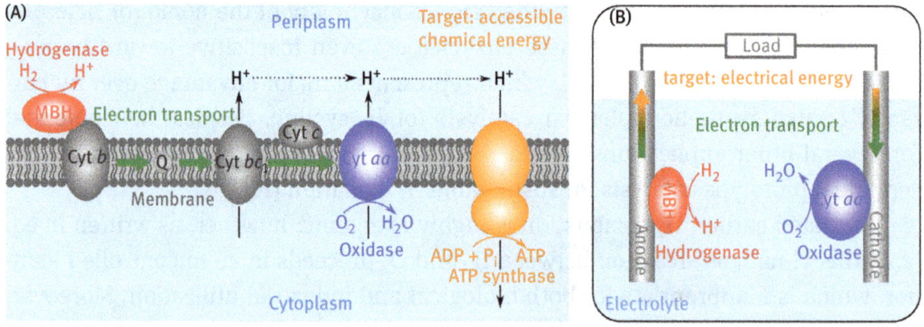

Figure 2.17: Schematic representations of biological and electrochemical energy conversion involving (O_2-tolerant) [NiFe] hydrogenases. (A) Aerobic respiration utilizing H_2 as a source of electrons and O_2 as an electron acceptor. Anode (hydrogenase) and cathode (oxidase) reactions are coupled by a simplified electron transport chain, and the resulting chemiosmotic potential drives the synthesis of ATP, catalyzed by ATP synthase. (B) General design of a H_2/O_2-driven biofuel cell for the generation of electrical energy. Here the oxidation of H_2 is coupled with the reduction of O_2 by wiring a hydrogenase with an oxidase. Abbreviations used: MBH, membrane-bound hydrogenase; Cyt *b*, cytochrome *b*; Q, quinone; Cyt bc_1, cytochrome bc_1 complex (respiratory complex III); Cyt *c*, cytochrome *c*; Cyt aa_3, cytochrome aa_3 complex (respiratory complex IV); ADP, adenosine diphosphate; P_i, inorganic phosphate; ATP, adenosine triphosphate. Due to the complexity of the respiration process, all stoichiometric coefficients have been left out for the sake of simplicity.

2.2.3 Enzymatic structure and function

Given the challenges associated with H_2 activation under ambient conditions, relations between enzymatic structure and function have been of major interest since the earliest days of [NiFe] hydrogenase research. Besides a fundamental scientific interest, research on this topic has been largely motivated by the aim to design H_2 oxidation catalysts that are superior in terms of overall performance, size or adaptability to special applications and general industrial requirements (see Sections 2.2.5.3 and 2.2.6). Rational approaches to meet these requirements require a detailed understanding of the native enzyme's physiological function, the catalytic mechanism and the underlying structural determinants.

While the function and performance of [NiFe] hydrogenases can be well explored by biochemical assays and protein electrochemistry [1, 26, 29–32], the investigation of structural and electronic features of this diverse class of enzymes has proven to be more difficult. While crystal structures are available for a number of enzymes [1], these data provide little insight into mechanistically relevant details and dynamic properties. Moreover, several subclasses of [NiFe] hydrogenases have so far refused crystallization. As a consequence, various spectroscopic techniques have played a central role in the characterization of the [NiFe] active site and other cofactors found in these enzymes [1, 13, 33–36]. While an exhaustive overview of the applied techniques is clearly beyond the scope of this chapter, we will shortly refer to the most important ones wherever it is required for the understanding of the outlined structural properties. For more detailed insights into the quoted methods, the reader is referred to part IV of this book.

Despite considerable insights gained from spectroscopic analyses, this approach is naturally limited since, typically, only selected cofactors or redox-structural states can be studied in detail [1, 34]. As a consequence, a multitude of techniques is often required to obtain comprehensive insights [34], and the overall structure and cofactor composition of several [NiFe] hydrogenases can be deduced from homology modeling and sequence comparison with other, well-characterized enzymes only [2, 3, 37–39]. Moreover, several catalytically relevant processes proceed on timescales that are, with few exceptions [40, 41], inaccessible by most techniques applied so far. As a consequence, numerous aspects of the proposed catalytic cycle for H_2 oxidation by [NiFe] hydrogenases built on steady state or non-turnover studies [1, 29–34, 42], quantum mechanical calculations [18, 43–45] and analogies to easily accessible small-molecule analogues (see Section 2.2.5.3) [1, 19, 46–51].

2.2.3.1 Classification and cellular tasks

[NiFe] hydrogenases can be found in a wide range of bacteria and archaea, where they perform a number of important cellular tasks [2, 3]. While [NiFe] hydrogenases can be clearly distinguished from [FeFe] and [Fe] hydrogenases in terms of active site structure and phylogeny, they are far from being uniform. Based on comprehensive phylogenetic analyses, [NiFe] hydrogenases are classified into four groups, which differ with respect to subunit composition, cofactor content, cellular location and physiological function (Figure 2.18) [2, 3]. In the following, we will shortly introduce these different groups laying emphasis on those that have been characterized in most detail so far.

Group 1 consists of *membrane-bound H_2-uptake [NiFe] hydrogenases* (Figure 2.18, top left), which catalyze H_2 oxidation for energy conversion, as sketched in Section 2.2.2. Members of this group are generally membrane-associated, typically located on the periplasmic side of the membrane, and linked to additional electron transfer and/or transmembrane proteins [2, 3]. Numerous group 1 hydrogenases have been extensively studied for almost three decades, and some of these enzymes, designated as "standard [NiFe] hydrogenases," have been widely perceived as canonical [NiFe] hydrogenases (see Section 2.2.3.2). [1, 26, 29, 35–37, 52, 53]. However, group 1 also contains members of the subclass of [NiFeSe] hydrogenases [54] and the recently discovered actinobacterial-like [NiFe]-hydrogenases [2, 55, 56]. In contrast to all other group 1 hydrogenases, the latter enzymes are located in the cytoplasm, and they typically exhibit an extraordinary affinity toward H_2, which allows the utilization of trace H_2 present in the atmosphere [55–58].

Cytosolic H_2-uptake [NiFe] hydrogenases of group 2 (Figure 2.18, center left) may fulfill different, partly unknown cellular functions [3]. Prominent examples are cyanobacterial uptake and H_2-sensing regulatory [NiFe] hydrogenases. While the latter are soluble cytoplasmic proteins, the former may also be associated with the photosynthetic membrane [3]. Similar to a few members of group 1, cyanobacterial uptake hydrogenases catalyze the oxidation of H_2 that is produced, for example, as a side product of nitrogenase-mediated dinitrogen fixation [2]. H_2-sensing regulatory [NiFe] hydrogenases differ considerably from most other [NiFe] hydrogenases by their deliberately poor activity toward H_2 oxidation and evolution [1, 26]. This characteristic owes to their cellular function, namely H_2 sensing rather than rapid turnover. Sensory [NiFe] hydrogenases are constituents of multicomponent signaling cascades that regulate the biosynthesis of energy-converting [NiFe] hydrogenases according to the availability of H_2 [2, 3, 26].

Group 3 comprises *cytosolic bidirectional [NiFe] hydrogenases* (Figure 2.18, bottom left) that are modular multisubunit enzymes exhibiting pronounced similarity to respiratory complex I (see Section 2.2.5; Figure 2.18, bottom right) [2,3,37–39]. Many of these enzymes are soluble proteins, but some members may also be associated with the cytoplasmic or thylakoid membrane [37]. In general, group 3 [NiFe]

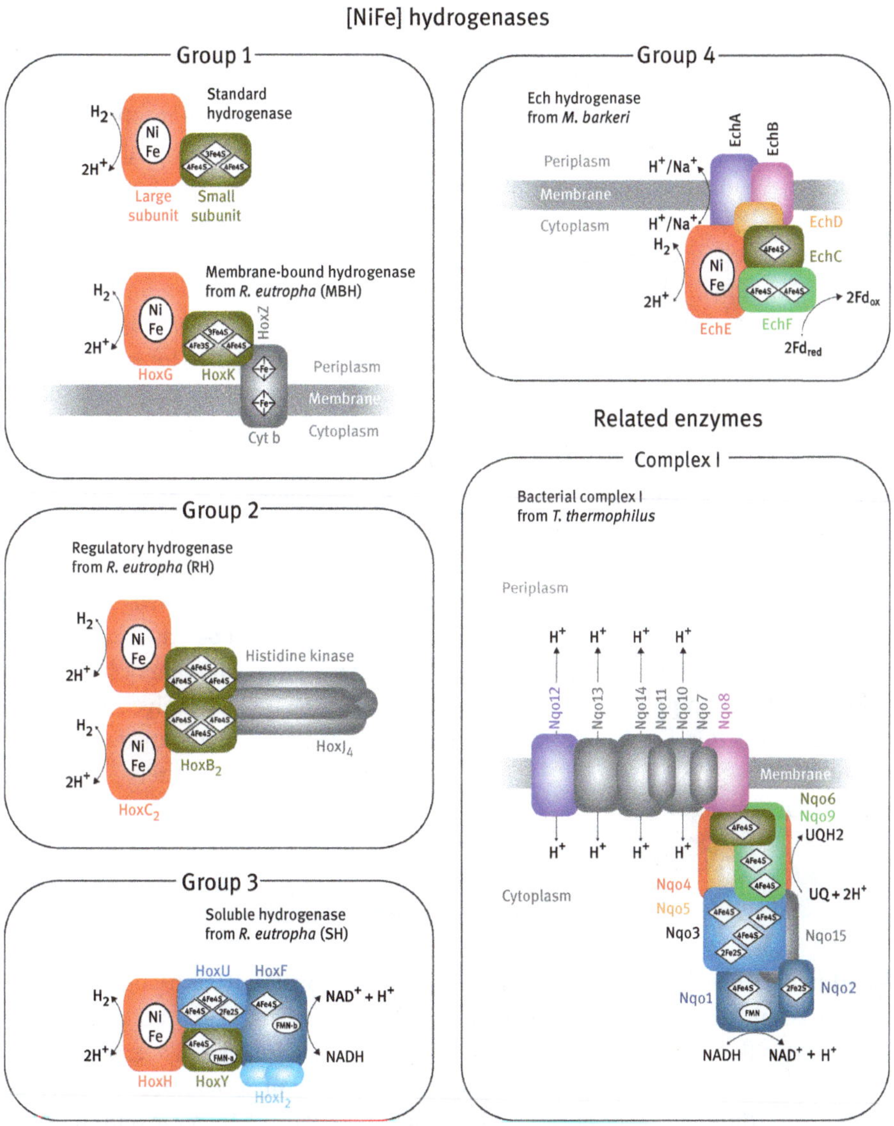

Figure 2.18: Subunit composition and cofactor content of selected [NiFe] hydrogenases from the four different groups and respiratory complex I. Homologous subunits are color-coded. Abbreviations used: Cyt *b*, cytochrome *b*; FMN, flavin mononucleotide; NAD^+/NADH, oxidized/reduced nicotinamide adenine dinucleotide; Fd_{ox}/Fd_{red}, oxidized/reduced ferredoxin; UQ/UQH_2, oxidized/reduced ubiquinone. *R.*, *Ralstonia*; *M.*, *Methanosarcina*; *T.*, *Thermus*. For further details, see Sections 2.2.3.2, 2.2.4 and 2.2.5.1.

hydrogenases couple the cleavage of H_2 to the reduction of soluble cofactors, for example, nicotinamide adenine dinucleotide (NAD), its phosphorylated congener (NADP) or 8-hydroxy-5-deazaflavin (cofactor F_{420}) [2,3]. Members of this group are termed "bidirectional," because they may also catalyze the reverse reaction, that is, H_2 evolution and oxidation of the soluble cofactor, under physiological conditions [3,37]. Due to this trait and their bifunctional character, these enzymes can fulfill several metabolic tasks, and their potential for biotechnological applications has been extensively explored (see Section 2.2.6) [2,3,37,59]. The cellular function of group 3 [NiFe] hydrogenases depends on the individual enzyme, the host organism and the current metabolic situation [3,37]. On the one hand, they may enable H_2 evolution as a means to dispose of excess electrons deriving from, for example, fermentation or photosynthesis. On the other hand, they allow utilizing H_2 as a source of energy or reducing equivalents for biosyntheses, including carbon fixation.

Similar to members of group 3, *membrane-bound H_2-evolving [NiFe] hydrogenases* of group 4 (Figure 2.19; top right) are large multimeric enzymes consisting of six or more subunits [2,3]. They also share some similarity with group 1 hydrogenases as they are membrane-associated and involved in energy conversion. Group 4 hydrogenases, however, appear to be located at the cytoplasmic side of the plasma membrane, and energy conversion involving these enzymes does not utilize H_2 as an electron donor [2,3]. In contrast, energy is gained by the oxidation of low-potential one-carbon compounds, for example, CO or $HCOO^-$, and H_2 evolves as a result of proton reduction [3]. In other words, the H_2/H^+ redox couple is involved in the cathode rather than the anode reaction in catabolic processes that rely on group 4 [NiFe] hydrogenases.

2.2.3.2 Consensus features and "standard [NiFe] hydrogenases"

As adumbrated earlier, many [NiFe] hydrogenases consist of numerous subunits, and the resulting quaternary structures are often arranged in superordinate homo- or heteromultimeric ensembles [2, 3]. In addition to the [NiFe] active site responsible for H_2 conversion, these protein complexes can contain several other cofactors, for example, different types of FeS clusters, flavin derivatives, heme groups and single alkaline earth metal or iron ions (Figure 2.18) [1–3, 26, 37, 52]. Despite their structural complexity and considerable differences between members of the four groups, all [NiFe] hydrogenases exhibit distinct consensus features that appear to represent common prerequisites for H_2 cycling by these enzymes. These features will be outlined in the following, while their explicit functional roles will be analyzed in more detail in Section 2.2.3.4.

[NiFe] hydrogenases are composed of at least one large subunit and one small subunit (Figure 2.19) [1–3, 6, 7, 37]. The large subunit contains four strictly conserved cysteines involved in the formation of the deeply buried [NiFe] active site

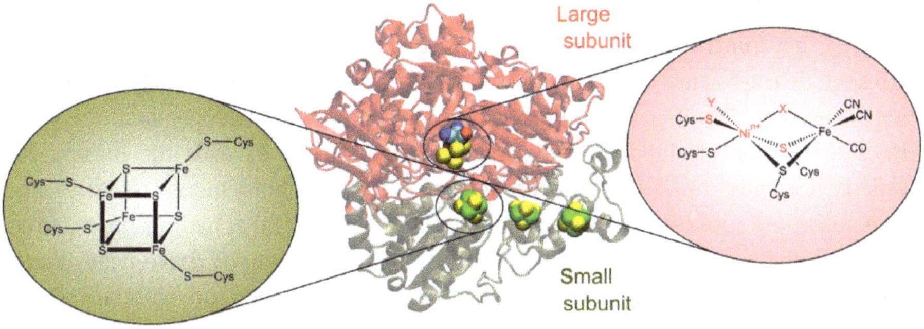

Figure 2.19: Crystal structure representation of a standard [NiFe] hydrogenase (from *Desulfovibrio gigas*; PDB entry: 1YQ9) [89]. Metal cofactors are depicted as colored spheres. Skeletal formulas are shown for the [NiFe] active site and a nearby (proximal) [4Fe–4S] cluster. Molecular sites involved in redox-structural changes of the [NiFe] active site are highlighted in red. For details, see Sections 2.2.3.3 and 2.2.3.4 as well as Figure 2.21.

(see Section 2.2.3.4), while the small subunit harbors at least one [4Fe–*X*S] cluster ("*X*" represents the number of inorganic sulfide anions), which is located in electron-transfer distance to the active site [1–3, 6, 7, 37]. Interestingly, this minimum set of cofactors is reminiscent of [FeFe] hydrogenases (see Section 2.1), which also contain a [4Fe–4S] cluster closely connected to the unique [FeFe] site [1, 9, 10]. This suggests that the combination of a bimetallic center and a nearby FeS cluster represents a universal minimum motif for biological H_2 conversion according to eq. (2.2). In contrast to [FeFe] hydrogenases, however, the two cofactor species of [NiFe] hydrogenases are not covalently linked by a single cysteine side chain but separated by a few ångströms and distributed over two separate subunits (Figure 2.19) [1, 6, 7]. Moreover, while native [FeFe] hydrogenases containing *only* the minimum set of cofactors are known and well characterized [1–3], no such enzyme from the [NiFe] hydrogenase class has been reported so far. Thus, the outlined consensus features should be perceived as minimum requirements that are necessary but generally not sufficient for catalytic H_2 cycling by these latter enzymes.

Group 1 [NiFe] hydrogenases from anaerobic organisms of the genera *Desulfovibrio* and *Allochromatium* were among the first that have been extensively studied, and, thus, enzymes of this type have been frequently termed "standard [NiFe] hydrogenases" [1, 26, 29, 35–37, 52, 53]. Within the scope of this chapter, these O_2-sensitive enzymes represent valuable model systems for several reasons: (1) Standard [NiFe] hydrogenases can be purified as compact globular proteins that contain no subunits other than the essential large and small subunits, and their cofactor content is small compared to most other energy-converting [NiFe] hydrogenases (Figure 2.18) [1–3, 6, 7, 37]. Apart from the [NiFe] active site, the large subunit of these enzymes harbors only a remote Mg^{2+} ion, while the small subunit contains a chain of three FeS clusters, consisting of a [4Fe–4S] cluster "proximal" to the active site, a "medial" [3Fe4S] cluster, and a "distal" [4Fe–4S] cluster (Figure 2.19) [1, 6, 7]. (2) Standard [NiFe] hydrogenases are stable but inactive under aerobic oxidizing conditions [1, 26]. Consequently, they represent ideal reference systems to elucidate structural peculiarities of O_2-tolerant hydrogenases that allow H_2 cycling in the presence of air. (3) Knowledge about the structure and function of standard [NiFe] hydrogenases clearly exceeds the information available for most other [NiFe] hydrogenases [1]. While fundamental features and mechanisms are supposed to be essentially uniform among [NiFe] hydrogenases, structural and functional details, as outlined in the following sections, have been largely derived from experimental and theoretical studies performed on standard [NiFe] hydrogenases.

2.2.3.3 The H_2-converting active site

While the catalytic function of [NiFe] hydrogenases depends on a large number of partly unexplored structural features (see Section 2.2.3.4), the fundamental processes of H_2 cleavage and evolution take place at a unique heterobimetallic cofactor [1]. This active site contains the two metal ions, Ni and Fe, which are coordinated by four strictly conserved cysteines (Figure 2.19) [1–3, 6]. Two of these sulfur donors are found in bridging positions between the two metals, while the other two serve as terminal ligands to the nickel. In a subclass of [NiFe] hydrogenases, one of the (terminal) cysteines is replaced by a selenocysteine, which results in a number of specialties that are beyond the scope of this chapter [1, 54, 60]. Remarkably, the Fe ion of all [NiFe] hydrogenases is furthermore coordinated by three nonproteic ligands, one CO and two CN^- [1, 7, 61–64]. As a matter of fact, these biologically uncommon ligands have not been observed for any enzyme other than hydrogenases so far. Consequently, their first infrared (IR) spectroscopic detection in a [NiFe] hydrogenase came as a big surprise to the biochemical community, since both CO and CN^- were traditionally known as toxic compounds and metalloprotein inhibitors only. Later, both types of ligands were detected in [FeFe] hydrogenases as well [9,10,65], and also [Fe] hydrogenases contain CO ligands [8, 66], indicating a special function in biological H_2 cycling (see Section 2.2.3.5) [14, 19]. Even in the absence of these diatomic ligands, the active site of [NiFe] hydrogenases would differ considerably from those of other Ni-containing enzymes, most of which catalyze reactions involving gases as well. Known members of this diverse group contain either mononuclear Ni sites with a nitrogen-rich first coordination sphere or complex [NiFeS] clusters, as observed for CO dehydrogenase or acetyl-CoA synthase (see chapters 3 and 7) [67]. Similar to [NiFe] hydrogenases, the C-cluster of CO dehydrogenase exhibits an $S_2Ni(\mu_2\text{-}S)_2Fe$ motif as well; however, the active site of [NiFe] hydrogenases differs from this and other metal-sulfur clusters by the complete absence of inorganic sulfide anions. Consequently, it represents a unique heterobimetallic cofactor "tailored" for biological H_2 cycling.

Under catalytic turnover, the [NiFe] active site reacts with either H_2 or protons and electrons. Likewise, inhibitors, including O_2, reactive oxygen species and CO, may interact with this cofactor as well. Thus, depending on the current conditions, the general architecture, electronic structure and geometry of the [NiFe] active site may be subject to considerable variations: (1) Inspecting the structural depiction of the active site in Figure 2.19, two vacant coordination sites can be recognized, one between the two metals (X) and one at a terminal position of the Ni (Y). Both appear to be relevant for enzymatic function (see Section 2.2.3.4) [44, 45] and especially the bridging site has long been known to readily bind several substrate- or inhibitor-derived ligands [1, 33–36]. (2) Since interactions with H_2 and O_2 represent redox reactions, diverse combinations of oxidation states may be observed for the two metal ions of the [NiFe] active site. In fact, the Ni is redox-active [1, 33, 36], while the Fe persists in the low-spin ($S = 0$) ferrous state under all conditions explored so far [16, 68–70]. (3) In a similar sense, the four cysteine thiolates ($Cys–S^-$) may act as redox noninnocent ligands under aerobic conditions [50], leading to possible oxidation products including sulfenate ($Cys–SO^-$) [71–73] or sulfinate ($Cys–SOO^-$) [74] species. (4) Terminal cysteine ligands may also participate in acid–base reactions resulting in protonation, that is, thiol (Cys–SH) formation [1, 33, 41, 44, 45, 75–79]. (5) Last but not least, structural changes in the vicinity of the active site may affect its electronic and geometric properties in a complex manner.

In the subsequent paragraphs, we will illustrate these possible variations by describing typical states of the [NiFe] active site observed under different redox conditions (Figure 2.20). Some of these states have been structurally explored by X-ray crystallography, but most information is based on spectroscopic studies, involving especially IR and electron paramagnetic resonance (EPR) techniques [1, 13, 33, 35, 36]. The latter provide selective insights into paramagnetic species and the coordination environment of the Ni, while the former is able to explore all redox-structural states by probing the stretching vibrations of the CO and CN^- ligands. These normal modes are highly sensitive toward electron density changes at the Fe, so that detailed insights into diverse structural variations at the [NiFe] site are available [33, 34, 63, 71, 80]. Historically, different states of the [NiFe] active site have first been assigned on the basis of characteristic Ni-related signals in the EPR spectra [13, 36, 81–85]. Thus, all redox-structural states are typically denoted as Ni_X-Y, where X and Y are descriptors whose meaning will be explained in the following.[2]

In the absence of H_2, O_2 and CO, both vacant coordination sites of the [NiFe] center are likely unoccupied (Figure 2.19) [70, 78]. Moreover, in the absence of powerful redox agents, the Ni ion is expected to be found in the +II oxidation state, which is most common in both biological and synthetic Ni compounds. This configuration may be perceived as the resting state of the [NiFe] active site, which is likely

2 Note that slightly different nomenclatures are used in parts of the literature.

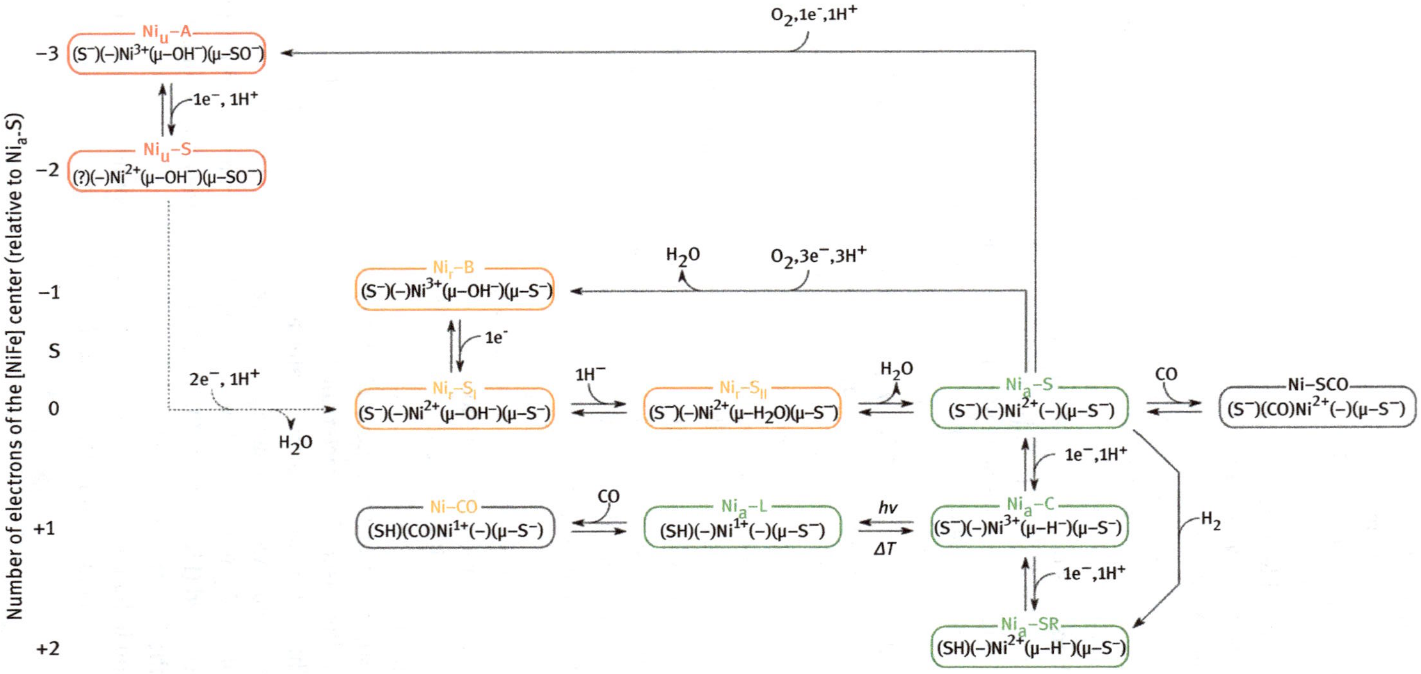

Figure 2.20: Characteristics and reactions of typical redox-structural states of the [NiFe] active site, as observed experimentally for several (standard) [NiFe] hydrogenases. Unready, ready and active states are highlighted in red, orange and green, respectively. Structural properties of each state are indicated according to the general scheme (1)(2)Ni^{n+}(3)(4). Here, $n+$ is the formal oxidation state of the Ni ion, and the other descriptors are defined as follows: (1) protonation state of one of the terminal cysteines; (2) ligand bound to the vacant terminal coordination site of the Ni ion; (3) ligand bound to the vacant bridging site between Ni and Fe; and (4) presence or absence of sulfoxygenated cysteine at one of the two bridging positions. Note that alternative structural interpretations have been proposed for some of the depicted states. For details, see text and Figure 2.20.

to be most eager to react with any of the above gaseous substrates. Generally, this resting state is referred to as Ni_a-S (Figure 2.20), where the label "S" indicates an EPR-"silent" (nondetectable) form, while the subscript "a" flags it as "active" (an intermediate of the catalytic cycle) [1, 33, 34, 78, 86]. Under aerobic conditions, this state may react with O_2, and, thus, "as-isolated" [NiFe] hydrogenases typically exhibit a mixture of EPR-detectable inactive states (Ni^{III}, $S = 1/2$), denoted as Ni_u-A and Ni_r-B (Figure 2.20) [1, 36, 81–83]. The latter is considered a "ready" state (subscript "r") as it is kinetically competent and easily activated under reducing conditions [1, 36, 82, 87, 88]. In contrast, Ni_u-A is "unready" in the sense that reductive activation takes a long time, up to hours [1, 36, 82, 87, 88]. Differences in the activation kinetics of the two states have been ascribed to their individual structures. While Ni_r-B is generally accepted to carry a bridging hydroxo ligand [1, 36, 89, 90], the structure of the Ni_u-A state is not entirely clear. Former studies indicated the presence of a bridging hydroperoxo ligand [72, 89], while more recent investigations point toward the presence of a bridging hydroxo ligand and sulfenate formation at one of the two bridging cysteines (Figure 2.20) [73, 91]. Notably, the two proposed structures can be perceived as tautomers, and the sulfoxygenated variant may be easily formed from the hydroperoxo form [71, 73, 92]. Irrespective of its exact structure, Ni_u-A is two electrons more oxidized than Ni_r-B, and, consequently, Ni_u-A and Ni_r-B are generally accepted to be formed, for example, upon reaction with O_2 under electron-poor and electron-rich conditions, respectively [30, 32, 93, 94].

Starting from Ni_u-A and Ni_r-B, the [NiFe] active site can be reduced by consecutive $1e^-/1H^+$ steps [1, 35, 36, 95, 96]. The first of these steps yields the unready Ni_u-S and the ready Ni_r-S state, both of which are catalytically inactive and EPR-silent (Ni^{II}). The exact structure of these states and the fate of the proton are not unambiguously known, but in a simplified scheme the overall structures of Ni_u-A and Ni_r-B can be assumed to be roughly retained upon one-electron reduction (Figure 2.20) [1, 36]. Specifically, the Ni_r-S state was reported to exist in two pH-dependent subforms carrying either a hydroxo or an aqua ligand in the bridging position between the two metals [86]. The latter appears to be weakly bound and easily lost, thereby yielding the Ni_a-S state as an entry point to the catalytic cycle. Formation of this species from Ni_u-S appears to be the rate-limiting step in the re-activation of Ni_u-A [95], but the exact mechanism is not known.

Further one-electron reduction of Ni_a-S yields the paramagnetic Ni_a-C state, which carries a hydrido ligand in the bridging position between the two metals (Figure 2.20) [1, 36, 81, 84, 85, 97, 98]. Consequently, this species is generally considered a central intermediate of the catalytic cycle. Interestingly, this ligand can be photo-dissociated and the resulting tautomeric form Ni_a-L, characterized by a formal N^I ($S = 1/2$) oxidation state and a protonated terminal cysteine, is stabilized at low temperatures (Figure 2.20) [1, 33, 36, 75, 76, 78, 79, 84]. Ni_a-L can be inhibited by CO and exist in up to three sub-forms, whose structural differences are not entirely clear [1, 33, 97, 99]. Further one-electron reduction of Ni_a-C yields the fully reduced Ni_a-SR state of typical [NiFe] hydrogenases, which may also exist in up to three so-far unexplored sub-forms [1, 33, 35, 36]. At least one sub-species of this EPR-silent (N^{II}, $S = 0$) catalytic intermediate appears to comprise a protonated terminal cysteine in addition to a bridging hydrido ligand (Figure 2.20) [77, 100, 101].

Besides reacting with the substrates (H_2, H^+, e^-) or the inhibitor O_2, the Ni_a-S state of the [NiFe] active site can also bind extrinsic CO, yielding the EPR-silent Ni-SCO state (Ni^{II}) [1, 36, 64, 102, 103]. In contrast to O_2- and H_2-related ligands, CO appears to bind to the terminal vacant coordination site at the Ni rather than the bridging location between the two metals (Figure 2.20) [64, 102, 103]. In most [NiFe] hydrogenases, this reaction inhibits catalytic H_2 cycling [1, 104], providing an indirect indication for the proposed involvement of this terminal binding site in the catalytic cycle (see Section 2.2.3.4) [44, 45].

As illustrated in the previous paragraphs, several (catalytically relevant) states of the [NiFe] active site exhibit a formal Ni^{II} oxidation state, as *inter alia* deduced from the absence of active site-related signals in the EPR spectra. The spin state of the Ni ion, however, has been a matter of debate, since neither the low-spin ($S = 0$) nor the high-spin ($S = 1$) configuration is observable by "conventional" perpendicular-mode EPR spectroscopy. Consequently, numerous spectroscopic and quantum mechanical studies have dealt with this question, and controversial results have been published [44, 45, 68, 78, 105, 106–112]. However, all recent studies tend to favor the low-spin over the high-spin configuration [44, 78, 109–112], and differences between the two may be highly important for catalysis (see Section 2.2.3.4).

2.2.3.4 Catalytic mechanism

Depending on the net driving force, [NiFe] hydrogenases can in principle catalyze both H_2 cleavage and H_2 evolution, but the underlying enzymatic reaction pathways are not necessarily identical [107]. As a matter of fact, efficiencies for the two catalytic directions vary significantly across [NiFe] hydrogenases (see Section 2.2.3.1), and most of them – including standard [NiFe] hydrogenases – tend to be biased toward H_2 cleavage [1, 26, 29]. As a consequence, the catalytic mechanism of H_2 oxidation has been most thoroughly explored, as detailed in the following.

The active site of [NiFe] hydrogenases is deeply buried within the protein matrix, so that H_2 transformation involves four fundamental processes (Figure 2.21A): (1) the educt, H_2, has to be transferred to the [NiFe] active site in an efficient and controlled manner. According to crystal structure data and molecular dynamic simulations, this process is accomplished by dynamic hydrophobic gas channels that connect the protein surface with the [NiFe] center [113, 114]. (2) H_2 has to be cleaved at the active site, which involves the crucial step of dihydrogen bond activation, as outlined in detail below. (3) The products, two electrons and two protons, have to be transferred from the [NiFe] active site toward the protein surface. While electron

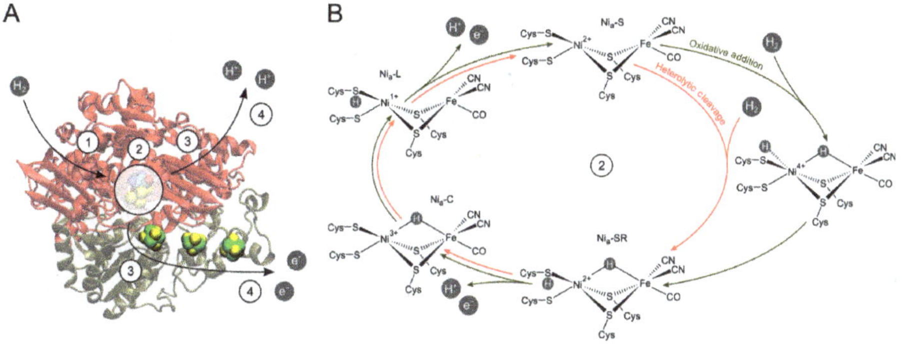

Figure 2.21: (A) Overall process of H_2 conversion by [NiFe] hydrogenases involving the intramolecular transport (1) and cleavage (2) of the educt, H_2, as well as the intramolecular transport (3) and release (4) of the products, two protons and two electrons. (B) Possible catalytic mechanisms of H_2 cleavage at the [NiFe] active site. For details, see text.

transfer proceeds *via* a sequence of FeS clusters, proton transfer is generally accepted to involve ionizable amino acid side chains and structural water molecules embedded in the protein matrix [1, 33]. (4) Finally, the products of H_2 cleavage have to be released from the enzyme. While protons are taken up by the buffered aqueous solvent, electrons are transferred to a suitable electron acceptor. *In vivo*, this process requires either a second metalloprotein or a soluble low-molecular-weight cofactor, whose reduction can involve a second catalytic reaction [1–3, 26, 37]. *In vitro*, electrons may also be accepted by electrode or chemical oxidants such as redox dyes.

Given the physiological importance of H_2 conversion, the overall architecture of [NiFe] hydrogenases is subject to considerable evolutionary pressure and therefore optimized for maximum performance. Depending on the relative efficiencies, each of the aforementioned four processes may in principle become rate-limiting during cellular H_2 cycling. While substrate conversion and interfacial electron transfer are mandatory tasks common to all types of redox enzymes, intramolecular educt and product transport are specifically dictated by the surface-remote location of the deeply buried [NiFe] active site. This situation suggests that the encapsulation of this cofactor is beneficial for the overall efficiency and/or robustness of cellular H_2 cycling. Obviously, an enclosed catalytic center represents a simple means to limit side reactions with potential inhibitors or undesired substrates. Indeed, standard [NiFe] hydrogenases have not been reported to catalyze any side reaction, in sharp contrast to other metalloenzymes such as nitrogenase (see chapter 4). Moreover, encapsulation of the [NiFe] active site inside the protein matrix allows to precisely tune its environment in order to optimize ground and transition state energies along and off the reaction coordinate. In line with this statement, biomimetic compounds lacking this trait are typically instable and poor H_2-cycling catalysts [1, 46], and results from computational studies on [NiFe] hydrogenase function were found to critically depend on protein matrix modeling [43–45, 110–112, 115]. These findings highlight the functional role of the active site environment as a multi-layered anisotropic "solvent."

Following the sketched reaction sequence outlined earlier, both substrate access and product release are clearly separated from H_2 conversion at the [NiFe] active site in a temporal and/or spatial manner. Since details of these processes are also neither conserved among different [NiFe] hydrogenase nor intimately linked to the principles of biological H_2 cycling, we will not dwell on these aspects in the following. Instead, possible mechanisms of H_2 conversion at the [NiFe] active site will be described in more detail, laying emphasis on fundamental reaction schemes, proposed elementary steps and the functional relevance of concomitant intramolecular proton and electron transfer events.

Due to its inherent strength and apolarity [18], breaking the dihydrogen bond requires prior activation, which could in principle proceed *via* three fundamental mechanisms, that is, homolysis, heterolysis or a *generalized* oxidative addition step that is neither homo- nor heterolytic in a strict sense [19, 44, 45, 116, 117]. Within the framework of transition metal-based (bio)catalysis, these mechanisms can be illustrated by the following reaction schemes:[3]

Homolysis:

$$M^X + H_2 \rightarrow M^X(H_2) \rightarrow M^{X+2}(H^-)_2 \tag{2.5a}$$

$$2M^X + H_2 \rightarrow M^X(H_2)M^X \rightarrow 2M^{X+1}(H^-)_2 \tag{2.5b}$$

Heterolysis:

$$M^X + B + H_2 \rightarrow M^X(H_2) + B \rightarrow M^X(H^-) + B(H^+) \tag{2.6}$$

Oxidative addition:

$$M^X + H_2 \rightarrow M^X(H_2) \rightarrow M^{X+2}(H_a{}^-)(H_b{}^-) \tag{2.7}$$

3 The usage of these termini is not uniform throughout literature, and the terminology used here has been explicitly chosen to illustrate the differences between the possible mechanisms. Clearly, reactions (2.5a) and (2.5b) can also be considered special cases of oxidative addition. In this respect, reactions of type (2.5a) have also been designated as (*cis*) oxidative addition, while phrases like homolysis or homolytic addition/activation/cleavage are sometimes reserved for reactions involving two metal ions (2.5b). The mechanism reflected by reaction (2.7) is uncommon for synthetic transition metal compounds and thus no commonly accepted designation has been established. Semantic subtleties aside, the reader is encouraged to carefully analyze reaction mechanisms described by others rather than relying on the termini used.

Here, M^X is a transition metal ion in oxidation state X and B is a base. The anticipated formation of a metal–dihydrogen adduct intermediate is common to all three mechanisms; however, the exact structures and the resulting products differ considerably.

H_2 homolysis at transition metal sites involves the two-electron oxidation of one metal ion (eq. (2.5a)) or two metal ions that formally donate each one electron (eq. (2.5b)). After bond cleavage, this process stabilizes the resulting hydrogen species in the form of metal–hydrides. By definition, homolysis requires that electronic charge derived from dihydrogen bond cleavage is equally distributed between the two resulting hydrogen species. This requirement clearly conflicts with the highly asymmetric structure of the [NiFe] active site, and, consistently, catalytic H/D exchange experiments revealed the formation of two different hydrogen species in the course of dihydrogen bond cleavage [42, 116, 117]. As a consequence, homolytic cleavage of H_2 by [NiFe] hydrogenases can be excluded.

H_2 heterolysis by [NiFe] hydrogenase (eq. (2.6)) is assumed to proceed *via* polarization of the dihydrogen bond facilitated by a (frustrated) Lewis acid–base pair. Here, one of the two metal ions would serve as an initial hydride acceptor (Lewis acid), while a nearby ionizable amino acid could act as a primary proton acceptor (Lewis base). Obviously, this mechanism results in two chemically distinguishable hydrogen species, in line with the observed H/D exchange behavior [42, 116, 117]. Therefore, heterolysis has long been considered the only plausible mechanism for H_2 cleavage by [NiFe] hydrogenases.

Only recently, quantum mechanical studies have proposed oxidative addition to a single metal ion (eq. (2.7)) as an alternative reaction mechanism [44, 45]. Dihydrogen bond activation *via* this reaction channel involves a pronounced population of the antibonding σ^* orbital of H_2, and subsequent two-electron oxidation of the metal ion results in the formation of a metal–dihydride product. In contrast to metal-based homolysis, which involves oxidative addition as well, this mechanism yields two hydrogen species, labeled H_a and H_b in eq. (2.7), which are *not* identical due to an asymmetric environment. As a consequence, this generalized oxidative addition mechanism represents an alternative model for H_2 cleavage by [NiFe] hydrogenases that complies with observations from experiments on catalytic H/D exchange [42, 116, 117].[4]

So far, the earliest intermediates of biological H_2 oxidation have not been monitored experimentally, indicating that they are both kinetically and thermodynamically instable. Therefore, neither heterolytic cleavage nor oxidative addition can be clearly ruled out, and two possible catalytic cycles that illustrate these fundamental

4 Notably, the common mechanistic interpretation of catalytic H/D exchange experiments assumes that the observed effects relate to the *immediate* products of H_2 cleavage. This is not necessarily the case, stressing the possibility of reaction mechanisms other than heterolytic cleavage.

mechanisms will be described in the following (Figure 2.21B) [44, 45]. In both cases, H_2 has to bind to the Ni_a-S state of the [NiFe] active site prior to cleavage. This process is widely assumed to involve the formation of a metal–dihydrogen σ-bond complex as the first intermediate [19]. Since no experimental insights into this side-on (three-center) adduct have been reported so far, the initial binding site of H_2 is not entirely clear. Based on former theoretical studies and reports on synthetic σ-bond complexes (see Section 2.2.5.3) [18, 19, 106, 108, 118, 119], H_2 was widely believed to bind to the Fe ion of the [NiFe] active site, but more recent quantum mechanical calculations suggest H_2 binding to the Ni ion (Figure 2.22A) [44, 111, 112]. This suggestion is in line with the Ni ion's location at the end of the H_2 channel, the presence of a vacant terminal coordination site at this metal and the observation of H_2 cycling catalyzed by synthetic Ni compounds and Ni-substituted rubredoxin (see Section 2.2.5.3) [1, 46–49, 51, 113, 120, 121]. Theoretical studies on [NiFe] hydrogenases also indicate that spin state and geometry of Ni_a-S are highly important for thermodynamically favorable H_2 binding [44]. Typically, low-spin and high-spin ground states of four-coordinate d8 transition metal compounds favor square planar and tetrahedral geometries, respectively (Figure 2.22). In [NiFe] hydrogenases, however, the Ni^{II} site of Ni_a-S adopts a distinct seesaw-shaped geometry that is dictated by the arrangement of the four active site cysteines. Interestingly, energetically favorable H_2 binding appears to strictly depend on this forced geometry and a Ni^{II} low-spin configuration (Figure 2.22B). This finding demonstrates the importance of proteic constraints imposed *via* the unique all-cysteine coordination pattern, which may explain the absence of sulfido ligands that dominate other polynuclear metal–sulfur sites in biology (see Section 2.2.3.3).

Once H_2 is bound to the [NiFe] active site, dihydrogen bond cleavage could proceed *via* two different routes. Heterolytic cleavage likely involves the formation of a terminal Ni-bound hydride [44], which is immediately transferred to the bridging position between the two metals (Figure 2.21B) [44, 45]. In this reaction, Ni would act as an acid, but the corresponding base of the (frustrated) Lewis pair has not been unambiguously assigned [33]. Both a conserved arginine close to the [NiFe] center [122, 123] and one of the terminal active site cysteines [18, 43–45, 79, 119, 124] have been proposed as the initial proton acceptors, and the latter seems to provide a first stable relay site [77,100,101]. Alternatively, Ni-bound H_2 could also be cleaved in an oxidative addition mechanism, which would yield a transient species with a formal Ni^{IV} site and each one hydride bound to the vacant bridging and terminal positions of the [NiFe] center (Figure 2.21B) [44]. A subsequent reductive elimination step could enable re-formation of Ni^{II} and protonation of one of the terminal cysteines at the expense of the terminal hydride. Despite mechanistic differences, both pathways depend on bridging and terminal vacant sites, and two chemically distinct hydrogen species are formed upon H_2 splitting (*vide supra*). Moreover, the experimentally observed Ni_a-SR state is formed as the first stable intermediate in both reaction schemes, so that all subsequent steps are identical (Figure 2.21B).

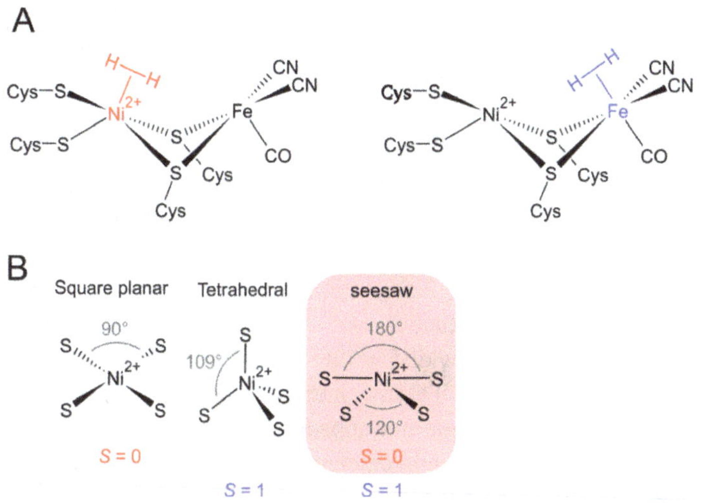

Figure 2.22: Structural aspects of H_2 binding to the [NiFe] active site. (A) Structural depictions of possible σ-bond complexes involving either of the two metal ions. (B) Possible spin states and idealized geometries of four-coordinate Ni^{II} sites, relevant for the [NiFe] center. For details, see text.

Notably, heterolytic cleavage could also take place at the Fe ion, while oxidative addition is limited to the Ni site with its two vacant coordination sites.

After cleavage of the dihydrogen bond, a total of two protons and two electrons have to be released from the [NiFe] site in a kinetically efficient manner. This process is expected to involve the proximal $[4Fe–4S]^{1+/2+}$ cluster and, probably, a conserved glutamate residue as initial electron and proton acceptors [1, 33, 35, 36, 125], which excludes $2H^+/2e^-$ reactions. Thus, Ni_a-SR is first oxidized to the Ni_a-C state in a net $1H^+/1e^-$ reaction (Figure 2.21B) [44, 45], which likely proceeds in a sequential manner [40]. The Ni_a-C species was proposed to subsequently transform to the Ni_a-L state or a similar tautomeric species with a formal Ni^I oxidation state and a protonated terminal cysteine [18, 33, 34, 40, 41, 75, 76, 78, 79, 107]. This species can be converted to Ni_a-S in a concerted $1H^+/1e^-$ step [41], thereby closing the catalytic cycle (Figure 2.21B).

Despite little knowledge about the details of subsequent proton and electron transfer processes, a few generalized statements can be made. As stated earlier, each of the two $1H^+/1e^-$ reactions at the [NiFe] site is assumed to require the presence of an oxidized proximal [4Fe–4S] cluster (and a deprotonated glutamate side chain), which implies that intramolecular charge transport and active site reactions cannot be clearly separated. Consequently, both processes are integral parts of H_2 conversion, and the overall enzymatic efficiency may depend on their well-tuned orchestration rather than a mere optimization of their individual kinetics. In other words, the driving force and efficiency of any elementary step may depend on the current state of other reaction sites that are not directly involved in this process. In addition, electron transport chain architectures of [NiFe] hydrogenases may also explain deviations from an idealized catalyst, namely the typical bias toward H_2 cleavage and the requirement of overpotentials observed for certain oxygen-tolerant exemplars (see Section 2.2.4) [1, 26, 29, 30]. At physiological pH, the reduction potential of the H_2/H^+ couple is −413 mV, and at 1 bar H_2 an idealized catalyst should enable H_2 oxidation (evolution) at any potential higher (lower) than this value. For standard [NiFe] hydrogenases, however, all FeS clusters seem to possess higher reduction potentials [126], and electron transfer toward the active site includes at least two sequential intramolecular endergonic steps [126, 127]. This situation may diminish the driving force for electron flow toward the active site, thereby disfavoring H_2 evolution. FeS cluster potentials of O_2-tolerant membrane-bound [NiFe] hydrogenases (Section 2.2.4) are even higher [128–130], and consistently these enzymes appear to be even more strictly biased toward H_2 oxidation [1, 26, 29, 30]. At mild oxidizing potentials, electron release from the high-potential distal cluster of these enzymes is also endergonic, which could explain why H_2 oxidation by these enzymes requires considerable overpotentials [1, 26, 29, 30]. Besides these basic aspects related to catalytic H_2 cycling, properties of the electron transport chain also play a central role in the reversible inhibition and oxygen tolerance of certain [NiFe] hydrogenases, as outlined in Section 2.2.4 [1, 26, 34, 52, 53, 71, 131–134].

2.2.3.5 CO and CN^-: exploiting toxic ligands

The presence of CO and CN^- ligands at the active site of [NiFe] hydrogenases represents one of the most puzzling features of these enzymes. On the one hand, both diatomics inadvertently bind to metal sites of, for example, heme proteins, thereby severely interfering with cellular vitality. On the other hand, metalloenzymes containing intrinsic CO and CN^- ligands were unknown prior to the structural exploration of [NiFe] and [FeFe] hydrogenases [1, 7, 9, 10, 61–65]. [Fe] hydrogenases contain CO ligands as well [8, 66], and a conserved $Fe(CO)_m(RS^-)_n$ motif is found in all known types of hydrogenases – although these three classes are phylogenetically unrelated [2, 3, 14]. As a consequence, CN^- and, especially, CO likely play a crucial role in biological H_2 transformation.

Elucidating the exact function of CO and CN^- ligands requires a closer inspection of [NiFe] hydrogenase catalysis. According to the possible reaction mechanisms (see Section 2.2.3.4; Figure 2.21B), H_2 binding and bond breakage represent the two elementary steps that are unique and potentially rate-limiting with respect to the overall reaction taking place at the active site. Thus, the architecture of the [NiFe] cofactor is likely optimized to facilitate these processes, and the functional role of CO and CN^- ligands may be revealed by exploring their potential contribution to this task. In this respect, several aspects have to be considered, specifically kinetic competency of all intermediates and (moderate) exergonicity along the reaction coordinate as well as kinetic and thermodynamic stabilities with respect to undesired reaction channels. The possible contributions CO and CN^- to H_2 cycling will therefore depend on the details of the catalytic mechanism, especially the site of H_2 binding and conversion. In the following, we will shortly discuss essential properties of CO and CN^- and their relevance for possible ways of H_2 activation at either Ni and Fe.

Synthetic $M(CO)_m(L)_n(H_2)$ complexes have been known to organometallic chemists since the first identification of a metal–dihydrogen σ-bond complex, $W(CO)_3(PR_3)_2(H_2)$, more than 30 years ago [135]. As a consequence, the influence of CO ligands on H_2 binding and cleavage at the *same* metal ion has been rigorously explored, and a number of valuable statements regarding putative H_2 transformation at the Fe site of [NiFe] hydrogenases can be made [19]. In general, H_2 binds to low-valent metal ions as a moderate σ-donor ligand, but the stability of the complex can be strongly increased by backdonation from a filled metal d orbital toward the antibonding σ* orbital of H_2 (Figure 2.23A). This process, however, needs to be finely tuned, since backdonation also results in a gradual elongation of the dihydrogen bond, which may result in the stabilization of an undesired (compressed) dihydride complex (Figure 2.23B). Somewhat similar to H_2, CO binds to low-valent metal ions as an amphoteric ligand as well by acting both as a σ-donor and a strong π-acceptor (Figure 2.23C). The latter property decreases the electron density at the metal, thereby limiting backdonation to H_2. This effect is even more pronounced, if CO binds *trans* to H_2, and, consistently, such a configuration is observed in many organometallic metal–dihydrogen complexes and anticipated for both [NiFe] and [FeFe] hydrogenases (assuming H_2 binding to Fe in both cases). In [Fe] hydrogenases, both CO ligands are bound in *cis*-positions, but this potentially unfavorable architecture might be compensated by the *trans*-guanylylpyridinol ligand of these enzymes [14]. Interestingly, binding to an electron-poor metal ion may also decrease the pK_a of H_2 by up to 55 units, thereby strongly favoring heterolytic cleavage. Besides tuning the properties of Fe–H_2 and Fe–H^- adducts, decreased metal electron density inferred from a (*trans*) CO ligand may also prevent the competitive binding of N_2, which is a much poorer σ-donor than H_2. CN^- ligands appear to play a lesser role in H_2 binding and activation by synthetic organometallic complexes. Nonetheless, these strong-field σ donor ligands may be important in stabilizing the Fe ion of hydrogenases in its low-spin state, which is a prerequisite for strong CO binding [136]. In addition, CN^- ligands may further tune the electronic structure of the active site, possibly facilitated by their pronounced capability to form electrostatic and hydrogen bonding interactions with the protein matrix.

Elucidating the functional role of CO and CN^- ligands is less straightforward, if H_2 activation is assumed to take place at Ni rather than Fe. In this case, the impact of these ligands on substrate binding and activation would be indirect and, thus, more elusive. Such a mechanism would also imply a situation where the site of H_2 binding and cleavage is not directly tuned by one or more CO ligands at the same

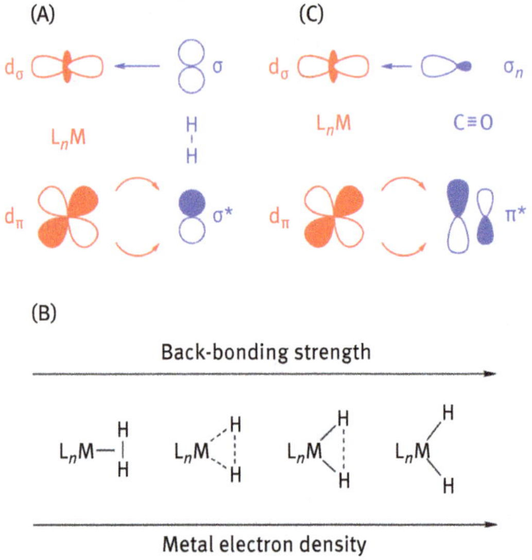

Figure 2.23: Interactions of (A) H_2 and (C) CO with a transition metal site L_nM. Complexes of both ligands are stabilized by a synergistic interplay of σ donation from a bonding (σ) or nonbonding (σ_n) ligand orbital to an empty d_σ orbital of the metal ion (top) and π back-donation from an occupied metal d_π orbital to an antibonding (σ* or π*) ligand orbital (bottom). Increased back-bonding due to a higher electron density at the metal ion stabilizes the H_2 adduct, but the H_2 bond is gradually elongated and finally broken upon dihydride formation (B).

metal, which would be in contrast to both [Fe] and [FeFe] hydrogenases. On the other hand, the latter enzyme also contains CO and CN^- ligands at the so-called proximal Fe ion, which is not directly involved in H_2 binding and transformation. This situation supports the idea of a functional role of CO and CN^- ligands that goes beyond a direct tuning of the H_2-binding metal ion. In fact, the $Fe(CO)(CN^-)_2$ moiety of [NiFe] hydrogenases was reported to facilitate Ni-based substrate conversion by increasing H_2 affinity [44]. Similar effects may also be important for tuning metal–metal bonding as well as ground and transition state energies associated with (bridging) metal–hydride intermediates. However, no systematic studies on these aspects are available so far. Thus, elucidating the exact role of CO and CN^- ligands remotely from the putative H_2 binding site remains an important task in the investigation of both [NiFe] and [FeFe] hydrogenases.

2.2.4 Coping with O_2: inhibition and counterstrategies

While [Fe] hydrogenase is a hydride-transferring enzyme, both [NiFe] and [FeFe] hydrogenases catalyze the complete reversible cleavage of H_2 according to eq. (2.1) [1]. Therefore, their active sites are designed to enable low-potential redox reactions in order to exchange electrons with the $H_2/2H^+$ couple. As a consequence of this trait, however, both enzymes can also easily react with O_2 and reactive oxygen species in a thermodynamically feasible manner, which results in an inactivation or degradation of the (catalytic) cofactor(s).[5] This situation represents a severe limitation for potential applications of most hydrogenases and bioinspired H_2-cycling catalysts alike (see Sections 2.2.5.3 and 2.2.6). Therefore, understanding the reaction(s) of hydrogenases with O_2 is of major interest, both from a fundamental and application point of view. In this section, we will describe interactions of O_2 with the active site of [NiFe] hydrogenases and known biological strategies that limit or reverse these reactions.

In contrast to [FeFe] hydrogenases, [NiFe] hydrogenases are usually not irreversibly damaged upon interaction with atmospheric O_2, but still most of these enzymes are catalytically inactive under aerobic conditions [1, 26]. [NiFe] hydrogenases from aerobic organisms, however, are catalytically active even in the presence of atmospheric O_2 levels [1, 26]. Notably, these O_2-tolerant enzymes usually do not form the inactive Ni_u-A state observed for oxidized standard [NiFe] hydrogenases under aerobic conditions (see Section 2.2.3.3). Since this redox-structural species is characterized by a very slow activation [1, 36, 82, 87, 88], its prevention during H_2 cycling in the presence of O_2 may be a fundamental key to O_2 tolerance.

In principle, a functional transition metal site will be inactivated by reaction with O_2 if all following conditions are fulfilled: (1) the reaction with O_2 is favorable both from a kinetic and thermodynamic point of view, (2) the resulting product is unable to participate in the molecule's usual function and (3) there are no reaction channels that allow to regenerate the functional form from the O_2-inhibited state on reasonable timescales. The second aspect is a trivial condition, since an O_2-inhibited transition metal site is very unlikely to be functionally unimpaired. For instance, the Ni_u-A and Ni_r-B states of O_2-inhibited standard [NiFe] hydrogenases are clearly unable to participate in H_2 cycling since the bridging coordination site between the two metals is occupied by an O_2-derived ligand [1, 72, 73, 89]. As a consequence, we are left with two variables governing O_2 inhibition that represent potential targets for the design of O_2-tolerant H_2-cycling catalysts. In the following, we will explore nature's approach to this task by illustrating structural and functional peculiarities of different prototypical O_2-tolerant [NiFe] hydrogenases.

So far, O_2-tolerant [NiFe] hydrogenases have been identified in groups 1, 2 and 3, (see Section 2.2.3.1) [1, 26]. Given the structural and functional differences between enzymes from these three groups [1–3], O_2 tolerance strategies are likely to be nonuniform but explicitly related to the physiological function of each individual [NiFe] hydrogenase. *Ralstonia eutropha* is a "Knallgas" bacterium that is able to thrive on H_2, O_2 and CO_2 [138]. This ability relies on the presence of four distinct O_2-

5 [Fe] hydrogenases are also inhibited by O_2, but the underlying processes and the exact role of the redox-inactive iron ion are not entirely clear.

tolerant [NiFe] hydrogenases from different groups [26]. Therefore, this organism represents an ideal model system to explore different O_2 tolerance strategies and their relation to enzymatic structure and physiological function.

R. eutropha contains three [NiFe] hydrogenases that contribute to energy conversion by catalyzing H_2 cleavage [26]. In addition, this organism also harbors a group 2 regulatory hydrogenase (RH), which acts as a cytoplasmic H_2 sensor [26, 139–144]. Similar to standard [NiFe] hydrogenases, this enzyme consists of a large subunit (HoxC) containing the [NiFe] active site and a small subunit (HoxB) harboring three FeS clusters [70, 145, 146]. In combination with a histidine protein kinase (HoxJ), the RH forms a cytoplasmic H_2-sensing complex with an overall $(HoxBC)_2(HoxJ)_4$ stoichiometry (Figure 2.18, center left) [26, 139–144]. This complex detects H_2 and initiates a signaling cascade in order to adjust the synthesis of energy-converting hydrogenases of *R. eutropha* according to the availability of H_2. In line with its task as a molecular switch, the RH exhibits low catalytic activity [26, 141, 146] and only two dominant redox-structural states of the [NiFe] active site, namely Ni_a-C in the presence and Ni_a-S in the absence of H_2 [78, 145, 146].

In contrast to standard [NiFe] hydrogenases, RH and similar enzymes from other organisms appear to exhibit a modified intramolecular gas channel that contains two bulky amino acids (isoleucine and phenylalanine) close to the active site [141, 147]. These two residues are supposed to form a gate preventing gases larger than H_2 from accessing the [NiFe] center (Figure 2.24A) [147, 148]. In line with this proposal, RH is also insensitive toward CO inhibition [146], and exchange of the bulky amino acids by smaller ones severely limits O_2 tolerance [147]. As proposed in the literature, this mechanism holds no implications regarding the thermodynamic feasibility of O_2 or CO binding to the RH active site, which may be as strong as in standard [NiFe] hydrogenases. Thus, O_2 tolerance of RH should be perceived as a kinetic protection that builds on slowing down the macroscopic rate at which O_2 binds to the active site, apparently, to time scales beyond the lifetime of the enzyme. Despite the plausibility of this mechanism, the molecular details of the proposed gating are so far unclear, since no crystal structure is available for RH or other H_2-sensing hydrogenases. Nonetheless, some important features can be deduced from fundamental considerations as well as biochemical, spectroscopic, and electrochemical studies. Given the general flexibility of the protein matrix, the performance of any gating can be affected by stationary and/or dynamic structural changes. As a consequence, the O_2 tolerance mechanism proposed for RH may not categorically exclude O_2 or CO inhibition under all conditions, in line with observations from spectroelectrochemical studies [149]. In addition, O_2 protection achieved by means of a limited gas access also holds a potential risk of slowing down H_2 transport and, thus, catalytic turnover. As stated above, RH is indeed a very poor H_2-cycling catalyst, and catalytic H/D exchange experiments suggest an unusual retention of H_2 molecules close to the active site [26, 141, 146]. While such side effects are probably negligible or even favorable for an H_2-sensing "enzyme," they may

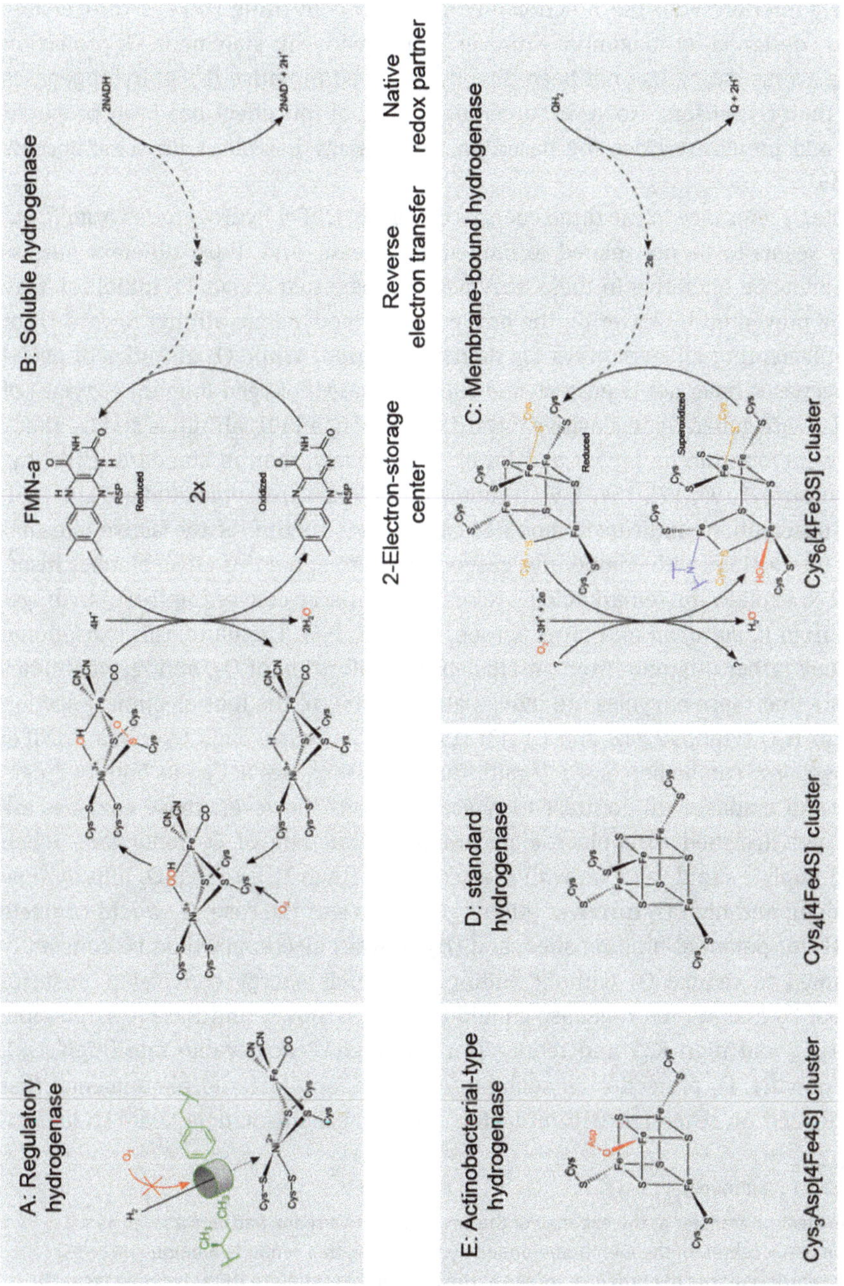

Figure 2.24: Structural and mechanistic aspects related to the O_2 tolerance of [NiFe] hydrogenases from *R. eutropha*. (A) Prevention of O_2 attack by a narrowed gas channel, as suggested for the regulatory hydrogenase. (B and C) Reductive detoxification of O_2, as anticipated for the soluble hydrogenase (B) and the membrane-bound hydrogenase (C). Both mechanisms B and C involve

severely interfere with the functionality of energy-converting [NiFe] hydrogenases that are designed for maximum turnover. In line with this statement, O_2 protection by gas access gating has not been directly observed in native [NiFe] hydrogenases other than H_2 sensors. However, a certain impact of this effect has been proposed [114], and protein engineering based on this strategy has met with some success [150–152].

The O_2 tolerance of the three energy-converting [NiFe] hydrogenases from *R. eutropha* seems to be not related to limited gas access, and, thus, different mechanisms must be operative in these enzymes. As suggested above, O_2 inhibition may also be prevented by lowering the enzyme's thermodynamic affinity toward O_2 or by implementing efficient ways for its detoxification. While O_2 affinities of [NiFe] hydrogenases have not been systematically evaluated, oxygen-tolerant enzymes of this class are generally insensitive toward CO as well, and H_2 affinities and O_2 attack rates were found to be higher and lower, respectively, than in standard [NiFe] hydrogenases [27, 93, 153, 154, 155]. In line with EPR spectroscopic studies [128, 156], these observations indicate a modified electronic structure of the active site that limits interactions with competitive gaseous inhibitors, but this effect alone is insufficient to explain the remarkable O_2 tolerance of energy-converting [NiFe] hydrogenases from *R. eutropha* and other aerobes [93, 153, 155]. Therefore, this exceptional trait may rather originate from an efficient detoxification of O_2, and recent studies indicate that these enzymes are indeed able to catalyze the four-electron reduction of O_2 to H_2O (Figure 2.24B and C) [131, 157]. In this sense, such O_2-tolerant [NiFe] hydrogenases can be perceived as multifunctional enzymes acting as both hydrogenases and oxidases. In contrast to typical oxidases, however, these enzymes are likely not designed to achieve a maximum overall rate of O_2 reduction, which would imply a rapid reaction with the active site (high k_{bind}), high O_2 affinity (low K_m value), and high O_2 turnover (high k_{cat}). If this was the case, O_2 would compete with H_2 for potential binding sites, and (H_2-derived) electrons would be constantly consumed to reduce O_2 without adding to cellular energy conversion. Instead, an ideal O_2-tolerant hydrogenase should bind O_2 as slowly and loosely as possible (low k_{bind} and high K_m) and reduce it at the highest achievable rate (high k_{cat}). Consequently, O_2 protection in such enzymes is a multifactorial phenomenon that may depend on several structural factors. Since mechanistic details are so far lim-

Figure 2.24 (continued)

reverse electron transfer at the expense of the enzymes' native redox partners as well as a 2-electron redox center. In the membrane-bound hydrogenase, this center is a unique [4Fe–3S] cluster, which replaces the proximal [4Fe–4S] cluster found in standard [NiFe] hydrogenases (D). The actinobacterial-type hydrogenase contains a modified [4Fe–4S] cluster in this position as well (E), but its functional relevance is so far unknown. Structural peculiarities of these two unusual FeS-cluster are highlighted in color. Abbreviations used: FMN, flavin mononucleotide; R5P, ribose 5-phophate; NAD^+/NADH, oxidized/reduced nicotinamide adenine dinucleotide; Q/QH_2, oxidized/reduced quinone. For details, see text.

ited for the actinobacterial-type hydrogenase from *R. eutropha* (AH) [58, 158], these aspects will be exemplarily illustrated for the other two energy-converting [NiFe] hydrogenases from this organism (Figure 2.24B and C).

The first of these enzymes is a periplasmic membrane-bound [NiFe] hydrogenase (MBH) belonging to group 1 [26, 52, 53]. Similar to other enzymes of this class, MBH is composed of a large subunit (HoxG) harboring the [NiFe] active site and a small subunit (HoxK) containing three FeS clusters (Figure 2.18, top left) [133]. *In vivo*, this enzyme is linked to the quinone pool of the electron transport chain *via* an additional di-heme cytochrome *b* subunit (HoxZ) [159], and multimerization of the heterotrimeric enzyme has been described [160]. While active site structure and overall architecture of MBH are similar to other group 1 hydrogenases, crystal structure and spectroscopic analyses have revealed considerable differences in the electron transport chain [132–134]. Specifically, the proximal [4Fe–4S] cluster found in standard [NiFe] hydrogenases is replaced by a unique [4Fe–3S] cluster, which lacks one of the inorganic sulfides (Figure 2.24C and D). This unusual cluster is coordinated by six cysteines, two of which are absent in standard [NiFe] hydrogenases but conserved among oxygen-tolerant MBH-like enzymes [1, 26, 52, 132]. One of these cysteines substitutes the missing sulfide of the inorganic cluster core while the other one serves as an additional terminal ligand of one of the Fe ions. This coordination pattern implies two $Fe(Cys)_2$ sites, which establish a distorted non-cubane structure lacking one of the Fe–S bonds present in canonical [4Fe–4S] clusters (Figure 2.24C).

This unusual [4Fe–3S] center is able to undergo two one-electron transitions at physiological redox potentials [132, 134, 161], which is in sharp contrast to almost all other known FeS clusters. While the reduced and semioxidized forms exhibit similar overall geometries, the superoxidized state features additional Fe coordination by a backbone amide N atom, and another Fe–S bond of the cluster core is broken [52, 134]. This structural reorganization results in an even more distorted geometry, and an additional OH^- ligand may stabilize this highly oxidized state (Figure 2.24C) [134]. Notably, both redox transitions exhibit rather high reduction potentials in the physiologically relevant range, that is, close to and slightly above that of the quinone pool, respectively [3, 52, 134]. Thus, two electrons derived from H_2 cleavage can be "stored" on the cluster and rapidly supplied to the [NiFe] active site in order to catalytically reduce O_2, thereby preventing the formation of Ni_u-A or similar "unready" states (Figure 2.24C). The remaining two electrons necessary for this reaction may derive from the [NiFe] center and the medial [3Fe4S] cluster, which exhibits a higher reduction potential in MBH-like enzymes as well [128–130]. In total, nature has established an elegant mechanism for catalytic O_2 detoxification, which is remarkable for two major reasons. First, the anticipated mechanism involves a finely tuned redox center that is able to participate both in O_2 reduction and the enzyme's primary task, that is, H_2 oxidation for energy conversion. Second, the properties of this exceptional [4Fe–3S] cofactor are inherently governed by extended interactions with the protein matrix involving coordination by up to seven amino acid donor atoms (Figure 2.24C). This observation is reminiscent of the catalytic role of $(Cys)_4$ coordination at the [NiFe] active site (see Section 2.2.3.3), highlighting the exceptional relevance of proteic ligands for the functional tuning of transition metal cofactors in biology.

The third energy-converting [NiFe] hydrogenase from *R. eutropha* is a cytoplasmic, bidirectional biocatalyst from group 3, which is generally denoted as soluble hydrogenase (SH) [26, 37]. This enzyme couples the reversible cleavage of H_2 to the redox conversion of NAD^+, thereby contributing to respiratory energy conversion, carbon fixation and redox homeostasis [37, 138, 162]. In contrast to AH, RH and MBH, the overall architecture of SH differs strongly from standard [NiFe] hydrogenases (Figure 2.18, bottom left), and pronounced sequence identity indicates structural similarity to respiratory complex I (Figure 2.18, bottom right) [2, 3, 37–39]. Although *R. eutropha* SH has been studied for more than 40 years [37, 163], crystal structure data are so far unavailable. Nonetheless, a crystal structure of a closely related enzyme has been recently reported [164], and considerable insights into structure and function of the SH from *R. eutropha* are available from biochemical and spectroscopic studies [34, 37, 71, 131, 165–172].

In total, SH is composed of six subunits, which are arranged in two functional modules (Figure 2.18, bottom left) [37]. Apparently, the truncated small subunit (HoxY) of the hydrogenase module contains only a single [4Fe–4S] cluster and a flavin mononucleotide cofactor (FMN-a) [37, 166, 167, 173–175], while the large subunit (HoxH) harbors the [NiFe] active site [37, 176]. The catalytic center exhibits a typical structure under catalytically relevant conditions [37, 71, 165, 170, 177], but spectroscopic and theoretical studies indicate the reversible formation of cysteine sulfenates in the presence of O_2 (Figure 2.24B) [34, 71, 171, 172]. The second module is an NAD^+ reductase (diaphorase), which also consists of two subunits, HoxF and HoxU [37, 169]. HoxF contains the active site for NAD^+ conversion, that is, a nucleotide binding site with a nearby FMN cofactor (FMN-b) and a [4Fe–4S] cluster [37, 166, 169, 173–175, 177–182]. HoxU harbors a [2Fe2S] cluster and two further [4Fe–4S] clusters, all of which are assumed to mediate electron transfer between the two catalytic sites [37, 166, 169, 173, 177–180, 182]. In addition, the NAD^+ reductase module carries two copies of the HoxI subunit [183], which has been described to have a regulatory function.

Notably, SH exhibits extraordinary O_2 tolerance, and oxidase activity of a hydrogenase has first been experimentally proven for this enzyme [131]. Similar to MBH, SH and related enzymes contain additional cysteines and histidines that may infer O_2 tolerance by modifying the properties of the "proximal" FeS cluster in HoxY. However, exchange of these amino acids yields only faint effects, indicating that the mechanism underlying O_2 resistance is distinct from that of MBH, in line with the different architectures and physiological context of the two enzymes [168]. Since O_2 tolerance is likely related to the prevention of Ni_u-A formation or persistence, catalytic reduction of an inhibited state at the H_2O_2 redox level likely represents the critical step in O_2 detoxification. If no special precautions are taken, such a peroxidase reaction likely starts with the tautomerization of an initial hydroperoxo adduct to a thermodynamically more stable species containing a cysteine sulfenate [92], as assumed for oxidized *R. eutropha* SH and the Ni_u-A state of standard [NiFe] hydrogenases (Figure 2.24B) [34, 71, 73, 91]. Once formed, this sulfoxygenated intermediate (S^0) has to be re-reduced to a thiolate (S^{2-}), which requires (1) constant access to low-potential electrons, (2) a suitable cofactor chain for electron transfer toward the active site and (3) an electron relay close to the [NiFe] center that enables a two-electron elementary step [34, 71]. Obviously, standard [NiFe] hydrogenases are incompatible with these requirements, and, thus, sulfoxygenation *per se* yields an "unready" inactivated state, Ni_u-A. Even O_2-tolerant MBH enzymes do not fulfil all requirements, indicating that these enzymes have to prevent the formation of sulfoxygenated species as well. In contrast [34, 71], SH (1) has access to a sufficiently strong reductant, NADH [37, 163, 169, 183], (2) contains a chain of FeS clusters optimized for reversible transfer of low-potential electrons [37, 138, 162, 166, 167, 169, 177–180, 182], and (3) may allow two-electron reactions at the active site due to a nearby FMN cofactor [167, 174, 175]. Thus, sulfoxygenated active site species can be activated by the enzyme's native substrate, which represents a fundamental difference to standard [NiFe] hydrogenases exhibiting a structurally similar Ni_u-A state [34, 71, 91]. As a consequence, SH may allow O_2-tolerant H_2 cycling by catalyzing sulfur-centered peroxide reduction (Figure 2.24B), similar to other nonheme peroxidases that utilize NADH, FMN and redox-active cysteines as well [92, 184].

In total, both MBH and SH prevent inhibition of the [NiFe] active site by catalyzing the reductive detoxification of O_2. In compliance with their distinct physiological context, this reaction is accomplished by different mechanisms, both of which build on native redox partners and cofactors of the enzymes (Figure 2.24B and C). In each case, Ni_u-A or similar "unready" states are prevented from interfering with catalytic H_2 cycling, albeit based on different strategies. In MBH, structural formation of such species is prevented by rapid electron transfer to the active site and, probably, other unknown factors. In contrast, sulfoxygenated [NiFe] intermediates, similar to Ni_u-A, are likely formed during O_2 detoxification by SH, however, without affecting enzymatic H_2 cycling. Another common feature of both mechanisms is the importance of proteic ligands, mainly cysteines, and the considerable structural plasticity of the involved cofactors (Figure 2.24B and C). This indicates an extraordinary role of the protein matrix, which has to tune ground and transition state energies of the individual species in order to allow their interconversions to proceed in a rapid and thermodynamically feasible manner. Moreover, O_2 detoxification is in both cases dependent on at least one non-[NiFe] cofactor, which enables the supply of electrons at adequate potentials and rates (Figure 2.24B and C). In the future, these findings and the general statements on idealized O_2-tolerant hydrogenases (see above) may guide the design of synthetic catalysts for H_2 cycling under aerobic conditions [34, 71, 172].

2.2.5 Formation and creation

Chemical structure as a molecule's central property may be analyzed with respect to two basic aspects, each of which is important both from a fundamental and application point of view. On the one hand, structure determines interactions with other chemical and physical entities and, thus, the molecule's potential with respect to a predefined function, for example, catalysis of a certain reaction. These structure–function relationships have been thoroughly outlined in Sections 2.2.3 and 2.2.4 of this chapter. On the other hand, molecular structure itself is, generally speaking, the result of a complex sequence of chemical and physical events. In the current section, we will discuss these events, specifically those processes that amount to the formation and creation of natural and synthetic H_2-cycling [NiFe] catalysts, respectively. We will start this treatise with a short survey on the emergence and evolution of [NiFe] hydrogenases, followed by a more detailed description of their biosynthesis in extant organisms as well as their intellectual transformation for the design of bioengineered and synthetic H_2-cycling catalysts.

2.2.5.1 Chemical and biological evolution

The three classes of hydrogenases are phylogenetically unrelated [2, 3], and structural similarities represent an intriguing example of convergent evolution rather than a common origin. In fact, [NiFe] hydrogenases have been reported to be phylogenetically older than both [FeFe] and [Fe] hydrogenases [2, 4, 23, 185]. Remarkably, these heterobimetallic enzymes can be traced back to the last universal common ancestor (LUCA), which represents life on the Earth at the branching point of the universal tree of life ca. 3.5 billion years ago [23, 185]. Thus, [NiFe] hydrogenases first appeared in the mysterious past of the young earth, which is difficult to explore by phylogenetic analyses. Nonetheless, evolutionary precursors of [NiFe] hydrogenases and other energy-converting LUCA enzymes (see chapters 3b and 4b) share a number of distinct features [23, 186], all of which are consistent with a chemoautotrophic origin of life and its emergence from autocatalytic reaction cycles at volcanic sites or hydrothermal vents [23, 24, 186–189]. First, all of these enzymes catalyze redox reactions involving inorganic substrates by using transition metal-based cofactors [23, 186]. Moreover, their active sites reflect motifs of simple inorganic and organometallic compounds that have been proposed as crucial players in prebiotic catalysis and abiogenesis [23, 24, 186–188]. Specifically, active sites of several [NiFeS] enzymes (see also chapters 3b and 4b) resemble the structures of catalytic minerals, for example, nickelian mackinawite in the case of [NiFe] hydrogenases (Figure 2.25) [23, 186]. In addition, both CO and CN^- have been proposed as substrates and ligands of transition metal catalysts in the primordial metabolism of a putative pioneering organism [24, 187, 188], and, indeed, organometallic compounds like $Fe_2(RS^-)_2(CO)_6$ are spontaneously formed under conditions mimicking the anticipated habitat of this early life form [190]. From this point of view, the presence of the diatomic ligands at the active site of [NiFe] hydrogenases appears less exotic, as their incorporation may simply reflect chemical evolution within the reaction environment faced by the first H_2-cycling organisms or their prebiotic progenitors. One may also speculate that the emergence of prototypical [NiFe] hydrogenases dates back to the earliest days of life on the Earth, which is in line with the high availability of H_2 at that time [3, 24], especially at the proposed sites of abiogenesis [23, 189].

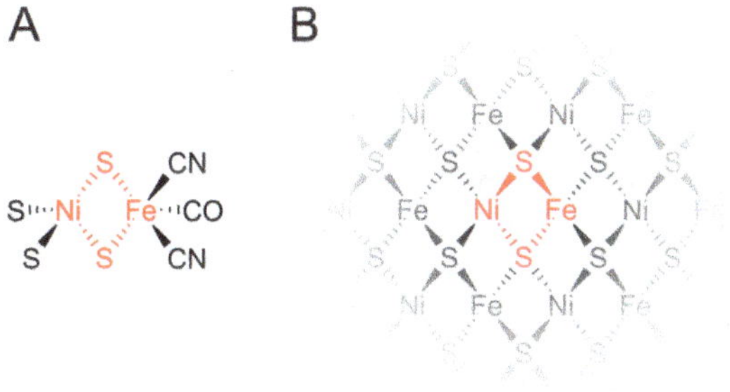

Figure 2.25: Structural comparison of (A) the [NiFe] active site of hydrogenases and (B) a slice through an idealized nickelian mackinawite crystal. Note that typical mackinawite contains more Fe than Ni.

Another interesting aspect of [NiFe] hydrogenases is their phylogenetic relationship to other energy-converting enzymes, especially respiratory complex I (Figure 2.18, bottom right) [2, 3, 37–39, 176], which is an NADH dehydrogenase acting both as an electron entry point to the electron transport chain (peripheral arm) and a redox-driven proton pump (membrane integral part) [38, 39]. Thus, complex I is crucial for ATP generation by chemiosmotic energy conversion in almost all eukaryotes and many bacteria and archaea. All [NiFe] hydrogenases contain two subunits that are homologous to parts of complex I [176], but similarities are most pronounced for members from groups 3 and 4 (Figure 2.18) [2, 3, 37–39]. Group 3 [NiFe] hydrogenases strongly resemble the peripheral arm of complex I including the NADH binding site, its FMN cofactor and large parts of the FeS cluster chain for electron transfer (Figure 2.18, bottom) [37, 38]. The [NiFe] center for H_2 cycling, however, is replaced by a quinone binding site in complex I [38]. Mechanistically, group 4 [NiFe] hydrogenases are even more similar to complex I in so far as they contain homologous membrane-spanning, ion-translocating subunits as well (Figure 2.19 right) [2, 3, 38, 39]. These biocatalysts vary in terms of overall complexity, and the smallest exemplars can be perceived as today's minimal enzymes for redox-coupled ion translocation involved in chemiosmosis (Figure 2.18, top right) [38]. While the exact evolutionary relationships between complex I and today's [NiFe] hydrogenases are not entirely clear [39, 191], an ancient soluble [NiFe] hydrogenase consisting of one large and one small subunit has been proposed as an ancestor of dimeric [NiFe] hydrogenases and a primordial membrane-bound hydrogenase complex [39]. The latter may have served as a prototypical proton pump for chemiosmotic energy conversion in early prokaryotes and, thus, represent an ancestor of complex I and group 4 (as well as group 3) [NiFe] hydrogenases [39].

2.2.5.2 Active site biogenesis

While the evolutionary emergence of the [NiFe] active site is largely obscure, its biosynthesis in extant organisms represents a complex but rather well-studied process [1, 4, 24, 192]. Cytoplasmic maturation of the heterodimeric [NiFe] hydrogenase module involves the following fundamental steps: (1) biosynthesis of the diatomic ligands and assembly of the [NiFe] active site in the apo-form of the large subunit, (2) insertion of FeS clusters into the small subunit and (3) oligomerization of the two subunits. In a fourth step, the hydrogenase module is – in case of periplasmic enzymes – translocated through the cytoplasmic membrane and/or connected to further electron-accepting/donating subunits that are either periplasmic or membrane integral. In the following, we will focus on the first step of active site assembly, which is unique but uniform in all [NiFe] hydrogenases (Figure 2.26). Maturation of the holo-form of the large hydrogenase subunit can be further divided into four substeps, namely (1) synthesis of the CO and CN^- ligands, (2) assembly and insertion of the $Fe(CO)(CN^-)_2$ moiety, (3) Ni insertion and (4) active site encapsulation within the protein framework, which is in most cases accompanied by endoproteolytic cleavage of a C-terminal extension of the large subunit. Apart from few exceptions, this maturation process requires six dedicated accessory proteins, designated HypA–HypF, and a hydrogenase-specific endoprotease, as will be outlined in the following [1, 4, 24, 192].

Each of the two CN^- ligands is synthesized from carbamoyl phosphate *via* two ATP-dependent reactions involving a complex of the accessory proteins HypE and HypF (Figure 2.26A) [193–200]. First, carbamoyl phosphate is adenylated by HypF [198], followed by transfer of the carboxamide moiety to a C-terminal cysteine of HypE [195–197]. In a subsequent ATP-dependent reaction, the resulting *S*-thiocarbamate is dehydrated to thiocyanate [195–197, 199, 200], which serves as the direct precursor for the CN^- ligand. The synthesis of the CO ligand is more elusive, since CO may, in principle, derive from both atmospheric and metabolic precursors. While CO levels well above the atmospheric CO concentration have been shown to be sufficient for gaseous CO being incorporated into the [NiFe] active site, this process is likely irrelevant under most if not all physiological conditions [201]. CO_2 has also been suggested as a potential precursor of CO [202, 203], but this scenario is rather unlikely according to a previous physiological study [194]. In any case, the active site CO ligand seems not to be formed from atmospheric precursors, and two metabolic pathways for its generation have been proposed (Figure 2.26B) [201]. The first mechanism is essentially unknown but likely common to all [NiFe] hydrogenases and apparently operative under anaerobic conditions [193, 194, 201, 204]. The second pathway is relevant for CO synthesis under aerobic conditions, and the underlying biochemical machinery is only found in organisms that produce O_2-tolerant [NiFe] hydrogenases (see Section 2.2.4) [205]. Specifically, this mechanism depends on an additional accessory protein, HypX, which consists of an N-terminal N^{10}-formyl-tetrahydrofolate hydrolase and a C-terminal hydratase moiety. It has been shown that the formyl group of N^{10}-formyl-tetrahydrofolate – an intermediate of the cellular one-carbon metabolism – is converted by HypX into CO, which, in turn, ends up as a ligand of O_2-tolerant [NiFe] hydrogenase.

Once CO and CN^- have been synthesized, the $Fe(CO)(CN^-)_2$ moiety can be assembled on an Fe-loaded scaffold provided by a complex of two further accessory proteins, HypC and HypD (Figure 2.26) [202, 206, 207]. So far, it is unclear how the correct stoichiometry of one CO and two CN^- ligands is achieved, and the exact sequence of their incorporation has been discussed controversially [206, 208]. Most likely, CN^- ligands are transferred to the Fe ion prior to CO binding [206], which is in line with synthetic routes for model compounds (see Section 2.2.5.3) and the anticipated functional role of CN^- ligands in [NiFe] hydrogenases (see Section 2.2.3.5) [19, 136]. According to this model, two cyanated HypE proteins interact successively with the HypCD complex in order to transfer a total of two CN^- ligands to the Fe ion jointly bound by HypC and HypD [199, 202, 206, 209, 210]. Each of these reactions requires two electrons [195], which are likely delivered *via* a [4Fe–4S] cluster in HypD [199, 210]. After subsequent transfer of CO to the HypCD–$Fe(CN^-)_2$ complex, the entire $Fe(CO)(CN^-)_2$ unit is translocated to the hydrogenase large subunit (Figure 2.27). This process likely involves additional interactions with HypC [209, 211, 212], which may also induce conformational changes that prime the large subunit for subsequent Ni insertion [192, 213]. Notably, the apo-large subunit also interacts with HypC alone [212], but it is unknown whether this interaction is physiologically relevant. For some O_2-tolerant [NiFe] hydrogenases, $Fe(CO)(CN^-)_2$ synthesis and translocation involves extra maturation factors that are supposed to protect maturation intermediates from O_2 attack [52, 53, 214].

The premature active site is completed by insertion of nickel, which is assisted by at least two accessory proteins, HypA and HypB (Figure 2.26). Nickel insertion may be considered a delicate step in [NiFe] hydrogenase maturation, since Ni is potentially toxic and, thus, assumed to be barely accumulated within the cell [192]. Therefore, trafficking and storage of this potentially limiting nutrient are critical

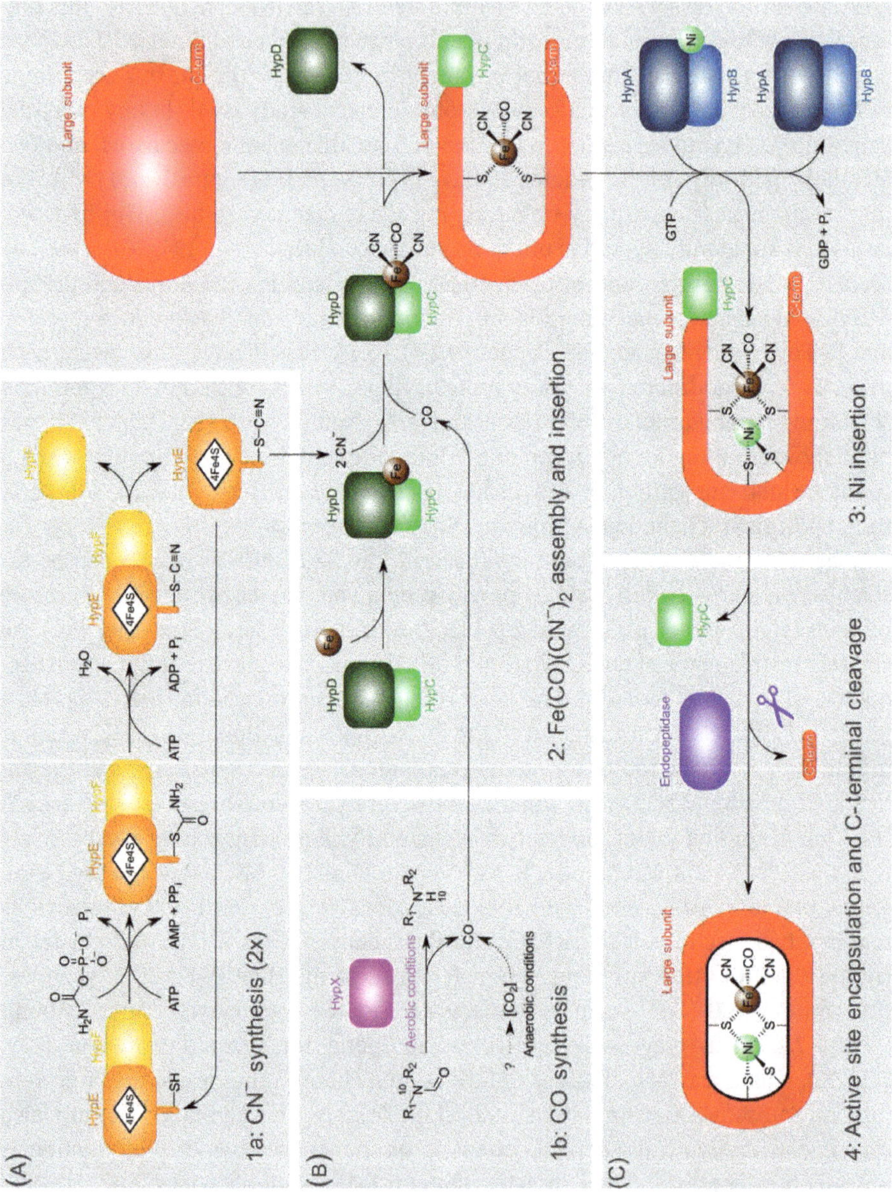

Figure 2.26: Active site biosynthesis of [NiFe] hydrogenases. Abbreviations used: ATP, adenosine triphosphate; AMP, adenosine monophosphate; PP_i, pyrophosphate; ADP, adenosine diphosphate; P_i, inorganic phosphate; GTP, guanosine triphosphate; GDP, guanosine diphosphate; C-term, C-terminus of the large subunit. For details, see text.

cellular processes that are not entirely uniform among different organisms [192]. Indeed, structures of HypA and HypB vary among species, and different pathways involving additional metallochaperones may be operative concomitantly or under distinct conditions [192]. In general, HypA appears to bind a single Ni ion, and a second, Zn-binding domain may be important for inter- and intramolecular communication or Ni transfer regulation [192, 215–217]. This protein is able to interact both with HypB and the hydrogenase large subunit, which indicates a central role in Ni delivery and the formation of a cooperative Ni-transferring complex [192, 213, 217, 218]. HypB generally exhibits the capability to bind and hydrolyze GTP or ATP [192], which is likely important for conformational regulation and Ni delivery [216, 219, 220]. Consistently, most HypB proteins also exhibit at least one metal binding motif, and additional coordination sites for Ni transfer and storage are present in HypB proteins from many species [192]. For instance, HypB from *Escherichia coli* contains two metal binding sites that exhibit high and low Ni affinities, respectively [221]. Upon complex formation, Ni may be selectively transferred from the HypB low-affinity site to HypA, especially, if HypB resides in its nucleotide-bound state [219]. Alternatively, Ni bound to the high-affinity site of HypB may also be delivered toward the hydrogenase large subunit [192]. While the exact mechanism of the latter pathway is not entirely clear [192], efficient Ni release from this site appears to require nucleotide binding and hydrolysis as well as – in the case of *E. coli* – complexation with an additional metallochaperone, SlyD [219, 222].

After incorporation of all nonproteic components into the hydrogenase large subunit, the C-terminus is cleaved off by a specific endoprotease (Figure 2.26) [223]. This process appears to induce a notable change in the tertiary structure [224], which is generally assumed to represent a central checkpoint in [NiFe] hydrogenase maturation. Specifically, it may allow oligomerization with the small subunit and subsequent cellular re-localization of the hydrogenase module [225]. Endoproteolytic cleavage, however, depends on the complete incorporation of the [NiFe] active site, which is possibly signaled by the dissociation of HypC from the large subunit upon Ni insertion. In this way, the presence of a fully matured [NiFe] active site can be guaranteed for all synthesized hydrogenases. It should be noted, however, that the C-terminal extension is missing in some precursors of [NiFe] hydrogenase large subunits [3], although these proteins undergo the same [NiFe] site assembly process as canonical hydrogenases. Thus, the exact role of the C-terminal processing is therefore not entirely clear.

2.2.5.3 Synthetic chemistry and biology

For decades, [NiFe] hydrogenases have served as prototypes for the creation of ideal molecular catalysts for H_2 cycling, and numerous attempts have been made to adapt their extraordinary properties for applications (see also Section 2.2.6). In this respect, three fundamentally different strategies can be distinguished, namely (1) optimization and extension of native [NiFe] hydrogenases or other suitable proteins, (2) synthesis of H_2-cycling catalysts that emulate the structural or mechanistic traits of their biological blueprints and (3) a hybrid approach aiming at the design of novel types of artificial Ni-containing hydrogenases. In addition, (4) [NiFe] hydrogenase may serve as an inspiration for heterogeneous catalysts with local structural properties optimized for H_2 cycling.

(1) Depending on the envisaged goal, attempts to modify native [NiFe] hydrogenases may take one of the following three approaches. First, the overall activity of these enzymes could be optimized by exchanging individual metal ions, nonproteic ligands or amino acids. Due to the excellent performance of [NiFe] hydrogenases, however, few attempts have been made to improve H_2 cycling, and site-directed mutagenesis has been mainly employed to identify structural determinants of enzymatic function [1, 33, 52, 122, 132, 147, 168]. Nonetheless, amino acid exchanges close to the gas and electron transfer pathways were shown to yield engineered [NiFe] hydrogenases with an improved tolerance toward O_2 and an altered catalytic bias [150–152, 226]. These findings demonstrate that site-specific structural changes may allow improving native [NiFe] hydrogenases with respect to general functional aspects.

Similarly, these enzymes can also be customized for specific biotechnological applications, for example light-driven H_2 production (see Section 2.2.6). This concept is nicely illustrated by a modular fusion protein that was constructed from an O_2-tolerant [NiFe] hydrogenase and photosystem I (PS I), a central part of the photosynthetic apparatus [227–230]. Initial studies demonstrated the catalytic activity of this hybrid enzyme, thereby paving the way for sustainable H_2 evolution and artificial modular enzymes in general.

Last but not least, [NiFe] hydrogenases may also be modified to diminish their large size, which represents an inherent disadvantage of biocatalysts. In contrast to synthetic molecular catalysts, most enzymes contain thousands of atoms, only few of which appear to be directly involved in catalytically relevant processes. This situation represents an undesirable trait, since the synthesis of excessively large macromolecules implies a waste of matter and energy. Furthermore, large molecular volumes impede the efficient packing of catalysts in bioelectronic devices (see Section 2.2.2), which results in a small number of active sites per unit area or volume [1]. In principle, these obstacles can be overcome through the design of minimal [NiFe] hydrogenases by removing unnecessary parts of the protein matrix. However, such an approach is complicated by the complex interplay between the "bulk" protein matrix and catalytically relevant reaction and relay sites. A mere removal of apparently unnecessary "ballast" is therefore unlikely to yield highly active minimum enzymes, and no successful attempts have been reported so far. An alternative approach to this challenge aims at the design of minimal hydrogenases from other, smaller, proteins. The Ni-substituted form of rubredoxin, a small electron transport protein [231], represents a prominent example, which catalyzes H_2 cycling by a tetrahedral $Ni(Cys)_4$ center resembling the Ni site of [NiFe] hydrogenases (Figure 2.27) [120, 121, 232, 233]. Although the catalytic properties of this model enzyme are inferior to those of native [NiFe] hydrogenases, it represents a potential starting point for designing minimal analogues that may also help to explore the functional role of the $Fe(CO)(CN^-)_2$ moiety in more detail (see Section 2.2.3.5).

(2) The construction of size-reduced variants of [NiFe] hydrogenases or other proteins can be perceived as a top-down approach to the creation of minimal H_2-cycling catalysts. In this sense, the design of small synthetic analogues that emulate native [NiFe] hydrogenases represents a complementary bottom-up strategy. Since the earliest insights into the active site structure of these enzymes, numerous attempts have been made to precisely reproduce the sophisticated architecture of this cofactor. These studies have focused on the first coordination sphere of the [NiFe] active site [46], and the resulting complexes are generally referred to as biomimetic compounds (Figure 2.28, top). Besides their potential relevance as size-reduced catalysts, such complexes primarily represent simplified model systems that provide valuable insights into structural, electronic and biosynthetic aspects of the enzymatic active site. Due to intrinsic difficulties associated with the synthesis of heterobimetallic complexes, many early attempts focused on the design of homonuclear compounds mimicking either of the two metal sites of the [NiFe] center [46]. These efforts led to a plethora of $Ni_l(SR^-)_m(L)_n$ and $Fe(CO)_x(CN^-)_y(L)_z$ model complexes of varying complexity and accuracy (Figure 2.28) [46, 234]. While some of these compounds reproduce selected spectroscopic properties of [NiFe] hydrogenases quite well [80, 235], none of them is catalytically active in terms of H_2 cleavage or H^+ reduction. Thus, besides these simple complexes, many sophisticated di- and poly-heteronuclear [NiFe] compounds have been synthesized as well [46, 236]. Several of these contain Ni and Fe in a 1:1 stoichiometry, bridging (and terminal) thiolate donors and CO/CN^- ligands (Figure 2.28). Despite their remarkable resemblance of the overall active site architecture, most of these complexes are catalytically inactive as well. Exceptions are often restricted to H^+ reduction [1, 46, 236, 237], and so far none of them is able to compete with native [NiFe] hydrogenases in terms of catalytic performance. Taking into account the biocatalytic relevance of subtle structural details as well as interactions with the protein matrix and other cofactors, this observation is probably not surprising (see Section 2.2.3.4). Given the complex interplay of these factors and the intrinsic difficulties of precise biomimicry, the contribution of this approach to the design of minimal H_2-cycling catalysts is therefore somewhat limited.

Due to the challenges associated with the development of functional biomimetic compounds, the design of bioinspired catalysts has been established as an alternative strategy. While biomimicry aims to explicitly reproduce structural aspects of functional biomolecules, the bioinspired approach is directed toward an

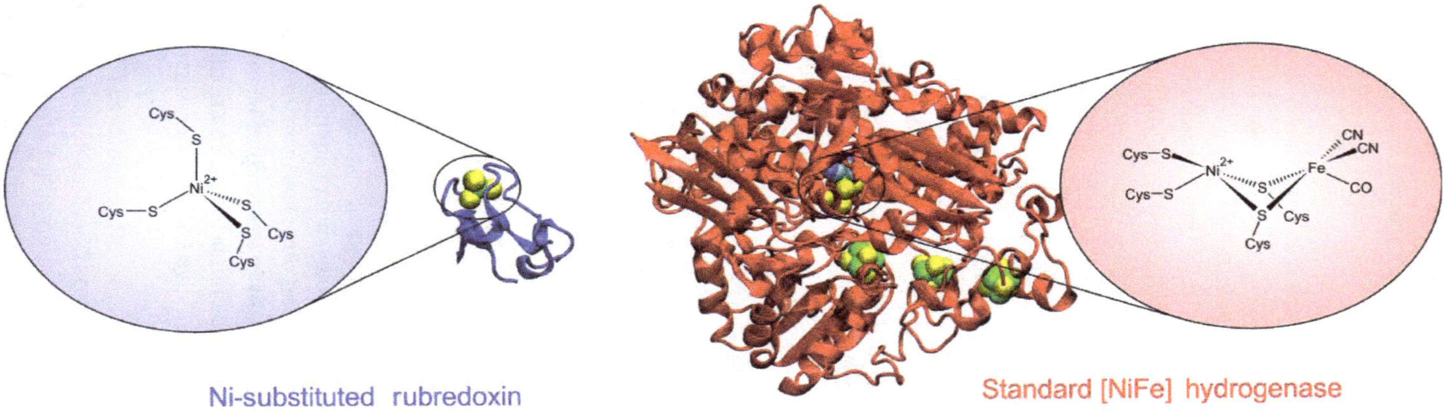

Figure 2.27: Structural comparison of Ni-substituted rubredoxin and a standard [NiFe] hydrogenase (from *D. gigas*; PDB entry: 1YQ9) [89]. The molecular mass of rubredoxin (5 kDa) is considerably smaller than that of the hydrogenase (90 kDa). Protein matrices of both enzymes are shown in cartoon representation, while metal cofactors are depicted as colored spheres. Skeletal formulas are shown for the respective catalytic sites. Note that the depicted crystal structure of rubredoxin refers to the native, Fe-containing form of the protein (from *D. desulfuricans*; PDB entry: 6RXN) [231].

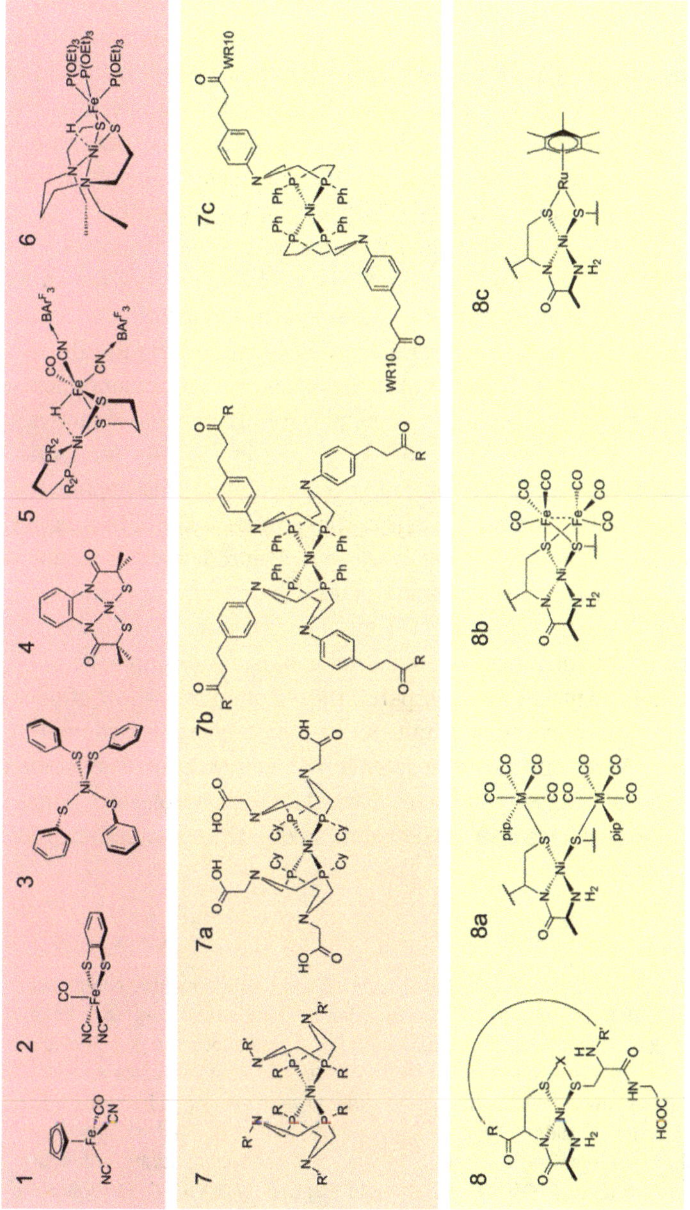

Figure 2.28: Selected biomimetic (red background), bioinspired (orange background) and biohybrid (yellow background) models of the [NiFe] active site of hydrogenases. **1** and **2** display mononuclear biomimetic models of the Fe-moiety that resemble its IR spectroscopic properties and structural features, respectively [46]. Mononuclear biomimetic models of the Ni-moiety have been synthesized from monodentate (**3**) and polydentate thiolato ligands (**4**) [234]. This approach has led to homoleptic (**3**) and heteroleptic (**4**) compounds mimicking, for example, tetrahedral (**3**) or square planar (**4**) geometries that had been anticipated for the [NiFe] active site prior to the

adaption of mechanistic concepts through abstraction and emulation of generalized functional features of enzymes. From a technical point of view, this approach is less complicated, since the design of the envisaged compounds is not restricted to the exact reproduction of biological motifs, which are often difficult to synthesize. On the other hand, the development of bioinspired compounds is conceptually more demanding, as the underlying biocatalytic principles have to be understood on a fundamental level. Bioinspired chemistry therefore represents a more rational approach to the design of minimal enzyme-derived catalysts, which, however, is strictly dependent on a thorough and probably lengthy investigation of the target enzymes. With respect to [NiFe] hydrogenases, these aspects and the differences between biomimetic and bioinspired strategies are nicely illustrated by the DuBois-type catalysts (Figure 2.28(1–4)) [48, 49, 51]. Simply speaking, these mononuclear complexes are designed to emulate basic principles of hydrogenase catalysis, namely the presence of a redox-active metal ion, a free coordination site for hydride binding and a base acting as a primary proton acceptor/donor. This implies that both first and second coordination sphere effects have to be taken into account, and the DuBois-type catalysts fulfill this requirement by combining an earth-abundant base metal ion, mainly Ni, with pendant amines. Obviously, these complexes only vaguely resemble the active site of [NiFe] hydrogenases, and similarities are even less pronounced in derivatives containing other metals, for example, Fe or Co. However, similar to their enzymatic counterparts, these complexes allow efficient catalysis by tuning the above-mentioned traits so that high-energy reaction steps are avoided. Consequently, decent turnover rates for H^+ reduction or H_2 oxidation can be obtained for several of these complexes. Notably, H_2 evolution rates outperforming native hydrogenases have been observed at moderate overpotentials [238],

---

Figure 2.28 (continued)

availability of crystal structure data (see also Section 2.2.3.4). **5** and **6** depict hydride-containing biomimetic models of the entire heterodinuclear [NiFe] site that are both structurally representative and functionally relevant (R is typically a phenyl substituent) [237d,e]. **7** shows the generalized architecture of a DuBois-type bioinspired Ni catalyst with variable pairs of substituents R and R′ [48, 49, 51]. Biohybrid compounds with improved water solubility were developed from this type of complex by the inclusion of amino acids, for example, glycine **(7a)** [241]. Extended variants containing 3-(4-aminophenyl)propionic acid substituents were also used as platforms for binding amino acids and their ester variants, R, **(7b)** as well as the 10-mer peptide WR10 **(7c)** [241]. Both latter approaches led to complexes with improved catalytic activities. **8** depicts the basic architecture of a set of Ni-containing biohybrid compounds, consisting of a short Ni-binding peptide, ACDLPCG, derived from Ni-superoxide dismutase and various organometallic building blocks X [243]. Only the underlined amino acids of the peptide are shown, while the remainder is indicated by the curved segment R–R′. **8a–8b** represent different variants of this compound, where M is either W or Mo [243]. Abbreviations used: BAr^F_3, tris-[3,5-bis(trifluoromethyl)phenyl]borane; P(OEt)$_3$, triethylphosphite; Cy, cyclohexyl; Ph, phenyl. Note that charges are not indicated since some of the depicted complexes may adopt multiple redox states.

and bidirectional systems catalyzing reversible H_2 cycling have been reported as well [239].

(3) So far, most attempts to design compact H_2-cycling catalysts have followed one of the above two routes in a rather strict manner so that they can be easily categorized as top-down bioengineering or, more often, bottom-up synthetic chemistry approaches. In addition, recent studies describe hybrid strategies that combine concepts of synthetic chemistry with the design of protein-derived or tailor-made matrices that resemble the protein environment of native enzymes [240]. Some of these efforts have focused on the extension of existing synthetic catalysts, for example, the amendment of DuBois-type complexes with amino acids, aminoester derivatives or small peptides (Figure 2.28) [241]. These compounds often show superior catalytic activity due to increased active site rigidity, beneficial electrostatic interactions and/or optimized proton relay functionalities. In addition, such complexes are typically soluble in water or aqueous solvent mixtures, which is beneficial for industrial applications and in sharp contrast to the typical synthetic compounds discussed earlier.

While these attempts may still be perceived as a mere extension of classical inorganic or organometallic strategies, others represent biosynthetic hybrid approaches in the strictest sense. For instance, a compact biohybrid device for light-driven H_2 evolution has been derived by combining Ni-substituted rubredoxin (Figure 2.27) with a synthetic photosensitizer [121, 233]. Likewise, a highly efficient device for light-driven H_2 production was derived by combining a DuBois-type Ni catalyst (Figure 2.28) and cyanobacterial PS I [242]. Conceptually, these modular constructs represent hybrid analogues of the hydrogenase-PS I fusion protein described earlier, and H_2 evolution rates of the latter are considerably higher than those observed for comparable DuBois-derived systems using synthetic photosensitizers. To further optimize this DuBois-PS I system, the synthetic Ni complex was subsequently incorporated into the FMN-free apo-form of flavodoxin, a physiological electron acceptor of PS I. This strategy enabled tight interaction of the semiartificial H_2-evolving catalyst with the electron release site of PS I, which facilitates interfacial electron transfer and hence light-driven H_2 production. In principle, this strategy can be applied to other synthetic catalysts as well, thereby creating a modular platform for biosynthetic hybrid constructs enabling light-driven reduction reactions.

Another interesting approach to the development of hybrid H_2-cycling catalysts is the *de novo* design of minimal enzymes on the basis of metal-binding peptides and synthetic building blocks. This strategy was used to create di- and trinuclear heterometallic compounds that represent potential starting points for the construction of artificial [NiM_x] hydrogenases without biological counterparts (Figure 2.28) [243]. These complexes were built from a modified N-terminal peptide of Ni superoxide dismutase that coordinates Ni by two nitrogen and two sulfur donor sites. The latter ligands served as anchoring points for the attachment of mono- or dinuclear building blocks that were derived from synthetic precursor complexes with labile ligands (Figure 2.28). From a synthetic point of view, this strategy is similar to those applied to the construction of biomimetic heterobimetallic [NiFe] models (see above). The design of artificial [NiM_x] hydrogenases, however, offers free choice of metal building blocks and the possibility to tune first and subsequent coordination spheres by modifying and extending the peptide fragment in a bioinspired fashion. In summary, the design of artificial H_2-cycling catalysts by this rational hybrid approach is probably both promising and highly demanding.

(4) All previous approaches aim at designing molecular H_2-cycling catalysts from [NiFe] hydrogenase prototypes. While this approach provides extraordinary control of structural and functional properties, homogeneous catalysts generally suffer from poor stability and a low density of active sites. In principle, this drawback could be overcome by designing functional materials whose local structure resembles the geometrical and electronic properties of the [NiFe] active site. Considering the proposed emergence of the [NiFe] active site from prebiotic catalytic minerals (see Section 2.2.5.1; Figure 2.25) [23, 24, 186–188], this approach may add another interesting aspect to the fruitful interplay between biological and synthetic inorganic chemistry.

2.2.6 Application and perspectives

As stated at the outset of this chapter, the widespread interest in [NiFe] hydrogenases is largely related to their potential relevance for biotechnological applications. Thus, we will close our treatise on these enzymes by returning to this initial research motivation. In this respect, both currently explored applications and future perspectives will be presented.

Given their diverse biological functions (see Section 2.2.3.1), various approaches for the technological utilization of [NiFe] hydrogenases have been proposed. Apart from bioelectronic H_2 sensors [244], envisaged applications are primarily aiming at either (1) sustainable energy conversion or (2) the utilization of H_2 as a cheap, clean and powerful reductant [1, 26]. The qualities of H_2 as an ideally clean fuel and prerequisites for its usage have already been outlined together with advantages of enzymatic fuel cells employing [NiFe] hydrogenases as H_2 oxidation catalysts (see Section 2.2.2). In addition, these enzymes can also be used as H^+ reduction catalysts for the sustainable production of H_2 *via* three different approaches [245]. First, H_2 may be produced *in vivo* by the dark fermentation of organic low-cost or waste products. In nature, several fermentation products are released, which lowers the H_2 evolution yield. In a simplified scheme, however, H_2 evolution from the fermentation of glucose is accompanied by the formation of acetic acid and carbon dioxide ($n = 0$) or formic acid ($n = 2$) only:

$$C_6H_{12}O_6 + 2H_2O \rightarrow 2C_2H_4O_2 + 2CH_nO_2 + (4-n)\ H_2 \tag{2.8}$$

Alternatively, H_2 evolution may also be achieved by using isolated enzymes or whole cells to catalyze the water–gas shift reaction:

$$CO + H_2O \rightarrow CO_2 + H_2 \tag{2.9}$$

In vitro, this can be accomplished by conductive co-immobilization of two [NiFe] enzymes, namely hydrogenase and CO dehydrogenase (see Chapter 3) [246]. In contrast to typically applied catalysts, these enzymes operate at ambient temperatures where the exothermic water–gas shift reaction is thermodynamically more favorable. Last but not least, H_2 may also be produced in a light-driven fashion, and many attempts have focused on this strategy [1, 245, 247, 248]. In principle, numerous photosynthetic organisms provide the molecular machinery for photocatalytic water splitting that may be utilized for human demands:

$$2H_2O \rightarrow 2H_2 + O_2 \tag{2.10}$$

On the one hand, the employment of living cells from these species is favorable, since the intrinsic light-induced damage of the photosynthetic apparatus can be easily compensated by the cellular repair machinery. On the other hand, net H_2 production by these organisms is rather low, and their native hydrogenases are typically sensitive toward O_2, which is produced in large amounts during oxygenic photosynthesis. These challenges may be overcome by genetic engineering, optimized cultivations strategies and/or the employment of heterologous O_2-tolerant [NiFe] hydrogenases (see Section 2.2.4) [247]. Alternatively, light-driven H_2 production may also be accomplished *in vitro* by combining isolated [NiFe] hydrogenases with photosynthetic or artificial photosensitizers [1, 248]. Different approaches have been proposed, some of which employ H_2O as a source of electrons (eq. (2.9)), while others use sacrificial or electrochemical electron donors [1]. In this respect, hybrid constructs of PSI and [NiFe] hydrogenases or bioinspired H_2-cycling catalysts have also been explored as H_2 evolution devices (see Section 2.2.5.3) [227–230, 242].

Besides being a valuable fuel, H_2 also represents a clean reductant that provides low-potential electrons for a wide range of biotechnologically relevant reactions, given that H_2 activation can be accomplished on reasonable timescales. As powerful H_2 oxidation catalysts, [NiFe] hydrogenases are predestined for numerous applications that benefit from using H_2 as a carbon-free reducing agent. For instance, these enzymes have been demonstrated to catalyze chromate reduction, which may be utilized for environmental decontamination [249]. Most envisaged applications, however, aim at electron delivery for the enzyme-catalyzed production of industrially relevant chemicals, for example enantiopure drug precursors. In principle, this goal may be achieved by co-immobilizing a [NiFe] hydrogenase with another oxidoreductase catalyzing the desired reaction (Figure 2.29). While the general feasibility of this approach has been demonstrated [250], many reactions of

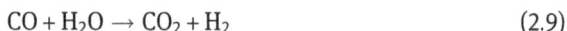

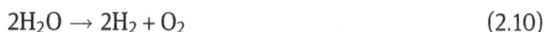

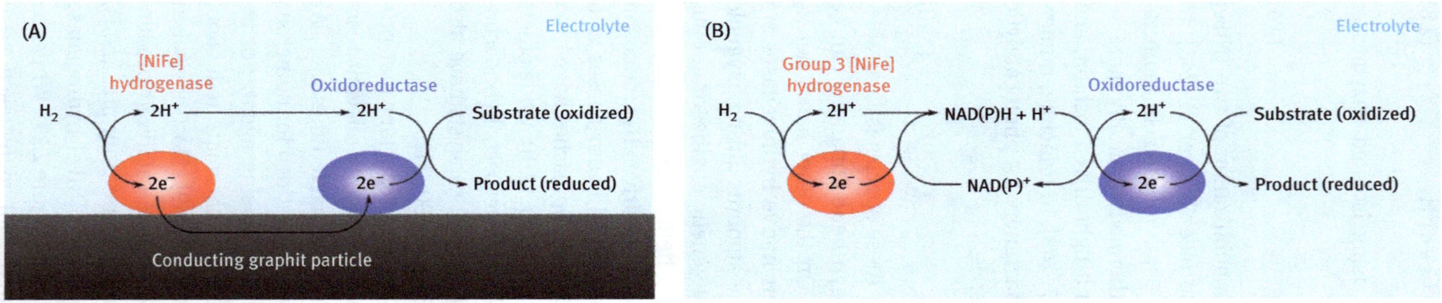

Figure 2.29: Schematic representation of biotechnological strategies for utilizing H_2 as a clean, carbon-free reductant in redox biotransformations. (A) An oxidoreductase that catalyzes the desired two-electron reduction is co-immobilized with a [NiFe] hydrogenase. H_2 is split by the hydrogenase, and the released electrons are delivered to the oxidoreductase *via* the conductive support. (B) An NAD(P)H-dependent oxidoreductase is combined with a NAD(P)$^+$-reducing [NiFe] hydrogenase from group 3 (see Sections 2.2.3.1 and 4). The latter transfers H_2-derived electrons to NAD(P)$^+$, and the resulting NAD(P)H is subsequently used by the oxidoreductase as a specific reductant. In case of both strategies, protons released into the aqueous solution are typically used up by the oxidoreductase reaction, so that no side products are generated. Abbreviations used: NAD(P)$^+$/NAD(P)H, oxidized/reduced nicotinamide adenine dinucleotide (phosphate).

biotechnological interest are catalyzed by enzymes that specifically depend on electron supply in the form of NAD(P)H. Continuous supply of these nucleotide reductants is economically unfavorable so that oxidized $NAD(P)^+$ has to be constantly regenerated in a kinetically and thermodynamically feasible manner [59]. This can be accomplished by using H_2 as a reductant and certain (group 3) [NiFe] hydrogenases as bifunctional biocatalysts for H_2 oxidation and $NAD(P)^+$ reduction (Figure 2.29B) [37, 59]. The feasibility of this approach has been demonstrated both *in vitro* and *in vivo*, and the overall performance was shown to be superior to many other (established) regeneration systems [59].

In summary, [NiFe] hydrogenases are promising prototype catalysts for sustainable energy conversion and biosynthesis approaches. This is particularly true for those that are able to catalyze H_2 cycling at ambient O_2 levels (see Section 2.2.4). To fully exploit their extraordinary traits for applications, however, several technological and scientific challenges have yet to be met. In particular, rational strategies are required in order to adapt the structural and functional properties of these enzymes for sustainable utilization. In principle, this may be accomplished by the modification of native [NiFe] hydrogenases, the creation of bioinspired catalysts or the construction of biosynthetic hybrid devices (see Section 2.2.5.3). However, all these approaches depend on a detailed understanding of enzymatic H_2 cycling and the underlying structural determinants. Considerable insights into these aspects have been obtained from experimental and theoretical studies, but several important details are still matters of debate (see Section 2.2.3). In particular, initial reaction steps and processes on short timescales are scarcely explored, and their investigation by ultrafast spectroscopic techniques remains as a central challenge for the future.

References

[1] Lubitz, W., Ogata, H., Rüdiger, O., and Reijerse, E. Hydrogenases. Chem Rev 2014, 114, 4081.

[2] Greening, C., Biswas, A., Carere, C. R., Jackson, C. J., Taylor, M. C., Stott, M. B., Cook, G. M., and Morales, S. E. Genomic and metagenomic surveys of hydrogenase distribution indicate H_2 is a widely utilised energy source for microbial growth and survival. ISME J 2015, 10, 761–777.

[3] Vignais, P. M., and Billoud, B. Occurrence, classification, and biological function of hydrogenases: An overview. Chem Rev 2007, 107, 4206.

[4] Peters, J. W., Schut, G. J., Boyd, E. S., Mulder, D. W., Shepard, E. M., Broderick, J. B., King, P. W., and Adams, M. W.W. [FeFe]- and [NiFe]-hydrogenase diversity, mechanism, and maturation. BBA – Mol Cell Res 2015, 1853, 1350.

[5] a) Bone, D. H. Localization of hydrogen activating enzymes in *Pseudomonas saccharophila*. Biochem Biophys Res Commun 1960, 3, 211. ; b) D. H. Bone, S. Bernstein, W. Vishniac, Purification and some properties of different forms of hydrogen dehydrogenase. Biochim. Biophys. Acta 1963, 67, 581.

[6] Volbeda, A., Charon, M. H., Piras, C., Hatchikian, E. C., Frey, M., and Fontecilla-Camps, J. C. Crystal structure of the nickel-iron hydrogenase from *Desulfovibrio gigas*. Nature 1995, 373, 580.

[7] Volbeda, A., Garcin, E., Piras, C., De Lacey, A. L., Fernandez, V. M., Hatchikian, E. C., Frey, M., and Fontecilla-Camps, J. C. Structure of the [NiFe] Hydrogenase active site: Evidence for biologically uncommon Fe ligands. J Am Chem Soc 1996, 118, 12989.
[8] Shima, S., Pilak, O., Vogt, S., Schick, M., Stagni, M. S., Meyer-Klaucke, W., Warkentin, E., Thauer, R. K., and Ermler, U. The crystal structure of [Fe]-Hydrogenase reveals the geometry of the active site. Science 2008, 321, 572.
[9] Peters, J. W., Lanzilotta, W. N., Lemon, B. J., and Seefeldt, L. C. X-ray crystal structure of the Fe-only hydrogenase (CpI) from *Clostridium pasteurianum* to 1.8 Angstrom resolution. Science 1998, 282, 1853.
[10] Nicolet, Y., Piras, C., Legrand, P., Hatchikian, C. E., and Fontecilla-Camps, J. C. *Desulfovibrio desulfuricans* iron hydrogenase: The structure shows unusual coordination to an active site Fe binuclear center. Structure 1999, 7, 13.
[11] Stephenson, M., and Stickland, L. H. Hydrogenase: A bacterial enzyme activating molecular hydrogen: The properties of the enzyme. Biochem J 1931, 25, 205.
[12] Kaserer, H. Die Oxydation des Wasserstoffes durch Mikroorganismen. Zent Bakt Par II 1906, 16, 681.
[13] Albracht, S. P. J. Nickel hydrogenases: In search of the active site. Biochim Biophys Acta 1994, 1188, 167.
[14] Armstrong, F. A., and Fontecilla-Camps, J. C. A natural choice for activating hydrogen. Science 2008, 321, 498.
[15] a) Shafiee, S., and Topal, E. When will fossil fuel reserves be diminished? Energy Policy 2009, 37, 181. b) IPCC. Climate Change 2014: Synthesis Report. Contribution of Working Groups I, II and III to the Fifth Assessment Report of the Intergovernmental Panel on Climate Change, Geneva, Switzerland, 2014.
[16] R. Cammack, M. Frey, Robson R., Eds., Hydrogen as a Fuel: Learning from Nature, Taylor and Francis, London, 2001.
[17] Crabtree, G. W., Dresselhaus, M. S., and Buchanan, M. V. The hydrogen economy. Physics Today 2004, 57, 39.
[18] Siegbahn, P. E., Tye, J. W., and Hall, M. B. Computational studies of [NiFe] and [FeFe] hydrogenases. Chem Rev 2007, 107, 4414.
[19] Kubas, G. J. Fundamentals of H_2 binding and reactivity on transition metals underlying hydrogenase function and H_2 production and storage. Chem Rev 2007, 107, 4152.
[20] Tye, J. W., Hall, M. B., and Darensbourg, M. Y. Better than platinum? Fuel cells energized by enzymes. Proc Natl Acad Sci U.S.A 2005, 102, 16911.
[21] O'M, J., and Bockris, M. A Hydrogen Economy. Science 1972, 176, 1323.
[22] a) Haldane, J. B. S. Daedalus; or, Science and the Future, E. P. Dutton and Company, Inc, Boston, Massachusetts, United States, 1924, b) J. Verne, L'Ile mysterieuse, Pierre-Jules Hetzel, France, 1874.
[23] Nitschke, W., McGlynn, S. E., Milner-White, E. J., and Russell, M. J. On the antiquity of metalloenzymes and their substrates in bioenergetics. Biochim Biophys Acta 2013, 1827, 871.
[24] McGlynn, S. E., Mulder, D. W., Shepard, E. M., Broderick, J. B., and Peters, J. W. Hydrogenase cluster biosynthesis: Organometallic chemistry nature's way. Dalton Trans 2009, 4274.
[25] Jones, A. K., Sillery, E., Albracht, S. P. J., and Armstrong, F. A. Direct comparison of the electrocatalytic oxidation of hydrogen by an enzyme and a platinum catalyst. Chem Commun 2002, 866.
[26] Lenz, O., Lauterbach, L., Frielingsdorf, S., and Friedrich, B. Biohydrogen, Ed., Matthias Rögner, Walter de Gruyter GmbH, Berlin, Germany, 2015, 61–96.

[27] Vincent, K. A., Cracknell, J. A., Lenz, O., Zebger, I., Friedrich, B., and Armstrong, F. A. Electrocatalytic hydrogen oxidation by an enzyme at high carbon monoxide or oxygen levels. Proc Natl Acad Sci U.S.A 2005, 102, 16951.
[28] Vincent, K. A., Cracknell, J. A., Clark, J. R., Ludwig, M., Lenz, O., Friedrich, B., and Armstrong, F. A. Electricity from low-level H_2 in still air–an ultimate test for an oxygen tolerant hydrogenase. Chem Commun 2006, 5033.
[29] Vincent, K. A., Parkin, A., and Armstrong, F. A. Investigating and exploiting the electrocatalytic properties of hydrogenases. Chem Rev 2007, 107, 4366.
[30] Armstrong, F. A., Evans, R. M., Hexter, S. V., Murphy, B. J., Roessler, M. M., and Wulff, P. Guiding principles of hydrogenase catalysis Instigated and clarified by protein film electrochemistry. Acc Chem Res 2016, 49, 884.
[31] Sensi, M., Del Barrio, M., Baffert, C., Fourmond, V., and Léger, C. New perspectives in hydrogenase direct electrochemistry. Curr Opin Electrochem 2017, 5, 135.
[32] Armstrong, F. A., Belsey, N. A., Cracknell, J. A., Goldet, G., Parkin, A., Reisner, E., Vincent, K. A., and Wait, A. F. Dynamic electrochemical investigations of hydrogen oxidation and production by enzymes and implications for future technology. Chem Soc Rev 2009, 38, 36.
[33] Ash, P. A., Hidalgo, R., and Vincent, K. A. Proton transfer in the catalytic cycle of [NiFe] hydrogenases, insight from vibrational spectroscopy. ACS Catal 2017, 7, 2471.
[34] Horch, M., Hildebrandt, P., and Zebger, I. Concepts in bio-molecular spectroscopy: Vibrational case studies on metalloenzymes. Phys Chem Chem Phys 2015, 17, 18222.
[35] De Lacey, A. L., Fernandez, V. M., Rousset, M., and Cammack, R. Activation and inactivation of hydrogenase function and the catalytic cycle: Spectroelectrochemical studies. Chem Rev 2007, 107, 4304.
[36] Lubitz, W., Reijerse, E., and van Gastel, M. [NiFe] and [FeFe] hydrogenases studied by advanced magnetic resonance techniques. Chem Rev 2007, 107, 4331.
[37] Horch, M., Lauterbach, L., Lenz, O., Hildebrandt, P., and Zebger, I. NAD(H)-coupled hydrogen cycling-structure-function relationships of bidirectional [NiFe] hydrogenases. FEBS Lett 2012, 586, 545.
[38] Efremov, R. G., and Sazanov, L. A. The coupling mechanism of respiratory complex I – a structural and evolutionary perspective. Biochim Biophys Acta 2012, 1817, 1785.
[39] Thorsten, F., and Dierk, S. The respiratory complex I of bacteria, archaea and eukarya and its module common with membrane-bound multisubunit hydrogenases. FEBS Lett 2000, 479, 1.
[40] Greene, B. L., Wu, C.-H., McTernan, P. M., Adams, M. W. W., and Dyer, R. B. Proton-coupled electron transfer dynamics in the catalytic mechanism of a [NiFe]-Hydrogenase. J Am Chem Soc 2015, 137, 4558.
[41] Greene, B. L., Vansuch, G. E., Wu, C.-H., Adams, M. W. W., and Dyer, R. B. Glutamate gated proton-coupled electron transfer activity of a [NiFe]-hydrogenase. J Am Chem Soc 2016, 138, 13013.
[42] Vignais, P. M. H/D exchange reactions and mechanistic aspects of the hydrogenases. Coord Chem Rev 2005, 249, 1677.
[43] Bruschi, M., Zampella, G., Fantucci, P., and de Gioia, L. DFT investigations of models related to the active site of [NiFe] and [Fe] hydrogenases. Coord Chem Rev 2005, 249, 1620.
[44] Bruschi, M., Tiberti, M., Guerra, A., and de Gioia, L. Disclosure of key stereoelectronic factors for efficient H_2 binding and cleavage in the active site of [NiFe]-hydrogenases. J Am Chem Soc 2014, 136, 1803.
[45] Lill, S. O., and Siegbahn, P. E. An autocatalytic mechanism for NiFe-hydrogenase: reduction to Ni(I) followed by oxidative addition. Biochemistry 2009, 48, 1056.
[46] Tard, C., and Pickett, C. J. Structural and functional analogues of the active sites of the [Fe]-, [NiFe]-, and [FeFe]-hydrogenases. Chem Rev 2009, 109, 2245.

[47] Schilter, D., Camara, J. M., Huynh, M. T., Hammes-Schiffer, S., and Rauchfuss, T. B. Hydrogenase enzymes and their synthetic models, the role of metal hydrides. Chem Rev 2016, 116, 8693.
[48] Shaw, W. J., Helm, M. L., and DuBois, D. L. A modular, energy-based approach to the development of nickel containing molecular electrocatalysts for hydrogen production and oxidation. BBA – Bioenergetics 2013, 1827, 1123.
[49] DuBois, D. L., and Bullock, R. M. Molecular electrocatalysts for the oxidation of hydrogen and the production of hydrogen – the role of pendant amines as proton relays. Eur J Inorg Chem 2011, 2011, 1017.
[50] Darensbourg, M. Y., and Weigand, W. Sulfoxygenation of active site models of [NiFe] and [FeFe] hydrogenases – a commentary on possible chemical models of hydrogenase enzyme oxygen sensitivity. Eur J Inorg Chem 2011, 2011, 994.
[51] Rakowski DuBois, M., and DuBois, D. L. The roles of the first and second coordination spheres in the design of molecular catalysts for H_2 production and oxidation. Chem Soc Rev 2009, 38, 62.
[52] Fritsch, J., Lenz, O., and Friedrich, B. Structure, function and biosynthesis of O_2-tolerant hydrogenases. Nat Rev Microbiol 2013, 11, 106.
[53] Lenz, O., Ludwig, M., Schubert, T., Burstel, I., Ganskow, S., Goris, T., Schwarze, A., and Friedrich, B. H_2 conversion in the presence of O_2 as performed by the membrane-bound [NiFe]-hydrogenase of *Ralstonia eutropha*. Chem Phys Chem 2010, 11, 1107.
[54] Baltazar, C. S. A., Marques, M. C., Soares, C. M., De Lacey, A. L., Pereira, I. A. C., and Matias, P. M. Nickel -Iron -Selenium Hydrogenases – An overview. Eur J Inorg Chem 2011, 2011, 948.
[55] Constant, P., Chowdhury, S. P., Hesse, L., Pratscher, J., and Conrad, R. Genome data mining and soil survey for the novel group 5 [NiFe]-hydrogenase to explore the diversity and ecological importance of presumptive high-affinity H_2-oxidizing bacteria. Appl Environ Microbiol 2011, 77, 6027.
[56] Constant, P., Chowdhury, S. P., Pratscher, J., and Conrad, R. Streptomycetes contributing to atmospheric molecular hydrogen soil uptake are widespread and encode a putative high-affinity [NiFe]-hydrogenase. Environ Microbiol 2010, 12, 821.
[57] Schäfer, C., Friedrich, B., and Lenz, O. Novel, oxygen-insensitive group 5 [NiFe]-hydrogenase in *Ralstonia eutropha*. Appl Environ Microbiol 2013, 79, 5137.
[58] Schäfer, C., Bommer, M., Hennig, S. E., Jeoung, J.-H., Dobbek, H., and Lenz, O. Structure of an actinobacterial-type [NiFe]-hydrogenase reveals insight into O_2-tolerant H_2 oxidation. Structure 2016, 24, 285.
[59] Lauterbach, L., Lenz, O., and Vincent, K. A. H_2-driven cofactor regeneration with $NAD(P)^+$-reducing hydrogenases. FEBS J 2013, 280, 3058.
[60] Shafaat, H. S., Rüdiger, O., Ogata, H., and Lubitz, W. [NiFe] hydrogenases: a common active site for hydrogen metabolism under diverse conditions. Biochim Biophys Acta 2013, 1827, 986.
[61] Happe, R. P., Roseboom, W., Pierik, A. J., Albracht, S. P. J., and Bagley, K. A. Biological activation of hydrogen. Nature 1997, 385, 126.
[62] Pierik, A. J., Roseboom, W., Happe, R. P., Bagley, K. A., and Albracht, S. P. J. Carbon monoxide and cyanide as intrinsic ligands to iron in the active site of [NiFe]-hydrogenases. J Biol Chem 1999, 274, 3331.
[63] Bagley, K. A., Duin, E. C., Roseboom, W., Albracht, S. P. J., and Woodruff, W. H. Infrared-detectable groups sense changes in charge density on the nickel center in hydrogenase from *Chromatium vinosum*. Biochemistry 1995, 34, 5527.
[64] Bagley, K. A., van Garderen, C. J., Chen, M., Duin, E. C., Albracht, S. P. J., and Woodruff, W. H. Infrared studies on the interaction of carbon monoxide with divalent nickel in hydrogenase from *Chromatium vinosum*. Biochemistry 1994, 33, 9229.

[65] van der Spek, T. M., Arendsen, A. F., Happe, R. P., Yun, S., Bagley, K. A., Stufkens, D. J., Hagen, W. R., and Albracht, S. P. J. Similarities in the architecture of the active sites of Ni-hydrogenases and Fe-hydrogenases detected by means of infrared spectroscopy. Eur J Biochem 1996, 237, 629.

[66] Lyon, E. J., Shima, S., Boecher, R., Thauer, R. K., Grevels, F. W., Bill, E., Roseboom, W., and Albracht, S. P. J. Carbon monoxide as an intrinsic ligand to iron in the active site of the iron-sulfur-cluster-free hydrogenase H_2-forming methylenetetrahydromethanopterin dehydrogenase as revealed by infrared spectroscopy. J Am Chem Soc 2004, 126, 14239.

[67] a) Boer, J. L., Mulrooney, S. B., and Hausinger, R. P. Nickel-dependent metalloenzymes. Arch Biochem Biophys 2013, b) S. W. Ragsdale, Nickel-based enzyme systems. J. Biol. Chem. 2009, 284, 18571.

[68] Dole, F., Fournel, A., Magro, V., Hatchikian, E. C., Bertrand, P., and Guigliarelli, B. Nature and electronic structure of the Ni-X dinuclear center of *Desulfovibrio gigas* hydrogenase. Implications for the enzymatic mechanism. Biochemistry 1997, 36, 7847.

[69] a) Surerus, K. K., Chen, M., van der Zwaan, J. W., Rusnak, F. M., Kolk, M., Duin, E. C., Albracht, S. P. J., and Munck, E. Further characterization of the spin coupling observed in oxidized hydrogenase from *Chromatium vinosum*. A Mössbauer and multifrequency EPR study. Biochemistry 1994, 33, 4980. b) J. E. Huyett, M. Carepo, A. Pamplona, R. Franco, I. Moura, J. J. G. Moura, B. M. Hoffman, ^{57}Fe Q-Band Pulsed ENDOR of the hetero-dinuclear site of nickel hydrogenase: Comparison of the NiA, NiB, and NiC states. J. Am. Chem. Soc. 1997, 119, 9291.

[70] Roncaroli, F., Bill, E., Friedrich, B., Lenz, O., Lubitz, W., and Pandelia, M.-E. Cofactor composition and function of a H_2-sensing regulatory hydrogenase as revealed by Mössbauer and EPR spectroscopy. Chem Sci 2015, 6, 4495.

[71] Horch, M., Lauterbach, L., Mroginski, M. A., Hildebrandt, P., Lenz, O., and Zebger, I. Reversible active site sulfoxygenation can explain the oxygen tolerance of a NAD^+-reducing [NiFe] hydrogenase and its unusual infrared spectroscopic properties. J Am Chem Soc 2015, 137, 2555.

[72] Ogata, H., Hirota, S., Nakahara, A., Komori, H., Shibata, N., Kato, T., Kano, K., and Higuchi, Y. Activation process of [NiFe] hydrogenase elucidated by high-resolution X-ray analyses: Conversion of the ready to the unready state. Structure 2005, 13, 1635.

[73] Volbeda, A., Martin, L., Barbier, E., Gutiérrez-Sanz, O., De Lacey, A. L., Liebgott, P.-P., Dementin, S., Rousset, M., and Fontecilla-Camps, J. C. Crystallographic studies of [NiFe]-hydrogenase mutants: Towards consensus structures for the elusive unready oxidized states. J Biol Inorg Chem.2015, 20, 11.

[74] Marques, M. C., Coelho, R., De Lacey, A. L., Pereira, I. A., and Matias, P. M. The three-dimensional structure of [NiFeSe] hydrogenase from *Desulfovibrio vulgaris* Hildenborough: A hydrogenase without a bridging ligand in the active site in its oxidised, "as-isolated" state. J Mol Biol 2010, 396, 893.

[75] Fichtner, C., van Gastel, M., and Lubitz, W. Wavelength dependence of the photo-induced conversion of the Ni-C to the Ni-L redox state in the [NiFe] hydrogenase of *Desulfovibrio vulgaris* Miyazaki F. Phys Chem Chem Phys 2003, 5, 5507.

[76] Siebert, E., Horch, M., Rippers, Y., Fritsch, J., Frielingsdorf, S., Lenz, O., Velazquez Escobar, F., Siebert, F., Paasche, L., Kuhlmann, U., Lendzian, F., Mroginski, M. A., Zebger, I., and Hildebrandt, P. Resonance Raman spectroscopy as a tool to monitor the active site of hydrogenases. Angew Chem Int Ed 2013, 52, 5162.

[77] Ogata, H., Nishikawa, K., and Lubitz, W. Hydrogens detected by subatomic resolution protein crystallography in a [NiFe] hydrogenase. Nature 2015, 520, 571.

[78] Horch, M., Schoknecht, J., Mroginski, M. A., Lenz, O., Hildebrandt, P., and Zebger, I. Resonance Raman spectroscopy on [NiFe] hydrogenase provides structural insights into catalytic intermediates and reactions. J Am Chem Soc 2014, 136, 9870.

[79] Dong, G., and Ryde, U. Protonation states of intermediates in the reaction mechanism of [NiFe] hydrogenase studied by computational methods. J Biol Inorg Chem 2016, 21, 383.
[80] Darensbourg, M. Y., Lyon, E. J., and Smee, J. J. The bio-organometallic chemistry of active site iron in hydrogenases. Coord Chem Rev 2000, 206–207, 533.
[81] LeGall, J., Ljungdahl, P. O., Moura, I., Peck Jr, H. D., Xavier, A. V., Moura, J. J., Teixera, M., Huynh, B. H., and DerVartanian, D. V. The presence of redox-sensitive nickel in the periplasmic hydrogenase from *Desulfovibrio gigas*. Biochem Biophys Res Commun 1982, 106, 610.
[82] Cammack, R., Fernandez, V. M., and Schneider, K. Activation and active sites of nickel-containing hydrogenases. Biochimie 1986, 68, 85.
[83] Teixeira, M., Moura, I., Xavier, A. V., Huynh, B. H., DerVartanian, D. V., Peck, H. D., LeGall, Jr., J., and Moura, J. J. Electron paramagnetic resonance studies on the mechanism of activation and the catalytic cycle of the nickel-containing hydrogenase from *Desulfovibrio gigas*. J Biol Chem 1985, 260, 8942.
[84] van der Zwaan, J. W., Albracht, S. P. J., Fontijn, R. D., and Slater, E. C. Monovalent nickel in hydrogenase from *Chromatium vinosum*. Light sensitivity and evidence for direct interaction with hydrogen. FEBS Lett 1985, 179, 271.
[85] Moura, J. J., Moura, I., Huynh, B. H., Kruger, H. J., Teixeira, M., DuVarney, R. C., DerVartanian, D. V., Xavier, A. V., Peck, H. D., and LeGall, Jr., J. Unambiguous identification of the nickel EPR signal in ^{61}Ni-enriched *Desulfovibrio gigas* hydrogenase. Biochem Biophys Res Commun 1982, 108, 1388.
[86] Bleijlevens, B., van Broekhuizen, F. A., De Lacey, A. L., Roseboom, W., Fernandez, V. M., and Albracht, S. P. J. The activation of the [NiFe]-hydrogenase from *Allochromatium vinosum*. An infrared spectro-electrochemical study. J Biol Inorg Chem 2004, 9, 743.
[87] Fernandez, V. M., Hatchikian, E. C., and Cammack, R. Properties and reactivation of two different deactivated forms of *Desulfovibrio gigas* hydrogenase. Biochim Biophys Acta 1985, 832, 69.
[88] Fernandez, V. M., Rao, K. K., Fernandez, M. A., and Cammack, R. Activation and deactivation of the membrane-bound hydrogenase from *Desulfovibrio desulfuricans*, Norway strain. Biochimie 1986, 68, 43.
[89] Volbeda, A., Martin, L., Cavazza, C., Matho, M., Faber, B. W., Roseboom, W., Albracht, S. P. J., Garcin, E., Rousset, M., and Fontecilla-Camps, J. C. Structural differences between the ready and unready oxidized states of [NiFe] hydrogenases. J Biol Inorg Chem 2005, 10, 239.
[90] a) van Gastel, M., Stein, M., Brecht, M., Schröder, O., Lendzian, F., Bittl, R., Ogata, H., Higuchi, Y., and Lubitz, W. A single-crystal ENDOR and density functional theory study of the oxidized states of the [NiFe] hydrogenase from *Desulfovibrio vulgaris* Miyazaki F. J Biol Inorg Chem 2006, 11, 41. b) H. Ogata, S. Hirota, A. Nakahara, H. Komori, N. Shibata, T. Kato, K. Kano, Y. Higuchi, Activation process of [NiFe] Hydrogenase elucidated by high-resolution X-ray analyses: Conversion of the ready to the unready State. Structure 2005, 13, 1635.
[91] Breglia, R., Ruiz-Rodriguez, M. A., Vitriolo, A., Gonzàlez-Laredo, R. F., de Gioia, L., Greco, C., and Bruschi, M. Theoretical insights into [NiFe]-hydrogenases oxidation resulting in a slowly reactivating inactive state. J Biol Inorg Chem 2017, 22, 137.
[92] Gupta, V., and Carroll, K. S. Sulfenic acid chemistry, detection and cellular lifetime. BBA – General Subjects 2014, 1840, 847.
[93] Cracknell, J. A., Wait, A. F., Lenz, O., Friedrich, B., and Armstrong, F. A. A kinetic and thermodynamic understanding of O_2 tolerance in [NiFe]-hydrogenases. Proc Natl Acad Sci U. S.A. 2009, 106, 20681.
[94] Breglia, R., Greco, C., Fantucci, P., de Gioia, L., and Bruschi, M. Theoretical investigation of aerobic and anaerobic oxidative inactivation of the [NiFe]-hydrogenase active site. Phys Chem Chem Phys 2018, 20, 1693.

[95] De Lacey, A. L., Hatchikian, E. C., Volbeda, A., Frey, M., Fontecilla-Camps, J. C., and Fernandez, V. M. Infrared-spectroelectrochemical characterization of the [NiFe] hydrogenase of *Desulfovibrio gigas*. J Am Chem Soc 1997, 119, 7181.

[96] Davidson, G., Choudhury, S. B., Gu, Z., Bose, K., Roseboom, W., Albracht, S. P. J., and Maroney, M. J. Structural examination of the nickel site in *Chromatium vinosum* hydrogenase: Redox state oscillations and structural changes accompanying reductive activation and CO binding. Biochemistry 2000, 39, 7468.

[97] Brecht, M., van Gastel, M., Buhrke, T., Friedrich, B., and Lubitz, W. Direct detection of a hydrogen ligand in the [NiFe] center of the regulatory H_2-sensing hydrogenase from *Ralstonia eutropha* in its reduced state by HYSCORE and ENDOR spectroscopy. J Am Chem Soc 2003, 125, 13075.

[98] a) Förster, S., Stein, M., Brecht, M., Ogata, H., Higuchi, Y., and Lubitz, W. Single crystal EPR studies of the reduced active site of [NiFe] hydrogenase from *Desulfovibrio vulgaris* Miyazaki F. J Am Chem Soc 2003, 125, 83. b) S. Förster, M. van Gastel, M. Brecht, W. Lubitz, An orientation-selected ENDOR and HYSCORE study of the Ni-C active state of *Desulfovibrio vulgaris* Miyazaki F hydrogenase. J. Biol. Inorg. Chem. 2005, 10, 51.

[99] a) Pandelia, M. E., Ogata, H., Currell, L. J., Flores, M., and Lubitz, W. Inhibition of the [NiFe] hydrogenase from *Desulfovibrio vulgaris* Miyazaki F by carbon monoxide: An FTIR and EPR spectroscopic study. Biochim Biophys Acta 2010, 1797, 304. b) R. P. Happe, W. Roseboom, S. P. J. Albracht, Pre-steady-state kinetics of the reactions of [NiFe]-hydrogenase from *Chromatium vinosum* with H_2 and CO. Eur. J. Biochem. 1999, 259, 602; c) F. Dole, M. Medina, C. More, R. Cammack, P. Bertrand, B. Guigliarelli, Spin-spin interactions between the Ni site and the [4Fe-4S] centers as a probe of light-induced structural changes in active *Desulfovibrio gigas* hydrogenase. Biochemistry 1996, 35, 16399; d) M. Medina, E. Claude Hatchikian, R. Cammack, Studies of light-induced nickel EPR signals in hydrogenase: Comparison of enzymes with and without selenium. BBA – Bioenergetics 1996, 1275, 227.

[100] Ogata, H., Krämer, T., Wang, H., Schilter, D., Pelmenschikov, V., van Gastel, M., Neese, F., Rauchfuss, T. B., Gee, L. B., Scott, A. D., Yoda, Y., Tanaka, Y., Lubitz, W., and Cramer, S. P. Hydride bridge in [NiFe]-hydrogenase observed by nuclear resonance vibrational spectroscopy. Nat Commun 2015, 6, 7890.

[101] Wang, H., Yoda, Y., Ogata, H., Tanaka, Y., and Lubitz, W. A strenuous experimental journey searching for spectroscopic evidence of a bridging nickel-iron-hydride in [NiFe] hydrogenase. J Synchrotron Rad 2015, 22, 1334.

[102] van der Zwaan, J. W., Coremans, J. M., Bouwens, E. C., and Albracht, S. P. J. Effect of $^{17}O_2$ and ^{13}CO on EPR spectra of nickel in hydrogenase from *Chromatium vinosum*. Biochim Biophys Acta 1990, 1041, 101.

[103] Ogata, H., Mizoguchi, Y., Mizuno, N., Miki, K., Adachi, S., Yasuoka, N., Yagi, T., Yamauchi, O., Hirota, S., and Higuchi, Y. Structural studies of the carbon monoxide complex of [NiFe] hydrogenase from *Desulfovibrio vulgaris* Miyazaki F: Suggestion for the initial activation site for dihydrogen. J Am Chem Soc 2002, 124, 11628.

[104] Purec, L., Krasna, A. I., and Rittenberg, D. The inhibition of hydrogenase by carbon monoxide and the reversal of this inhibition by light. Biochemistry 1962, 1, 270.

[105] a) Fan, H. J., and Hall, M. B. High-spin Ni(II), a surprisingly good structural model for [NiFe] hydrogenase. J Am Chem Soc 2002, 124, 394. b) M. Bruschi, G. L. De, G. Zampella, M. Reiher, P. Fantucci, M. Stein, A theoretical study of spin states in Ni-S_4 complexes and models of the [NiFe] hydrogenase active site. J. Biol. Inorg. Chem. 2004, 9, 873; c) A. T. Kowal, I. C. Zambrano, I. Moura, J. J. G. Moura, J. LeGall, M. K. Johnson, Electronic and magnetic properties of nickel-substituted rubredoxin: A variable-temperature magnetic circular dichroism study. Inorg. Chem. 1988, 27, 1162; d) C. P. Wang, R. Franco, J. J. Moura, I. Moura, E. P. Day, The nickel site in active *Desulfovibrio baculatus* [NiFeSe] hydrogenase is diamagnetic. Multifield saturation

magnetization measurement of the spin state of Ni(II). J. Biol. Chem. 1992, 267, 7378; e) H. Wang, C. Y. Ralston, D. S. Patil, R. M. Jones, W. Gu, M. Verhagen, M. Adams, P. Ge, C. Riordan, C. A. Marganian, P. Mascharak, J. Kovacs, C. G. Miller, T. J. Collins, S. Brooker, p. D. Croucher, K. Wang, E. I. Stiefel, S P. Cramer, Nickel L-edge soft X-ray spectroscopy of nickel–iron hydrogenases and model Compounds – evidence for high-spin nickel(II) in the active enzyme. J. Am. Chem. Soc. 2000, 122, 10544; f) H. Wang, D. S. Patil, W. Gu, L. Jacquamet, S. Friedrich, T. Funk, S. P. Cramer, L-edge X-ray absorption spectroscopy of some Ni enzymes: Probe of Ni electronic structure. J. Electron Spectrosc. Relat. Phenom. 2001, 114, 855; g) S. Li, M. B. Hall, Modeling the active sites of metalloenzymes. 4. Predictions of the unready states of [NiFe] *Desulfovibrio gigas* hydrogenase from density functional theory. Inorg. Chem. 2001, 40, 18; h) J. van Elp, S. J. George, J. Chen, G. Peng, C. T. Chen, L. H. Tjeng, G. Meigs, H. J. Lin, Z. H. Zhou, M. W. Adams, Soft X-ray magnetic circular dichroism: A probe for studying paramagnetic bioinorganic systems. Proc. Natl. Acad. Sci. U.S.A. 1993, 90, 9664.

[106] Jayapal, P., Robinson, D., Sundararajan, M., Hillier, I. H., and McDouall, J. High level ab initio and DFT calculations of models of the catalytically active Ni-Fe hydrogenases. J. Phys Chem Chem Phys 2008, 10, 1734.

[107] Pardo, A., De Lacey, A. L., Fernandez, V. M., Fan, H. J., Fan, Y., and Hall, M. B. Density functional study of the catalytic cycle of nickel-iron [NiFe] hydrogenases and the involvement of high-spin nickel(II). J Biol Inorg Chem 2006, 11, 286.

[108] Wu, H., and Hall, M. B. Density functional theory on the larger active site models for [NiFe] hydrogenases: Two-state reactivity? C R Chim 2008, 11, 790.

[109] Kaliakin, D. S., Zaari, R. R., and Varganov, S. A. Effect of H_2 binding on the nonadiabatic transition probability between singlet and triplet States of the [NiFe]-hydrogenase active site. J Phys Chem A 2015, 119, 1066.

[110] Delcey, M. G., Pierloot, K., Phung, Q. M., Vancoillie, S., Lindh, R., and Ryde, U. Accurate calculations of geometries and singlet-triplet energy differences for active-site models of [NiFe] hydrogenase. Phys Chem Chem Phys 2014, 16, 7927.

[111] Dong, G., Phung, Q. M., Hallaert, S. D., Pierloot, K., and Ryde, U. H_2 binding to the active site of [NiFe] hydrogenase studied by multiconfigurational and coupled-cluster methods. Phys Chem Chem Phys 2017, 19, 10590.

[112] Dong, G., Ryde, U., Aa. Jensen, H. J., and Hedegard, E. D. Exploration of H_2 binding to the [NiFe]-hydrogenase active site with multiconfigurational density functional theory. Phys Chem Chem Phys 2018, 20, 794.

[113] Montet, Y., Amara, P., Volbeda, A., Vernede, X., Hatchikian, E. C., Field, M. J., Frey, M., and Fontecilla-Camps, J. C. Gas access to the active site of Ni-Fe hydrogenases probed by X-ray crystallography and molecular dynamics. Nat Struct Biol 1997, 4, 523.

[114] Kalms, J., Schmidt, A., Frielingsdorf, S., van der Linden, P., von Stetten, D., Lenz, O., Carpentier, P., and Scheerer, P. Krypton derivatization of an O_2-tolerant membrane-bound [NiFe] hydrogenase reveals a hydrophobic tunnel network for gas transport. Angew Chem Int Ed 2016, 55, 5586.

[115] Rippers, Y., Horch, M., Hildebrandt, P., Zebger, I., and Mroginski, M. A. Revealing the absolute configuration of the CO and CN^- ligands at the active site of a [NiFe] hydrogenase. Chem Phys Chem 2012, 13, 3852.

[116] Krasna, A. I., and Rittenberg, D. The mechanism of action of the enzyme hydrogenase 1. J Am Chem Soc 1954, 76, 3015.

[117] Rittenberg, D., and Krasna, A. I. Interaction of hydrogenase with hydrogen. Discuss Faraday Soc 1955, 20, 185.

[118] Pavlov, M., Blomberg, M. R. A., and Siegbahn, P. E. M. New aspects of H_2 activation by nickel–iron hydrogenase. Int J Quantum Chem 1999, 73, 197.

[119] Niu, S., Thomson, L. M., and Hall, M. B. Theoretical characterization of the reaction intermediates in a model of the nickel–iron hydrogenase of *Desulfovibrio gigas*. J Am Chem Soc 1999, 121, 4000.
[120] Saint-Martin, P., Lespinat, P. A., Fauque, G., Berlier, Y., LeGall, J., Moura, I., Teixeira, M., Xavier, A. V., and Moura, J. J. Hydrogen production and deuterium-proton exchange reactions catalyzed by *Desulfovibrio* nickel(II)-substituted rubredoxins. Proc Natl Acad Sci U.S.A 1988, 85, 9378.
[121] Stevenson, M. J., Marguet, S. C., Schneider, C. R., and Shafaat, H. S. Light-driven hydrogen evolution by nickel-substituted rubredoxin. Chem Sus Chem 2017, 10, 4424.
[122] Evans, R. M., Brooke, E. J., Wehlin, S. A. M., Nomerotskaia, E., Sargent, F., Carr, S. B., Phillips, S. E. V., and Armstrong, F. A. Mechanism of hydrogen activation by [NiFe] hydrogenases. Nat Chem Biol 2015, 12, 4.
[123] Szőri-Dorogházi, E., Maróti, G., Szőri, M., Nyilasi, A., Rákhely, G., and Kovács, K. L. Analyses of the large subunit histidine-rich motif expose an alternative proton transfer pathway in [NiFe] hydrogenases. PLoS ONE 2012, 7, e34666.
[124] Niu, S., and Hall, M. B. Modeling the active sites in metalloenzymes 5. The heterolytic bond cleavage of H_2 in the [NiFe] hydrogenase of *Desulfovibrio gigas* by a nucleophilic addition mechanism. Inorg Chem 2001, 40, 6201.
[125] Dementin, S., Burlat, B., De Lacey, A. L., Pardo, A., Adryanczyk-Perrier, G., Guigliarelli, B., Fernandez, V. M., and Rousset, M. A glutamate is the essential proton transfer gate during the catalytic cycle of the [NiFe] hydrogenase. J Biol Chem 2004, 279, 10508.
[126] Teixeira, M., Moura, I., Xavier, A. V., Moura, J. J., LeGall, J., DerVartanian, D. V., Peck, H. D., and Huynh, B. H. Redox intermediates of *Desulfovibrio gigas* [NiFe] hydrogenase generated under hydrogen. Mössbauer and EPR characterization of the metal centers. J Biol Chem 1989, 264, 16435.
[127] Page, C. C., Moser, C. C., Chen, X., and Dutton, P. L. Natural engineering principles of electron tunnelling in biological oxidation–reduction. Nature 1999, 402, 47.
[128] Roessler, M. M., Evans, R. M., Davies, R. A., Harmer, J., and Armstrong, F. A. EPR spectroscopic studies of the Fe-S clusters in the O_2-tolerant [NiFe]-hydrogenase Hyd-1 from *Escherichia coli* and characterization of the unique [4Fe-3S] cluster by HYSCORE. J Am Chem Soc 2012, 134, 15581.
[129] Pandelia, M. E., Nitschke, W., Infossi, P., Giudici-Orticoni, M. T., Bill, E., and Lubitz, W. Characterization of a unique [FeS] cluster in the electron transfer chain of the oxygen tolerant [NiFe] hydrogenase from *Aquifex aeolicus*. Proc Natl Acad Sci U.S.A 2011, 108, 6097.
[130] Knuettel, K., Schneider, K., Erkens, A., Plass, W., Mueller, A., Bill, E., and Trautwein, A. X. Redox properties of the metal centers in the membrane-bound hydrogenase from *Alcaligenes eutrophus* CH34. Bull Pol Acad Sci Chem 1994, 42, 495.
[131] Lauterbach, L., and Lenz, O. Catalytic production of hydrogen peroxide and water by oxygen-tolerant [NiFe]-hydrogenase during H_2 cycling in the presence of O_2. J Am Chem Soc 2013, 135, 17897.
[132] Goris, T., Wait, A. F., Saggu, M., Fritsch, J., Heidary, N., Stein, M., Zebger, I., Lendzian, F., Armstrong, F. A., Friedrich, B., and Lenz, O. A unique iron-sulfur cluster is crucial for oxygen tolerance of a [NiFe]-hydrogenase. Nat Chem Biol 2011, 7, 310.
[133] Fritsch, J., Scheerer, P., Frielingsdorf, S., Kroschinsky, S., Friedrich, B., Lenz, O., and Spahn, C. M. The crystal structure of an oxygen-tolerant hydrogenase uncovers a novel iron-sulphur centre. Nature 2011, 479, 249.
[134] Frielingsdorf, S., Fritsch, J., Schmidt, A., Hammer, M., Löwenstein, J., Siebert, E., Pelmenschikov, V., Jaenicke, T., Kalms, J., Rippers, Y., Lendzian, F., Zebger, I., Teutloff, C., Kaupp, M., Bittl, R., Hildebrandt, P., Friedrich, B., Lenz, O., and Scheerer, P. Reversible [4Fe-3S] cluster morphing in an O_2-tolerant [NiFe] hydrogenase. Nat Chem Biol 2014, 10, 378.

[135] Kubas, G. J., Ryan, R. R., Swanson, B. I., Vergamini, P. J., and Wasserman, H. J. Characterization of the first examples of isolable molecular hydrogen complexes, $M(CO)_3(PR_3)_2(H_2)$ (M = molybdenum or tungsten; R = Cy or isopropyl). Evidence for a side-on bonded dihydrogen ligand. J Am Chem Soc 1984, 106, 451.
[136] Rauchfuss, T. B., Contakes, S. M., Hsu, S. C., Reynolds, M. A., and Wilson, S. R. The influence of cyanide on the carbonylation of iron(II): Synthesis of Fe-SR-CN-CO centers related to the hydrogenase active sites. J Am Chem Soc 2001, 123, 6933.
[137] Huang, G., Wagner, T., Ermler, U., Bill, E., Ataka, K., and Shima, S. Dioxygen sensitivity of [Fe]-hydrogenase in the presence of reducing substrates. Angew Chem Int Ed 2018, 57, 4917.
[138] Cramm, R. Genomic view of energy metabolism in *Ralstonia eutropha* H16. J Mol Microbiol Biotechnol 2009, 16, 38.
[139] Lenz, O., Bernhard, M., Buhrke, T., Schwartz, E., and Friedrich, B. The hydrogen sensing apparatus in *Ralstonia eutropha*. J Mol Microbiol Biotechnol 2002, 4, 255.
[140] Buhrke, T., Lenz, O., Porthun, A., and Friedrich, B. The H_2-sensing complex of *Ralstonia eutropha*: Interaction between a regulatory [NiFe] hydrogenase and a histidine protein kinase. Mol Microbiol 2004, 51, 1677.
[141] Kleihues, L., Lenz, O., Bernhard, M., Buhrke, T., and Friedrich, B. The H_2 sensor of *Ralstonia eutropha* is a member of the subclass of regulatory [NiFe] hydrogenases. J Bacteriol 2000, 182, 2716.
[142] Friedrich, B., Buhrke, T., Burgdorf, T., and Lenz, O. A hydrogen-sensing multiprotein complex controls aerobic hydrogen metabolism in *Ralstonia eutropha*. Biochem Soc Trans 2005, 33, 97.
[143] Lenz, O., Strack, A., Tran-Betcke, A., and Friedrich, B. A hydrogen-sensing system in transcriptional regulation of hydrogenase gene expression in *Alcaligenes* species. J Bacteriol 1997, 179, 1655.
[144] Lenz, O., and Friedrich, B. A novel multicomponent regulatory system mediates H_2 sensing in *Alcaligenes eutrophus*. Proc Natl Acad Sci U.S.A 1998, 95, 12474.
[145] Pierik, A. J., Schmelz, M., Lenz, O., Friedrich, B., and Albracht, S. P. J. Characterization of the active site of a hydrogen sensor from *Alcaligenes eutrophus*. FEBS Lett 1998, 438, 231.
[146] Bernhard, M., Buhrke, T., Bleijlevens, B., De Lacey, A. L., Fernandez, V. M., Albracht, S. P. J., and Friedrich, B. The H_2 sensor of *Ralstonia eutropha*. Biochemical characteristics, spectroscopic properties, and its interaction with a histidine protein kinase. J Biol Chem 2001, 276, 15592.
[147] Buhrke, T., Lenz, O., Krauss, N., and Friedrich, B. Oxygen tolerance of the H_2-sensing [NiFe] hydrogenase from *Ralstonia eutropha* H16 is based on limited access of oxygen to the active site. J Biol Chem 2005, 280, 23791.
[148] Volbeda, A., Montet, Y., Vernède, X., Hatchikian, E. C., and Fontecilla-Camps, J. C. High-resolution crystallographic analysis of *Desulfovibrio fructosovorans* [NiFe] hydrogenase. Int J Hydrog Energy 2002, 27, 1449.
[149] Ash, P. A., Liu, J., Coutard, N., Heidary, N., Horch, M., Gudim, I., Simler, T., Zebger, I., Lenz, O., and Vincent, K. A. Electrochemical and infrared spectroscopic studies provide insight into reactions of the NiFe regulatory hydrogenase from *Ralstonia eutropha* with O_2 and CO. J Phys Chem B 2015, 119, 13807.
[150] Dementin, S., Leroux, F., Cournac, L., De Lacey, A. L., Volbeda, A., Leger, C., Burlat, B., Martinez, N., Champ, S., Martin, L., Sanganas, O., Haumann, M., Fernandez, V. M., Guigliarelli, B., Fontecilla-Camps, J. C., and Rousset, M. Introduction of methionines in the gas channel makes [NiFe] hydrogenase aero-tolerant. J Am Chem Soc 2009, 131, 10156.
[151] Liebgott, P. P., Leroux, F., Burlat, B., Dementin, S., Baffert, C., Lautier, T., Fourmond, V., Ceccaldi, P., Cavazza, C., Meynial-Salles, I., Soucaille, P., Fontecilla-Camps, J. C., Guigliarelli, B., Bertrand, P., Rousset, M., and Leger, C. Relating diffusion along the substrate tunnel and oxygen sensitivity in hydrogenase. Nat Chem Biol 2010, 6, 63.

[152] Liebgott, P. P., De Lacey, A. L., Burlat, B., Cournac, L., Richaud, P., Brugna, M., Fernandez, V. M., Guigliarelli, B., Rousset, M., Leger, C., and Dementin, S. Original design of an oxygen-tolerant [NiFe] hydrogenase: Major effect of a valine-to-cysteine mutation near the active site. J Am Chem Soc 2011, 133, 986.
[153] Luo, X., Brugna, M., Tron-Infossi, P., Giudici-Orticoni, M. T., and Lojou, E. Immobilization of the hyperthermophilic hydrogenase from *Aquifex aeolicus* bacterium onto gold and carbon nanotube electrodes for efficient H_2 oxidation. J Biol Inorg Chem 2009, 14, 1275.
[154] a) Leger, C., Dementin, S., Bertrand, P., Rousset, M., and Guigliarelli, B. Inhibition and aerobic inactivation kinetics of *Desulfovibrio fructosovorans* NiFe hydrogenase studied by protein film voltammetry. J Am Chem Soc 2004, 126, 12162. b) M. Ludwig, J. A. Cracknell, K. A. Vincent, F. A. Armstrong, O. Lenz, Oxygen-tolerant H_2 oxidation by membrane-bound [NiFe] hydrogenases of *Ralstonia* species. Coping with low level H_2 in air. J. Biol. Chem. 2009, 284, 465.
[155] Lukey, M. J., Parkin, A., Roessler, M. M., Murphy, B. J., Harmer, J., Palmer, T., Sargent, F., and Armstrong, F. A. How *Escherichia coli* is equipped to oxidize hydrogen under different redox conditions. J Biol Chem 2010, 285, 3928.
[156] Pandelia, M. E., Infossi, P., Stein, M., Giudici-Orticoni, M. T., and Lubitz, W. Spectroscopic characterization of the key catalytic intermediate Ni-C in the O_2-tolerant [NiFe] hydrogenase I from *Aquifex aeolicus*: Evidence of a weakly bound hydride. Chem Commun 2012, 48, 823.
[157] Wulff, P., Day, C. C., Sargent, F., and Armstrong, F. A. How oxygen reacts with oxygen-tolerant respiratory [NiFe]-hydrogenases. Proc Natl Acad Sci U.S.A 2014, 111, 6606.
[158] Schäfer, C., Friedrich, B., and Lenz, O. Characteristics of a novel, oxygen-insensitive group 5 [NiFe]-hydrogenase in *Ralstonia eutropha*. Appl Environ Microbiol 2013.
[159] Frielingsdorf, S., Schubert, T., Pohlmann, A., Lenz, O., and Friedrich, B. A trimeric supercomplex of the oxygen-tolerant membrane-bound [NiFe]-hydrogenase from *Ralstonia eutropha* H16. Biochemistry 2011, 50, 10836.
[160] Radu, V., Frielingsdorf, S., Evans, S. D., Lenz, O., and Jeuken, L. J. C. Enhanced oxygen-tolerance of the full heterotrimeric membrane-bound [NiFe]-hydrogenase of *Ralstonia eutropha*. J Am Chem Soc 2014, 136, 8512.
[161] Schneider, K., Patil, D. S., and Cammack, R. ESR properties of membrane-bound hydrogenases from aerobic hydrogen bacteria. BBA – Protein Struct Mol Enzym 1983, 748, 353.
[162] Kuhn, M., Steinbüchel, A., and Schlegel, H. G. Hydrogen evolution by strictly aerobic hydrogen bacteria under anaerobic conditions. J Bacteriol 1984, 159, 633.
[163] Schneider, K., and Schlegel, H. G. Purification and properties of soluble hydrogenase from *Alcaligenes eutrophus* H 16. Biochim Biophys Acta 1976, 452, 66.
[164] Shomura, Y., Taketa, M., Nakashima, H., Tai, H., Nakagawa, H., Ikeda, Y., Ishii, M., Igarashi, Y., Nishihara, H., Yoon, K.-S., Ogo, S., Hirota, S., and Higuchi, Y. Structural basis of the redox switches in the NAD^+-reducing soluble [NiFe]-hydrogenase. Science 2017, 357, 928.
[165] Horch, M., Lauterbach, L., Saggu, M., Hildebrandt, P., Lendzian, F., Bittl, R., Lenz, O., and Zebger, I. Probing the active site of an O_2-tolerant NAD^+-reducing [NiFe]-hydrogenase from *Ralstonia eutropha* H16 by in situ EPR and FTIR spectroscopy. Angew Chem Int Ed 2010, 49, 8026.
[166] Lauterbach, L., Wang, H., Horch, M., Gee, L. B., Yoda, Y., Tanaka, Y., Zebger, I., Lenz, O., and Cramer, S. P. Nuclear resonance vibrational spectroscopy reveals the FeS cluster composition and active site vibrational properties of an O_2-tolerant NAD^+-reducing [NiFe] hydrogenase. Chem Sci 2015.
[167] Lauterbach, L., Liu, J., Horch, M., Hummel, P., Schwarze, A., Haumann, M., Vincent, K. A., Lenz, O., and Zebger, I. The hydrogenase subcomplex of the NAD^+-reducing [NiFe] hydrogenase from *Ralstonia eutropha* – insights into catalysis and redox interconversions. Eur J Inorg Chem 2011, 2011, 1067.

[168] Karstens, K., Wahlefeld, S., Horch, M., Grunzel, M., Lauterbach, L., Lendzian, F., Zebger, I., and Lenz, O. Impact of the iron–sulfur cluster proximal to the active site on the catalytic function of an O_2-tolerant NAD^+-reducing [NiFe]-hydrogenase. Biochemistry 2015, 54, 389.
[169] Lauterbach, L., Idris, Z., Vincent, K. A., and Lenz, O. Catalytic properties of the isolated diaphorase fragment of the NAD-reducing [NiFe]-hydrogenase from *Ralstonia eutropha*. PLoS One 2011, 6, e25939.
[170] Horch, M., Rippers, Y., Mroginski, M. A., Hildebrandt, P., and Zebger, I. Combining spectroscopy and theory to evaluate structural models of metalloenzymes: A case study on the soluble [NiFe] hydrogenase from *Ralstonia eutropha*. Chem Phys Chem 2013, 14, 185.
[171] Löwenstein, J., Lauterbach, L., Teutloff, C., Lenz, O., and Bittl, R. Active site of the NAD^+-reducing hydrogenase from *Ralstonia eutropha* studied by EPR Spectroscopy. J Phys Chem B 2015, 119, 13834.
[172] Lindenmaier, N. J., Wahlefeld, S., Bill, E., Szilvási, T., Eberle, C., Yao, S., Hildebrandt, P., Horch, M., Zebger, I., and Driess, M. An S-oxygenated [NiFe] complex modelling sulfenate intermediates of an O_2-tolerant hydrogenase. Angew Chem Int Ed 2017, 56, 2208.
[173] Pilkington, S. J., Skehel, J. M., Gennis, R. B., and Walker, J. E. Relationship between mitochondrial NADH-ubiquinone reductase and a bacterial NAD-reducing hydrogenase. Biochemistry 1991, 30, 2166.
[174] Schneider, K., and Schlegel, H. G. Identification and quantitative determination of the flavin component of soluble hydrogenase from *Alcaligenes eutrophus*. Biochem Biophys Res Commun 1978, 84, 564.
[175] van der Linden, E., Faber, B. W., Bleijlevens, B., Burgdorf, T., Bernhard, M., Friedrich, B., and Albracht, S. P. J. Selective release and function of one of the two FMN groups in the cytoplasmic NAD^+-reducing [NiFe]-hydrogenase from *Ralstonia eutropha*. Eur J Biochem 2004, 271, 801.
[176] Albracht, S. P. J. Intimate relationships of the large and the small subunits of all nickel hydrogenases with two nuclear-encoded subunits of mitochondrial NADH: ubiquinone oxidoreductase. Biochim Biophys Acta 1993, 1144, 221.
[177] Erkens, A., Schneider, K., and Müller, A. The NAD-linked soluble hydrogenase from *Alcaligenes eutrophus* H16: Detection and characterization of EPR signals deriving from nickel and flavin. J Biol Inorg Chem 1996, 1, 99.
[178] Happe, R. P., Roseboom, W., Egert, G., Friedrich, C. G., Massanz, C., Friedrich, B., and Albracht, S. P. J. Unusual FTIR and EPR properties of the H_2-activating site of the cytoplasmic NAD-reducing hydrogenase from *Ralstonia eutropha*. FEBS Lett 2000, 466, 259.
[179] Schneider, K., Cammack, R., Schlegel, H. G., and Hall, D. O. The iron-sulphur centres of soluble hydrogenase from *Alcaligenes eutrophus*. Biochim Biophys Acta 1979, 578, 445.
[180] van der Linden, E., Burgdorf, T., De Lacey, A. L., Buhrke, T., Scholte, M., Fernandez, V., Friedrich, B., and Albracht, S. P. J. An improved purification procedure for the soluble [NiFe]-hydrogenase of *Ralstonia eutropha*: New insights into its (in)stability and spectroscopic properties. J Biol Inorg Chem 2006, 11, 247.
[181] Patel, S. D., Aebersold, R., and Attardi, G. cDNA-derived amino acid sequence of the NADH-binding 51-kDa subunit of the bovine respiratory NADH dehydrogenase reveals striking similarities to a bacterial NAD^+-reducing hydrogenase. Proc Natl Acad Sci U.S.A 1991, 88, 4225.
[182] Long, M., Liu, J., Chen, Z., Bleijlevens, B., Roseboom, W., and Albracht, S. P. J. Characterization of a HoxEFUYH type of [NiFe] hydrogenase from *Allochromatium vinosum* and some EPR and IR properties of the hydrogenase module. J Biol Inorg Chem 2007, 12, 62.
[183] Burgdorf, T., van der Linden, E., Bernhard, M., Yin, Q. Y., Back, J. W., Hartog, A. F., Muijsers, A. O., de Koster, C. G., Albracht, S. P. J., and Friedrich, B. The soluble NAD^+-reducing [NiFe]-hydrogenase from *Ralstonia eutropha* H16 consists of six subunits and can be specifically activated by NADPH. J Bacteriol 2005, 187, 3122.

[184] a) Yeh, J. I., Claiborne, A., and Hol, W. G. J. Structure of the native cysteine-sulfenic acid redox center of enterococcal NADH peroxidase refined at 2.8 Å resolution. Biochemistry 1996, 35, 9951. b) L. B. Poole, A. Claiborne, The non-flavin redox center of the streptococcal NADH peroxidase. II. Evidence for a stabilized cysteine-sulfenic acid. J. Biol. Chem. 1989, 264, 12330.
[185] Boyd, E. S., Schut, G. J., Adams, M. W. W., and Peters, J. W. Hydrogen metabolism and the evolution of biological respiration. Microbe 2014, 9, 361.
[186] Volbeda, A., Fontecilla-Camps, J. C. Catalytic nickel–iron–sulfur clusters: From minerals to enzymes. In: Simonneaux, G., ed. Bioorganometallic Chemistry, Heidelberg, Germany, Springer, 2006, 57–82.
[187] Günter, Wächtershäuser. On the chemistry and evolution of the pioneer organism. Chem Biodivers 2007, 4, 584.
[188] Wächtershäuser, G. From volcanic origins of chemoautotrophic life to Bacteria, Archaea and Eukarya. Phil Trans R Soc B Biol Sci 2006, 361, 1787.
[189] Martin, W., Baross, J., Kelley, D., and Russell, M. J. Hydrothermal vents and the origin of life. Nat Rev Microbiol 2008, 6, 805.
[190] Cody, G. D., Boctor, N. Z., Filley, T. R., Hazen, R. M., Scott, J. H., Sharma, A., and Yoder, H. S. Primordial carbonylated iron-sulfur compounds and the synthesis of pyruvate. Science 2000, 289, 1337.
[191] a) Moparthi, V. K., and Hagerhall, C. The evolution of respiratory chain complex I from a smaller last common ancestor consisting of 11 protein subunits. J Mol Evol 2011, 72, 484. b) C. Mathiesen, C. Hägerhäll, The 'antiporter module' of respiratory chain Complex I includes the MrpC/NuoK subunit – a revision of the modular evolution scheme. FEBS Lett. 2003, 549, 7; c) B. C. Marreiros, A. P. Batista, A. M. Duarte, M. M. Pereira, A missing link between complex I and group 4 membrane-bound [NiFe] hydrogenases. Biochim. Biophys. Acta 2013, 1827, 198.
[192] Lacasse, M. J., and Zamble, D. B. [NiFe]-hydrogenase maturation. Biochemistry 2016, 55, 1689.
[193] Forzi, L., Hellwig, P., Thauer, R. K., and Sawers, R. G. The CO and CN^- ligands to the active site Fe in [NiFe]-hydrogenase of *Escherichia coli* have different metabolic origins. FEBS Lett 2007, 581, 3317.
[194] Lenz, O., Zebger, I., Hamann, J., Hildebrandt, P., and Friedrich, B. Carbamoylphosphate serves as the source of CN^-, but not of the intrinsic CO in the active site of the regulatory [NiFe]-hydrogenase from *Ralstonia eutropha*. FEBS Lett 2007, 581, 3322.
[195] Reissmann, S., Hochleitner, E., Wang, H., Paschos, A., Lottspeich, F., Glass, R. S., and Bock, A. Taming of a poison: Biosynthesis of the NiFe-hydrogenase cyanide ligands. Science 2003, 299, 1067.
[196] Blokesch, M., Paschos, A., Bauer, A., Reissmann, S., Drapal, N., and Bock, A. Analysis of the transcarbamoylation-dehydration reaction catalyzed by the hydrogenase maturation proteins HypF and HypE. Eur J Biochem 2004, 271, 3428.
[197] Rangarajan, E. S., Asinas, A., Proteau, A., Munger, C., Baardsnes, J., Iannuzzi, P., Matte, A., and Cygler, M. Structure of [NiFe] hydrogenase maturation protein HypE from *Escherichia coli* and its interaction with HypF. J Bacteriol 2008, 190, 1447.
[198] Paschos, A., Bauer, A., Zimmermann, A., Zehelein, E., and Bock, A. HypF, a carbamoyl phosphate-converting enzyme involved in [NiFe] hydrogenase maturation. J Biol Chem 2002, 277, 49945.
[199] Watanabe, S., Matsumi, R., Arai, T., Atomi, H., Imanaka, T., and Miki, K. Crystal structures of [NiFe] hydrogenase maturation proteins HypC, HypD, and HypE: Insights into cyanation reaction by thiol redox signaling. Mol Cell 2007, 27, 29.
[200] Shomura, Y., Komori, H., Miyabe, N., Tomiyama, M., Shibata, N., and Higuchi, Y. Crystal structures of hydrogenase maturation protein HypE in the Apo and ATP-bound forms. J Mol Biol 2007, 372, 1045.

[201] Bürstel, I., Hummel, P., Siebert, E., Wisitruangsakul, N., Zebger, I., Friedrich, B., and Lenz, O. Probing the origin of the metabolic precursor of the CO ligand in the catalytic center of [NiFe] hydrogenase. J Biol Chem 2011, 286, 44937.
[202] Soboh, B., Stripp, S. T., Muhr, E., Granich, C., Braussemann, M., Herzberg, M., Heberle, J., and Sawers, R. G. [NiFe]-hydrogenase maturation: Isolation of a HypC-HypD complex carrying diatomic CO and CN^- ligands. FEBS Lett 2012, 586, 3882.
[203] Soboh, B., Stripp, S. T., Bielak, C., Lindenstrauss, U., Braussemann, M., Javaid, M., Hallensleben, M., Granich, C., Herzberg, M., Heberle, J., and Sawers, R. G. The [NiFe]-hydrogenase accessory chaperones HypC and HybG of *Escherichia coli* are iron- and carbon dioxide-binding proteins. FEBS Lett 2013, 587, 2512.
[204] Roseboom, W., Blokesch, M., Bock, A., and Albracht, S. P. J. The biosynthetic routes for carbon monoxide and cyanide in the Ni-Fe active site of hydrogenases are different. FEBS Lett 2005, 579, 469.
[205] Bürstel, I., Siebert, E., Frielingsdorf, S., Zebger, I., Friedrich, B., and Lenz, O. CO synthesized from the central one-carbon pool as source for the iron carbonyl in O_2-tolerant [NiFe]-hydrogenase. Proc Natl Acad Sci U.S.A 2016, 113, 14722.
[206] Bürstel, I., Siebert, E., Winter, G., Hummel, P., Zebger, I., Friedrich, B., and Lenz, O. A universal scaffold for synthesis of the $Fe(CN)_2(CO)$ moiety of [NiFe] hydrogenase. J Biol Chem 2012, 287, 38845.
[207] a) Soboh, B., Lindenstrauss, U., Granich, C., Javed, M., Herzberg, M., Thomas, C., and Stripp, S. T. [NiFe]-hydrogenase maturation: Analysis of the roles of the HybG and HypD accessory proteins. Biochem J 2014, 464, 169. b) S. T. Stripp, B. Soboh, U. Lindenstrauss, M. Braussemann, M. Herzberg, D. H. Nies, R. G. Sawers, J. Heberle, HypD is the scaffold protein for $Fe\text{-}(CN)_2CO$ cofactor assembly in [NiFe]-hydrogenase maturation. Biochemistry 2013, 52, 3289.
[208] Stripp, S. T., Lindenstrauss, U., Granich, C., Sawers, R. G., and Soboh, B. The influence of oxygen on [NiFe]–hydrogenase cofactor biosynthesis and how ligation of carbon monoxide precedes cyanation. PLoS ONE 2014, 9, e107488.
[209] Blokesch, M., and Bock, A. Maturation of [NiFe]-hydrogenases in *Escherichia coli*: The HypC cycle. J Mol Biol 2002, 324, 287.
[210] Blokesch, M., Albracht, S. P. J., Matzanke, B. F., Drapal, N. M., Jacobi, A., and Bock, A. The complex between hydrogenase-maturation proteins HypC and HypD is an intermediate in the supply of cyanide to the active site iron of [NiFe]-hydrogenases. J Mol Biol 2004, 344, 155.
[211] a) Blokesch, M., Magalon, A., and Bock, A. Interplay between the specific chaperone-like proteins HybG and HypC in maturation of hydrogenases 1, 2, and 3 from *Escherichia coli.* J Bacteriol 2001, 183, 2817. b) G. Butland, J. W. Zhang, W. Yang, A. Sheung, P. Wong, J. F. Greenblatt, A. Emili, D. B. Zamble, Interactions of the Escherichia coli hydrogenase biosynthetic proteins: HybG complex formation. Interactions of the *Escherichia coli* hydrogenase biosynthetic proteins: HybG complex formation. FEBS Lett. 2006, 580, 677; c) A. K. Jones, O. Lenz, A. Strack, T. Buhrke, B. Friedrich, NiFe hydrogenase active site biosynthesis: Identification of Hyp protein complexes in *Ralstonia eutropha*. Biochemistry 2004, 43, 13467.
[212] Drapal, N., and Bock, A. Interaction of the hydrogenase accessory protein HypC with HycE, the large subunit of *Escherichia coli* hydrogenase 3 during enzyme maturation. Biochemistry 1998, 37, 2941.
[213] Chan Chung, K. C., and Zamble, D. B. Protein interactions and localization of the *Escherichia coli* accessory protein HypA during nickel insertion to [NiFe] hydrogenase. J Biol Chem 2011, 286, 43081.
[214] a) Fritsch, J., Lenz, O., and Friedrich, B. The maturation factors HoxR and HoxT contribute to oxygen tolerance of membrane-bound [NiFe] hydrogenase in *Ralstonia eutropha* H16. J Bacteriol 2011, 193, 2487. b) M. Ludwig, T. Schubert, I. Zebger, N. Wisitruangsakul, M.

Saggu, A. Strack, O. Lenz, P. Hildebrandt, B. Friedrich, Concerted action of two novel auxiliary proteins in assembly of the active site in a membrane-bound [NiFe] hydrogenase. J. Biol. Chem. 2009, 284, 2159.

[215] a) Xia, W., Li, H., Sze, K. H., and Sun, H. Structure of a nickel chaperone, HypA, from *Helicobacter pylori* reveals two distinct metal binding sites. J Am Chem Soc 2009, 131, 10031. b) S. Watanabe, T. Arai, R. Matsumi, H. Atomi, T. Imanaka, K. Miki, Crystal structure of HypA, a nickel-binding metallochaperone for [NiFe] hydrogenase maturation. J. Mol. Biol. 2009, 394, 448; c) D. C. Kennedy, R. W. Herbst, J. S. Iwig, P. T. Chivers, M. J. Maroney, A dynamic Zn site in *Helicobacter pylori* HypA – a potential mechanism for metal specific protein activity. J. Am. Chem. Soc. 2008, 129, 16; d) M. Rowinska-Zyrek, S. Potocki, D. Witkowska, D. Valensin, H. Kozlowski, The zinc-binding fragment of HypA from *Helicobacter pylori*: A tempting site also for nickel ions. Dalton Trans. 2013, 42, 6012.

[216] Watanabe, S., Kawashima, T., Nishitani, Y., Kanai, T., Wada, T., Inaba, K., Atomi, H., Imanaka, T., and Miki, K. Structural basis of a Ni acquisition cycle for [NiFe] hydrogenase by Ni-metallochaperone HypA and its enhancer. Proc Natl Acad Sci U.S.A 2015, 112, 7701.

[217] Herbst, R. W., Perovic, I., Martin-Diaconescu, V., O'Brien, K., Chivers, P. T., Pochapsky, S. S., Pochapsky, T. C., and Maroney, M. J. Communication between the zinc and nickel sites in dimeric HypA: Metal recognition and pH sensing. J Am Chem Soc 2010, 132, 10338.

[218] a) Atanassova, A., and Zamble, D. B. *Escherichia coli* HypA is a zinc metalloprotein with a weak affinity for nickel. J Bacteriol 2005, 187, 4689. b) D. Sasaki, S. Watanabe, T. Kanai, H. Atomi, T. Imanaka, K. Miki, Characterization and *in vitro* interaction study of a [NiFe] hydrogenase large subunit from the hyperthermophilic archaeon *Thermococcus kodakarensis* KOD1. Biochem. Biophys. Res. Commun. 2012, 417, 192.

[219] Douglas, C. D., Ngu, T. T., Kaluarachchi, H., and Zamble, D. B. Metal transfer within the *Escherichia coli* HypB-HypA complex of hydrogenase accessory proteins. Biochemistry 2013, 52, 6030.

[220] Sydor, A. M., Lebrette, H., Ariyakumaran, R., Cavazza, C., and Zamble, D. B. Relationship between Ni(II) and Zn(II) coordination and nucleotide binding by the *Helicobacter pylori* [NiFe]-hydrogenase and urease maturation factor HypB. J Biol Chem 2014, 289, 3828.

[221] a) Leach, M. R., Sandal, S., Sun, H., and Zamble, D. B. Metal binding activity of the *Escherichia coli* hydrogenase maturation factor HypB. Biochemistry 2005, 44, 12229. b) A. V. Dias, C. M. Mulvihill, M. R. Leach, I. J. Pickering, G. N. George, D. B. Zamble, Structural and biological analysis of the metal sites of *Escherichia coli* hydrogenase accessory protein HypB. Biochemistry 2008, 47, 11981.

[222] a) Kaluarachchi, H., Zhang, J. W., and Zamble, D. B. *Escherichia coli* SlyD, more than a Ni(II) reservoir. Biochemistry 2011, 50, 10761. b) M. R. Leach, J. W. Zhang, D. B. Zamble, The role of complex formation between the *Escherichia coli* hydrogenase accessory factors HypB and SlyD. J. Biol. Chem. 2007, 282, 16177; c) H. Kaluarachchi, M. Altenstein, S. R. Sugumar, J. Balbach, D. B. Zamble, C. Haupt, Nickel binding and [NiFe]-hydrogenase maturation by the metallochaperone SlyD with a single metal-binding site in *Escherichia coli*. J. Mol. Biol. 2012, 417, 28; d) H. Kaluarachchi, J. F. Siebel, S. Kaluarachchi-Duffy, S. Krecisz, D. E. Sutherland, M. J. Stillman, D. B. Zamble, Metal selectivity of the *Escherichia coli nickel* metallochaperone, SlyD. Biochemistry 2011, 50, 10666; e) H. Kaluarachchi, D. E. Sutherland, A. Young, I. J. Pickering, M. J. Stillman, D. B. Zamble, The Ni(II)-binding properties of the metallochaperone SlyD. J. Am. Chem. Soc. 2009, 131, 18489.

[223] a) Theodoratou, E., Paschos, A., Mintz, W., and Bock, A. Analysis of the cleavage site specificity of the endopeptidase involved in the maturation of the large subunit of hydrogenase 3 from *Escherichia coli*. Arch Microbiol 2000, 173, 110. b) E. Theodoratou, A. Paschos, A. Magalon, E. Fritsche, R. Huber, A. Bock, Nickel serves as a substrate recognition

motif for the endopeptidase involved in hydrogenase maturation. Eur. J. Biochem. 2000, 267, 1995; c) E. Theodoratou, R. Huber, A. Bock, [NiFe]-hydrogenase maturation endopeptidase: Structure and function. Biochem. Soc. Trans. 2005, 33, 108.
[224] Senger, M., Stripp, S. T., and Soboh, B. Proteolytic cleavage orchestrates cofactor insertion and protein assembly in [NiFe]-hydrogenase biosynthesis. J Biol Chem 2017.
[225] a) Magalon, A., and Bock, A. Dissection of the maturation reactions of the [NiFe] hydrogenase 3 from *Escherichia coli* taking place after nickel incorporation. FEBS Lett 2000, 473, 254. b) C. Thomas, E. Muhr, R. G. Sawers, Coordination of synthesis and assembly of a modular membrane-associated [NiFe]-hydrogenase is determined by cleavage of the C-terminal peptide. J. Bacteriol. 2015, 197, 2989.
[226] a) Abou, H. A., Dementin, S., Liebgott, P. P., Gutierrez-Sanz, O., Richaud, P., Lacey, A. L. De., Rousset, M., Bertrand, P., Cournac, L., and Leger, C. Understanding and tuning the catalytic bias of hydrogenase. J Am Chem Soc 2012, 134, 8368. b) S. Dementin, V. Belle, P. Bertrand, B. Guigliarelli, G. Adryanczyk-Perrier, A. L. De Lacey, V. M. Fernandez, M. Rousset, C. Leger, Changing the ligation of the distal [4Fe4S] cluster in NiFe hydrogenase impairs inter- and intramolecular electron transfers. J. Am. Chem. Soc. 2006, 128, 5209.
[227] Ihara, M., Nakamoto, H., Kamachi, T., Okura, I., and Maeda, M. Photoinduced hydrogen production by direct electron transfer from photosystem I cross-linked with cytochrome c_3 to [NiFe]-hydrogenase. Photochem Photobiol 2006, 82, 1677.
[228] Ihara, M., Nishihara, H., Yoon, K. S., Lenz, O., Friedrich, B., Nakamoto, H., Kojima, K., Honma, D., Kamachi, T., and Okura, I. Light-driven hydrogen production by a hybrid complex of a [NiFe]-hydrogenase and the cyanobacterial photosystem I. Photochem Photobiol 2006, 82, 676.
[229] Krassen, H., Schwarze, A., Friedrich, B., Ataka, K., Lenz, O., and Heberle, J. Photosynthetic hydrogen production by a hybrid complex of photosystem I and [NiFe]-hydrogenase. ACS Nano 2009, 3, 4055.
[230] Schwarze, A., Kopczak, M. J., Rogner, M., and Lenz, O. Requirements for construction of a functional hybrid complex of photosystem I and [NiFe]-hydrogenase. Appl Environ Microbiol 2010, 76, 2641.
[231] Stenkamp, R. E., Sieker, L. C., and Jensen, L. H. The structure of rubredoxin from *Desulfovibrio desulfuricans* strain 27774 at 1.5 Å resolution. Proteins 1990, 8, 352.
[232] a) Mus-Veteau, I., Diaz, D., Gracia-Mora, J., Guigliarelli, B., Chottard, G., and Bruschi, M. Spectroscopic studies of the nickel-substituted *Desulfovibrio vulgaris* Hildenborough rubredoxin: Implication for the nickel site in hydrogenases. BBA – Bioenergetics 1991, 1060, 159. b) Y. H. Huang, J. B. Park, M. W. W. Adams, M. K. Johnson, Oxidized nickel-substituted rubredoxin as a model for the Ni-C EPR signal of NiFe hydrogenases. Inorg. Chem. 1993, 32, 375.
[233] Slater, J. W., and Shafaat, H. S. Nickel-substituted rubredoxin as a minimal enzyme model for hydrogenase. J Phys Chem Lett 2015, 6, 3731.
[234] Bouwman, E., and Reedijk, J. Structural and functional models related to the nickel hydrogenases. Coord Chem Rev 2005, 249, 1555.
[235] a) Darensbourg, D. J., Reibenspies, J. H., Lai, C.-H., Lee, W.-Z., and Darensbourg, M. Y. Analysis of an organometallic iron site model for the heterodimetallic unit of [NiFe] hydrogenase. J Am Chem Soc 1997, 119, 7903. b) C.-H. Lai, W.-Z. Lee, M. L. Miller, J. H. Reibenspies, D. J. Darensbourg, M. Y. Darensbourg, Responses of the $Fe(CN)_2(CO)$ unit to electronic changes as related to its role in [NiFe] hydrogenase. J. Am. Chem. Soc. 1998, 120, 10103.
[236] Ohki, Y., and Tatsumi, K. Thiolate-bridged iron–nickel models for the active site of [NiFe] hydrogenase. Eur J Inorg Chem 2010, 2011, 973.
[237] a) Barton, B. E., and Rauchfuss, T. B. Hydride-containing models for the active site of the nickel-iron hydrogenases. J Am Chem Soc 2010, 132, 14877. b) W. Zhu, A. C. Marr, Q. Wang, F. Neese, D. J. Spencer, A. J. Blake, P. A. Cooke, C. Wilson, M. Schroder, Modulation of the

electronic structure and the Ni-Fe distance in heterobimetallic models for the active site in [NiFe] hydrogenase. Proc. Natl. Acad. Sci. U.S.A. 2005, 102, 18280; c) S. Canaguier, M. Field, Y. Oudart, J. Pecaut, M. Fontecave, V. Artero, A structural and functional mimic of the active site of NiFe hydrogenases. Chem. Commun. 2010, 46, 5876; d) S. Ogo, K. Ichikawa, T. Kishima, T. Matsumoto, H. Nakai, K. Kusaka, T. Ohhara, A functional [NiFe] hydrogenase mimic that catalyzes electron and hydride transfer from H_2. Science 2013, 339, 682; e) B. C. Manor, T. B. Rauchfuss, Hydrogen activation by biomimetic [NiFe]-hydrogenase model containing protected cyanide cofactors. J. Am. Chem. Soc. 2013, 135, 11895; f) T. Xu, D. Chen, X. Hu, Hydrogen-activating models of hydrogenases. Coord. Chem. Rev. 2015, 303, 32.
[238] Helm, M. L., Stewart, M. P., Bullock, R. M., DuBois, M. R., and DuBois, D. L. A synthetic nickel electrocatalyst with a turnover frequency above 100,000 s^{-1} for H_2 production. Science 2011, 333, 863.
[239] Smith, S. E., Yang, J. Y., DuBois, D. L., and Bullock, R. M. Reversible electrocatalytic production and oxidation of hydrogen at low overpotentials by a functional hydrogenase mimic. Angew Chem Int Ed 2012, 51, 3152.
[240] a) Caserta, G., Roy, S., Atta, M., Artero, V., and Fontecave, M. Artificial hydrogenases: Biohybrid and supramolecular systems for catalytic hydrogen production or uptake. Curr Opin Chem Biol 2015, 25, 36. b) A. Onoda, T. Hayashi, Artificial hydrogenase: Biomimetic approaches controlling active molecular catalysts. Curr. Opin. Chem. Biol. 2015, 25, 133.
[241] Ginovska-Pangovska, B., Dutta, A., Reback, M. L., Linehan, J. C., and Shaw, W. J. Beyond the active site: The impact of the outer coordination sphere on electrocatalysts for hydrogen production and oxidation. Acc Chem Res 2014, 47, 2621.
[242] Silver, S. C., Niklas, J., Du, P., Poluektov, O. G., Tiede, D. M., and Utschig, L. M. Protein delivery of a Ni catalyst to photosystem I for light-driven hydrogen production. J Am Chem Soc 2013, 135, 13246.
[243] Dutta, A., Hamilton, G. A., Hartnett, H. E., and Jones, A. K. Construction of heterometallic clusters in a small peptide scaffold as [NiFe]-hydrogenase models: Development of a synthetic methodology. Inorg Chem 2012, 51, 9580.
[244] Lutz, B. J., Fan, Z. H., Burgdorf, T., and Friedrich, B. Hydrogen sensing by enzyme-catalyzed electrochemical detection. Anal Chem 2005, 77, 4969.
[245] Benemann, J. Hydrogen biotechnology, progress and prospects. Nat Biotech 1996, 14, 1101.
[246] Lazarus, O., Woolerton, T. W., Parkin, A., Lukey, M. J., Reisner, E., Seravalli, J., Pierce, E., Ragsdale, S. W., Sargent, F., and Armstrong, F. A. Water-gas shift reaction catalyzed by redox enzymes on conducting graphite platelets. J Am Chem Soc 2009, 131, 14154.
[247] Prince, R. C., and Kheshgi, H. S. The photobiological production of hydrogen: Potential efficiency and effectiveness as a renewable fuel. Crit Rev Microbiol 2005, 31, 19.
[248] Winkler, M., Kawelke, S., and Happe, T. Light driven hydrogen production in protein based semi-artificial systems. Bioresour Technol 2011, 102, 8493.
[249] a) Chardin, B., Giudici-Orticoni, M. T., De, L. G., Guigliarelli, B., and Bruschi, M. Hydrogenases in sulfate-reducing bacteria function as chromium reductase. Appl Microbiol Biotechnol 2003, 63, 315. b) C. Michel, M.-T. Giudici-Orticoni, F. Baymann, M. Bruschi, Bioremediation of chromate by sulfate-reducing bacteria, cytochromes c_3 and hydrogenases. Wat Air Soil Pollut: Focus 2003, 3, 161.
[250] Vincent, K. A., Li, X., Blanford, C. F., Belsey, N. A., Weiner, J. H., and Armstrong, F. A. Enzymatic catalysis on conducting graphite particles. Nat Chem Biol 2007, 3, 761.

2.3 [Fe]-hydrogenase

Takashi Fujishiro

2.3.1 Introduction

Hydrogenases are metalloenzymes that reversibly catalyze conversion of molecular hydrogen (H_2) to protons and electrons. Based on the metallic active sites, three distinct classes of hydrogenases are known: [NiFe]-hydrogenase, [FeFe]-hydrogenase and [Fe]-hydrogenase [1, 2]. Three types of hydrogenases have different protein structures and metallic active sites, although they commonly have their ability to heterolytically cleave H_2 to a hydride and a proton, which is as the result of convergent evolution. [NiFe]- and [FeFe]-hydrogenases can catalyze both of heterolytic cleavage of H_2 to a proton and a hydride and further extraction of two electrons from a hydride. On the other hand, [Fe]-hydrogenase contains a mononuclear Fe as its active site, and directly uses a hydride for catalytic reduction of the substrate methenyl-tetrahydromethanopterin to methylene-tetrahydromethanopterin after heterolytic cleavage of H_2 at its Fe center requiring methenyl-tetrahydromethanopterin (Figure 2.30) [3, 4]. In other words, [Fe]-hydrogenase cannot heterolytically cleave H_2 without methenyl-tetrahydromethanopterin, although the other two classes of hydrogenases can with their dinuclear metallic active sites. Thus, [Fe]-hydrogenase takes its unique strategy for hydrogenase catalysis by using its unique structural and functional features, which are distinct from those of other two types of hydrogenases.

In this chapter, we explore the structure, function and catalytic mechanism of [Fe]-hydrogenase to understand its structure–function relationships. Furthermore, the unique metallic active site of [Fe]-hydrogenase, called the *iron-guanylylpyridinol* (FeGP) cofactor, is discussed from viewpoints ranging from synthetic model chemistry to biosynthesis toward the development of hydrogen-based catalytic systems.

2.3.2 Identification of [Fe]-hydrogenase (Hmd) and its paralogs

[Fe]-hydrogenase has been found only in hydrogenotrophic methanogenic archaea and is expressed under nickel-limiting conditions [6, 7], whereas the other two hydrogenases are widely distributed in various organisms [8–10]. First discovered in *Methanothermobacter marburgensis* [11], [Fe]-hydrogenase is also called *H_2-forming methylene-tetrahydromethanopterin dehydrogenase*, abbreviated as Hmd [12–14]. It catalyzes the conversion of methenyl-tetrahydromethanopterin to methylene-tetrahydromethanopterin ($\Delta G^{0\prime}$ = −5.5 kJ mol^{-1}) using a hydride generated by heterolytic cleavage of H_2, as well as the reverse reaction (Figure 2.30). This conversion is an intermediate step in methanogenesis, which is an energy-conservation process via multistep reductions of CO_2 to CH_4 in hydrogenotrophic methanogenic

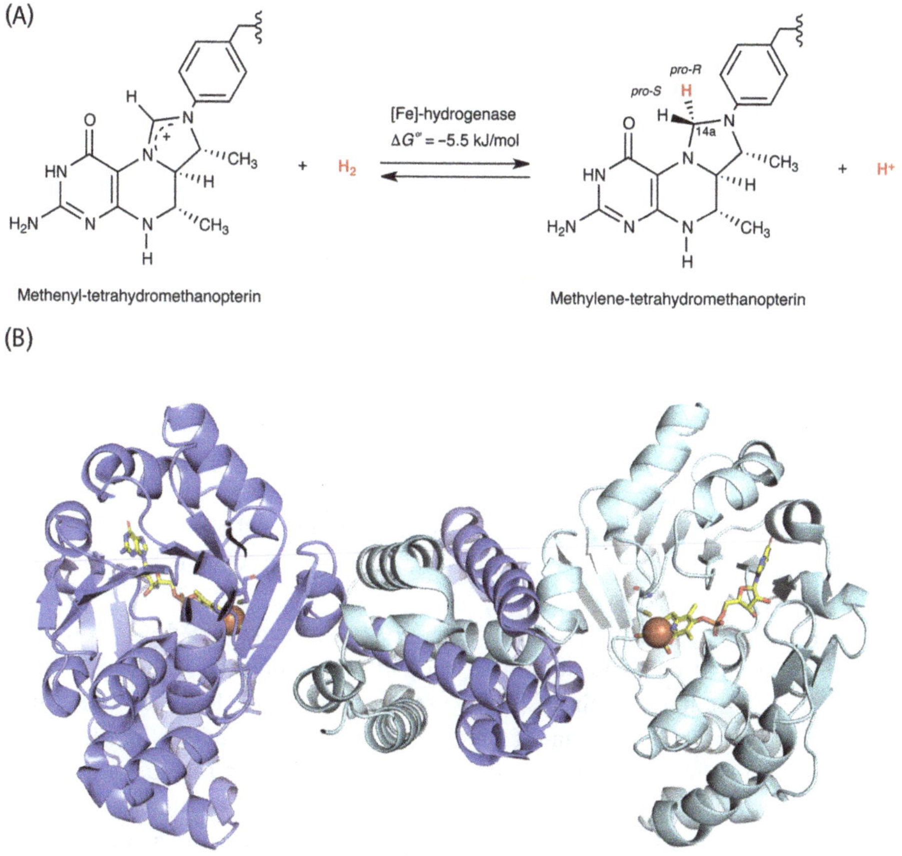

Figure 2.30: (A) [Fe]-hydrogenase-catalyzed reaction [3, 4]. Molecular hydrogen (H_2) is heterolytically cleaved to proton and hydride, and the hydride is transferred stereospecifically to the C14a position of methenyl-tetrahydromethanopterin to form methylene-tetrahydromethanopterin and proton. (B) Overall structure of holo-[Fe]-hydrogenase wild-type (PDB ID: 3F47) [5]. The iron-guanylylpyridinol (FeGP) cofactor is shown as a stick model, and the iron center is shown as a sphere model.

archaea [15, 16]. When the nickel in growth medium is not limited, F_{420}-dependent methylene-tetrahydromethanopterin dehydrogenase (Mtd) [17, 18] is in charge of the conversion of methenyl-tetrahydromethanopterin to methylene-tetrahydromethanopterin. Also, [NiFe]-hydrogenase (Frh), a group III family member of [NiFe]-hydrogenase [19, 20], is responsible for providing reduced coenzyme F_{420} toward Mtd-catalyzed conversion. Studies of [Fe]-hydrogenase expression in the cell found that the production of Frh is greatly decreased by Ni limitation, whereas Mtd is overproduced [6,7]. Thus, [Fe]-hydrogenase is also important in a coupling reaction with Mtd for production of reduced coenzyme F_{420} by Frh with H_2 under nickel-sufficient conditions.

About two decades ago, it was believed that [Fe]-hydrogenase contains no metals unlike the other two types of hydrogenases, so that [Fe]-hydrogenase was also once called "metal-free" or "iron-sulfur cluster-free hydrogenase" in contrast to [NiFe]- and [FeFe]-hydrogenases [21]. Later on, researchers found that [Fe]-hydrogenase does harbor a light-sensitive cofactor composed of iron, carbon monoxide (CO) and guanylylpyridinol [22–25]. Further detailed biochemical, spectroscopic and structural analyses unveiled its structural characteristics [26–30]. Finally, the structure of [Fe]-hydrogenase harboring an FeGP cofactor was determined to be its holo-form (Figure 2.31 and 2.32), and the geometry of its iron coordination structure with unique organometallic acyl–iron bonding was confirmed [5, 31, 32]. Afterward, the crystal structure of holo-[Fe]-hydrogenase C176A variant, which is an inactive form with a cysteinyl ligand replaced by alanine, was also determined to be in complex with methylene-tetrahydromethanopterin [33]. Notably, the FeGP cofactor in the structure of holo-[Fe]-hydrogenase C176A variant contains one dithiothreitol (DTT) molecule that came from the buffer solutions used. One thiolate group and one hydroxy group of DTT are at the equivalent positions to the conserved cysteinyl ligand and a water molecule in the structure of wild-type [Fe]-hydrogenase, respectively.

[Fe]-hydrogenase displays a homodimeric structure stabilized with a unique intertwined domain at the C-terminal region of each monomer [5, 31, 33] (Figure 2.31). The N-terminal domain of [Fe]-hydrogenase shows a Rossmann-like fold, which is a typical α-β-α-type motif capable of binding nucleotides. The FeGP cofactor is bound to each of the N-terminal domain to constitute the active site (Figure 2.32). Clearly, the guanosine mono-phosphate (GMP) moiety of the FeGP cofactor is anchored to the N-terminal Rossmann-like domain with several polar interactions, which is consistent with the known affinity of Rossmann-like fold to nucleotides. In addition to binding site of the GMP moiety, the N-terminal domain contains a key cysteine residue for coordinating to the iron center of the FeGP cofactor, as well as some hydrophobic residues for stabilizing the pyridinol moiety of the FeGP cofactor. These structural features of the N-terminal Rossmann-like domain are well suited for the FeGP-cofactor binding. More importantly, [Fe]-hydrogenase is expected to facilitate the N-terminal domain motions for its catalysis (Figure 2.31). By this conformational change of the N-terminal domain, [Fe]-hydrogenase can change the size of its active site pocket in either the open or closed form. Indeed, apo-[Fe]-hydrogenase (FeGP cofactor-free [Fe]-hydrogenase) shows a closed form, whereas holo-[Fe]-hydrogenase shows an open cleft between the N-terminal Rossman domain and C-terminal intertwined region (Figure 2.31). This possible conformational change is important for consideration of a hydrogen-activation mechanism by [Fe]-hydrogenase with methenyl-tetrahydromethanopterin. The structure-based catalytic mechanism of [Fe]-hydrogenase with methenyl-tetrahydromethanopterin is described in detail below.

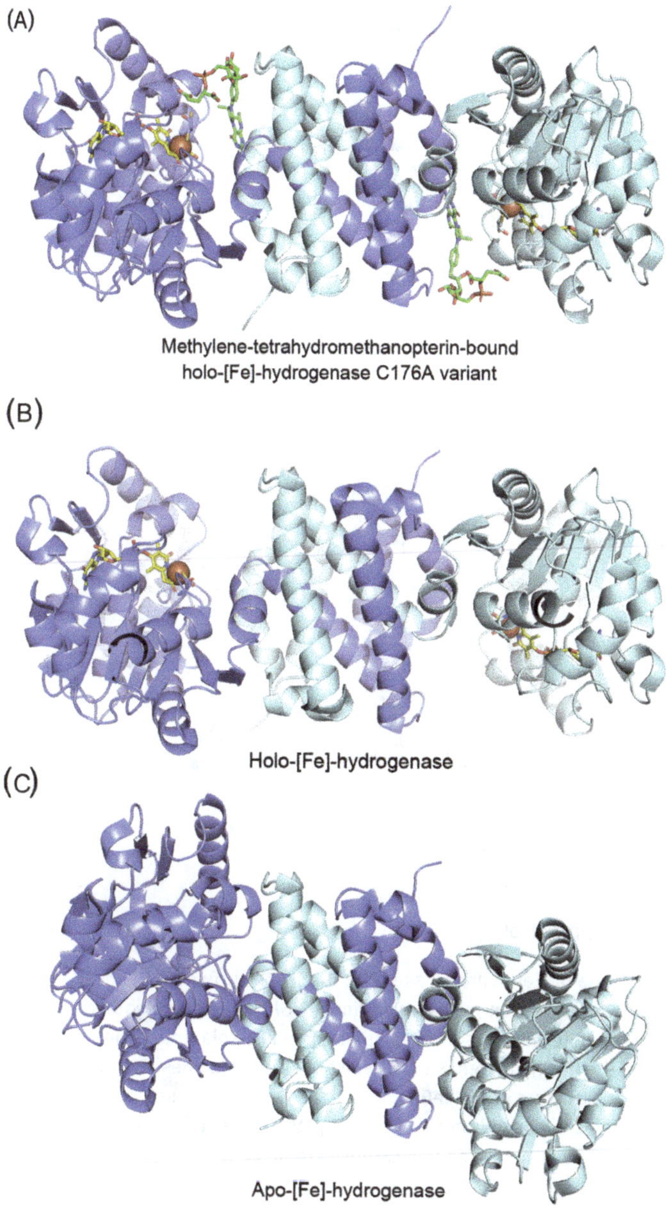

Figure 2.31: Comparison of closed and open conformations of [Fe]-hydrogenase. (A) Closed conformation of [Fe]-hydrogenase C176A variant in complex with methylene-tetrahydromethanopterin (PDB ID: 3H65) [33] and (B) [Fe]-hydrogenase wild-type (PDB ID: 3F47) [5]. (C) Open conformation of apo-[Fe]-hydrogenase (PDB ID: 2B0J) [28]. Organic moieties of the iron-guanylylpyridinol (FeGP) cofactor, dithiothreitol (DTT), and methylene-tetrahydromethanopterin are shown as stick models, [Fe]-hydrogenase homodimeric protein scaffolds are shown as ribbon models and Fe ions are shown as spheres.

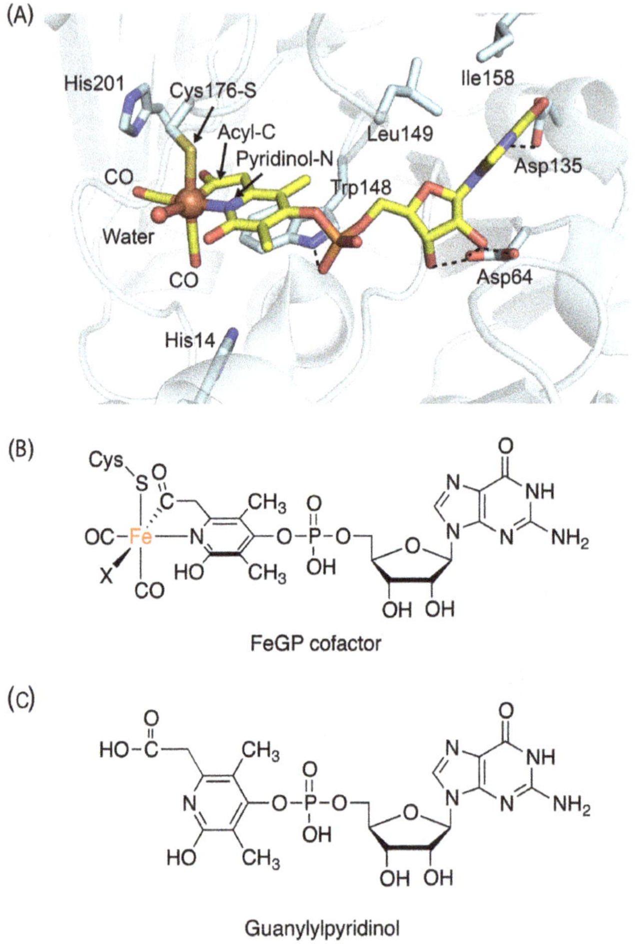

Figure 2.32: (A) Active site structure of [Fe]-hydrogenase (PDB ID: 3F47) [5]. Two CO ligands, a cysteinyl ligand and an organic moiety of iron-guanylylpyridinol (FeGP) cofactor are shown as stick models. The Fe ion and the water molecule are shown as sphere models. Polar interactions between the FeGP cofactor and the [Fe]-hydrogenase are depicted as dashed lines. Arrows indicate atoms ligated to Fe. (B) Chemical structure of the FeGP cofactor. (C) Chemical structure of guanylylpyridinol (GP), the light-decomposed iron-free FeGP cofactor under alkaline conditions. The acyl group of the FeGP cofactor was converted to a carboxy group [24].

Structural details of the [Fe]-hydrogenase protein scaffold show that it is well designed for utilizing the FeGP cofactor and the substrate methenyl-tetrahydromethanopterin. However, some (but not all) hydrogenotrophic methanogenic archaea having [Fe]-hydrogenase additionally possess two [Fe]-hydrogenase paralogs, named HmdII and HmdIII [3, 7, 34, 35]. These two paralogs are very similar to each other in their amino acid sequence (ca. 80%), but have low amino acid sequence identity (<20%) to [Fe]-hydrogenase. Furthermore, HmdII and HmdIII have no [Fe]-hydrogenase activity. Recently, three crystal structures of HmdII were determined as a holo-form, an apo-forms and a methylene-tetrahydromethanopterin-bound apo-form (Figure 2.33) [36]. [Fe]-hydrogenase and HmdII resemble each other in some parts of their three-dimensional structures: for example, both exhibit homodimer forms that contain N-terminal Rossmann-like domain connecting to C-terminal intertwined as a core dimer interface. However, HmdII shows extended C-terminal alpha-helices beside the N-terminal domain, which may prevent their N-terminal domain from conformational changes unlike the [Fe]-hydrogenase. More importantly, HmdII is likely to bind to either FeGP cofactor or methylene-tetrahydromethanopterin at the partially shared space of the cleft between the N-terminal and C-terminal domains. This eliminates the possible formation of a hydrogen-activating environment as in [Fe]-hydrogenase. Nevertheless, the FeGP cofactor coordination (i.e., a conserved

(A)

(B)

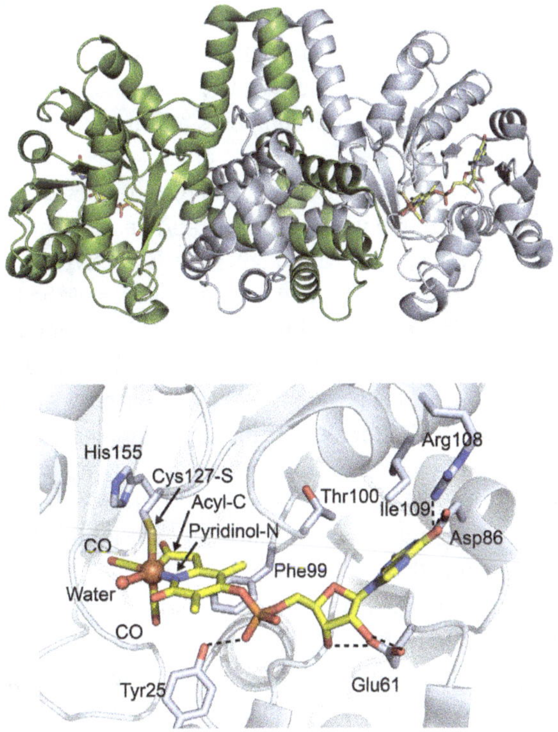

Figure 2.33: (A) Overall structure of holo-HmdII (PDB ID: 4YT4) [36], a [Fe]-hydrogenase paralog. (B) FeGP cofactor-binding site of HmdII. Two CO ligands, a cysteinyl ligand and an organic moiety of iron-guanylylpyridinol (FeGP) cofactor are shown as stick models. The Fe ion and the water molecule are shown as sphere models. Polar interactions between the FeGP cofactor and HmdII are depicted as dashed lines. Arrows indicate atoms ligated to Fe.

cysteinyl ligand, two cis-COs, an acyl–iron coordination and a water molecule) is consistent with [Fe]-hydrogenase. Based on this observation, it is assumed that HmdII may use the FeGP cofactor for a function other than [Fe]-hydrogenase activity, which should be further investigated.

2.3.3 Structure and property of FeGP cofactor

Biological organometallic complexes involving a metal–carbon bond are known to exist in not only the three distinct types of hydrogenases [1, 2], but also some metalloenzymes using metalloclusters [37–45], cobalamin-dependent enzymes [46] and nickel pincer-type complexes [47].

The FeGP cofactor is also an organometallic mononuclear iron-carbonyl complex bearing a six-coordinated octahedral geometry (Figure 2.33). The iron center of the FeGP cofactor has a low-spin Fe(II) state, as indicated by Mössbauer spectroscopy [26] and X-ray absorption spectroscopy [30]. Moreover, this iron center is redox-inactive unlike the other metallic active centers of [NiFe]- and [FeFe]-hydrogenases. Importantly, the FeGP cofactor is light sensitive [23, 24] like typical Fe-carbonyl complexes, and so it must be carefully handled in experiments under red light [48]. When irradiated by UV-A/blue light, it decomposes and its acyl group is converted to a carboxy group (Figure 2.33).

The ligands to the iron of the FeGP cofactor are two CO molecules [25, 29], one cysteine thiolate, one pyridinol-nitrogen, one water molecule and an acylmethyl ligand [4]. There are no cyanide (CN^-) ligands, although metallic active sites in [NiFe] and [FeFe]-hydrogenases have them [1, 2]. The two CO molecules of the FeGP cofactor are ligated to the iron in a *cis*-form: one *trans* to the cysteinyl ligand and the other *trans* to the pyridinol-nitrogen. The pyridinol ligand is highly substituted, with a hydroxyl group (pyridinol-OH) at the 2-position, two methyl groups at the 3- and 5-positions, a relatively hydrophilic GMP moiety at the 4-positionand the acylmethyl group at the 6-position. The 2-hydroxy group is crucial for activation of H_2, as demonstrated in the reconstitution of [Fe]-hydrogenase using FeGP cofactor model complexes [49]. The roles of the two methyl groups are still under debate, but they may be electron-donating substituents to the pyridinol ligand. This electron-donating property of the methyl groups may influence the electronic property of the pyridinol ligand and the iron thereafter. Additionally, these methyl groups could have hydrophobic or van der Waals interactions with their surroundings at the FeGP-cofactor-binding site of the N-terminal domain of [Fe]-hydrogenase [31, 36]. Supposedly, the methyl groups may help fixing the FeGP cofactor as an active site of [Fe]-hydrogenase. By contrast, the GMP moiety connected to the 4-position of the pyridinol ligand is clearly more important for anchoring the FeGP cofactor to the N-terminal Rossman-like domain with several polar interactions [31, 36].

However, the electronic effect of the phosphate group on the pyridinol is unclear at this moment.

The most unique and interesting ligand is the acylmethyl forming an organometallic iron–carbon bond *trans* to the open-site, where a water molecule is bound at the resting state. Because of its *trans* position to the open-site via the Fe, this acyl group is considered to be important for hydrogen activation catalyzed by [Fe]-hydrogenase. However, the FeGP cofactor alone and even the holo-[Fe]-hydrogenase alone have no hydrogenase activation property. This fact is completely distinct from the [NiFe]- and [FeFe]-hydrogenases. Nevertheless, this organometallic acyl–iron structure of FeGP cofactor is attractive because it is regarded as a stable organometallic M–C(=O)–CH_2-type complex in the "resting state," although a few M–C(=O)–R-type species (R=alkyl or O) have been reported as "catalytic intermediates" (e.g., a "bent" CO_2-bound Ni–C(=O)–O–Fe species of CO dehydrogenase [50], and a hypothetical Ni–C(=O)–CH_3 species of acetyl-CoA synthase [39]). As an overview, model synthesis and theoretical studies of the FeGP cofactor could help us understand why this acyl–iron can stably exist, and how the acyl group contributes to the [Fe]-hydrogenase catalyzed reaction.

An interesting property of the FeGP cofactor is that the intact FeGP cofactor can be extracted from holo-[Fe]-hydrogenase and reconstituted to apo-[Fe]-hydrogenase [48]. This technique was successfully applied to obtain the crystal structures of holo-[Fe]-hydrogenase [5, 31, 33]. In these cases, the FeGP cofactor extracted from native [Fe]-hydrogenase of *M. marburgensis* was reconstituted to recombinant *Methanocaldococcus jannaschii* apo-[Fe]-hydrogenase expressed in *Escherichia coli* cells. Also, the chemical reconstitution technique for FeGP cofactor was recently applied to investigate the roles of functional groups (e.g. 2-hydroxy group) of the FeGP cofactor using synthetic FeGP cofactor models, as described in details below [49].

2.3.4 Catalytic and inhibition mechanisms of [Fe]-hydrogenase

The hydrogen activation mechanism of [Fe]-hydrogenase is the most attractive, because the holo-[Fe]-hydrogenase requires methenyl-tetrahydromethanopterin as both a substrate and an activator. In other words, the [Fe]-hydrogenase catalytic mechanism should be considered in the ternary complex of the bound FeGP cofactor, methenyl-tetrahydromethanopterin and the surrounding amino acids of [Fe]-hydrogenase.

In the beginning of the mechanistic investigations, H/D exchange (H = hydrogen, D = deuterium) experiments were performed using native [Fe]-hydrogenase [51–54]. Also, it was found that [Fe]-hydrogenase transfers a hydride directly to the *pro-R* site of the C(14a) of methenyl-tetrahydromethanopterin by using NMR analysis of the substrate and the product with hydrogen isotopes in [Fe]-hydrogenase catalysis [55–57]. These studies demonstrated that [Fe]-hydrogenase works as a ternary complex for its H_2-utilizing catalysis with methenyl-tetrahydromethanopterin. In contrast, the [NiFe]- and [FeFe]-active bimetallic sites alone catalyze the proton and para/ortho-H_2 exchange reactions, reflecting a different mechanism as known the ping-pong mechanism [58]. (In enzymology, the ping-pong mechanism is that one product is formed and released from an enzyme before another substrate is bound to it.) In the ping-pong mechanism of [NiFe]- and [FeFe]-hydrogenases, extraction of two electrons from the hydride bound to the enzymes with electron acceptors (e.g. methyl viologen) proceeds not simultaneously, but successively.

The structure-based catalytic mechanism of [Fe]-hydrogenase for hydrogen activation has been discussed. The X-ray crystal structure of *M. jannaschii* holo-[Fe]-hydrogenase [5, 31] revealed the amino acid environments around the FeGP cofactor (Figure 2.32). Based on this structure, two conserved histidines, His14 and His201, were substituted by alanine to test their effects on [Fe]-hydrogenase activity. The substitution of His14 by Ala greatly decreased the activity, whereas that of His201 just showed minor effect. One hypothetical role of His14 is to be involved in hydrogen-bonding network, which may facilitate a proton mobilization. Next, the crystal structure of [Fe]-hydrogenase C176A variant with bound methylene-tetrahydromethanopterin [33] revealed that the *pro-R site* of C(14a) is certainly facing the open site of the FeGP cofactor, which explains the stereoselectivity in the hydrogenation. This methylene-tetrahydromethanopterin-bound [Fe]-hydrogenase C176A variant could be also used to consider a hydrogen-activating, closed form of [Fe]-hydrogenase with methenyl-tetrahydromethanopterin. The conformational change of [Fe]-hydrogenase to its hydrogen-activating state upon the binding of methenyl-tetrahydromethanopterin could also be supported by iron-chromophore circular dichroism spectroscopy [59].

Based on the biochemical and structural analyses, a [Fe]-hydrogenase catalytic mechanism is proposed [33]. First, the substrate methenyl-tetrahydromethanopterin can be bound at the cleft of an open form of holo-[Fe]-hydrogenase (Figure 2.32). Subsequently, a further conformational change possibly occurs to make the cleft size smaller, to result in a shorter distance (ca. 3 Å) between the carbocation (C14a) of methenyl-tetrahydromethanopterin and the iron site of FeGP cofactor. By this conformational change, the Fe site is turned to be a hydrogen-activating environment. In this environment, the iron site of the FeGP cofactor and the carbocation (C14a) of methenyl-tetrahydromethanopterin are reasonably arranged for hydrogen activation and a facile hydride transfer to the C14a of methenyl-tetrahydromethanopterin, to afford methylene-tetrahydromethanopterin in a stereoselective manner.

Experiments with inhibitors of [Fe]-hydrogenase also possibly give insights about its mechanism. For example, CO and CN^- bound to the open site can inhibit [Fe]-hydrogenase [25, 27, 30], although the inhibition is weak unlike the cases of [NiFe]- and [FeFe]-hydrogenases [60, 61]. In contrast, isocyanide shows much stronger inhibition than CO and CN^- [62]. X-ray crystal structures of isocyanide-inhibited [Fe]-hydrogenase revealed that the isocyanic-carbon is ligated to the iron at the open site, forming a five-membered-metalacyclic ring (Figure 2.34) [63]. This implies that the 2-hydroxy group of the pyridinol ligand has a nucleophilic or basic property, which may allow it to attack the H_2 to heterolytically cleave it. Cu (I) ion and H_2O_2 were also employed for inhibitory assays of [Fe]-hydrogenase [64]. Cu(I) was found to work as an inhibitor probably by affecting the thiolate ligand, which was also demonstrated by using synthetic FeGP cofactor models [65]. On the other hand, H_2O_2 inhibited the [Fe]-hydrogenase, but the inhibitory mechanism seemed to be different from that of Cu(I). H_2O_2 is likely to attack the iron at the open site rather than the thiolate, resulting in conversion of Fe(II) to Fe (III). Fe(II) ion was also tested for an inhibitory assay [64]. As a result, Fe(II) was found to be a reversible inhibitor with quite a low K_i (40 nM), which were different

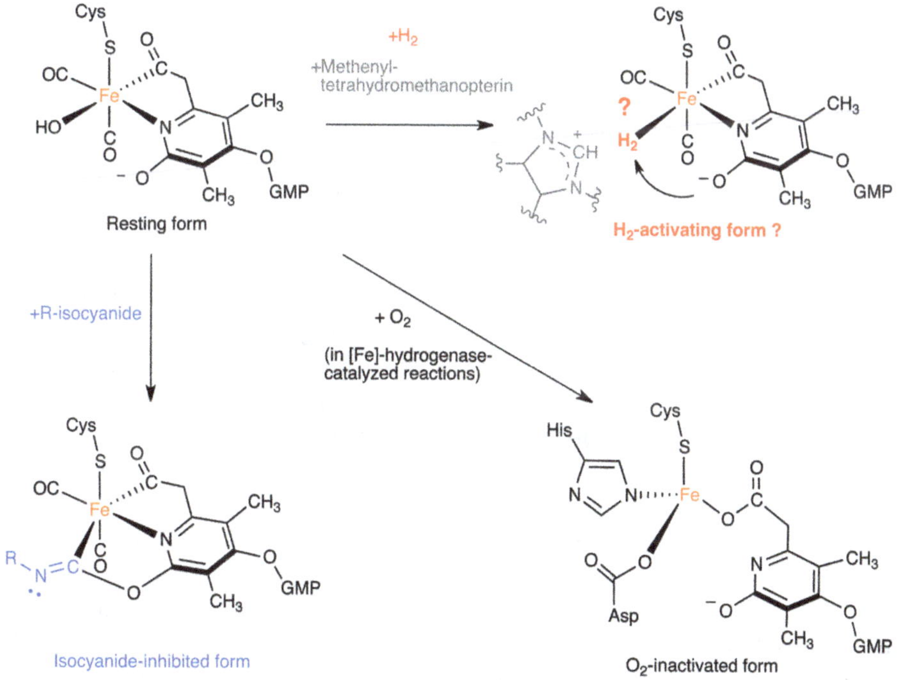

Figure 2.34: Inhibition of [Fe]-hydrogenase by isocyanide and inactivation by O_2 during [Fe]-hydrogenase-catalyzed reaction. Isocyanide inhibition involves a 2-hydroxy group of the pyridinol forming a metallacycle, which implies the possibility of the 2-hydroxy group as a nucleophile or a base [49]. The O_2-inactivation is driven by H_2O_2 formed by reduction of O_2 with a reducing equivalent (methylene-tetrahydromethanopterin or H_2/methenyl-tetrahydromethanopterin) during [Fe]-hydrogenase-catalyzed reactions [66]. This O_2-inactivation process forming H_2O_2 implies that a hydride is formed during [Fe]-hydrogenase reaction. In a proposed H_2-activating form of [Fe]-hydrogenase, H_2 could be first bound at the open site where a water molecule is bound in the resting form, and then H_2 should be heterolytically cleaved to a proton and a hydride with the support of the 2-hydroxy group at the open site.

from the fact that both Cu(I) and H_2O_2 are irreversible inhibitors (half-maximal inactivation rates were found in the cases of using 1 μM of Cu(I) and 20 μM of H_2O_2, respectively) [64]. More recently, inhibition of [Fe]-hydrogenase with O_2 has been intensively studied by using different gaseous conditions with H_2 and N_2 [66]. This study revealed that the O_2-induced inhibition of [Fe]-hydrogenase was caused by O_2 reduction to H_2O_2 (whose inhibition mechanism is explained above) [64] with an external reducing equivalent, most likely a hydride, rather than redox change in the iron of [Fe]-hydrogenase. Although direct evidence of the hydride is not available yet, further attempts to capture it on [Fe]-hydrogenase can be evoked by this O_2-inhibition study. Moreover, it is curious to compare the inhibitory mechanisms by reactive oxidative species (ROS) for the three distinct hydrogenases [67]. Unveiling the underlying mechanisms may be useful for the design of hydrogen-based metal catalysts that can tolerate ROS-induced inactivation.

2.3.5 Theoretical approaches for understanding [Fe]-hydrogenase catalytic mechanism

Theoretical chemistry has been used to examine the catalytic mechanism of [Fe]-hydrogenase. For example, density functional theory (DFT) calculation was conducted for the FeGP cofactor model, omitting the GMP and dimethyl substituents of the pyridinol [68]. Two possible pathways for hydrogen activation were demonstrated involving a thiolate ligand or a hydroxy group of the pyridinol as a base for heterolytic cleavage of H_2, although methenyl-tetrahydromethanopterin was found to be crucial for hydrogen activation. Another DFT calculation mentioned that protonation of the thiolate could influence ligand binding despite the lack of methenyl-tetrahydromethanopterin in the calculation [69]. Then, another research group introduced a dispersion interaction between methenyl-tetrahydromethanopterin and the FeGP cofactor model into the DFT calculation [70]. This resulted in a lower-energy barrier in the step of hydride transfer. The DFT calculation technique was used not only for understanding hydrogen-binding and activation mechanisms of [Fe]-hydrogenase, but also for the design of its mimetic catalysts these days [71]. Molecular dynamics (MD) calculation and/or quantum-mechanics/molecular-mechanics (QM/MM) calculation were alternative approaches that can be favorably applied to [Fe]-hydrogenase with methenyl-tetrahydromethanopterin, rather than just surroundings of the FeGP cofactor [72]. For example, MD and QM/MM calculations of [Fe]-hydrogenase demonstrated that His14 could serve as a base to deprotonate 2-hydroxy group of the pyridinol [72], which could be related to the largely decreased activity of [Fe]-hydrogenase H14A variant [31]. A study of theoretical Mössbauer spectroscopy was also reported to examine the possibility of discriminating different Fe states of some mononuclear Fe-carbonyl complexes as FeGP cofactor models [73].

These theoretical studies have given insights into the mechanism of [Fe]-hydrogenase. However, it is still elusive how the thiolate and/or 2-hydroxy group in the deprotonated/protonated forms are involved in hydrogen activation by [Fe]-hydrogenase. Further experimental approaches that have been utilized for [NiFe]-, [FeFe]-hydrogenases and other metalloproteins (e.g., subatomic high-resolution X-ray crystallography [74], neutron crystallography [75, 76] and advanced spectroscopy [77–80]) may be necessary for understanding the detailed mechanisms of [Fe]-hydrogenase by direct observation of a hydride, a proton and a hydrogen molecule bound to [Fe]-hydrogenase.

2.3.6 Biosynthesis of FeGP cofactor

Structural confirmation of the FeGP cofactor motivated biochemists to study how this cofactor, regarded as a naturally occurring organometallic compound, can be produced in a cellular aqueous environment by biosynthetic enzymes, because traditional organometallic compounds are unstable in water. The first attempt in studying the biosynthesis used stable isotope (^{13}C and ^{2}H) labeled precursors for the FeGP cofactor in *Methanothermobacter marburgensis* and *Methanobrevibacter smithii* [81]. The resulting labeling patterns in combination with a retrobiosynthetic approach gave clues to the FeGP cofactor biosynthetic pathways. For example, the pyridinol-ring should be uniquely biosynthesized by unknown reactions with hypothetical precursors, for example, conjugation/elimination of β-alanine or aspartate with 2,3-dihydroxy-4-oxo-pentanoate or 6-deoxy-5-ketoallulose-1-phosphate [82, 83] by unknown enzyme(s). The labeling pattern of the GMP moiety was identical to that found in the canonical biosynthetic pathway of methanogenic archaea. The CO ligands and the acyl carbon of the FeGP cofactor were labeled with CO_2 and CO, but not the C-1 and C-2 of acetate or pyruvate. This result can exclude at least one possibility that the CO ligands are derived from L-tyrosine, which comes from pyruvate, as in the case of [FeFe]-hydrogenase [84, 85]. However, more details about the origins of CO ligands and acyl carbon are still obscure because there remain several possibilities of CO formation [86] as well as possible scrambling of acyl- and CO ligands as model complexes [87]. More importantly, the 3-methyl substituent of the pyridinol came from L-methionine-(methyl-^{13}C) most probably via *S*-adenosylmethionine (SAM), suggesting that a SAM-utilizing enzyme could be involved in the FeGP cofactor biosynthesis.

The FeGP cofactor biosynthetic genes and enzymes have been sought for by comparison of genomes of methanogenic archaea, which resulted in the identification of seven conserved genes neighboring the [Fe]-hydrogenase-encoding gene, namely, *hmd* gene, in many of the methanogenic archaea (Figure 2.35) [16]. Thus, the seven conserved genes are designated *hmd*-co-occurring genes (*hcgA-G* genes) and considered to encode the biosynthetic enzymes. Genetic study by using gene

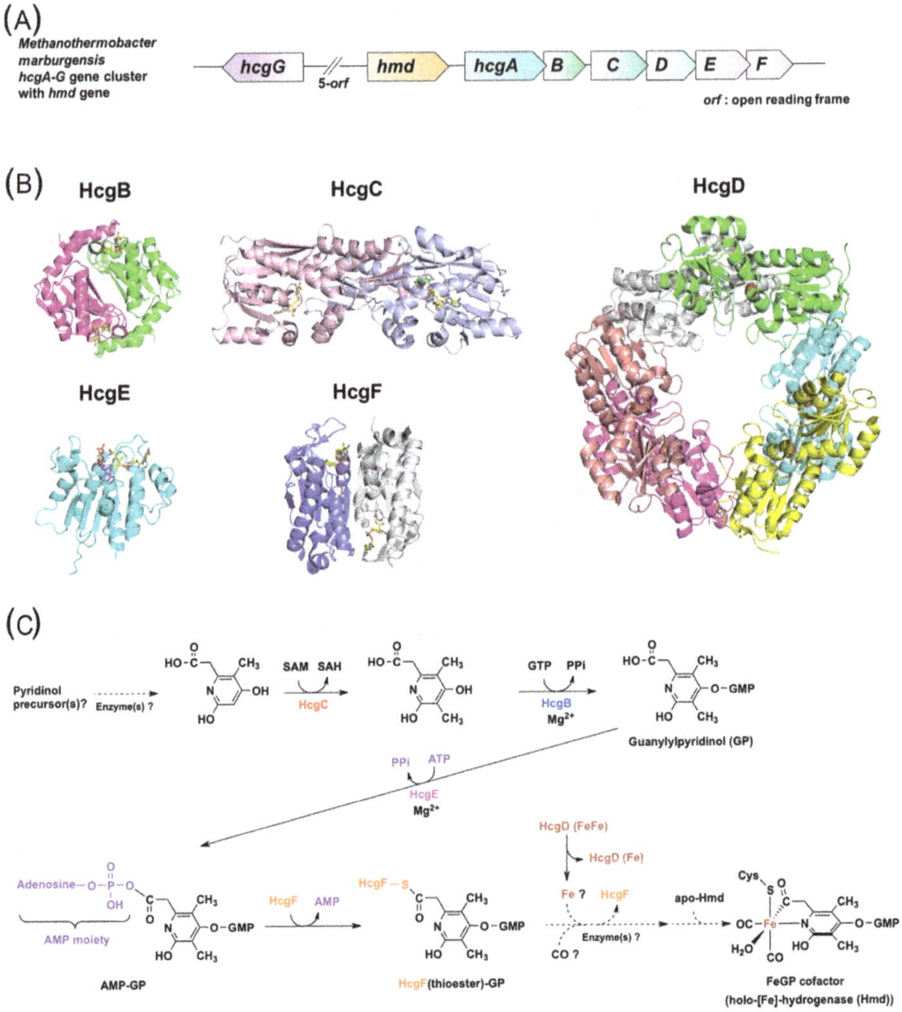

Figure 2.35: (A) The *hcgA-G* gene cluster and *hmd* gene in *Methanothermobacter marburgensis* [16]. (B) Structures of HcgB, HcgC, HcgD, HcgE and HcgF proteins (PDB ID: 3WB0, 5O4J, 3WSE, 3WV9 and 3WVC, respectively) [90, 93, 94, 96]. Guanylylpyridinol (GP) in HcgB, HcgE and HcgF; 6-carboxymethyl-2,4-dihydroxy-5-methylpyridine in HcgC and ATP in HcgE are shown as stick models. Fe ions in HcgD are shown as spheres. (C) A proposed biosynthetic pathway for the FeGP cofactor based on biochemical analyses of HcgB–HcgF. The structure-based functions of HcgB, HcgC, HcgE and HcgF have been characterized, whereas HcgA and HcgG are still under investigation. HcgD is likely to be an iron chaperon, which may provide one Fe ion to the FeGP cofactor biosynthesis. Also, there are still some uncharacterized biosynthetic substrates and reactions: for example, the pyridinol precursor(s) and a key Fe center formation reaction including CO formation. PPi indicates a pyrophosphate ion.

deletions of *hcgA-G*, *hmd* and *hmdII* gene (*hmdII* encodes HmdII, a [Fe]-hydrogenase paralog, described above) of *Methanococcus maripaludis* also demonstrated that *hcg* genes as well as *hmd* were important for the growth [35]. By contrast, the gene deletion of *hmdII* did not affect the growth.

HcgA is a member of radical-*S*-adenosylmethionine (radical-SAM) enzymes, which harbor at least one [4Fe-4S] cluster coordinated by the conserved Cys-rich motif (CX_3CX_2C or CX_4CX_2C motif) as their active sites [41]. However, HcgA contains a unique Cys-rich motif, CX_5CX_2C, which has not been found in the other radical SAM enzymes. An *E. coli* recombinant HcgA was used for conversion of SAM to 5′-deoxyadenosine in its [4Fe-4S] cluster-reconstituted form [88]. However, the function of HcgA in the FeGP cofactor biosynthesis is not known yet, although some radical SAM enzyme-catalyzed reactions in cofactor biosynthesis [89] might be useful when considering a possible HcgA function. HcgG, annotated as a fibrillarin-like protein [16], was also briefly mentioned in the HcgA study [88], because the *hcgA* and *hcgG* genes tend to be clustered beside *hmd* gene rather than the other *hcg* genes, although HcgG has not been structurally or functionally characterized so far.

A breakthrough approach, a so-called structure-to-function analysis, was used for studying Hcg proteins, in which three-dimensional structures of Hcg proteins are used to predict and then characterize their functions by referring to a structural database (Figure 2.35). Such functional analysis was first conducted on HcgB [90, 91]. HcgB was structurally characterized by a structural genomics project (PDB ID: 3BRC), although its function was unknown. However, HcgB was found to show some three-dimensional structural similarity to the active site of nucleoside-triphosphate phosphatases (NTPases) by searching for functionally characterized HcgB homologs. Such structural similarity suggests that HcgB could also use an NTP. Based on this idea and the fact that GMP moiety of the FeGP cofactor comes from the canonical pathway according to the stable isotope labeling experiment, HcgB has been applied for guanylyltransferase activity assay using GTP and several commercially available pyridinols. This biosynthetic model reaction by HcgB demonstrated that HcgB can catalyze guanylylpyridinol formation reaction using GTP and pyridinol [90, 91]. Afterward, it was demonstrated that the pyridinol derived from GP can be also a natural substrate for HcgB [91]. The X-ray crystal structures of HcgB with a pyridinol substrate and GP-type products confirmed that HcgB can bind to these ligands with specific interactions at its active site formed by the dimeric interface [90, 91].

This structure-guided functional analysis for HcgB was a successful example of using a combination of biochemistry, X-ray crystallography and structural bioinformatics. This strategy was applied to further characterize some other Hcg proteins. HcgC was predicted to have a Rossman-like domain, which is a well-known nucleotide-binding domain. Based on this information, the function of HcgC was investigated using nucleotide-containing compounds [91–93]. Consequently, HcgC functions as a SAM-dependent methyltransferase for stereoselective modification at the 3-position of the substrate pyridinol with the SAM-derived methyl group, which is consistent with a unique ^{13}C-labeling on 3-methyl group of the pyridinol [81]. HcgD is annotated as a member of the Nif3 protein superfamily. *E. coli* recombinant HcgD had two Fe ions at its metal-binding site [94]. An additional feature of HcgD is that one of its irons can be removed by treatment with chelating reagents. A current hypothetical role of HcgD is being an iron chaperon to deliver one iron to the FeGP cofactor biosynthesis, by considering the facts that (1) HcgD-type proteins as members of the Nif3 protein family [95] are widely found in other metabolic processes and (2) *hcgD* deletion is less critical than other *hcg* deletions in *Methanococcus maripaludis* [35].

More importantly, HcgE and HcgF have been characterized as key enzymes involved in the activation of the carboxy group of GP toward formation of a unique acyl-ligand of FeGP cofactor [96]. HcgE is homologous to the ATP-dependent ubiquitin-activating E1 adenylyltransferase enzyme (E1 enzyme) family. This type of enzymes usually catalyzes the attachment of AMP moiety to the carboxy group at the C-terminus of the protein substrates (ubiquitin-like small proteins) [97, 98]. However, there are no ubiquitin-like proteins among the Hcg proteins. Considering the adenylylation chemistry for the carboxy group by this type of enzymes [97, 98], an attempt was made to cocrystallize HcgE with both ATP and GP, which contains a carboxy group [96]. Surprisingly, both ATP and GP were bound adjacent in the HcgE-active site cleft, although the other members of the E1 enzyme family use ubiquitin-like proteins as substrates. Finally, biochemical assay using HcgE, ATP and GP confirmed that HcgE is an adenylyltransferase to form AMP-GP. HcgF was also studied by structure-based functional analysis, showing structural similarity to a nicotinamide-mononucleotide-utilizing PnnC enzyme [99]. The structure of HcgF and GP cocrystal revealed that the GP was bound to the HcgF cavity, and its carboxy group was unprecedently converted to a thioester with a covalent link to the conserved Cys9. This finding of the HcgF-driven thioester formation of GP as well as the HcgE-catalyzed adenylylation were important for revealing the acyl formation mechanism in FeGP cofactor biosynthesis. In fact, the acyl–iron structure of some synthetic FeGP cofactor models have been synthesized by oxidative addition of thioester to Fe(0) [100, 101]. If the same thioester chemistry occurs in the FeGP cofactor biosynthesis, it is important to confirm whether Fe(0), an extremely reduced state of Fe, could be really formed. In addition, one should also consider whether monoiron carbonyl species can exist in aqueous cellular environment, because these water-insoluble Fe–carbonyl species [49] may have to be handled in the proteins' hydrophobic environments. Clearly, the biosynthesis of FeGP cofactor includes novel bio-organometallic chemistry to achieve the challenging synthesis of its acyl–iron structure in aqueous environments.

2.3.7 Synthetic structural models of [Fe]-hydrogenase

Synthetic chemists have made great effort to prepare structural and/or functional models of the FeGP cofactor, not only for understanding its structure–function relationships but also for application of the models to H_2-based catalytic hydrogenation systems [102–104]. To the FeGP cofactor models, the unique iron six-coordination structure including a unique acyl–iron has been synthesized, except for the hydrophilic GMP moiety. Early works focusing on the acyl–iron structural models of the FeGP cofactor successfully demonstrated the following two synthetic strategies: syntheses of the acyl (–C(=O)–)-ligated Fe–carbonyl complex (which is not the "acylmethyl (–C(=O)–CH_2–)" that was also found in the FeGP cofactor) [100, 101, 105, 106], and the carbamoyl (–C(=O)–NH–)-ligated Fe–carbonyl complexes (Figure 2.36) [106–109]. The thioester-mediated acyl–iron formation for synthesizing the former complexes is biomimetic to formation of GP-thioester-HcgF in the FeGP cofactor biosynthesis [96]. The latter model is a facile way to form the five-membered-ring of Fe–C(=O)–NH–pyridinol, like the Fe–acylmethyl (Fe–C(=O)–CH_2–)–pyridinol in the FeGP cofactor. Curiously, a Fe-hydrido species with a carbamoyl ligand has recently been reported by using octahedral six-coordinated carbamoyl model with $NaHBEt_3$ [109]. This hydride formation is not biomimetic, but it may provide access to hydride-bound mononuclear iron complexes with different ligand sets in organometallic chemistry.

To create structures more similar to the original FeGP cofactor, a bidentate pyridinol ligand with an acylmethyl group has been introduced to the synthetic models (Figure 2.37) [65, 87, 104, 110–116]. In this type of FeGP cofactor structural models, not only the unique FeGP cofactor geometry, but also various 2-substituent R groups (R = OH, OCH_3, etc.) of the pyridinol ligand were achieved, and some of them were further used for reconstitution experiments to create "semisynthetic" [Fe]-hydrogenase (see below) [49]. Also, five-coordination FeGP cofactor models have been developed and studied in detail [65, 117–123]. These five-coordination structures were structurally more labile with regard to, for example, dimerization and ligand exchanges, but they are not solely reactive with H_2. This result was relevant to the active site environments where the FeGP cofactor, methenyl-tetrahydromethanopterin and amino acid residues are appropriately positioned for H_2 catalysis.

2.3.8 Synthetic functional [Fe]-hydrogenase models as hydrogenation catalysts with H_2

Based on the proposed catalytic mechanism of [Fe]-hydrogenase, synthetic chemists have also considered the synthesis of FeGP cofactor functional models (Figure 2.38). Although these synthetic models have low structural similarity or use metals other than iron, they retain the key feature of [Fe]-hydrogenase-like activation of H_2. For

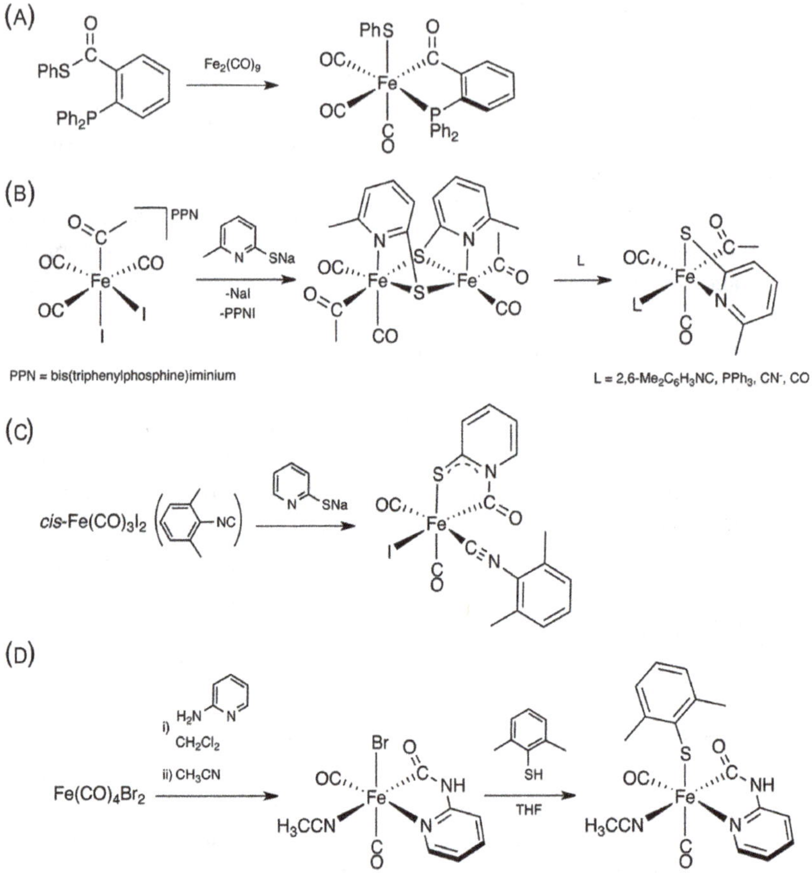

Figure 2.36: Synthesis of FeGP cofactor structural models: (A) with an acyl–iron structure via thioester activation [100, 101], (B) with an acetyl–iron structure [105] and (C) with an acyl–iron structure presumably via substitution of one CO by pyridinol-N, followed by insertion of the CO to pyridinol–N–Fe bond [106]. (D) A carbamoyl-containing model synthesis [107].

example, metal (e.g., Ir and Ru) complexes using a 2-hydroxypyridine moiety [71] were synthesized and used as hydrogen-utilizing catalysts [124–127]. These complexes can activate H_2 via their heterolytic cleavage at the spot between 2-OH and the metal center and catalyze reduction of CO_2 to HCOOH reversibly.

Another focus for the FeGP cofactor functional models is the role of methenyl-tetrahydromethanopterin as a Lewis acid in [Fe]-hydrogenase catalysis. Thus, intermolecular [Fe]-hydrogenase-like catalytic systems were demonstrated [128] based on a frustrated Lewis acid and base pair concept [129]. A "frustrated" word in this concept is derived from the fact that a formation of an adduct between Lewis acid and base in a classical donor–acceptor interaction can be sterically precluded by bulky groups of the Lewis acid and base. Moreover, this concept has also been

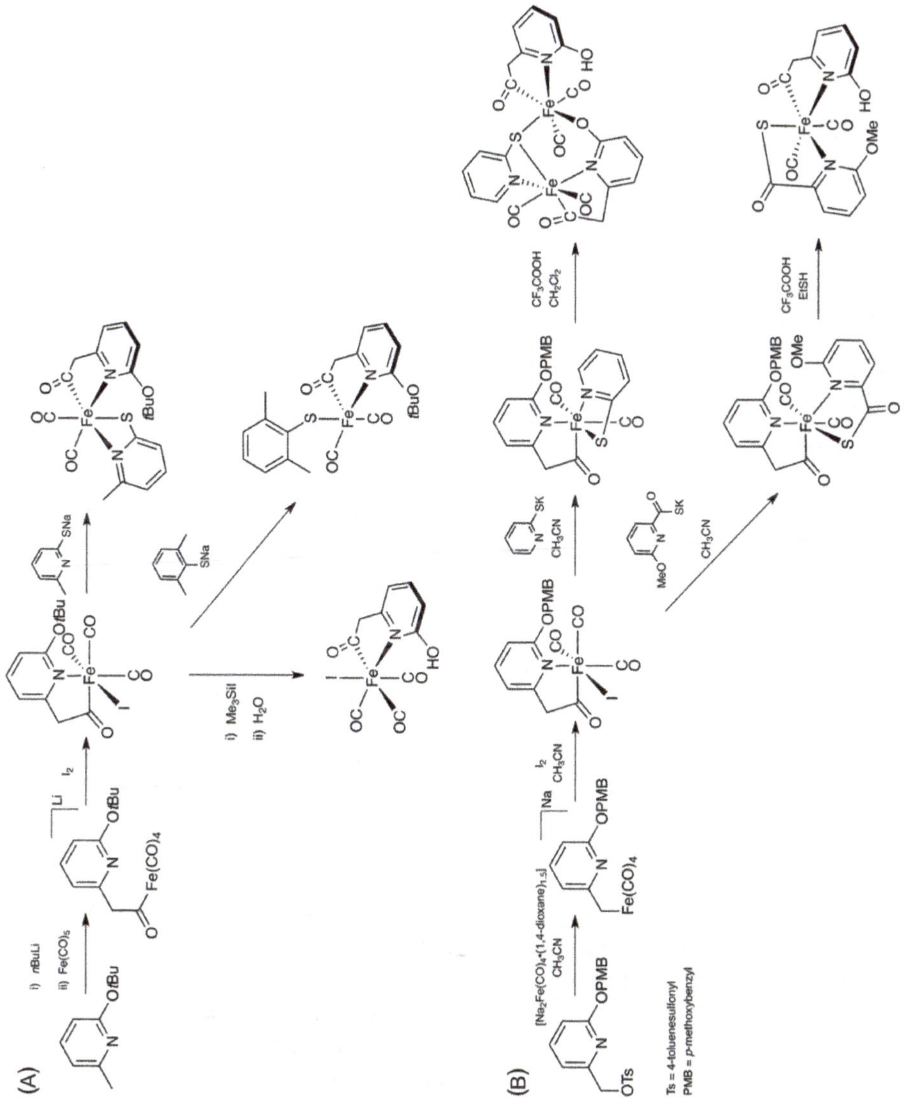

Figure 2.37: Two strategies for the synthesis of acylmethyl–pyridinol ligand-containing FeGP cofactor models by (A) using a neutral Fe-carbonyl with 6-methyl group of the pyridine and a strong base [87, 110] and (B) using an anionic Fe-carbonyl with 6-(4-toluenesulfonyl)methyl group of the pyridine [112, 113].

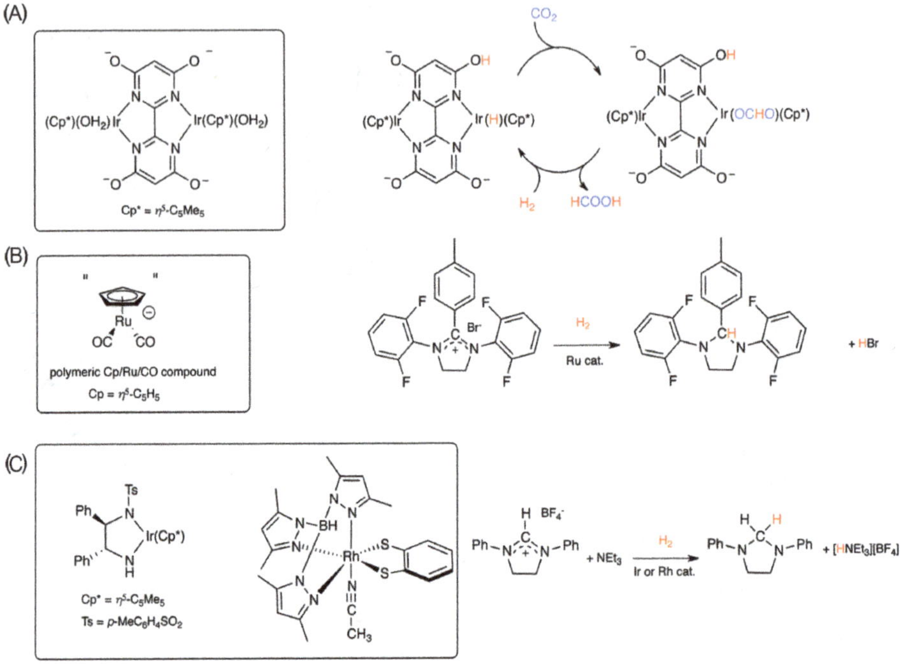

Figure 2.38: Three functional FeGP cofactor models. (A) A binuclear Ir complex showing 2-hydroxypyridine-N ligation to Ir [124]. Frustrated Lewis acid and base pair-based functional FeGP cofactor model systems using (B) an Ru compound [128] and (c) Ir or Rh complex [130]. These were inspired by a proposed methenyl-tetrahydromethanopterin-dependent [Fe]-hydrogenase mechanism for H_2 activation.

applied for making even "metal-free" hydrogen-activating catalysts in their inter- and/or intramolecular systems so far. Combination of $[RuCp(CO)_2]^-$ (Cp = C_5H_5) complex or its related polymeric Cp/Ru/CO compound with various imidazolinium salts were utilized as a Lewis acid and a base, respectively, to cause the heterolytic cleavage of H_2. This concept was further employed for pairs of different organometallic Ru or Ir complexes and an *N,N'*-diphenylimidazolinium cation [33, 130].

2.3.9 Synthetic structural and functional models of [Fe]-hydrogenase

Very recently, functional and structural models have been synthesized and demonstrated activity for heterolytic cleavage of H_2. A type of bidentate acylmethylpyridinol-containing FeGP cofactor model has been developed, which has a bidentate ligand $Et_2PCH_2N(Me)CH_2PEt_2$ (PNP) as a base for hydrogen catalysis (Figure 2.39A) [131, 132]. In this PNP-bridging model, the PNP-nitrogen and the 2-methoxy group sandwich one

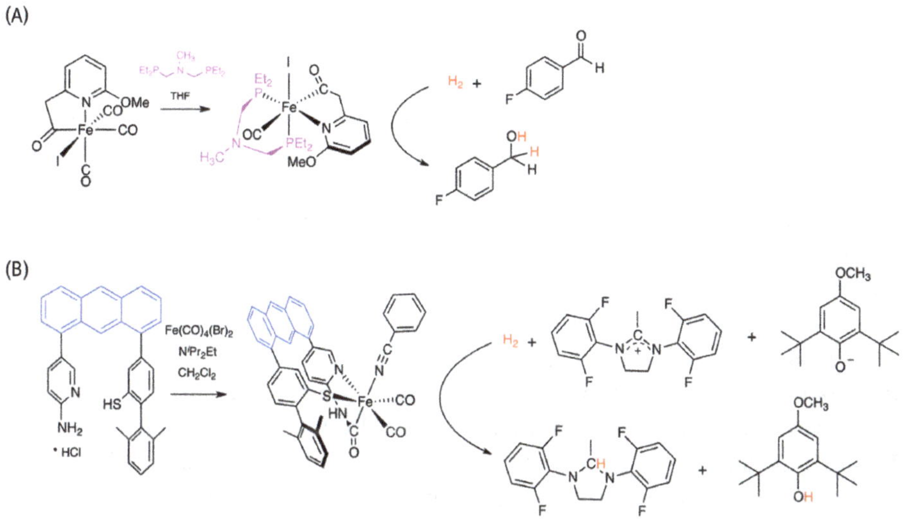

Figure 2.39: Structural and functional FeGP cofactor models. (A) Synthesis of an acylmethylpyridinol FeGP cofactor model featuring a $Et_2PCH_2N(Me)CH_2PEt_2$ (PNP) ligand [132]. This PNP- and acylmethylpyridinol-ligands-containing FeGP cofactor model catalyzes hydrogenation of aldehyde to alcohol with H_2. (B) Synthesis of a FeGP cofactor model having a carbamoylpyridine ligand featuring an anthracene scaffold [134]. This complex could activate H_2 in a manner similar to the functional Ru-complexes in a frustrated Lewis acid and base pair strategy [128, 130].

CO ligand *trans* to the acyl-ligand. This model can function in H_2/D_2 exchange and hydrogenation based on a proposed mechanism involving ligand-exchange processes and heterolytic cleavage of H_2 at the PNP-nitrogen site. The PNP of this FeGP cofactor model is somewhat structurally and functionally analogous to the bridging ligand (–S–CH_2–NH–CH_2–S–) of the H-cluster of [FeFe]-hydrogenase, despite the structural difference between *N*-methyl in this FeGP cofactor model and NH of the H-cluster [133]. Another structural and functional FeGP cofactor model has been just reported featuring a bulky anthracene moiety-containing carbamoyl-pyridine ligand (Figure 2.39B) [134]. This FeGP cofactor model could work as a hydrogen-activating catalyst with NEt_4 [MeO*t*Bu_2PhO] as an exogenous proton acceptor. The generated hydride on the Fe center was transferred to 2,6-difluoro(phenyl)-2-(4-methyl)imidazolium cation (CH_3Im^+), which is analogous to methenyl-tetrahydromethanopterin, as demonstrated in some non-Fe-type functional models using an imidazolinium cation [128, 130].

2.3.10 Semisynthetic [Fe]-hydrogenase using model complexes and its apo-protein

A hybrid of synthetic model complexes and apo-proteins has recently been attracting attention, because this so-called semisynthetic approach enables not only investigating the native functions of metallic active sites in metalloenzymes [133–137], but also the creation of artificial metalloenzymes [138–140]. For example, the bridging ligand of the H-cluster in [FeFe]-hydrogenase was studied by reconstitution of H-cluster-free-[FeFe]-hydrogenase with synthetic H-cluster models having a series of bridging ligands (–S–CH_2–X–CH_2–S–, X = CH_2, NH, O). Among them, only X = NH is an active form [133, 136]. Also, this approach has recently been applied to study the roles of the FeGP cofactor of [Fe]-hydrogenase (Figure 2.40) [49]. Reconstitution of apo-[Fe]-hydrogenase was performed with the synthetic FeGP cofactor models having the 6-acylmethyl-pyridine with 2-hydroxy or 2-methoxy group as a ligand. The reconstituted [Fe]-hydrogenases were investigated by catalytic activity assay as native [Fe]-hydrogenase, in order to identify the role of 2-substituent of the pyridine ligand. In the results, only the 2-hydroxy group-containing model showed activity (albeit being relatively low). This finding certainly indicated that the 2-hydroxy group is important for the catalysis, which has been suggested by model and theoretical studies. This study of chemical reconstitution of the synthetic FeGP cofactor models to apo-[Fe]-hydrogenase may help future investigation in the roles of other moieties of the FeGP cofactors, such as the two methyl groups at 3- and 5-positions and the 4-GMP moiety of the pyridine ligand.

2.3.11 Closing remarks

In this chapter, a wide range of topics about [Fe]-hydrogenase have been summarized from different perspectives, including organometallic chemistry and biochemistry. Many important and attractive features of [Fe]-hydrogenase have been explored and unveiled over the decades, especially in the last 10 years. Nevertheless, there are still questions about the mechanisms of [Fe]-hydrogenase: for example, how the hydrogen is activated and how the hydride-bound intermediate is formed. For the organometallic FeGP cofactor biosynthesis, issues remain about the biosynthesis of the unique organometallic acyl–iron as well as the pyridinol structures. These aspects may be further studied by focusing on Hcg proteins and/or seeking other new possible biosynthetic enzymes. Furthermore, there is the hope to apply synthetic FeGP cofactor models for H_2-utilizing catalytic systems, by referring to not only the revealed [Fe]-hydrogenase mechanisms but also strategies from other fields or combinations of different approaches like the “semisynthetic” approach. Bio-organometallic chemistry, a multidisciplinary research field integrating areas including biochemistry, organometallics, structural biology and so on, is going to promote a drastic breakthrough toward the comprehensive understanding of [Fe]-hydrogenase.

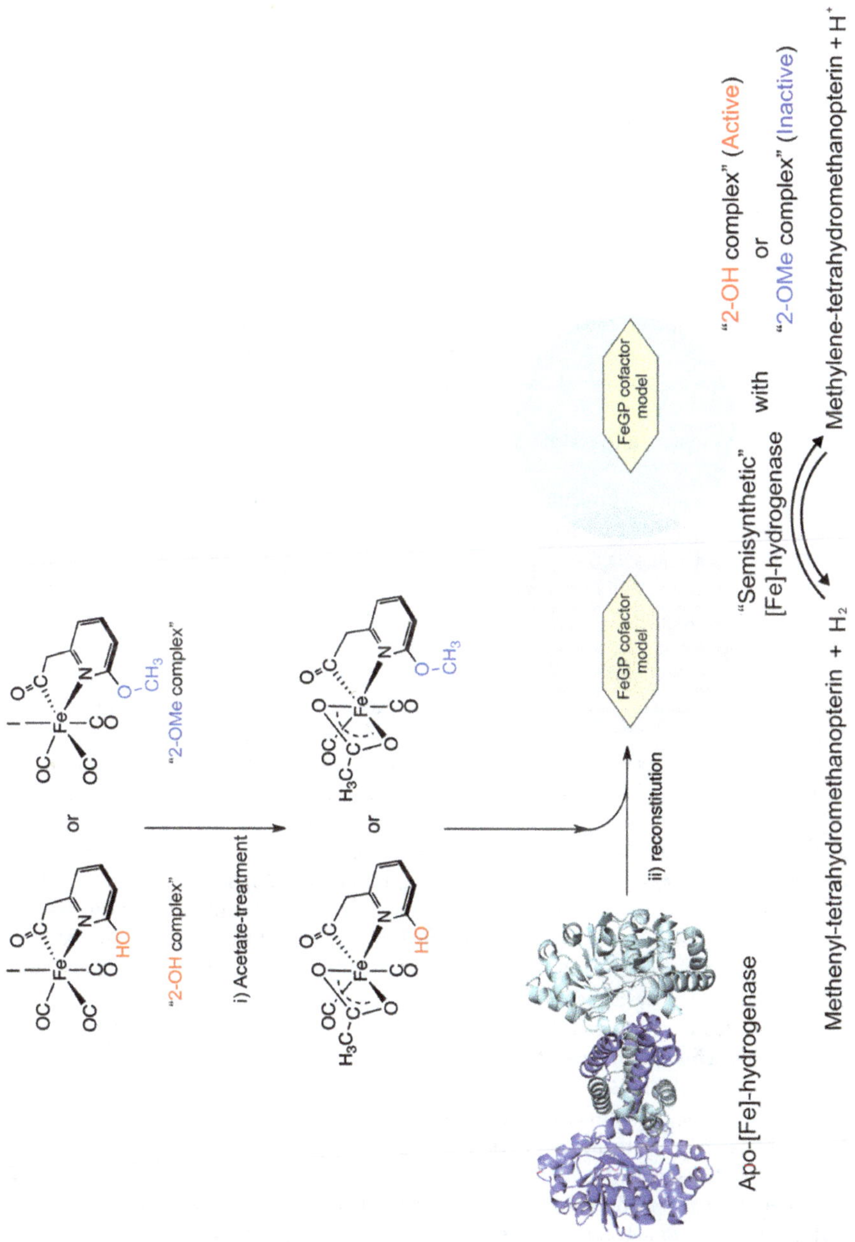

Figure 2.40: An approach for creating "semisynthetic" [Fe]-hydrogenase by reconstitution of apo-[Fe]-hydrogenase with FeGP cofactor models [49]. The model complexes were treated with acetate [32, 110] to make them water soluble. Then, the water-soluble ones were reconstituted to apo-[Fe]-hydrogenase produced in *Escherichia coli* [5, 31].

Acknowledgment: The author thanks Dr Seigo Shima for the opportunity to study [Fe]-hydrogenase and related subjects, especially the FeGP cofactor biosynthesis, during his early research career in his lab, with generous supports, inspiring discussions and fruitful suggestions. The author also thanks all the coworkers and colleagues during that time for their supports and kindness.

References

[1] Fontecilla-Camps, JC., Volbeda, A., Cavazza, C., and Nicolet, Y. Structure/function relationships of [NiFe]- and [FeFe]-hydrogenases. Chem Rev 2007, 107, 4273–303.

[2] Lubitz, W., Ogata, H., Rüdiger, O., and Reijerse, E. Hydrogenases. Chem Rev 2014, 114, 4081–4148.

[3] Shima, S., and Thauer, RK. A third type of hydrogenase catalyzing H_2 activation. Chem Rec 2007, 7, 37–46.

[4] Shima, S., and Ermler, U. Structure and function of [Fe]-hydrogenase and its iron-guanylylpyridinol (FeGP) cofactor. Eur J Inorg Chem 2011, 2011, 963–72.

[5] Hiromoto, T., Ataka, K., Pilak, O., Vogt, S., Salomone, M., Meyer-Klaucke, W., Warkentin, E., Thauer, RK., Shima, S., and Ermler, U. The crystal structure of C176A mutated [Fe]-hydrogenase suggests an acyl-iron ligation in the active site iron complex. FEBS Lett 2009, 583, 585–90.

[6] Afting, C., Hochheimer, A., and Thauer, RK. Function of H_2-forming methylenetetrahydromethanopterin dehydrogenase from *Methanobacterium thermoautotrophicum* in coenzyme F_{420} reduction with H_2. Arch Microbiol 1998, 169, 206–10.

[7] Afting, C., Kremmer, E., Brucker, C., Hochheimer, A., and Thauer, RK. Regulation of the synthesis of H_2-forming methylenetetrahydromethanopterin dehydrogenase (Hmd) and of HmdII and HmdIII in *Methanothermobacter marburgensis*. Arch Microbiol 2000, 174, 225–32.

[8] Vignais, PM., and Billoud, B. Occurrence, classification, and biological function of hydrogenases: an overview. Chem Rev 2007, 107, 4206–72.

[9] Greening, C., Biswas, A., CR, Carere., CJ, Jackson., Taylor, MC., MB, Stott., GM, Cook., and SE, Morales. Genomic and metagenomic surveys of hydrogenase distribution indicate H_2 is a widely utilised energy source for microbial growth and survival. ISME J 2016, 10, 761–77.

[10] Søndergaard, D., Pedersen, CNS., and Greening, C. HydDB: A web tool for hydrogenase classification and analysis. Sci Rep 2016, 6, 34212.

[11] Zirngibl, C., Hedderich, R., and Thauer, RK. N^5, N^{10}-Methylenetetrahydromethanopterin dehydrogenase from *Methanobacterium thermoautotrophicum* has hydrogenase activity. FEBS Lett 1990, 261, 112–6.

[12] Ma, K., Zirngibl, C., Linder, D., Stetter, KO., and Thauer, RK. N^5, N^{10}-Methylenetetrahydromethanopterin dehydrogenase (H_2-forming) from the extreme thermophile *Methanopyrus kandleri*. Arch Microbiol 1991, 156, 43–8.

[13] Schwörer, B., and Thauer, RK. Activities of formylmethanofuran dehydrogenase, methylenetetrahydromethanopterin dehydrogenase, methylenetetrahydromethanopterin reductase, and heterodisulfide reductase in methanogenic bacteria. Arch Microbiol 1991, 155, 459–65.

[14] Hartmann, GC., Santamaria, E., Fernández, VM., and Thauer, RK. Studies on the catalytic mechanism of H_2-forming methylenetetrahydromethanopterin dehydrogenase: *para-ortho* H_2 conversion rates in H_2O and D_2O. J Biol Inorg Chem 1996, 1, 446–50.

[15] Thauer, RK., Kaster, AK., Seedorf, H., Buckel, W., and Hedderich, R. Methanogenic archaea: ecologically relevant differences in energy conservation. Nat Rev Microbiol 2008, 6, 579–91.
[16] Thauer, RK., Kaster, AK., Goenrich, M., Schick, M., Hiromoto, T., and Shima, S. Hydrogenases from methanogenic archaea, nickel, a novel cofactor, and H_2 storage. Annu Rev Biochem 2010, 79, 507–36.
[17] Shima, S., Warkentin, E., Grabarse, W., Sordel, M., Wicke, M., Thauer, RK., and Ermler, U. Structure of coenzyme F_{420} dependent methylenetetrahydromethanopterin reductase from two methanogenic archaea. J Mol Biol 2000, 300, 935–50.
[18] Ceh, K., Demmer, U., Warkentin, E., Moll, J., Thauer, RK., Shima, S., and Ermler, U. Structural basis of the hydride transfer mechanism in F_{420}-dependent methylenetetrahydromethanopterin dehydrogenase. Biochemistry 2009, 48, 10098–105.
[19] Mills, DJ., Vitt, S., Strauss, M., Shima, S., and Vonck, J. De novo modeling of the F_{420}-reducing [NiFe]-hydrogenase from a methanogenic archaeon by cryo-electron microscopy. eLife 2013, 2, e00218.
[20] Vitt, S., Ma, K., Warkentin, E., Moll, J., Pierik, AJ., Shima, S., and Ermler, U. The F_{420}-reducing [NiFe]-hydrogenase complex from *Methanothermobacter marburgensis*, the first X-ray structure of a group 3 family member. J Mol Biol 2014, 426, 2813–26.
[21] Zirngibl, C., Van Dongen, W., Schwörer, B., Von Bünau, R., Richter, M., Klein, A., and Thauer, RK. H_2-forming methylenetetrahydromethanopterin dehydrogenase, a novel type of hydrogenase without iron-sulfur clusters in methanogenic archaea. Eur J Biochem 1992, 208, 511–20.
[22] Buurman, G., Shima, S., and Thauer, RK. The metal-free hydrogenase from methanogenic archaea: evidence for a bound cofactor. FEBS Lett 2000, 485, 200–4.
[23] Lyon, EJ., Shima, S., Buurman, G., Chowdhuri, S., Batschauer, A., Steinbach, K., and Thauer, RK. UV-A/blue-light inactivation of the 'metal-free' hydrogenase (Hmd) from methanogenic archaea – The enzyme contains functional iron after all. Eur J Biochem 2004, 271, 195–204.
[24] Shima, S., Lyon, EJ., Sordel-Klippert, M., Kauß, M., Kahnt, J., Thauer, RK., Steinbach, K., Xie, X., Verdier, L., and Griesinger, C. The cofactor of the iron-sulfur cluster free hydrogenase Hmd: structure of the light-inactivation product. Angew Chem Int Ed 2004, 43, 2547–51.
[25] Lyon, EJ., Shima, S., Boecher, R., Thauer, RK., Grevels, FW., Bill, E., Roseboom, W., and Albracht, SPJ. Carbon monoxide as an intrinsic ligand to iron in the active site of the iron-sulfur-cluster-free hydrogenase H_2-forming methylenetetrahydromethanopterin dehydrogenase as revealed by infrared spectroscopy. J Am Chem Soc 2004, 126, 14239–48.
[26] Shima, S., Lyon, EJ., Thauer, RK., Mienert, B., and Bill, E. Mössbauer studies of the iron-sulfur cluster-free hydrogenase: the electronic state of the mononuclear Fe active site. J Am Chem Soc 2005, 127, 10430–35.
[27] Korbas, M., Vogt, S., Meyer-Klaucke, W., Bill, E., Lyon, EJ., Thauer, RK., and Shima, S. The iron-sulfur cluster-free hydrogenase (Hmd) is a metalloenzyme with a novel iron binding motif. J Biol Chem 2006, 281, 30804–13.
[28] Pilak, O., Mamat, B., Vogt, S., Hagemeier, CH., Thauer, RK., Shima, S., Vonrhein, C., Warkentin, E., and Ermler, U. The crystal structure of the apoenzyme of the iron-sulphur cluster-free hydrogenase. J Mol Biol 2006, 358, 798–809.
[29] Guo, Y., Wang, H., Xiao, Y., Vogt, S., Thauer, RK., Shima, S., Volkers, PI., Rauchfuss, TB., Pelmentschikov, V., Case, DA., Alp, EE., Sturhahn, W., Yada, Y., and Cramer, SP. Characterization of the Fe site in iron-sulfur-cluster-free hydrogenase (Hmd) and of a model compound *via* nuclear resonance vibrational spectroscopy (NRVS). Inorg Chem 2008, 47, 3969–77.

[30] Salomone-Stagni, M., Stellato, F., Whaley, CM., Vogt, S., Morante, S., Shima, S., Rauchfuss, TB., and Meyer-Klaucke, W. The iron-site structure of [Fe]-hydrogenase and model systems: an X-ray absorption near edge spectroscopy study. Dalton Trans 2010, 39, 3057–64.
[31] Shima, S., Pilak, O., Vogt, S., Schick, M., Stagni, MS., Meyer-Klaucke, W., Warkentin, E., Thauer, RK., and Ermler, U. The crystal structure of [Fe]-hydrogenase reveals the geometry of the active site. Science 2008, 321, 572–5.
[32] Shima, S., Schick, M., Kahnt, J., Ataka, K., Steinbach, K., and Linne, U. Evidence for acyl-iron ligation in the active site of [Fe]-hydrogenase provided by mass spectrometry and infrared spectroscopy. Dalton Trans 2012, 41, 767–71.
[33] Hiromoto, T., Warkentin, E., Moll, J., Ermler, U., and Shima, S. The crystal structure of an [Fe]-hydrogenase-substrate complex reveals the framework for H_2 activation. Angew Chem Int Ed 2009, 48, 6457–60.
[34] Goldman, AD., Leigh, JA., and Samudrala, R. Comprehensive computational analysis of Hmd enzymes and paralogs in methanogenic Archaea. BMC Evol. Biol. 2009, 9, 199.
[35] Lie, TJ., Costa, KC., Pak, D., Sakesan, V., and Leigh, JA. Phenotypic evidence that the function of the [Fe]-hydrogenase Hmd in *Methanococcus maripaludis* requires seven *hcg* (*hmd* co-occurring genes) but not *hmdII*. FEMS Microbiol Lett 2013, 343, 156–60.
[36] Fujishiro, T., Ataka, K., Ermler, U., and Shima, S. Towards a functional identification of catalytically inactive [Fe]-hydrogenase paralogs. FEBS J 2015, 282, 3412–23.
[37] Ragsdale, SW., and Kumar, M. Nickel-containing carbon monoxide dehydrogenase/Acetyl-CoA synthase. Chem Rev 1996, 96, 2515–39.
[38] Fontecilla-Camps, JC., Amara, P., Cavazza, C., Nicolet, Y., and Volbeda, A. Structure-function relationships of anaerobic gas-processing metalloenzymes. Nature 2009, 460, 814–22.
[39] Boer, JL., Mulrooney, SB., and Hausinger, RP. Nickel-dependent metalloenzymes. Arch Biochem Biophys 2014, 544, 142–52.
[40] Can, M., FA, Armstrong., and SW, Ragsdale. Structure, function, and mechanism of the nickel metalloenzymes, CO dehydrogenase, and Acetyl-CoA synthase. Chem Rev 2014, 114, 4149–74.
[41] Broderick, JB., Duffus, BR., Duschene, KS., and Shepard, EM. Radical *S*-Adenosylmethionine enzymes. Chem Rev 2014, 114, 4229–317.
[42] Horitani, M., Shisler, K., Broderick, WE., Hutcheson, RU., Duschene, KS., Marts, AR., Hoffman, BM., and Broderick, JB. Radical SAM catalysis via an organometallic intermediate with an Fe-[5 '-C]-deoxyadenosyl bond. Science 2016, 352, 822–25.
[43] Dong, M., Kathiresan, V., Fenwick, MK., Torelli, AT., Zhang, Y., Caranto, JD., Dzikovski, B., Sharma, A., Lancaster, KM., Freed, JH., Ealick, SE., Hoffman, BM., and Lin, H. Organometallic and radical intermediates reveal mechanism of diphthamide biosynthesis. Science 2018, 359, 1247–50.
[44] Ribbe, MW., Hu, Y., Hodgson, KO., and Hedman, B. Biosynthesis of nitrogenase metalloclusters. Chem Rev 2014, 114, 4063–80.
[45] Sippel, D., and Einsle, O. The structure of vanadium nitrogenase reveals an unusual bridging ligand. Nat Chem Biol 2017, 13, 956–60.
[46] Brown, KL. Chemistry and enzymology of vitamin B_{12}. Chem Rev 2005, 105, 2075–149.
[47] Desguin, B., Zhang, T., Soumillion, P., Hols, P., Hu, J., and Hausinger, RP. A tethered niacin-derived pincer complex with a nickel-carbon bond in lactate racemase. Science 2015, 349, 66–9.
[48] Shima, S., Schick, M., and Tamura, H. Preparation of [Fe]-hydrogenase from methanogenic archaea. Method Enzymol 2011, 494, 119–37.
[49] Shima, S., Chen, D., Wodrich, MD., Fujishiro, T., Schultz, KM., Kahnt, J., Ataka, K., and Hu, X. Reconstitution of [Fe]-hydrogenase using model complexes. Nat Chem 2015, 7, 995–1002.
[50] Fesseler, J., Jeoung, JH., and Dobbek, H. How the $[NiFe_4S_4]$ cluster of CO dehydrogenase activates CO_2 and NCO^-. Angew Chem Int Ed 2015, 54, 8560–64.

[51] Schwörer, B., Fernandez, VM., Zirngibl, C., and Thauer, RK. H_2-forming N^5, N^{10}-methylenetetrahydromethanopterin dehydrogenase from *Methanobacterium-Thermoautotrophicum* – studies of the catalytic mechanism of H_2 formation using hydrogen isotopes. Eur J Biochem 1993, 212, 255–61.

[52] Klein, AR., Fernández, VM., and Thauer, RK. H_2-forming N^5, N^{10}-methylenetetrahydromethanopterin dehydrogenase: mechanism of H_2 formation analyzed using hydrogen isotopes. FEBS Lett 1995, 368, 203–6.

[53] Klein, AR., Hartmann, GC., and Thauer, RK. Hydrogen isotope effects in the reactions catalyzed by H_2-forming N^5, N^{10}-methylenetetrahydromethanopterin dehydrogenase from methanogenic Archaea. Eur J Biochem 1995, 233, 372–6.

[54] Vogt, S., Lyon, EJ., Shima, S., and Thauer, RK. The exchange activities of [Fe] hydrogenase (iron-sulfur-cluster-free hydrogenase) from methanogenic archaea in comparison with the exchange activities of [FeFe] and [NiFe] hydrogenases. J Biol Inorg Chem 2008, 13, 97–106.

[55] Schleucher, J., Griesinger, C., Schwörer, B., and Thauer, RK. H_2-Forming N^5, N^{10}-methylenetetrahydromethanopterin dehydrogenase from *methanobacterium-thermoautotrophicum* catalyzes a stereoselective hydride transfer as determined by two-dimensional NMR-spectroscopy. Biochemistry 1994, 33, 3986–93.

[56] Schleucher, J., Schwörer, B., Thauer, RK., and Griesinger, C. Elucidation of the stereochemical course of chemical reactions by magnetic labeling. J Am Chem Soc 1995, 117, 2941–2.

[57] Klein, AR., and Thauer, RK. *Re*-face specificity at C14a of methylenetetrahydromethanopterin and *Si*-face specificity at C5 of coenzyme F_{420} for coenzyme F_{420}-dependent methylenetetrahydromethanopterin dehydrogenase from methanogenic Archaea. Eur J Biochem 1995, 227, 169–74.

[58] Vignais, PM. H/D exchange reactions and mechanistic aspects of the hydrogenases. Coordin Chem Rev 2005, 249, 1677–90.

[59] Shima, S., Vogt, S., Göbels, A., and Bill, E. Iron-chromophore circular dichroism of [Fe]-hydrogenase: the conformational change required for H_2 activation. Angew Chem Int Ed 2010, 49, 9917–21.

[60] Adams, MWW., Mortenson, LE., and Chen, JS. Hydrogenase. Biochim Biophys Acta 1980, 594, 105–76.

[61] Arp, DJ., and Burris, RH. Kinetic mechanism of the hydrogen-oxidizing hydrogenase from soybean nodule bacteroids. Biochemistry 1981, 20, 2234–40.

[62] Shima, S., and Ataka, K. Isocyanides inhibit [Fe]-hydrogenase with very high affinity. FEBS Lett 2011, 585, 353–6.

[63] Tamura, H., Salomone-Stagni, M., Fujishiro, T., Warkentin, E., Meyer-Klaucke, W., Ermler, U., and Shima, S. Crystal structures of [Fe]-hydrogenase in complex with inhibitory isocyanides: implications for the H_2-activation site. Angew Chem Int Ed 2013, 52, 9656–9.

[64] Hidese, R., Ataka, K., Bill, E., and Shima, S. Cu^I and H_2O_2 inactivate and Fe^{II} inhibits [Fe]-hydrogenase at very low concentrations. ChemBioChem 2015, 16, 1861–5.

[65] Hu, B., Chen, D., and Hu, X. A pyridinol acyl cofactor in the active site of [Fe]-hydrogenase evidenced by the reactivity of model complexes. Chem Eur J 2012, 18, 11528–30.

[66] Huang, G., Wagner, T., Ermler, U., Bill, E., Ataka, K., and Shima, S. Dioxygen sensitivity of [Fe]-hydrogenase in the presence of reducing substrates. Angew Chem Int Ed 2018, 57, 4917–20.

[67] Stiebritz, MT., and Reiher, M. Hydrogenases and oxygen. Chem Sci 2012, 3, 1739–51.

[68] Yang, XZ., and Hall, MB. Monoiron hydrogenase catalysis: hydrogen activation with the formation of a dihydrogen, $Fe\text{-}H^{\delta-}\cdots H^{\delta+}\text{-}O$, bond and methenyl-$H_4MPT^+$ triggered hydride transfer. J Am Chem Soc 2009, 131, 10901–8.

[69] Dey, A. Density functional theory calculations on the mononuclear non-heme iron active site of Hmd hydrogenase: role of the internal ligands in tuning external ligand binding and driving H_2 heterolysis. J Am Chem Soc 2010, 132, 13892–901.
[70] Finkelmann, AR., Stiebritz, MT., and Reiher, M. Kinetic modeling of hydrogen conversion at [Fe] hydrogenase active-site models. J Phys Chem B 2013, 117, 4806–17.
[71] Moore, CM., Quist, DA., Kampf, JW., and Szymczak, NK. A 3-fold-symmetric ligand based on 2-hydroxypyridine: regulation of ligand binding by hydrogen bonding. Inorg Chem 2014, 53, 3278–80.
[72] Finkelmann, AR., Senn, HM., and Reiher, M. Hydrogen-activation mechanism of [Fe] hydrogenase revealed by multi-scale modeling. Chem Sci 2014, 5, 4474–82.
[73] Hedegård, ED., Knecht, S., Ryde, U., Kongsted, J., and Saue, T. Theoretical ^{57}Fe Mössbauer spectroscopy: isomer shifts of [Fe]-hydrogenase intermediates. Phys Chem Chem Phys 2014, 16, 4853–63.
[74] Ogata, H., Nishikawa, K., and Lubitz, W. Hydrogens detected by subatomic resolution protein crystallography in a [NiFe] hydrogenase. Nature 2015, 520, 571–4.
[75] Casadei, CM., Gumiero, A., Metcalfe, CL., Murphy, EJ., Basran, J., Concilio, MG., Teixeira, SCM., Schrader, TE., Fielding, AJ., Ostermann, A., Blakeley, MP., Raven, EL., and Moody, PCE. Neutron cryo-crystallography captures the protonation state of ferryl heme in a peroxidase. Science 2014, 345, 193–7.
[76] Kwon, H., Smith, O., EL, Raven., and PCE, Moody. Combining X-ray and neutron crystallography with spectroscopy. Acta Crystallogr D Struct Biol 2017, 73, 141–7.
[77] Ogata, H., Krämer, T., Wang, H., Schilter, D., Pelmenschikov, V., van Gastel, M., Neese, F., Rauchfuss, TB., Gee, LB., Scott, AD., Yoda, Y., Tanaka, Y., Lubitz, W., and Cramer, SP. Hydride bridge in [NiFe]-hydrogenase observed by nuclear resonance vibrational spectroscopy. Nat Commun 2015, 6, 7890.
[78] Reijerse, EJ., Pham, CC., Pelmenschikov, V., Gilbert-Wilson, R., Adamska-Venkatesh, A., Siebel, JF., Gee, LB., Yoda, Y., Tamasaku, K., Lubitz, W., Rauchfuss, TB., and Cramer, SP. Direct observation of an iron-bound terminal hydride in [FeFe]-hydrogenase by nuclear resonance vibrational spectroscopy. J Am Chem Soc 2017, 139, 4306–9.
[79] DW, Mulder., YS, Guo., MW, Ratzloff., and PW, King. Identification of a catalytic iron-hydride at the H-cluster of [FeFe]-hydrogenase. J Am Chem Soc 2017, 139, 83–6.
[80] Rumpel, S., Sommer, C., Reijerse, E., Farès, C., and Lubitz, W. Direct detection of the terminal hydride intermediate in [FeFe] hydrogenase by NMR spectroscopy. J Am Chem Soc 2018, 140, 3863–6.
[81] Schick, M., Xie, X., Ataka, K., Kahnt, J., Linne, U., and Shima, S. Biosynthesis of the iron-guanylylpyridinol cofactor of [Fe]-hydrogenase in methanogenic Archaea as elucidated by stable-isotope labeling. J Am Chem Soc 2012, 134, 3271–80.
[82] White, R.H. L-aspartate semialdehyde and a 6-deoxy-5-ketohexose 1-phosphate are the precursors to the aromatic amino acids in Methanocaldococcus jannaschii. Biochemistry-Us 2004, 43, 7618–7627.
[83] Porat, I., Sieprawska-Lupa, M., Teng, Q., Bohanon, FJ., White, RH., and Whitman, WB. Biochemical and genetic characterization of an early step in a novel pathway for the biosynthesis of aromatic amino acids and *p*-aminobenzoic acid in the archaeon *Methanococcus maripaludis*. Mol Microbiol 2006, 62, 1117–31.
[84] Shepard, EM., Duffus, BR., George, SJ., McGlynn, SE., Challand, MR., Swanson, KD., Roach, PL., Cramer, SP., Peters, JW., and Broderick, JB. [FeFe]-hydrogenase maturation: hydg-catalyzed synthesis of carbon monoxide. J Am Chem Soc 2010, 132, 9247–9.
[85] Dinis, P., Suess, DLM., Fox, SJ., Harmer, JE., Driesener, RC., De La Paz, L., Swartz, JR., Essex, JW., David Britt, R., and Roach, PL. X-ray crystallographic and EPR spectroscopic analysis of

HydG, a maturase in [FeFe]-hydrogenase H-cluster assembly. Proc Natl Acad Sci USA 2015, 112, 1362–7.
[86] Samuel, BS., Hansen, EE., Manchester, JK., Coutinho, PM., Henrissat, B., Fulton, R., Latreille, P., Kim, K., Wilson, RK., and Gordon, JI. Genomic and metabolic adaptations of *Methanobrevibacter smithii* to the human gut. Proc Natl Acad Sci USA 2007, 104, 10643–8.
[87] Chen, D., Scopelliti, R., and Hu, X. [Fe]-Hydrogenase models featuring acylmethylpyridinyl ligands. Angew Chem Int Ed 2010, 49, 7512–5.
[88] McGlynn, SE., Boyd, ES., Shepard, EM., Lange, RK., Gerlach, R., Broderick, JB., and Peters, JW. Identification and characterization of a novel member of the radical adomet enzyme superfamily and implications for the biosynthesis of the Hmd hydrogenase active site cofactor. J Bacteriol 2010, 192, 595–8.
[89] AP, Mehta., SH, Abdelwahed., Mahanta, N., Fedoseyenko, D., Philmus, B., LE, Cooper., Liu, Y., Jhulki, I., SE, Ealick., and TP, Begley. Radical S-adenosylmethionine (SAM) enzymes in cofactor biosynthesis: a treasure trove of complex organic radical rearrangement reactions. J Biol Chem 2015(290), 3980–6.
[90] Fujishiro, T., Tamura, H., Schick, M., Kahnt, J., Xie, X., Ermler, U., and Shima, S. Identification of the HcgB enzyme in [Fe]-hydrogenase-cofactor biosynthesis. Angew Chem Int Ed 2013, 52, 12555–8.
[91] Bai, LP., Fujishiro, T., Huang, G., Koch, J., Takabayashi, A., Yokono, M., Tanaka, A., Xu, T., Ermler, U., and Shima, S. Towards artificial methanogenesis: biosynthesis of the [Fe]-hydrogenase cofactor and characterization of the semisynthetic hydrogenase. Faraday Discuss 2017, 198, 37–58.
[92] Fujishiro, T., Bai, L., Xu, T., Xie, X., Schick, M., Kahnt, J., Rother, M., Hu, X., Ermer, U., and Shima, S. Identification of HcgC as a SAM-dependent pyridinol methyltransferase in [Fe]-hydrogenase cofactor biosynthesis. Angew Chem Int Ed 2016, 55, 9648–51.
[93] Bai, L., Wagner, T., Xu, T., Hu, X., Ermler, U., and Shima, S. A water-bridged h-bonding network contributes to the catalysis of the SAM-dependent C-methyltransferase HcgC. Angew Chem Int Ed 2017, 56, 10806–9.
[94] Fujishiro, T., Ermler, U., and Shima, S. A possible iron delivery function of the dinuclear iron center of HcgD in [Fe]-hydrogenase cofactor biosynthesis. FEBS Lett 2014, 588, 2789–93.
[95] Godsey, MH., Minasov, G., Shuvalova, L., Brunzelle, JS., Vorontsov, II., Collart, FR., and Anderson, WF. The 2.2 Å resolution crystal structure of *Bacillus cereus* Nif3-family protein YqfO reveals a conserved dimetal-binding motif and a regulatory domain. Protein Sci 2007, 16, 1285–93.
[96] Fujishiro, T., Kahnt, J., Ermler, U., and Shima, S. Protein-pyridinol thioester precursor for biosynthesis of the organometallic acyl-iron ligand in [Fe]-hydrogenase cofactor. Nat Commun 2015, 6, 6895.
[97] Maxwell Burroughs, A., Iyer, LM., and Aravind, L. Natural history of the E1-like superfamily: Implication for adenylation, sulfur transfer, and ubiquitin conjugation. Proteins 2009, 75, 895–910.
[98] Schulman, BA., and Wade Harper, J. Ubiquitin-like protein activation by E1 enzymes: the apex for downstream signalling pathways. Nat Rev Mol Cell Biol 2009, 10, 319–31.
[99] Galeazzi, L., Bocci, P., Amici, A., Brunetti, L., Ruggieri, S., Romine, M., Reed, S., Osterman, AL., Rodionov, DA., Sorci, L., and Raffaelli, N. Identification of nicotinamide mononucleotide deamidase of the bacterial pyridine nucleotide cycle reveals a novel broadly conserved amidohydrolase family. J Biol Chem 2011, 286, 40365–75.
[100] Royer, AM., Rauchfuss, TB., and Gray, DL. Oxidative addition of thioesters to iron(0): active-site models for Hmd, nature's third hydrogenase. Organometallics 2009, 28, 3618–20.

[101] Royer, AM., Salomone-Stagni, M., Rauchfuss, TB., and Meyer-Klaucke, W. Iron acyl thiolato carbonyls: structural models for the active site of the [Fe]-hydrogenase (Hmd). J Am Chem Soc 2010, 132, 16997–7003.
[102] Schilter, D., Camara, JM., Huynh, MT., Hammes-Schiffer, S., and Rauchfuss, TB. Hydrogenase enzymes and their synthetic models: the role of metal hydrides. Chem Rev 2016, 116, 8693–749.
[103] Wright, JA., Turrell, PJ., and Pickett, CJ. The third hydrogenase: more natural organometallics. Organometallics 2010, 29, 6146–56.
[104] Schultz, KM., Chen, D., and Hu, X. [Fe]-hydrogenase and models that contain iron–acyl ligation. Chem Asian J 2013, 8, 1068–75.
[105] Chen, D., Scopelliti, R., and Hu, X. Synthesis and reactivity of iron acyl complexes modeling the active site of [Fe]-hydrogenase. J Am Chem Soc 2010, 132, 928–9.
[106] Chen, D., Ahrens-Botzong, A., Schuemann, V., Scopelliti, R., and Hu, X. Synthesis and characterization of a series of model complexes of the active site of [Fe]-hydrogenase (Hmd). Inorg Chem 2011, 50, 5249–57.
[107] Turrell, PJ., Wright, JA., Peck, JNT., Oganesyan, VS., and Pickett, CJ. The third hydrogenase: a ferracyclic carbamoyl with close structural analogy to the active site of Hmd. Angew Chem Int Ed 2010, 49, 7508–11.
[108] Turrell, PJ., Hill, AD., Ibrahim, SK., Wright, JA., and Pickett, CJ. Ferracyclic carbamoyl complexes related to the active site of [Fe]-hydrogenase. Dalton Trans 2013, 42, 8140–6.
[109] Durgaprasad, G., Xie, Z., and Rose, MJ. Iron hydride detection and intramolecular hydride transfer in a synthetic model of mono-iron hydrogenase with a CNS chelate. Inorg Chem 2016, 55, 386–9.
[110] Hu, B., Chen, D., and Hu, X. Synthesis and reactivity of mononuclear iron models of [Fe]-hydrogenase that contain an acylmethylpyridinol ligand. Chem Eur J 2014, 20, 1677–82.
[111] Song, L., Xie, Z., Wang, M., Zhao, G., and Song, HB. Biomimetic models for the active site of [Fe]hydrogenase featuring an acylmethyl(hydroxymethyl)pyridine ligand. Inorg Chem 2012, 51, 7466–8.
[112] Song, L., Zhao, G., Xie, Z., and Zhang, J. A novel acylmethylpyridinol ligand containing dinuclear iron complex closely related to [Fe]-hydrogenase. Organometallics 2013, 32, 2509–12.
[113] Song, L., Hu, F., Zhao, G., Zhang, J., and Zhang, W. Several new [Fe]hydrogenase model complexes with a single fe center ligated to an acylmethyl(hydroxymethyl)pyridine or acylmethyl(hydroxy)pyridine ligand. Organometallics 2014, 33, 6614–22.
[114] Song, L., Hu, F., Wang, M., Xie, Z., Xu, K., and Song, H. Synthesis, structural characterization, and some properties of 2-acylmethyl-6-ester group-difunctionalized pyridine-containing iron complexes related to the active site of [Fe]-hydrogenase. Dalton Trans 2014, 43, 8062–71.
[115] Song, L., Xu, K., Han, X., and Zhang, J. Synthetic and structural studies of 2-Acylmethyl-6-R-difunctionalized pyridine ligand-containing iron complexes related to [Fe]-hydrogenase. Inorg Chem 2016, 55, 1258–69.
[116] Song, L., Zhu, L., Hu, F., and Wang, Y. Studies on chemical reactivity and electrocatalysis of two acylmethyl(hydroxymethyl)pyridine ligand-containing [Fe]-hydrogenase models (2-$COCH_2$-6-$HOCH_2C_5H_3N$)$Fe(CO)_2L$ ($L = \eta^1$-SCOMe, η^1-2-SC_5H_4N). Inorg Chem 2017, 56, 15216–30.
[117] Liu, T., Li, B., Popescu, CV., Bilko, A., Perez, LM., Hall, MB., and Darensbourg, MY. Analysis of a pentacoordinate iron dicarbonyl as synthetic analogue of the Hmd or mono-iron hydrogenase active site. Chem Eur J 2010, 16, 3083–9.
[118] Chen, D., Scopelliti, R., and Hu, X. A five-coordinate iron center in the active site of [Fe]-hydrogenase: hints from a model study. Angew Chem Int Ed 2011, 50, 5670–2.
[119] Chen, D., Scopelliti, R., and Hu, X. Reversible protonation of a thiolate ligand in an [Fe]-hydrogenase model complex. Angew Chem Int Ed 2012, 51, 1919–21.

[120] Hu, B., Chen, D., and Hu, X. Reversible dimerization of mononuclear models of [Fe]-hydrogenase. Chem Eur J 2013, 19, 6221–4.
[121] Wodrich, MD., and Hu, X. Electronic elements governing the binding of small molecules to a [Fe]-hydrogenase mimic. Eur J Inorg Chem 2013, 2013, 3993–9.
[122] Zhang, T., Sheng, L., Yang, Q., Jiang, S., Wang, Y., Jin, C., and Li, B. Synthesis, characterization and catalytic reactivity of pentacoordinate iron dicarbonyl as a model of the [Fe]-hydrogenase active site. Chinese J Catal 2015, 36, 2011–9.
[123] Jiang, S., Zhang, T., Zhang, X., Zhang, G., and Li, B. Nitrogen heterocyclic carbene containing pentacoordinate iron dicarbonyl as a [Fe]-hydrogenase active site model. Dalton Trans 2015, 44, 16708–12.
[124] Hull, JF., Himeda, Y., Wang, W., Hashiguchi, B., Periana, R., Szalda, DJ., Muckerman, JT., and Fujita, E. Reversible hydrogen storage using CO_2 and a proton-switchable iridium catalyst in aqueous media under mild temperatures and pressures. Nat Chem 2012, 4, 383–8.
[125] Wang, W., Hull, JF., Muckerman, JT., Fujita, E., and Himeda, Y. Second-coordination-sphere and electronic effects enhance iridium(III)-catalyzed homogeneous hydrogenation of carbon dioxide in water near ambient temperature and pressure. Energ Environ Sci 2012, 5, 7923–6.
[126] Wang, W., Muckerman, JT., Fujita, E., and Himeda, Y. Mechanistic insight through factors controlling effective hydrogenation of CO_2 catalyzed by bioinspired proton-responsive iridium(III) complexes. ACS Catal 2013, 3, 856–60.
[127] Siek, S., Burks, DB., Gelach, DL., Liang, G., Tesh, JM., Thompson, CR., Qu, F., Shankwitz, JE., Vasquez, RM., Chambers, N., Szulczewski, GJ., Grotjahn, DB., Webster, CE., and Papish, ET. Iridium and ruthenium complexes of *n*-heterocyclic carbene- and pyridinol-derived chelates as catalysts for aqueous carbon dioxide hydrogenation and formic acid dehydrogenation: the role of the alkali metal. Organometallics 2017, 36, 1091–106.
[128] Kalz, KF., Brinkmeier, A., Dechert, S., Mata, RA., and Meyer, F. Functional model for the [Fe] hydrogenase inspired by the frustrated lewis pair concept. J Am Chem Soc 2014, 136, 16626–34.
[129] Stephan, DW., and Erker, G. Frustrated lewis pairs: metal-free hydrogen activation and more. Angew Chem Int Ed 2010, 49, 46–76.
[130] Hatazawa, M., Yoshie, N., and Seino, H. Reversible hydride transfer to *N, N'*-diarylimidazolinium cations from hydrogen catalyzed by transition metal complexes mimicking the reaction of [Fe]-hydrogenase. Inorg Chem 2017, 56, 8087–99.
[131] Murray, KA., Wodrich, MD., Hu, X., and Corminboeuf, C. Toward functional type III [Fe]-hydrogenase biomimics for H_2 activation: insights from computation. Chem Eur J 2015, 21, 3987–96.
[132] Xu, T., Yin, CJM., Wodrich, MD., Mazza, S., Schultz, KM., Scopelliti, R., and Hu, X. A functional model of [Fe]-hydrogenase. J Am Chem Soc 2016, 138, 3270–3.
[133] Berggren, G., Adamska, A., Lambertz, C., Simmons, TR., Esselborn, J., Atta, M., Gambarelli, S., Mouesca, JM., Reijerse, E., Lubitz, W., Happe, T., Artero, V., and Fontecave, M. Biomimetic assembly and activation of [FeFe]-hydrogenases. Nature 2013, 499, 66–9.
[134] Kerns, SA., Magtaan, AC., Vong, PR., and Rose, MJ. Functional hydride transfer by a thiolate-containing model of mono-iron hydrogenase featuring an anthracene scaffold. Angew Chem Int Ed 2018, 57, 2855–8.
[135] Esselborn, J., Camilla, L., Adamska-Venkates, A., Simmons, T., Berggren, G., Noth, J., Siebel, J., Hemschemeier, A., Artero, V., Reijerse, E., Fontecave, M., Lubitz, W., and Happe, T. Spontaneous activation of [FeFe]-hydrogenases by an inorganic [2Fe] active site mimic. Nat Chem Biol 2013, 9, 607–9.
[136] Esselborn, J., Muraki, N., Klein, K., Engelbrecht, V., Metzler-Nolte, N., Apfel, UP., Hofmann, E., Kurisu, G., and Happe, T. A structural view of synthetic cofactor integration into [FeFe]-hydrogenases. Chem Sci 2016, 7, 959–68.

[137] Tanifuji, K., Lee, CC., Sickerman, NS., Tatsumi, K., Ohki, Y., Hu, Y., and Ribbe, MW. Tracing the 'ninth sulfur' of the nitrogenase cofactor via a semi-synthetic approach. Nat Chem 2018, 10, 568–72.
[138] Fruk, L., Kuo, CH., Torres, E., and Niemeyer, CM. Apoenzyme reconstitution as a chemical tool for structural enzymology and biotechnology. Angew Chem Int Ed 2009, 48, 1550–74.
[139] Happe, T., and Hemschemeier, A. Metalloprotein mimics – old tools in a new light. Trends Biotechnol 2014, 32, 170–6.
[140] Simmons, TR., Berggren, G., Bacchi, M., Fontecave, M., and Artero, V. Mimicking hydrogenases: from biomimetics to artificial enzymes. Coordin Chem Rev 2014, 270, 127–50.

Holger Dobbek

3 CO_2 Reduction

3.1 Carbon monoxide dehydrogenases

3.1.1 Introduction

Carbon monoxide (CO) is a colorless, odorless and flammable gas [1]. It has a similar molecular weight (M_r = 29) as air (M_r about 28) and readily mixes with it. In water, only 0.35 L of CO per L H_2O is soluble at 0 °C. The C–O bond length is 1.128 Å (gaseous CO), indicative of a CO-triple bond. The triple-bonded electronic structure of CO has formally a negative charge at C and a positive charge at O. However, the higher electronegativity of oxygen counterbalances the formal charges, making CO an unusual "carbonyl"-type compound with an electron-rich carbon atom. CO is isoelectronic with N_2 and CN^-, a property responsible for the CN^- inhibitory activity on CO-binding catalysts [2].

Natural and anthropogenic sources contribute to a constant production of CO, mostly by incomplete combustion of carbon-containing compounds. In the laboratory (small-scale production), CO can be generated by mixing concentrated sulfuric acid with formic acid (HCOOH), formally dehydrating formic acid to its anhydride CO. In technical and industrial settings (large-scale production), CO can be produced by coal gasification producing syngas, a mixture of CO, H_2, CO_2, CH_4 and water vapor [1].

Possible ways of interacting with CO are revealed from its frontier orbitals. The highest occupied molecular orbital (HOMO) of CO is centered at the carbon atom and CO therefore binds preferentially to metals through the carbon and not the electron lone pair at oxygen, which, as oxygen has a higher electronegativity than carbon, is in an orbital of lower energy. As CO is an unsaturated soft ligand, it is able to donate σ-electrons to the metal and to accept metal dπ electrons into its π*-orbital, a process termed as *back-bonding* (Figure 3.1). While electron donation along the σ-bond removes electron density from the carbon, the metal dπ electrons shifted to the ligand increase the electron density at both carbon and oxygen. Thus, when CO binds to a metal, the C-atom becomes more positive and the O-atom more negative, further polarizing the molecule. It thus becomes chemically activated, thereby increasing the tendency of its carbon to react with nucleophiles and its oxygen to react with electrophiles. This reactivity is especially pronounced when CO is bound to a metal of low π-basicity, such as Ni^{2+} [2]. CO also has the tendency to insert into metal alkyl bonds, forming metal acyl-compounds – a reactivity of no

Holger Dobbek, Humboldt-Universität zu Berlin

https://doi.org/10.1515/9783110496574-003

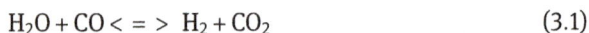

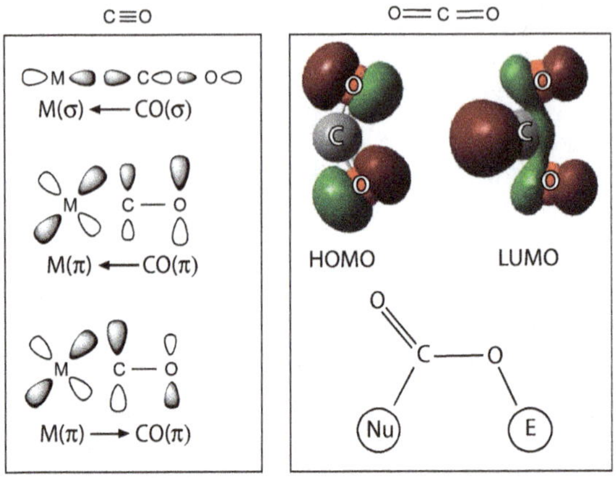

Figure 3.1: Chemistry of CO and CO_2.
Left side – CO: The interaction of CO with a metal is due to three electronic contributions: a σ-donor (top) and a π-donor bonding (middle) as well as a π-acceptor back-bonding interaction (bottom). The relative contribution of the three interactions determines the stability of the metal-carbonyl and the reactivity of bound CO toward nucleophiles. Right side – CO_2: The highest occupied molecular orbital (HOMO) activated (bent) CO_2 is centered at the nonbonding electron pairs of the oxygen atoms, favoring the interaction with electrophiles. In contrast, the lowest unoccupied molecular orbital (LUMO) is centered at the carbon atom, stabilizing its interaction with nucleophiles.

direct relevance for the enzymes discussed in this chapter, but important for acetyl-CoA synthases (see Chapter 7).

Closely related to the chemistry of carbon monoxide dehydrogenases (CODHs) is the industrial water gas shift reaction, which allows changing the CO:H_2 ratio in synthesis gas (syn-gas; Reaction (3.1)).

$$H_2O + CO <=> H_2 + CO_2 \tag{3.1}$$

Reaction (3.1) is catalyzed by various homogenous and heterogeneous transition metal-containing catalysts. The basic steps of the water gas shift reaction are of interest for us as they describe what a catalyst, such as a CODH, has to achieve: (1) CO has to bind to a metal, whereby CO becomes activated for the attack of a nucleophilic OH^- group. (2) The metalcarboxylic acid formed is decarboxylated either by a β-hydride shift (β-elimination) requiring a vacant site for the hydride at the metal or by a deprotonation of the carboxylate, followed by a reprotonation of the metal. (3) Protonation of the metal hydride liberates H_2 and regenerates the catalyst [2]. The last step, the formation of dihydrogen from a metal-bound hydride, does not occur with the enzymes, where protons and electrons are kept separate, as the electrons generated by CO oxidation are used as a source of energy by the bacteria.

3.1.2 The chemical CO_2

Carbon dioxide (CO_2) is a colorless, nonflammable gas, which does not support respiration or burning processes. With a molecular weight larger than that of air, it has a higher density and can accumulate near the ground. At room temperature about 0.9 L of gaseous CO_2 can be dissolved in 1 L of water. CO_2 is a linear molecule with formal C=O double bonds (C–O bond length of 1.17 Å). The molecule is overall nonpolar, but due to the difference in electronegativities of carbon and oxygen, such that the C-atom is positively polarized, while the O-atoms are negatively polarized [1].

The lowest unoccupied molecular orbital (LUMO) of CO_2 is localized at the carbon, making the molecule susceptible to attack of nucleophiles and to reduction (Figure 3.1). The typical first step in any reaction of CO_2 is its activation, involving a decrease in the C–O bond order, which is also apparent by a decrease of the O–C–O bond angle. Bending stabilizes and increases the carbon weight in the LUMO – in sum facilitating CO_2 reduction. The HOMO of CO_2 is localized at the oxygen atoms and favors their interaction with electrophiles [3]. In biological systems, activation of CO_2 typically occurs by a combined interaction of nucleophilic and electrophilic centers (Figure 3.1). In this bent state, CO_2 tends to undergo a two-electron reduction rather than a reduction by one electron [3].

3.1.3 Biology of CO/CO_2

Natural and anthropogenic processes produce CO typically by incomplete combustion. These sources are responsible for 2.500–2.600 Tg of CO emitted per year. Natural processes include atmospheric methane oxidation, natural hydrocarbon oxidation, volcanic activity, production by plants and photochemical degradation of organic matters, while anthropogenic processes are mostly due to the incomplete combustion of fossil fuels and various industrial processes, adding annually 1.200 Tg of CO to the atmosphere [4].

CO, a ubiquitous and toxic pollutant of the present atmosphere, probably enabled early life in a CO-containing primordial atmosphere. Ubiquitous volcanic emissions in the early earth atmosphere are believed to have raised the global CO concentration to 100 ppm. Nowadays, we find 0.05–0.35 ppm CO in nonurban environments, whereas dense urban regions with its abundant car traffic can raise CO concentrations to 1.30 ppm. Higher concentrations of up to 5,540 ppm can be encountered in volcanic environments, nurturing CO-consuming bacteria [4].

Microorganisms use CO as a source of energy and carbon. Soil and marine microbes together reduce the annual global budget of CO by 20%, to which soil microbes contribute with 200–600 Tg of CO [5]. CO-oxidizing bacteria can thrive in the top layer of burning charcoal piles from where several of these soil microorganisms have been isolated. On the other hand, microorganisms may also use biogenic CO, as is likely for pathogenic mycobacteria, like the tubercle bacillus *Mycobacterium tuberculosis*, which can grow on CO as sole source of carbon and energy [6]. Overall, biological processes keep CO at low trace gas levels in our atmosphere, for which oxygen-dependent and therefore CO-sensitive organisms like us can be grateful.

Considering the rising CO_2 concentrations in our atmosphere, biological pathways consuming CO_2 are of outstanding interest. Six pathways to assimilate CO_2 into biomass have been identified so far: 1) the Calvin cycle, also as an integral part of oxygenic photosynthesis, 2) the reductive citric acid cycle, 3) the 3-hydroxypropionate bicycle, 4) the hydroxypropionate-hydroxybutyrate cycle, 5) the dicarboxylate-hydroxybutyrate cycle and 6) the reductive acetyl-CoA pathway (Wood–Ljungdahl pathway) in which CODHs have a central role [7]. The reductive acetyl-CoA pathway is most likely the most ancient of the listed pathways and may be responsible for up to 20% of all conversion of atmospheric CO_2 into biomass [8]. The reductive acetyl-CoA pathway employs several dioxygen-sensitive enzymes and is therefore only found in anaerobic bacteria and archaea, such as sulfate reducers, hydrogenogens, acetogens and methanogens, where it generates biomass and energy from CO or CO_2 and H_2. The pathway consists of two branches, named the methyl and the carbonyl branch [9]. In the carbonyl branch, CODH catalyzes the reduction of CO_2 to CO, which is condensed in the second step with coenzyme A (CoA) and the methyl group generated in the methyl branch by acetyl-CoA synthase (ACS) to acetyl-CoA. The generated acetyl-CoA may serve as building block for cellular carbon compounds, to generate acetate (by acetogens) or as a source of energy (for acetoclastic methanogens). The methyl branch is a folate-dependent one-carbon pathway. Here, CO_2 is in a first step reduced to formate, which is subsequently coupled after an activation step to tetrahydrofolate (H4F) and is stepwise reduced to generate 5-methyl-H4F. The methyl group of methyl-H4F is passed via the corrinoid iron–sulfur protein (CoFeSP) to ACS, thereby connecting the methyl and carbonyl branches of the pathway [9].

3.1.4 Two types of carbon monoxide dehydrogenases

Two principal types of CODHs exist in nature. The enzymes use different cofactors, contain different metals and are distinguished by their capacity to reduce CO_2 as well as their stability in the presence of air. While the Cu,Mo-containing CODHs isolated from aerobic carboxydotrophic bacteria contain Cu and Mo in their active site, but do not catalyze reduction of CO_2 and are stable in the presence of air (under nonturnover conditions), the Ni,Fe-containing CODHs found in anaerobically living bacteria and archaea, contain Ni and Fe in their active site, efficiently reduce CO_2 but are sensitive to the presence of dioxygen [10]. Based on the changing bioavailability of the metals used by the two enzymes, as well as their occurrence in different organisms and their phylogeny, it is apparent that the Ni,Fe-containing CODHs are substantially older than the Cu,Mo-containing CODHs. Phylogenetic analysis indicate that Ni,Fe-containing CODHs existed already in the last universal common ancestor long before the oxygenation of the atmosphere, thus were present already more than 3.5 billion years ago [11]. The younger Cu,Mo-CODHs likely evolved from other Mo-containing hydroxylases by acquiring the ability to bind copper, an element that only became bioavailable after the oxygenation of the atmosphere [10].

3.2 Cu,Mo-containing carbon monoxide dehydrogenases

Cu,Mo-containing CODHs catalyze the oxidation of CO to CO_2 according to Reaction (3.2):

$$CO + H_2O \longrightarrow CO_2 + 2H^+ + 2e^- \tag{3.2}$$

This reaction occurs at the pH optimum of 7.2, with a k_{cat} of 93.3 s^{-1} and a K_m for CO of 10.7 μM at 25 °C. The rate-limiting step of the overall reaction is likely the reductive half-reaction with CO binding to the enzyme in a rapid initial step [12]. Cu,Mo-CODHs are, at least under in vitro conditions, also uptake hydrogenases and oxidize H_2 with a limiting-rate constant of 5.3 s^{-1} and a K_d for dihydrogen of 525 μM [13, 14]. A nonnative variant containing Ag (instead of Cu) has a limiting rate constant of 8.1 s^{-1} in CO oxidation [15]. Electrons gained from CO oxidation can be directly transferred to the quinone pool of the cytoplasmic membrane [16].

3.2.1 Structure of Cu,Mo-CODHs

Cu,Mo-CODHs are assembled from three protein subunits, which are encoded by three genes. The transcriptional order found in *Oligotropha carboxydovorans* is *coxM-coxS-coxL*, encoding subunits with 288, 166 and 809 amino acids, respectively [17]. The sequence of the largest subunit contains a signature for Cu,Mo-CODHs, the sequence motif VAYXCSFR found in the active site loop, where C denotes a cysteine residue binding the Cu ion. This sequence motif can safely be used to distinguish Cu,Mo-CODHs from the closely related molybdenum hydroxylases. Several sequences in databases annotated to be a Cu,Mo-CODH do not contain the sequence motif and are likely catalyzing the hydroxylation of (hetero)aromatic compounds and are not able to oxidize CO.

The structure of the *O. carboxydovorans* Cu,Mo-CODH has been resolved in different catalytic and noncatalytic (inactive) states and was refined to a highest resolution of 1.09 Å, while the structure of the enzyme from *H. pseudoflava* has been refined to a resolution of 2.2 Å [18, 19]. Among the Cu,Mo-CODH structures are three reduced states (reduced using H_2, CO and sodium dithionite) containing Mo in the Mo(+IV) state, all of which have identical crystal structures. Furthermore, structures for the oxidized state, an n-butylisocyanide and a CN-inactivated state, have been determined.

The overall structure of Cu,Mo-CODH resembles a butterfly consisting of a dimer of heterotrimers $(\alpha\beta\gamma)_2$ (Figure 3.2). Each of the three subunits embeds and stabilizes one type of cofactor. Cu,Mo-CODHs contain a pyranopterin cofactor (sometimes called *molybdopterin cofactor*, but it is also found in tungsten containing enzymes) in the active site and additionally employ two [2Fe2S] clusters and a flavin adenine dinucleotide (FAD) as cofactors. The large subunit (L subunit) carries the Mo-pyranopterin cytosine dinucleotide (MCD) cofactor, which is part of the active site. The small subunit (S subunit) harbors two [2Fe2S] clusters and the medium subunit (M subunit) binds an FAD molecule noncovalently. The cofactors are suitably arranged for electron transfer from the molybdenum ion via the [2Fe2S] clusters to the FAD, forming two independent electron transfer chains in the dimer with short cofactor–cofactor distances of up to 14 Å within each heterotrimer, allowing a rapid electron transfer (Figure 3.2) [20]. The arrangement of the cofactors as well as the fold of the three subunits is also similar to that of other molybdenum hydroxylases, such as the eukaryotic xanthine oxidases and aldehyde dehydrogenases as well as the bacterial hydroxylases [21].

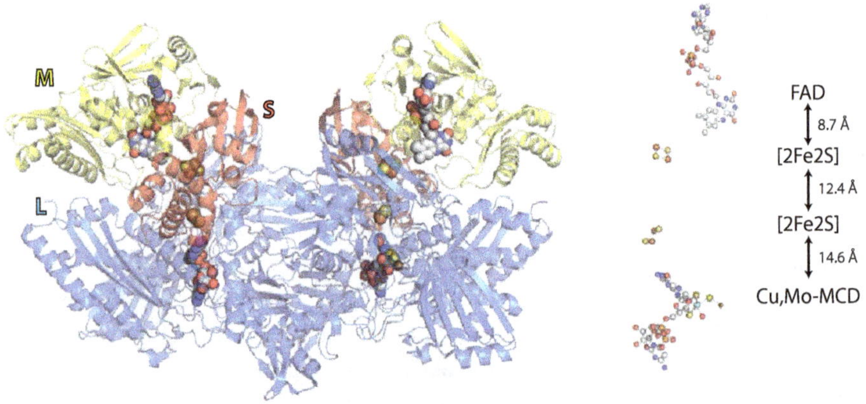

Figure 3.2: Overall structure of Cu,Mo-containing CODHs.
Left side: Cu,Mo-CODH is a dimer of heterotrimers, in which each subunit harbors one type of cofactor, the L-subunit (in blue) the Cu,Mo-MCD cofactor, the S-subunit (in red) two [2Fe2S]-cluster and the M-subunit (in yellow) FAD. Semitransparent cartoons display the protein fold surrounding the cofactors, which are shown as van der Waals spheres. Right side: Each LSM-heterotrimer arranges the cofactors to an electron-transfer chain through which electrons generated by CO oxidation at the Cu,Mo-active site are transferred via the [2Fe2S]-clusters to the FAD cofactor. Cofactors are shown in ball-and-stick mode.

The structure of the large subunit (L subunit) can be divided into two domains, together embracing the MCD cofactor in a dense network of hydrogen bonds, thereby burying the active site deeply inside the protein matrix [18]. The active site is accessible from the surface through a hydrophobic channel with a diameter of 6 Å, which ends at the [CuSMo] unit of the active site. The [2Fe2S] cluster proximal to the active site Mo is bound in a four-helix bundle domain, while the distal [2Fe2S] cluster is coordinated within a domain closely resembling the fold of a plant-type [2Fe2S]-ferredoxin. Both [2Fe2S] clusters are shielded from the solvent by the protein matrix. Finally, the FAD molecule is bound between the N-terminal and middle domain of the M subunit. Access to the redox-active N5 position of the isoalloxazine ring of FAD from the solvent is restricted by aromatic side chains, a tyrosine in the CODH of *O. carboxydovorans* and a tryptophan residue in the CODH of *H. pseudoflava* [18, 19].

The active site is located around the name-giving Cu- and Mo-ion, a cofactor assembly only found in CODHs. An enedithiolate moiety of the pyranopterin cofactor binds the Mo ion, while Cu is coordinated by the cysteine of the active site loop (Figure 3.3). A μ_2-sulfido-ligand bridges Cu and Mo, establishing a distance of 3.74 Å between Cu and Mo in the Mo(+VI) state. Reducing Mo to Mo(+IV) increases the Mo–Cu distance by 0.2 Å. The Mo-ligands are arranged in a distorted square pyramide, which may alternatively be described as a distorted tetrahedral coordination when treating the enedithiolate moiety as a single, bidentate ligand [22]. The enedithiolate moiety of the pyran ring, the bridging sulfido-ligand and one oxo/hydroxyo-ligand form the equatorial coordination sphere of the Mo ion and an additional oxo-ligand is found in the apical position. The second coordination sphere, whose importance to sustain catalysis has been shown for other molybdenum hydroxylases, is formed by a glutamate residue in *trans* to the apical oxo-ligand with a Glu-Mo distance of 3.14 Å and a glutamine residue in hydrogen-bonding distance to the apical oxo-ligand [18].

3.2.2 Spectral properties of the active site

In contrast to the [2Fe2S]-cluster and FAD, which absorb visible light and display characteristic redox-dependent UV/Vis spectra, the active site Cu,Mo lacks characteristic absorption in the UV/Vis region. However, the Mo(V) state is paramagnetic and can be inspected using EPR spectroscopy. Interestingly, the Mo(V) signal of Cu,Mo-CODH shows strong hyperfine coupling with the nuclear spin of Cu ($I = 3/2$), indicating a delocalization of the electron spin within the SOMO (singly occupied molecular orbital) of Mo(V) along the entire Mo(V)–S–Cu(I) moiety [12]. Upon addition of [^{13}C]CO, the Mo(V) signal splits, indicating that either the substrate [^{13}C]CO or the product [^{13}C]CO_2 is a part of the signal-giving species. Further analysis of the signal provided evidence for the presence of a

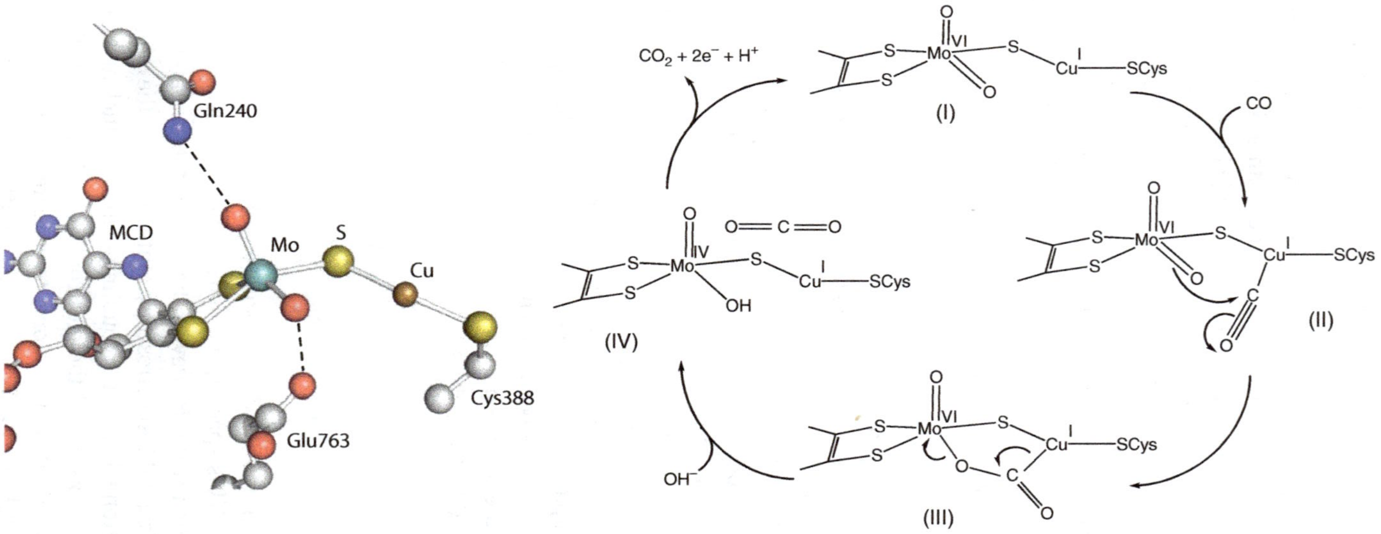

Figure 3.3: Active site structure and mechanism of Cu,Mo-containing CODHs.
Left side: Ball-and-stick presentation of the active site. Two amino acids, Gln240 and Glu763, are in hydrogen-bonding distance to the Mo-ion. Cu is linearly coordinated by the thiolate of Cys388 and a Cu,Mo-bridging sulfido-ligand. Right side: Catalytic cycle of CO oxidation. (I) The oxidized state contains Mo(+VI). (II) CO binds to the Cu(I) ion and reacts with the oxo-ligand of Mo. (III) A metallacycle, Mo–S–Cu–C–O is formed, which resolves with the oxidation of CO to CO_2 and the two-electron reduction of Mo (IV). The catalytic cycle closes with the release of CO_2 and the reoxidation of Mo(+IV) to Mo(+VI) by electron transfer via the electron transport chain.

copper–carbonyl intermediate in CO oxidation [23]. A change of the Mo(V) signal by the presence of substrate has also been observed with the alternative substrate H_2. When Cu,Mo-CODH reacts with H_2, a new EPR signal arises that has a larger g anisotropy and hyperfine coupling to the $^{63,65}Cu$ in the active site [14]. Complementary information on the structure of the active site can be gained by X-ray absorption spectroscopy (XAS). XAS confirmed the presence of Cu and Mo in the active site of Cu,Mo-CODH, but whereas the crystallographic studies indicated the presence of a hydroxyl-group in the equatorial plane, XAS studies are consistent with the presence of an oxo group [24]. This discrepancy may be due to the higher X-ray dose used in X-ray crystallography, which can lead to X-ray-induced photoreduction of the probe and a concomitant elongation of the Mo-O bond length [25].

3.2.3 Mechanism of Cu,Mo-CODHs

In addition to CO and H_2, Cu,Mo-CODHs react with various small molecules including inhibitory compounds. As these inhibitors typically test the chemistry of the active site, we can learn from them about the reactivity of CODHs. Several small-molecule inhibitors inactivate Cu,Mo-CODH irreversibly.

Cyanide (CN^-) is isoelectronic and isosteric with CO and inactivates (irreversibly inhibits) the oxidized enzyme with a half-life of about 30 min. Cyanolysis releases concomitantly up to 1 mol of thiocyanate and 1 mol of Cu per active site of Cu,Mo-CODH. The crystal structure and spectroscopic investigations indicate the formation of a Mo tri-oxo species upon cyanolysis, depleting the active site of Cu and the bridging sulfido-ligand [18]. The inactive Mo tri-oxo species also forms when active Cu,Mo-CODH reacts with CO under oxic conditions [16], showing that Cu,Mo-CODH is not completely protected from O_2-induced damage. Inactive Mo tri-oxo Cu,Mo-CODH can be reconstituted by adding sulfide and copper under anoxic, reducing conditions, resulting in about 50% functional enzyme [26].

Isocyanides and CO show a similar σ-donor and π-acceptor ligand character and are isoelectronic with a nonbonding pair of electrons in the p-orbital of the terminal carbon. Isocyanides inhibit Cu,Mo-CODH by inserting into the Cu–S bond. Inhibition is concurrent with oxidation of the inhibitor, which forms a thiocarbamate derivative after disrupting the Cu–S bond [18].

Based on the crystal structures, spectroscopic investigations, mechanisms of other Mo-containing hydroxylases, model calculations and chemical properties of its constituents, the following mechanisms for CO oxidation appears plausible (Figure 3.3): (I) CO binds to the Mo(VI)/Cu(I) state generating a Cu(I)–CO species. Binding to the electron-rich Cu(I) populates the π*-orbital of CO, thereby activating it for the nucleophilic attack of the equatorial Mo=O (or Mo–OH) oxygen on the carbon of Cu-bound CO. (II) CO oxidation likely leaves the Mo–S–Cu moiety intact and involves a Mo(VI)-S–Cu(I)-C–O-metallacycle in the next step, which breaks down to form CO_2 and Mo(IV). (III) The catalytic cycle is closed by electron transfer from the active site to external acceptors. Electrons generated by CO or H_2 oxidation are in a first step taken up by Mo reducing Mo(VI) to Mo(IV) and are subsequently transferred via the two different [2Fe-2S] clusters to the FAD, where they are used to reduce quinones, regenerating the enzyme for another turnover.

Despite different chemical characters of H_2 and CO, the conversion of both molecules depends on the presence of Cu(I); thus Cu(I) is most likely required to activate H_2. Catalytic H_2 oxidation was postulated to occur via formation of a copper hydride species in the active site, which is consistent with recent spectroscopic results on a Mo(V) species generated by H_2 incubation of Cu,Mo-CODH [14, 18].

Molybdenum hydroxylases hydroxylate C–H groups, whereas Cu,Mo-CODHs oxidize CO – two obviously different oxidation reactions. However, a comparison of the orbital contributions in both reactions found similar electronic structures for central intermediates in CO and C–H bond activation. Cu,Mo-CODHs stabilize the CO bound intermediate by $\mathrm{C-Cu}(\sigma) \rightarrow \mathrm{Mo-S}(\pi)$ and $\mathrm{Mo-S}(\pi) \rightarrow \mathrm{C-Cu}(\sigma)$ charge transfer, whereas C–H bond activation in xanthine oxidase involves $\mathrm{C-H}(\sigma) \rightarrow \mathrm{Mo-S}(\pi)$ and $\mathrm{Mo-S}\pi \rightarrow \mathrm{C-H}\sigma$ charge transfers. Thus, from the perspective of the orbital contributions, Cu,Mo-CODHs employ the Cu(I) ion as a substitute for the hydrogen bound to C in C–H groups, or in other words, CO–Cu(I) resembles a C–H unit activated for a nucleophilic attack [27]. This also explains why only a small change, the addition of a Cu(I)-ion, appeared to be sufficient to evolve a CODH from a progenitor molybdenum hydroxylase.

3.2.4 Model complexes of the Cu,Mo-CODH active site

No close mimics of the active site Cu–S–Mo unit are known to date. A first model, which appears still to be most close to the enzymes active site, contains a MoO (OAr)(μ_2-S) Cu core. The oxidation state of Mo is fixed as Mo(+V) and the EPR spectrum indicates that the SOMO, in principal localized at Mo(V), contains significant Cu character, explaining the large superhyperfine coupling observed [28]. Thus, the EPR characteristics of the Cu,Mo-CODH were well reproduced. However, the complex is not redox active and has not the dithiolene-type coordination of Mo deemed to be essential for its properties.

Other models, reproducing a dithiolene-type coordination of Mo, require supported Mo(μ_2–S)–Cu bridges with disulfido-bridged geometries. None of the available models shows CO-oxidizing activity.

3.3 Ni,Fe-containing carbon monoxide dehydrogenases

In 1979 the group of Rolf Thauer published the isolation of a CO-oxidizing, Ni-containing enzyme, the discovery of Ni,Fe-CODHs [29], and in 1980 the group of Harland Wood described the first bifunctional Ni-dependent CODH [30]; since than CODHs have been investigated by various techniques.

Depending on physiological conditions and source organism, Ni,Fe-CODHs are found either alone (monofunctional CODHs), in complex with acetyl-CoA synthase (ACS; condensing CO, CH_3^+ and CoA to acetyl-CoA) (bifunctional CODHs) or as a component of an even larger complex of approximately 2 MDa where CODH is not only associated with ACS, but also the corrinoid-iron/sulfur protein-carrying methyltransferase activity (CoFeSP/MeTr; CH_3-transfer to ACS; multifunctional CODH) [10].

Different physiological functions of Ni,Fe-CODHs are also reflected in the genome of bacteria living with CO, such as *C. hydrogenoformans*, where at least four different gene cluster contain structural genes encoding Ni,Fe-CODHs [31]. Two monofunctional Ni,Fe-CODHs ($CODHI_{Ch}$ and $CODHII_{Ch}$) catalyze the oxidation of CO, whereas a third is a bifunctional Ni,Fe-CODH ($CODHIII_{Ch}$) found in a stable complex with ACS, supporting autotrophic carbon assimilation, in which the physiological role of the CODH component is to reduce CO_2 to CO [32]. A fourth Ni,Fe-CODH ($CODHIV_{Ch}$) is found associated with enzymes for the detoxification of reactive oxygen species and catalyzes CO oxidation with high affinity, while being less susceptible to inactivation by the presence of air [33].

Ni,Fe-CODHs catalyze the reversible oxidation of CO (Reaction (3.3)):

$$CO + H_2O <-> CO_2 + 2H^+ + 2e^- \tag{3.3}$$

CO oxidation by $CODHII_{Ch}$ at 70 °C occurs with a k_{cat} of 31,000 s^{-1} and a K_M for CO of 18 μM [34]. CO oxidation by $CODHIV_{Ch}$ is strongly temperature dependent and both k_{cat} and K_M increase with rising temperature. The reason for this correlation of both constants is that k_{eff} (k_{cat}/K_M) is already limited by diffusion (3×10^8 to 10^9 M^{-1} s^{-1}) over the complete temperature range. At 25 °C CO oxidation occurs with a K_M of 47 nM; thus $CODHIV_{Ch}$ appears to have an exceptionally high affinity for CO, allowing *C. hydrogenoformans* to use CO even at low ppm concentrations [33].

3.3.1 Structure of Ni,Fe-CODHs

Ni,Fe-CODHs show different degrees of complexity and may be distinguished according to subunit composition into four classes [35]. Class I and II Ni,Fe-CODHs are large multifunctional complexes, also termed acetyl-CoA decarbonylase/synthases (ACDS) and are found in methanogenic and some sulfate-reducing archaea. Class I Ni, Fe-CODHs are used by obligate autotrophic methanogens such as *Methanobacterium thermoautotrophicum* to generate acetyl-CoA from CO_2, CoASH and H_2. Class II Ni, Fe-CODHs are used by facultative chemo-autotrophic methanogens, which are able to catabolize acetate generating CO_2, CoASH and a methyl group bound to tetrahydrosarcinapterin, a tetrahydrofolate analogue used by archaea. Both classes I and II Ni,Fe-CODH are composed of five different subunits (α, β, γ, δ and ε): CODH (α subunit), ACS (β subunit), CoFeSP/MeTr (γ and δ subunits) and an ε subunit whose function is still unclear. Class III Ni,Fe-CODHs are found in acetogens and are associated with ACS. Class IV Ni,Fe-CODHs are the monofunctional CODH homodimers.

Crystal structures have been determined for several monofunctional Ni, Fe-CODHs isolated from *C. hydrogenoformans* ($CODHII_{Ch}$ and $CODHIV_{Ch}$) and *R. rubrum* ($CODH_{Rr}$) [33, 36, 37]. Furthermore, crystal structures have been determined for bi- and multifunctional CODHs – the bifunctional CODH/ACS of *Moorella thermoacetica* as well as the CODH component ($\alpha_2\varepsilon_2$) of a class I/II Ni,Fe-CODH [38, 39]. The latter is part of the multifunctional ACDS complex and differs from the monofunctional bacterial enzymes by having two additional [$4Fe_4S$] clusters (clusters E and F) in each subunit. As the active site for CO/CO_2 conversion, as most likely the reaction mechanisms are conserved in Ni,Fe-CODHs, we will concentrate here on the more simple monofunctional CODHs.

The overall structure of Ni,Fe-CODH is that of a mushroom-shaped homodimer with five metal clusters, of which three are cubane-type [$4Fe_4S$]-clusters (termed cluster B and cluster D) and two are the active site cluster C (Figure 3.4) [36]. Each

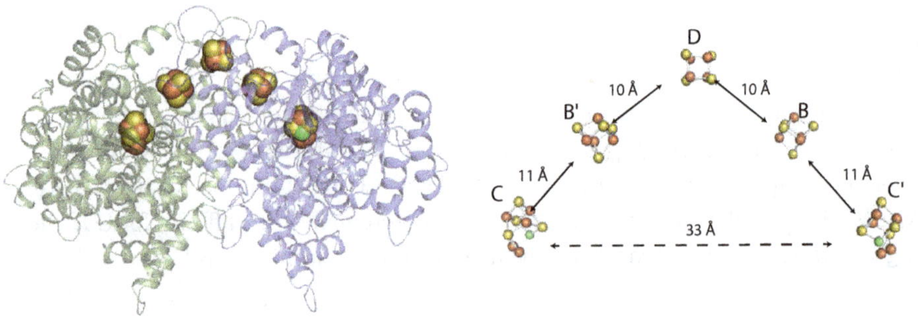

Figure 3.4: Overall structure of Ni,Fe-containing CODHs.
Left side: Dimeric structure of Ni,Fe-CODH, in which the secondary structure elements are depicted by semitransparent cartoons and the cofactors as van der Waals spheres. Right side: Ball-and-stick presentation of the V-shaped electron transfer chain containing the active site clusters C and C′ (the prime denotes the cluster bound to the second subunit), the [4Fe4S]-clusters B and B′ and the subunit bridging [4Fe4S]-cluster D.

subunit consists of three domains: the N-terminal domain (residues 3–237; numbering as in $CODHII_{Ch}$) is mostly α-helical and harbors the binding regions for the B- and D-clusters in its first subdomain (3–72); the middle (238–406) and the C-terminal (407–636) domain show a Rossman-fold topology.

The protein matrix forms conduits for the substrates: CO and CO_2; water as well as protons and electrons must be able to rapidly reach and egress from the active site to allow the high turnover observed with Ni,Fe-CODHs. Four different paths reach out from cluster C: a gas channel (CO/CO_2), a proton relay, a water network and an electron transfer chain.

Monofunctional Ni,Fe-CODHs employ two gas channels. The first channel is conserved and coincides with the tunnel connecting cluster C of Ni,Fe-CODH with the active site cluster A of ACS in bifunctional Ni,Fe-CODHs. The second hydrophobic channel is unique to monofunctional Ni,Fe-CODHs and is directed toward the solvent, allowing rapid pro- and egress of CO/CO_2 from the active site toward the solvent [38, 40].

The presence of a dynamically formed gas channel, through which CO_2 may diffuse from the solvent to cluster C in the bifunctional CODH/ACS complex, was recently suggested by a molecular dynamics study. The simulations also indicate that upon CO_2 reduction, the extended hydrogen network prohibits CO leakage through the dynamic gas channel [41].

Transfer of water as well as protons is most likely guided by charged and hydrophilic residues that form a continuous network from the protein surface to cluster C. Protons are thought to enter and leave the active site through His93, which is close to three consecutive His-residues (96, 99 and 102 in $CODHII_{Ch}$), of which the last residue of the chain is in contact with the solvent.

The protein matrix also aligns the redox-active metal clusters, forming a conduit for the electrons (Figure 3.4). Electrons may be transferred along the cascade of FeS-clusters (C <-> B <-> D) before reducing external electron acceptors. The participation of cluster D in electron transfer has not been established yet, as its reduction potential may be very negative [42]. Cluster D is coordinated by two Cys residues of each subunit at the dimer interface covalently linking both subunits. The two clusters B are positioned within typical biological electron transfer distances of 10 Å from cluster D and 11 Å from the active site cluster C.

The structure of cluster C in its more oxidized catalytic state has been deduced from the crystal structure of $CODHII_{Ch}$, poised to a redox potential of −320 mV [43], revealing a $[NiFe_4S_4{-}OH]$ composition (Figure 3.5). At this potential, the cluster adopts the oxidation state able to oxidize CO (below termed C_{red1}). Cluster C consists of a distorted $NiFe_3S_4$ heterocubane connected to an Fe(II) ion in *exo*. The Fe (II) ion (Fe1), also known as ferrous component II (FCII), is coordinated by a hydroxyl-ligand. Ni has an unusual coordination, as it is coordinatively unsaturated with a distorted planar T-shaped arrangement of two μ_3-S and one Cys thiolate

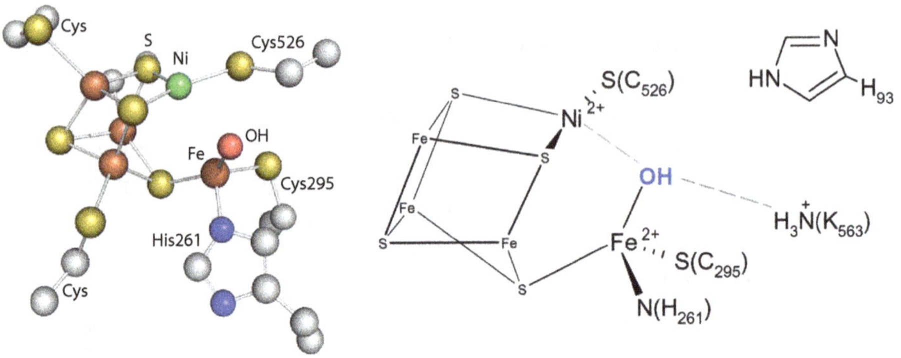

Figure 3.5: Active site structure of Ni,Fe-containing CODHs.
Left side: Ball-and-stick presentation of the active site cluster C. Only the side chains of coordinating amino acids are shown. Right side: Schematic presentation of cluster C in the same orientation as on the left side.

ligand. The hydroxyl-ligand of Fe1 may be regarded as a weak fourth ligand of Ni, with an Ni–OH distance of 2.7 Å, which would complete the square-planar coordination expected from a d^8 transition metal such as Ni^{2+} [2].

Structural insights into CO/CO_2-activation were gained from a CO_2-bound structure [43]. When crystals of CODHII$_{Ch}$ are incubated in a solution containing Ti(III)-citrate, a strong reducing agent employed to generate a reduction potential of −600 mV and 45 mM $NaHCO_3$, equivalent to a concentration of solvated CO_2 of 0.45 mM at pH 8.0, turnover conditions are generated in the crystal. Crystals were allowed to react in the solution for several minutes before being shock-frozen.

The structure of this CO_2-incubated state, finally resolved at a resolution of 1.03 Å, unraveled a [$NiFe_4S_4(CO_2)$]-cluster, in which a carboxylate ligand bridges Ni and Fe1 acting as a μ_2-η^2-ligand [44]. The carbon atom of CO_2 completes the square-planar coordination of Ni (η^1-CO_2 mode) and one of the oxygens of CO_2 replaces the hydroxo/water ligand at Fe1 (η^1-OCO mode). CO_2 is in an activated state as evident from an O–C–O–bending angle of approximately 117°, resembling a metal-bound carboxylate. The O–C–O angle and C–O bond length (1.30 and 1.32 Å) indicate that the metal-bound CO_2 is reduced by two electrons. The short Ni–C bond of 1.8 Å reveals substantial π-back-bonding and resembles a Ni–carbon complex. The Ni,Fe-bound CO_2 is additionally stabilized by hydrogen bonds to a lysine and a histidine residue. The overall structure of the Ni/Fe-sulfide frame of cluster C is practically unchanged by CO_2 binding, with only a slight movement (0.2 Å) of Ni, despite the additional ligand.

3.3.2 Spectral properties of the active site

According to various spectroscopic investigations, primarily relying on EPR spectroscopy, we can distinguish at least four different oxidation states of cluster C (C_{ox}, C_{red1}, C_{int}, C_{red2}). In the active state for CO oxidation (C_{red1}), cluster C has a hydroxyl-ligand bound to ferrous Fe1 [43]. C_{red1} is paramagnetic (S=1/2 spins) with EPR signals at around g_{av} = 1.87. The proposed oxidation state of the C_{red1} state is $\{[Ni^{2+}Fe^{2+}]:[Fe_3S_4]^-\}$, which can be reconciled with the S=1/2 spin state and Mössbauer spectra [42].

The midpoint potential of C_{red2} ($E^{o\prime}$ = −530 mV) coincides with the value obtained for the CO_2/CO couple ($E^{o\prime}$= −558 mV). C_{red2} shows a minor shift in the paramagnetic resonance spectrum compared to C_{red1} with a g_{av}-value of 1.86 (1.97, 1.87 and 1.75 for g_1, g_2 and g_3). The diamagnetic and catalytically inactive, oxidized C_{ox}-state can be reductively activated at potentials below −200 mV, regenerating the one-electron reduced C_{red1} state. C_{int} is formed by one-electron oxidation of C_{red2}. It produces no paramagnetic signal and likely has an integer spin state [45].

3.3.3 Mechanism of Ni,Fe-CODHs

As in the case of Cu,Mo-CODH, small-molecule inhibitors are able to reveal different aspects of the reactivity of the active site and cluster C and the mechanism of Ni,Fe-CODH. Ni,Fe-CODHs are inhibited or inactivated by several small molecules, binding and altering cluster C. These inhibitors include cyanide, cyanate, isocyanides, nitrous oxide and dioxygen [10].

Cyanide is isoelectronic to CO and acts as a slow-binding, competitive inhibitor. Binding and interaction of cyanide with cluster C has been investigated by kinetic, structural as well as spectroscopic methods, revealing the details of the inhibited state and indicating a possible model for CO-binding to cluster C. Cyanide acts as a slow binding, competitive inhibitor. For competitive inhibitors the inhibition constant K_i is equivalent to the dissociation constant of the enzyme inhibitor complex and have been determined to 8.5 µM (CODH of *R. rubrum)* and 21.7 µM ($CODHII_{Ch}$) [46, 47]. CN^- binds specifically to the C_{red1} state, changing the EPR signal. Both the specificity for C_{red1} as well as the K_i, which is in the same region as the K_m of CODHs for CO, show clear parallels between the interaction of CN^- and CO with Ni, Fe-CODHs.

In the structure of $CODHII_{Ch}$ with CN^-, cyanide binds to the open square-planar coordination site of Ni in an approximately linear manner (Ni–C–N angle of 168°). The OH-ligand on Fe1 is lost and Fe1 shifts from position A (Fe1A) to an alternative position B (Fe1B), more close to Ni. Cyanide binding is stabilized by hydrogen bonds with Lys563 and His93 [48]. A crystal structure of the bifunctional CODH/ACS complex from *Moorella thermoacetica* indicated a bent conformation for CN^-

with a distorted tetrahedral coordination of Ni [49]. IR spectroscopy on CN-inhibited samples revealed a band at 2,110 cm^{-1}, which was used as sensor for the environment and together with model calculations indicated that upon CN-binding the hydroxyl-ligand at Fe1 is lost and likely protonated by a nearby lysine residue [50]. The data are also consistent with a square planar coordination of Ni^{2+} and a tilted CN-ligand.

Cyanate inhibits CO_2 reduction and binds specifically to C_{red2}, stabilizing the oxidation state. Cyanate binds to cluster C in a bent conformation, strongly resembling the binding of CO_2. Just like CO_2, it is reduced by two electrons in the structure and corresponds to a carbamoyl-group stabilized by strong π-back-bonding [44]. Surprisingly, cyanate is most likely not itself the inhibiting species in solution, as it was shown to be reduced to the inhibitory CN [51].

When a crystal of $CODHII_{Ch}$ was incubated with *n*-butyl isocyanide (nBIC), a new variant of inhibition became obvious. nBIC acts as a rapid binding competitive inhibitor and slow-turnover substrate of Ni,Fe-CODH. In CODH crystals, nBIC reacted with the OH-ligand of Fe1, yielding a product-bound state in which the oxidation product *n*-butyl isocyanate is bound to the Ni ion of cluster C [40].

The spectroscopic investigations, the crystal structures with bound substrates and inhibitors as well as model calculations, agree with the following mechanism (Figure 3.6): (I) In the oxidized state (C_{red1}) Ni^{2+} has a T-shaped coordination with a weak interaction with the OH-ligand at the neighboring Fe^{2+} ion. (II) CO_2 is assumed to bind first to the Ni ion, before it replaces the hydroxyl-ligand at the Fe^{2+} ion, acting as a bridging ligand between Ni and Fe (III). The carboxylate is stabilized by electrostatic and H-bonding interactions with Lys563 and His93. The next step is most likely a protonation the carboxylate at the Fe–O position, initiating C–O bond cleavage, resulting in (IV). By formulating the in principal separated steps of the two-electron reduction of the cluster and the CO_2-binding in a single step between (I) and (II), it is not specified where the two electrons needed to reduce CO_2 are residing in cluster C. The location of the two electrons is one of the central unresolved issues in this mechanism, as the difference in the electronic state of C_{red1} and C_{red2} is unresolved – an issue of obvious importance as C_{red2} is the CO_2-binding and reducing state of the cluster. While being two electrons more reduced than the C_{red1} state, it is unclear where the two electrons reside. Thus, an electronic description of this important state is still under debate and involvement of Ni^0, a hydride-bound Ni^{2+} and an Ni–Fe bond (dative metal–metal bond) were proposed [52, 53].

Seen from the direction of CO oxidation, CO binds to (I), where it may either bind in the apical coordination site of nickel (Ni^{2+}), relocating to the equatorial site at the Ni ion, or it binds directly to the equatorial position. Bound CO is likely stabilized by a hydrogen bond with His93, increasing the polarization of the Ni-bound carbonyl and preparing the carbon atom of CO for a nucleophilic attack by the Fe1 bound OH-group.

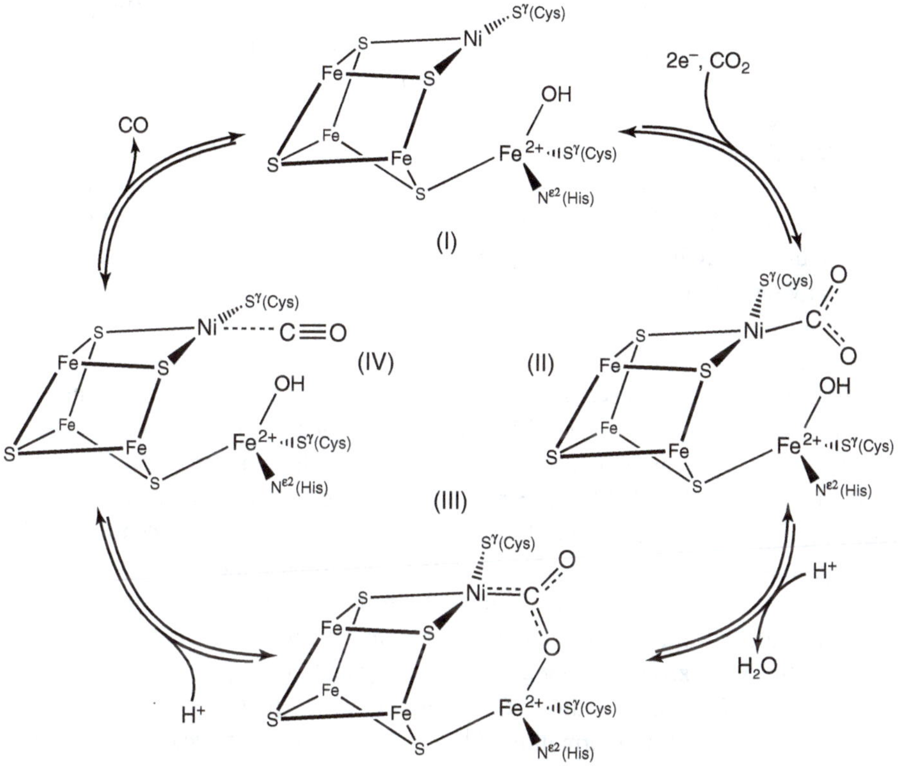

Figure 3.6: Mechanism of Ni,Fe-containing CODHs.
Schematic presentation of the catalytic cycle. For details see the main text.

3.3.4 Model complexes of the Ni,Fe-CODH active site

Synthetic catalysts able to reduce CO_2 are numerous and include various homo- and heterogeneous systems. In the following, a few cases that are either structurally or functionally related to Ni,Fe-CODHs will be discussed.

As cluster C allows CO_2 reduction with a minimal overpotential, synthesizing compounds of the structure of cluster C would be highly attractive. The core motif of cluster C is that of an $NiFe_3S_4$ heterocubane cluster, which can be synthesized. However, the step from an $NiFe_3S_4$ heterocubane cluster holds synthetic challenges: (I) a square-planar Ni(II) coordination unit has to be installed in a $NiFe_3S_4$ core; (II) an *exo*-Fe(II), bridged to the $NiFe_3S_4$ core has to be stabilized [54]. The first challenge has been overcome, the second not yet.

{$NiFe_3S_4$}-heterocubanes are known for some time. Using one of this {$NiFe_3S_4$}-heterocubanes as starting material, a tetrahedral to square-planar Ni(II) conversion was invoked by using a ligand with a strong in-plane ligand field. Introduction of the ligand also breaks an Ni–S bond of the former heterocubane cluster. The resulting $(Et_4N)[(dmpe)NiFe_3S_4(LS)_3]$ complex with a square planar Ni(II) incorporated into the heterocubane resembles the core of cluster C closely [54]. Unfortunately, this and similar complexes lack the *exo*-Fe and do not catalyze CO oxidation or CO_2 reduction.

Interestingly, some minerals share the basic composition of cluster C. Greigite, consisting of Fe_3S_4 units, has some CO_2-reducing activity, which increases when Ni is present (Ni-greigite $(Fe_3Ni)S_4$). Violarite, which has a higher Ni content ($FeNi_2S_4$), has a substantially higher efficiency in CO_2 reduction, indicating the beneficial effect of having both metals in the sulfide-rich materials [3].

Metal complexes with little resemblance of cluster C are not directly comparable to the enzyme, but can inform us about the principal challenges of CO_2 reduction and the individual mechanistic steps. Here only a few well-studied complexes will be mentioned. $[Ni(II)\ cyclam]^{2+}$ is a small complex containing Ni, interesting mostly because the catalytic mechanism has been studied in some detail revealing product specificity [55]. $[Ni(II)\ cyclam]^{2+}$ on an Hg-working electrode has high CO_2-reducing activity, in which most likely formation of a Ni(I) state is central for the activity. One challenge in CO_2 reduction by two electrons is the choice between the two possible products – CO versus formate. Investigation of the $[Ni(II)cyclam]^{+}$ complex indicates that the strong preference for formation of CO versus formate of this catalyst observed experimentally derives from the initial complex formed with CO_2 [55]. While the formation of the η^1-CO_2 complex, which preferentially gives CO, is nearly thermoneutral, the formation of the η^1-OCO complex, typically favoring formate production, is endergonic – thus the first step in the reaction sequence decides the reaction product. The overall mechanism may be described as concerted proton–electron transfer and C–O bond cleavage. In another Ni-complex a PNP-pincer ligand was used in combination with a CO-ligand [56]. When the Ni(II) complex is stepwise reduced to Ni(0), the complex changes from square-planar to distorted tetrahedral coordination, whereby the CO-ligand obviously support generation of Ni(0). This geometry change opens a coordination site for CO_2, which binds upon expulsion of CO. The resulting Ni(II) complex contains a Ni(η^1-CO_2) moiety resembling the CO_2-bound intermediate found in Ni,Fe-CODH.

CO_2 reduction is supported by several different mono- and bimetallic complexes. Fe-tetraphenylporphyrins have been thoroughly investigated using electrochemistry. A tetraphenylporphyrin variant, in which all phenyl-rings carry phenolic protons at the ortho and ortho' positions, achieves especially rapid turnover [57, 58]. The active species is Fe(0), which reacts with CO_2 to Fe(II)-CO, thereby circumventing the high-energy CO_2^{*-} radical anion. CO is generated with more than 90% Faraday efficiency and remaining electrons most likely reduce protons. The catalyst achieves 50 million turnovers in 4 h at a moderate overpotential (0.465 V) and shows no degradation. The increased efficiency of the catalyst brought about by the phenolic hydroxyl-groups is likely not solely due to the increased local proton concentration, but also due to a stabilization of the initial Fe(0)-CO_2 adduct – two effects that are also found in the active site of Ni,Fe-CODHs. A variant of this catalyst, in which two of the four phenyl-rings are perfluorinated, achieves near quantitative Faradaic efficiency, producing an efficient CO_2-to-CO reduction catalyst.

Pd–phosphine complexes are not similar to cluster C, but share several parallels with CODHs in their CO_2-reducing mechanism. The mechanism of CO_2 reduction at the mononuclear Pd-complex, [Pd(triphosphine)(solvent)]$^{2+}$, has been studied in some detail [3]. In the first step, Pd(II) is reduced to Pd(I), where in the second step CO_2 binds, forming a metal carboxylate. CO_2 gets reduced by one, not two electrons in the first step, and all subsequent steps include regulated proton/electron addition. Thus, before the one-electron-reduced CO_2 receives a second electron, it becomes protonated. A second protonation follows and initiates C–OH_2 bond cleavage whose products, CO and water remain bound to Pd(II) and the catalyst is regenerated by product release. The rate-limiting step appears to be the reaction of mononuclear Pd(I) with CO_2, which has been improved in a bimetallic Pd–phosphine complex, which facilitates formation of the CO_2-bound adduct and uses a similar way of CO_2 activation as Ni,Fe-CODH. Analogous to the above-mentioned mechanism, the following steps likely include reduction of Pd(II) to a Pd(I)-Pd(I) complex, formation of a diPd(II)-carboxylate complex, protonation of the carboxylate to promote C–O bond cleavage and release of CO and H_2O. A problem of this complex is its tendency to inactivate after few turnovers, likely due to Pd–Pd bond formation. This is likely less of a problem in Ni,Fe-CODH, as the two catalytic metals have different reduction potentials.

In sum, the model complexes underline some features found in the active site of Ni,Fe-CODHs, such as the importance of metal–metal interactions, a possible, biologically relevant Ni(I) state, the possibility that bound CO may support CO_2 reduction at cluster C and finally the relevance of proton-transfer-enabling PCET.

Acknowledgments: Work on CODHs in the lab was supported by the German funding agency DFG (DO-785) and the excellence initiative (EXC 314 – "Unifying concepts in catalysis, UniCat"). The author thanks all his past and present coworkers and collaborators in the field.

References

[1] Holleman, AF., Nils Wiberg, E., and Fischer, G. Lehrbuch der Anorganischen Chemie, Walter de Gruyter, Berlin New York, 2007.

[2] Crabtree, RH. The organometallic chemistry of the transition metals, John Wiley & Sons, Inc, Hoboken, NJ, USA, 2005.

[3] Appel, AM., Bercaw, JE., Bocarsly, AB., Dobbek, H., DuBois, DL., Dupuis, M. et al.. Frontiers, opportunities, and challenges in biochemical and chemical catalysis of CO_2 fixation. Chem Rev 2013, 113, 6621–6658.

[4] Khalil, MAK., Pinto, JP., and Shearer, MJ. Atmospheric carbon monoxide. Chemosphere – Global Change Sci 1999, 1, ix–xi.
[5] King, GM. Characteristics and significance of atmospheric carbon monoxide consumption by soils. Chemosphere – Global Change Sci 1999, 1, 53–63.
[6] King, GM., and Weber, CF. Distribution, diversity and ecology of aerobic CO-oxidizing bacteria. Nat Rev Micro 2007, 5, 107–118.
[7] Fuchs, G. Alternative pathways of carbon dioxide fixation: insights into the early evolution of life?. Annu Rev Microbiol 2011, 65, 631–658.
[8] Ljungdahl, LG. A life with acetogens, thermophiles, and cellulolytic anaerobes. Annu Rev Microbiol 2009, 63, 1–25.
[9] Ragsdale, SW., and Pierce, E. Acetogenesis and the wood-ljungdahl pathway of CO_2fixation. Biochimica et Biophysica Acta (BBA) – Proteins & Proteomics 2008, 1784, 1873–1898.
[10] Jeoung, J-H., Fesseler, J., Goetzl, S., and Dobbek, H. Carbon monoxide. Toxic gas and fuel for anaerobes and aerobes: carbon monoxide dehydrogenases. Met Ions Life Sci 2014, 14, 37–69.
[11] Adam, PS., Borrel, G., and Gribaldo, S. Evolutionary history of carbon monoxide dehydrogenase/acetyl-CoA synthase, one of the oldest enzymatic complexes. Proc Natl Acad Sci U.S.A 2018, 1, 201716667–201716668.
[12] Zhang, B., Hemann, CF., and Hille, R. Kinetic and spectroscopic studies of the molybdenum-copper CO dehydrogenase from *Oligotropha carboxidovorans*. J Biol Chem 2010, 285, 12571–12578.
[13] Santiago, B., and Meyer, O. Characterization of hydrogenase activities associated with the molybdenum CO dehydrogenase from *Oligotropha carboxidovorans*. FEMS Microbiol Lett 1996, 136, 157–162.
[14] Wilcoxen, J., and Hille, R. The hydrogenase activity of the molybdenum/copper-containing carbon monoxide dehydrogenase of *oligotropha carboxidovorans*. J Biol Chem 2013, 288, 36052–36060.
[15] Wilcoxen, J., Snider, S., and Hille, R. Substitution of silver for copper in the binuclear Mo/Cu center of carbon monoxide dehydrogenase from *Oligotropha carboxidovorans*. J Am Chem Soc 2011, 133, 12934–12936.
[16] Wilcoxen, J., Zhang, B., and Hille, R. Reaction of the molybdenum- and copper-containing carbon monoxide dehydrogenase from *Oligotropha carboxydovorans* with quinones. Biochemistry 2011, 50, 1910–1916.
[17] Santiago, B., Schübel, U., Egelseer, C., and Meyer, O. Sequence analysis, characterization and CO-specific transcription of the cox gene cluster on the megaplasmid pHCG3 of *Oligotropha carboxidovorans*. Gene 1999, 236, 115–124.
[18] Dobbek, H., Gremer, L., Kiefersauer, R., Huber, R., and Meyer, O. Catalysis at a dinuclear [CuSMo(==O)OH] cluster in a CO dehydrogenase resolved at 1.1-A resolution.. Proc Natl Acad Sci U.S.A 2002, 99, 15971–15976.
[19] Hänzelmann, P., Dobbek, H., Gremer, L., Huber, R., and Meyer, O. The effect of intracellular molybdenum in *Hydrogenophaga pseudoflava* on the crystallographic structure of the seleno-molybdo-iron-sulfur flavoenzyme carbon monoxide dehydrogenase. J Mol Biol 2000, 301, 1221–1235.
[20] Page, CC., Moser, CC., Chen, X., and Dutton, PL. Natural engineering principles of electron tunnelling in biological oxidation-reduction. Nature 1999, 402, 47–52.
[21] Dobbek, H. Structural aspects of mononuclear Mo/W-enzymes. Coord Chem Rev Elsevier B.V 2011, 255, 1104–1116.
[22] Dobbek, H., and Huber, R. The molybdenum and tungsten cofactors: a crystallographic view. Met Ions Biol Syst 2002, 39, 227–263.

[23] Shanmugam, M., Wilcoxen, J., Habel-Rodriguez, D., Cutsail, GE., Kirk, ML., Hoffman, BM. et al.. (13)C and (63,65)Cu ENDOR studies of CO dehydrogenase from *Oligotropha carboxidovorans*. Experimental evidence in support of a copper-carbonyl intermediate. J Am Chem Soc 2013, 135, 17775–17782.

[24] Gnida, M., Ferner, R., Gremer, L., Meyer, O., Meyer-Klaucke, W., and Novel Binuclear, A. [CuSMo] cluster at the active site of carbon monoxide dehydrogenase: characterization by X-ray absorption spectroscopy †. Biochemistry 2003, 42, 222–230.

[25] Yano, J., Kern, J., Irrgang, K-D., Latimer, MJ., Bergmann, U., Glatzel, P. et al.. X-ray damage to the Mn_4Ca complex in single crystals of photosystem II: a case study for metalloprotein crystallography. Proc Natl Acad Sci U.S.A 2005, 102, 12047–12052.

[26] Resch, M., Dobbek, H., and Meyer, O. Structural and functional reconstruction in situ of the [$CuSMoO_2$] active site of carbon monoxide dehydrogenase from the carbon monoxide oxidizing eubacterium *Oligotropha carboxidovorans*. J Biol Inorg Chem 2005, 10, 518–528.

[27] Stein, BW., and Kirk, ML. Orbital contributions to CO oxidation in Mo–Cu carbon monoxide dehydrogenase. Chem Commun 2013, 50, 1104.

[28] Gourlay, C., Nielsen, DJ., White, JM., Knottenbelt, SZ., Kirk, ML., and Young, CG. Paramagnetic active site models for the molybdenum-copper carbon monoxide dehydrogenase. J Am Chem Soc 2006, 128, 2164–2165.

[29] Diekert, GB., Graf, EG., and Thauer, RK. Nickel requirement for carbon monoxide dehydrogenase formation in *Clostridium pasteurianum*. Archives of Microbiol Springer-Verlag 1979, 122, 117–120.

[30] Drake, HL., Hu, SI., and Wood, HG. Purification of carbon monoxide dehydrogenase, a nickel enzyme from *Clostridium thermoaceticum*. J Biol Chem 1980, 255, 7174–7180.

[31] Wu, M., Ren, Q., Scott Durkin, A., Daugherty, SC., Brinkac, LM., Dodson, RJ. et al.. Life in hot carbon monoxide: The complete genome sequence of *Carboxydothermus hydrogenoformans* Z-2901. PLoS Genet 2005, 1, 563–574.

[32] Svetlitchnyi, V., Dobbek, H., Meyer-Klaucke, W., Meins, T., Thiele, B., Römer, P. et al.. A functional Ni-Ni-[4Fe-4S] cluster in the monomeric acetyl-CoA synthase from *Carboxydothermus hydrogenoformans*. Proc Natl Acad Sci U.S.A 2004, 101, 446–451.

[33] Domnik, L., Merrouch, M., Goetzl, S., Jeoung, J-H., Léger, C., Dementin, S. et al.. CODH-IV: a high-efficiency CO-scavenging CO dehydrogenase with resistance to O_2. Angew Chem Int Ed 2017, 56, 15466–15469.

[34] Svetlitchnyi, V., Peschel, C., Acker, G., and Meyer, O. Two membrane-associated NiFeS-carbon monoxide dehydrogenases from the anaerobic carbon-monoxide-utilizing eubacterium *Carboxydothermus hydrogenoformans*. J Bacteriol 2001, 183, 5134–5144.

[35] Lindahl, PA., and Chang, B. The evolution of acetyl-CoA synthase. Orig Life Evol Biosph 2001, 31, 403–434.

[36] Dobbek, H., Svetlitchnyi, V., Gremer, L., Huber, R., and Meyer, O. Crystal structure of a carbon monoxide dehydrogenase reveals a [Ni-4Fe-5S] cluster. Science. American Association for the Advancement of Science 2001, 293, 1281–1285.

[37] Drennan, CL., Heo, J., Sintchak, MD., Schreiter, E., and Ludden, PW. Life on carbon monoxide: X-ray structure of *Rhodospirillum rubrum* Ni-Fe-S carbon monoxide dehydrogenase. Proc Natl Acad Sci U.S.A 2001, 98, 11973–11978.

[38] Doukov, TI., Iverson, TM., Seravalli, J., Ragsdale, SW., and Drennan, CL. A Ni-Fe-Cu center in a bifunctional carbon monoxide dehydrogenase/acetyl-CoA synthase. Science 2002, 298, 567–572.

[39] Gong, W., Hao, B., Wei, Z., Ferguson, DJ., Tallant, T., Krzycki, JA. et al.. Structure of the $\alpha 2\varepsilon 2$ Ni-dependent CO dehydrogenase component of the *Methanosarcina barkeri* acetyl-CoA decarbonylase/synthase complex. Proc Natl Acad Sci 2008, 105, 9558–9563.

[40] Jeoung, J-H., and Dobbek, H. n-Butyl isocyanide oxidation at the [$NiFe_4S_4OH_x$] cluster of CO dehydrogenase. J Biol Inorg Chem 2012, 17, 167–173.
[41] Wang, P-H., Bruschi, M., De Gioia, L., and Blumberger, J. Uncovering a dynamically formed substrate access tunnel in carbon monoxide dehydrogenase/acetyl-CoA synthase. J Am Chem Soc 2013, 135, 9493–9502.
[42] Lindahl, P. The Ni-containing carbon monoxide dehydrogenase family: light at the end of the tunnel?. Biochemistry 2002, 41, 2097–2105.
[43] Jeoung, J-H., and Dobbek, H. Carbon dioxide activation at the Ni,Fe-cluster of anaerobic carbon monoxide dehydrogenase. Science 2007, 318, 1461–1464.
[44] Fesseler, J., Jeoung, J-H., and Dobbek, H. How the [$NiFe_4S_4$] cluster of CO dehydrogenase activates CO_2 and NCO^-. Angew Chem Int Ed Engl 2015, 54, 8560–8564.
[45] Anderson, ME., and Lindahl, PA. Spectroscopic states of the CO oxidation/CO_2 reduction active site of carbon monoxide dehydrogenase and mechanistic implications. Biochemistry 1996, 35, 8371–8380.
[46] Ensign, SA., Hyman, MR., and Ludden, PW. Nickel-specific, slow-binding inhibition of carbon monoxide dehydrogenase from *Rhodospirillum rubrum* by cyanide. Biochemistry 1989, 28, 4973–4979.
[47] Ha, S-W., Korbas, M., Klepsch, M., Meyer-Klaucke, W., Meyer, O., and Svetlitchnyi, V. Interaction of potassium cyanide with the [Ni-4Fe-5S] active site cluster of CO dehydrogenase from *Carboxydothermus hydrogenoformans*. J Biol Chem 2007, 282, 10639–10646.
[48] Jeoung, J-H., and Dobbek, H. Structural basis of cyanide inhibition of Ni, Fe-containing carbon monoxide dehydrogenase. J Am Chem Soc 2009, 131, 9922–9923.
[49] Kung, Y., Doukov, TI., Seravalli, J., Ragsdale, SW., and Drennan, CL. Crystallographic snapshots of cyanide- and water-bound C-clusters from bifunctional carbon monoxide dehydrogenase/acetyl-CoA synthase. Biochemistry 2009, 48, 7432–7440.
[50] Ciaccafava, A., Tombolelli, D., Domnik, L., Fesseler, J., Jeoung, J-H., Dobbek, H. et al.. When the inhibitor tells more than the substrate: the cyanide-bound state of a carbon monoxide dehydrogenase. Chem Sci 2016, 7, 3162–3171.
[51] Ciaccafava, A., Tombolelli, D., Domnik, L., Jeoung, J-H., Dobbek, H., Mroginski, MA. et al.. Carbon monoxide dehydrogenase reduces cyanate to cyanide. Angew Chem Int Ed Engl 2017, 56, 7398–7401.
[52] Amara, P., Mouesca, J-M., Volbeda, A., and Fontecilla-Camps, JC. Carbon monoxide dehydrogenase reaction mechanism: a likely case of abnormal CO_2 insertion to a Ni-H^- bond. Inorg Chem 2011, 50, 1868–1878.
[53] Lindahl, PA. Metal-metal bonds in biology. J Inorg Biochem 2012, 106, 172–178.
[54] Panda, R., Zhang, Y., McLauchlan, CC., Venkateswara Rao, P., Tiago de Oliveira, FA., Münck, E. et al. Initial structure modification of tetrahedral to planar nickel(II) in a nickel-iron-sulfur cluster related to the C-cluster of carbon monoxide dehydrogenase. J Am Chem Soc 2004, 126, 6448–6459.
[55] Song, J., Klein, EL., Neese, F., and Ye, S. The mechanism of homogeneous CO_2 reduction by Ni (cyclam): product selectivity, concerted proton-electron transfer and C-O bond cleavage. Inorg Chem 2014, 53, 7500–7507.
[56] Sahoo, D., Yoo, C., and Lee, Y. Direct CO_2 addition to a Ni(0)-CO species allows the selective generation of a nickel(II) carboxylate with expulsion of CO. J Am Chem Soc 2018, 140, 2179–2185.
[57] Costentin, C., Drouet, S., Robert, M., and Savéant, J-M. A local proton source enhances CO2 electroreduction to CO by a molecular Fe catalyst. Science 2012, 338, 90–94.
[58] Costentin, C., Passard, G., Robert, M., and Savéant, J-M. Ultraefficient homogeneous catalyst for the CO_2-to-CO electrochemical conversion. Proc Natl Acad Sci 2014, 111, 14990–14994.

Andrew Jasniewski, Caleb Hiller, Yilin Hu and Markus Ribbe

4 The study of nitrogen reduction by nitrogenase

4.1 General introduction

The major component (~78%) of Earth's atmosphere is the relatively inert molecular form of nitrogen, called N_2 or dinitrogen. N_2 is a stable molecule due to the strong orbital overlap between the nitrogen atoms, resulting in a diatomic triple bond with a bond dissociation energy of 945 kJ/mol [1]. Most organisms require atomic nitrogen (N) to generate fundamental cellular building blocks such as amino acids for protein synthesis and nucleobases for the synthesis of ribonucleic acid (RNA) and deoxyribonucleic acid (DNA). Thus, the cleavage of the N_2 triple bond and conversion into more accessible molecules such as ammonia (NH_3), a process known as nitrogen fixation, represents one of the most important and challenging chemical transformations in Nature. Humans have devised industrial methods to generate NH_3 from N_2, which is used in the production of fertilizers to support food production, and this has allowed for the ever-increasing world population [2, 3]. This "artificial" nitrogen fixation is known as the Haber–Bosch process and is represented in the following equation:

$$N_{2(g)} + 3H_{2(g)} \underset{400-500^{\circ}C}{\overset{15-25\,\text{Mpa}}{\rightleftharpoons}} 2NH_{3(g)}$$

where H_2 is hydrogen gas. This reaction is carried out with high temperature and pressure using a series of transition metal catalysts to overcome the reversible nature of the reaction, and therefore is an energy-intensive process [3]. Additionally, specialized high-pressure fittings and piping, reaction vessels and safety equipment are necessary to effectively carry out the Haber–Bosch reactions.

In Nature, diazotrophic organisms have developed biological machinery to facilitate the nitrogen fixation reaction in the form of the enzyme system nitrogenase [4, 5]. Nitrogenase produces NH_3 from N_2 by the following reaction:

$$N_2 + 8H^+ + 16MgATP + 8e^- \rightarrow 2NH_3 + H_2 + 16MgADP + 16P_i$$

where H^+ is a proton, MgATP is magnesium adenosine triphosphate, MgADP is magnesium adenosine diphosphate and P_i is inorganic phosphate. There are several advantages of biological nitrogen fixation compared to the artificial process,

Andrew Jasniewski, Yilin Hu, Department of Molecular Biology and Biochemistry, University of California, Irvine, CA, USA
Caleb Hiller, Markus Ribbe, Department of Chemistry, University of California, Irvine, CA, USA

https://doi.org/10.1515/9783110496574-004

the primary being that in the biological system ambient temperature and pressure is used, which greatly reduces the energy required for the scission of the N_2 bond. Additionally, water serves as the source of hydrogen for the reaction in the form of protons in nitrogenase, whereas hydrogen gas is required for Haber–Bosch, and H_2 is often generated industrially from the steam reformation of light hydrocarbons [6, 7]. However, biological nitrogen fixation also requires a large amount of the cellular energy "currency" MgATP, 16 molecules per N_2 molecule, which does not exactly reflect an efficient atom economy for the reaction. Regardless, N_2 reduction by nitrogenase has captured the attention of research groups interested in the important role the enzyme plays in the global nitrogen cycle since at least the early 1900s [8]. The reactivity of nitrogenase is facilitated by a series of complex FeS clusters that both transfer the multiple electron equivalents required for the chemical transformation, as well as bind N_2 directly for substrate reduction [4, 5, 9]. Over many decades, nitrogenase has also been found to reduce additional small molecules, including H^+ alone, N_3^-, C_2H_2, CO and CO_2, though the physiological importance of these reactions, if any, is not well understood [10–13].

Several forms of nitrogenase have been identified in diazotrophs, and these proteins are differentially expressed depending on the availability of transition metals during cell growth [14, 15]. The molybdenum-dependent nitrogenase (Mo-nitrogenase) is considered the most efficient at N_2 reduction and is the enzyme that is favorably expressed in bacterial and archaeal organisms [4, 5, 14]. Mo-nitrogenase from the soil bacterium *Azotobacter vinelandii* has been the most extensively characterized nitrogenase system [4, 5, 9], and is expressed and regulated by genes of the nitrogen fixation (*nif*) operon, some of which are shown in Table 4.1. In the absence of molybdenum,

Table 4.1: Designation and functions of the *nif*-gene products.

Gene	Protein code	Function	Reference
nifU	NifU	NifU is a scaffold protein involved in the assembly of small FeS clusters.	[16–18]
nifS	NifS	NifS is involved in sulfur mobilization to NifU for the assembly of small FeS clusters.	[16–21]
nifB	NifB	NifB mediates a radical SAM-dependent carbide insertion concomitant to L-cluster formation.	[22]
nifE, N	NifEN	NifEN is a scaffold protein used for M-cluster maturation.	[23–26]
nifH	NifH	NifH has three physiological roles: it mediates ATP-dependent electron transfer to the catalytic component, is involved in P-cluster biosynthesis, and acts as a Mo and homocitrate insertase in M-cluster maturation.	[4, 27–32]
nifZ	NifZ	NifZ is involved in the stepwise maturation of P-cluster.	[33, 30]
nifD, K	NifDK	NifDK catalyzes the reduction of N_2. Hydrocarbon formation is likely adventitious.	[4, 5]

This table is adapted from reference [9].

the vanadium-dependent (V-nitrogenase) or iron-only nitrogenase (Fe-nitrogenase) enzymes, termed "alternative nitrogenases," can be expressed in *A. vinelandii*, and are regulated by the *vnf* and *anf* genes, respectively [14, 15]. While not as efficient at N_2 reduction as the Mo-dependent nitrogenase, the alternative nitrogenases show interesting reactivity toward hydrocarbon substrates that are not observed in the Mo-containing system.

This chapter covers many of the important aspects of nitrogenase research, including protein structure, cofactor assembly and characterization, as well as the reactivity and mechanistic discussions.

4.1.1 Characterization of nitrogenases – introduction of characterization methods

Nitrogenase research is poised at the interface of several fields of study, including molecular biology, biochemistry, inorganic chemistry and spectroscopy. A multifaceted approach is necessary to best address many of the research questions that have yet to be answered, often employing techniques that span scientific disciplines. This requires a working knowledge of several characterization techniques that are critical to the understanding of problems related to the study of nitrogenase. This section briefly describes important methods and how the collected data can be used in nitrogenase research.

4.1.1.1 Anaerobic handling and gaseous substrates

The native substrate for nitrogenase is gaseous N_2 which is abundant in the atmosphere; however, the proteins and FeS clusters of nitrogenase are not tolerant of dioxygen (O_2) that is also found in the atmosphere. This poses a serious challenge for the handling of these proteins and the assessment of their reactivity. Fortunately, anaerobic handling techniques have been developed that allow for study of nitrogenase without interference from dioxygen.

Nitrogenase proteins are generally harvested from their native organisms, and these bacteria and archaea have methods to deal with the presence of O_2. The issues arise when these organisms are broken open to yield their contents, which potentially can expose the sensitive proteins to oxidative stress. During the protein purification process, the harvested cells are put under vacuum and back-filled with an inert gas such as argon to remove O_2 from the buffer-cell suspension. Glass or metal vacuum–gas manifolds, known as Schlenk lines, readily allow for this vacuum–gas exchange. In addition, a sacrificial oxygen scavenger such as sodium dithionite ($Na_2S_2O_4$) is added to the cell suspension before and after the cells are lysed to protect against degradation. The subsequent purification steps are performed under an inert gas environment, commonly in the presence of dithionite, and pure protein is frozen and stored in liquid nitrogen to mitigate exposure to oxygen.

To assess the reactivity of nitrogenase, the various protein components, including the catalytic and reductase components, are thawed in an inert atmosphere using a Schlenk line. To measure nitrogen reduction, a controlled quantity of N_2 is introduced to a sealed O_2-free reaction vessel, usually in a mixture of N_2 and argon. The nitrogenase proteins are then transferred using gastight syringes into the vessel with all the required components for catalysis. The reaction vessel can then be incubated at the appropriate temperature following established protocols, and the production of NH_3 can be analyzed from the resulting solution. Other gaseous substrates, such as CO or C_2H_2, can be introduced analogously and the resulting gaseous products can be analyzed from the reaction vessel headspace using gas chromatography.

4.1.1.2 X-ray crystallography

The functional nitrogenase system is composed of an $\alpha_2\beta_2$ heterotetrameric protein that contains the metallocofactors necessary for catalysis, as well as a smaller γ_2 homodimeric reductase protein (Fe protein), both of which are required for activity [4, 5]. In addition to the nitrogenase proteins responsible for reactivity, there are other necessary proteins and chaperones that assist in the biosynthesis of the critical metallocofactors, eventually transferring the synthesized cofactors to the catalytic nitrogenase component, generating active enzyme [9]. One main goal in nitrogenase research is understanding how these various proteins fold and interact with each other to regulate and facilitate N_2 reduction. X-ray crystallography is one of the best tools available for acquiring a three-dimensional structure of proteins and better understanding their interactions.

X-ray crystallography of proteins, or protein crystallography, allows the user to determine the structure of a protein through a combination of data collection and modeling. High-quality crystals are first grown of the protein(s) of interest, which is a nontrivial process that generally involves acquiring highly purified protein and screening through many conditions including chemical additive concentrations, buffers, temperatures and methods of crystallization. Once crystals have been collected, they are scrutinized under a microscope for quality, then are mounted on a circular holder called a loop and generally frozen and stored in liquid nitrogen. The loop can be installed precisely into an X-ray diffractometer, an instrument that generates an intense beam of X-rays that hits the sample crystal as it is rotated, and this results in a series of spots known as a diffraction pattern. The diffraction pattern is collected by the instrument detector, and the data can be processed and analyzed to generate a three-dimensional representation of the electron density observed in the crystal. A computational model of the protein is then generated and compared to the experimental electron density, and iterations of refinement are performed until there is acceptable agreement between the experiment and the model. The model and density are used to represent the 3-D structure of the protein to varying degrees of precision. Commercial X-ray diffractometers provide the ability to screen crystals for the quality of diffraction and generally yield low-resolution structures. The best crystals are usually selected and sent to a synchrotron facility, which can generate X-rays that are orders of magnitude more intense, and this is where high-resolution data can be collected. The higher the resolution of the collected data, the more accurately the protein structure can be determined.

With high-resolution protein structures available, a variety of information becomes accessible, including the overall protein fold, location of metallocofactors, amino acid residues that bind to said cofactors and the potentially charged residues on the protein surface. If a direct crystal structure of a protein–protein interaction is not available, all these factors can be used to try and predict these interactions using computation methods.

4.1.1.3 X-ray absorption and emission spectroscopies

While overall protein structure can be ascertained from crystallography, the structure of the metallocofactors in nitrogenase is not necessarily as well defined from this technique. The multiatomic FeS clusters are some of the most complex clusters found in nature, and the sub-angstrom resolution required to accurately observe chemical bonds is not generally achieved in protein crystallography experiments. One method that can be used to obtain more precise bonding metrics as well as metal oxidation state information is X-ray absorption spectroscopy (XAS).

XAS is a technique that is carried out at a synchrotron facility and the method targets a specific element, such as Fe in the nitrogenase metalloclusters, and provides average metal–ligand and metal–metal distances with high precision from the bulk measurement of a protein sample. A concentrated solution of the target metal-containing nitrogenase protein is mixed with a cryoprotectant such as glycerol to prevent ice crystal formation, and the resulting solution is added to a sample cell and frozen in liquid nitrogen. At the synchrotron beam line, the sample is placed into a cryostat at cryogenic temperature (usually between 4 and 77 K) and irradiated with X-rays. The X-rays are absorbed by the target element that causes a core electron to be excited and leave the target atom. The absorption generates a photoelectron, which subsequently scatters off atoms nearby the absorbing atom. The energy of the X-rays is varied during the sample collection, and the absorption/scattering near the target element at each energy point can be detected in several ways depending on the configuration of the experiment. With multiple scans collected per sample, the data can be processed and analyzed, providing the desired precise bonding metrics for the metal-containing nitrogenase protein. However, there are several caveats to the interpretation of the XAS data. The result of XAS analysis does not provide a three-dimensional picture of the target metal species, as one might expect from small-molecule crystallography, only the metal–scatterer distances are obtained. These pieces of information must be carefully considered to reconstruct a plausible model that agrees with the experimental data. Therefore, XAS analysis often pairs well with spectroscopic experiments and crystallography that provide complimentary structural information. Additionally, the *average* information is obtained in an XAS experiment, so pure protein samples with limited contaminants are required to acquire meaningful results. As nitrogenase contains multiple metalloclusters, proper control experiments must be carried out in parallel to the desired experiments to assure fruitful analysis.

X-ray emission spectroscopy (XES) is related to XAS in that the technique is element specific; however instead of directly acquiring bonding metrics, information about the electronic configuration of the target element is obtained. There are several different "flavors" of XES, but in the most general sense a sample is irradiated with an X-ray beam from a synchrotron source and a core electron from the target element is ionized, leaving a "core hole." Other electrons in higher energy levels can drop down to fill the core hole, which generates a detectible fluorescence signal. Depending on how the experiment is conducted, and which type of electron fills the core hole, the fluorescence data can be analyzed and interpreted to provide information about the types of bonds that the target element makes, for example, the nature of the $Fe–N_2$ bond in an iron complex. The interpretation of XES data is nontrivial and involves a combination of model complex studies to generate a library of well-characterized species that can be compared to experimental data of an unknown, as well as theoretical methodologies. XES has previously been used to determine that the interstitial atom in a nitrogenase metallocofactor was indeed a carbide (C^{4-}) as opposed to a nitride (N^{3-}) or an oxide (O^{2-}) atom [34].

Despite the challenges, XAS and XES are powerful tools for understanding the complex cofactors found in nitrogenase, particularly when paired with complimentary structural information and computations.

4.1.1.4 Electron paramagnetic resonance spectroscopy

The metallocofactors in nitrogenase can cycle through multiple oxidation states and in doing so play an invaluable role in facilitating electron transfer (ET) reactions as well as multielectron chemical transformations, such as the reduction of N_2 to NH_3. The redox changes of the cofactors bound to nitrogenase are not easy to assess using optical spectroscopies because they have similar properties with overlapping spectroscopic features, but electron paramagnetic resonance (EPR) spectroscopy provides a facile method to observe redox changes in the metalloclusters of nitrogenase [35].

EPR spectroscopy detects the presence of unpaired electrons by placing a paramagnetic sample in an external magnetic field and the sample is irradiated with microwave radiation. As the magnetic field strength is varied, the unpaired electron in the sample will flip its spin at a particular magnetic field strength, achieving resonance, and this results in the observation of a signal. The measured signal is sensitive to the electronic and nuclear environment, so as the nitrogenase cofactors are cycled through redox states that change the unpaired electrons, different characteristic signals can be detected for the cluster species. These changes are observed in the line-shape of the collected spectrum and the associated *g*-values, and have been valuable tools for studying the metallocluster biosynthesis in nitrogenase [9].

EPR spectroscopy is not limited to the observation of unpaired electrons, but can also be used to more extensively probe electron–nuclear interactions in paramagnetic systems [36, 37]. In an electron–nuclear double resonance (ENDOR) experiment, an EPR spectrometer is fit with a radiofrequency generator and receiver that can interact with the paramagnetic sample. When resonance is achieved using conventional EPR, the radiofrequency is scanned to perform a nuclear magnetic resonance experiment, coupling a nuclear spin flip with the electron spin flip and probing the electron–nuclear interactions in the sample. ENDOR can help identify the atoms that bind to the paramagnetic metal center, and for nitrogenase this could be used to identify substrate or product bound states of the metallocofactor, such as $Fe–N_2$ or Fe–H interactions. When the nucleus of interest is weakly coupled, or more distant from the paramagnetic metal center(s) in a sample, electron spin echo envelope modulation (ESEEM) experiments can be performed [36, 37]. In ESEEM studies, a paramagnetic sample with magnetic nuclei is placed in an external magnetic field. A precisely timed sequence of microwave radiation pulses irradiate the sample, and this allows for electron–nuclear interactions to be probed. Analysis of the ESEEM data allows for the identity and sometimes quantity of nuclei that interact with the unpaired electron to be determined. ENDOR and ESEEM provide similar types of information, but the physical methods used to acquire the data are completely different.

4.1.1.5 Mössbauer spectroscopy

With the discovery of recoilless nuclear resonance absorption by Rudolf Mössbauer in 1958, the phenomenon termed the “Mössbauer effect” has been used to study the nuclear transitions of Mössbauer-active atoms, particularly ^{57}Fe [38, 39]. As all the metallocofactors in nitrogenase are FeS clusters, Mössbauer spectroscopy is a useful tool in probing the electronic environment of and extracting chemical information about the Fe atoms in the clusters.

Mössbauer spectroscopy provides information about the electronic configuration of the ^{57}Fe atoms present in a sample, quantified in the value of δ, or the isomer shift, and also probes the ligand environment and symmetry of the metal centers, quantified by ΔE_Q, or the quadrupole splitting [38, 39]. These parameters together can detail the oxidation state, spin state and local coordination environment of the enriched Fe centers, as well as some magnetic properties. To collect data, a ^{57}Fe-containing sample is frozen and put into a cryostat at cryogenic temperatures. A radioactive ^{75}Co γ-ray source is used to irradiate the sample, and the absorption of the radiation is detected, yielding a spectrum. These experiments can be done in the absence or presence of an applied magnetic field depending on what information is desired. Computations and simulations of the data often accompany the interpretation of Mössbauer data.

A challenging aspect of Mössbauer spectroscopy is that analysis of an Fe-containing sample requires the enrichment of the ^{57}Fe isotope, which is naturally found in ~2% abundance compared to ~90% for the primary isotope ^{56}Fe. For a synthetic compound, it is relatively easy to source an ^{57}Fe salt to metallate a complex directly or generate a sample of naturally abundant Fe complex with a high concentration such that the 2% of ^{57}Fe present in the sample is enough to conduct experiments. However, for nitrogenase, the cells need to be grown on ^{57}Fe-encriched media, so that isotopically labeled metalloclusters will be present in the matured proteins [40]. Additionally, there are protocols to selectively enrich one nitrogenase cofactor with ^{57}Fe while the other remains ^{56}Fe, which provides the opportunity to study each cluster separately using the Mössbauer technique [41, 42].

4.2 Functional components of nitrogenase

Nitrogenase is a two-component system that is composed of a catalytic component responsible for the reduction of N_2 to NH_3, and a reductase component that delivers electrons to the catalytic component and is critical for nitrogenase function [4, 5]. Over decades of research and discovery, the terminology used to refer to these proteins has evolved, which slightly complicates a review of the literature (Table 4.2). Initially, the catalytic and reductase components were designated as such, or as components 1 and 2, respectively. As the nitrogenase proteins from *A. vinelandii* (and other organisms) were increasingly studied, this led to the use of the organism's abbreviation such as *Av*I and *Av*II interchangeably with components 1 and 2 in literature. There were also several groups that referred to component 1 as dinitrogenase and component 2 as dinitrogenase reductase. Additionally, the Mo-dependent nitrogenase component 1 can be found in the literature as Mo-nitrogenase, Mo-protein, MoFe-protein or iterations thereof, and component 2 is generally described as the Fe-protein. The "alternative" V- and Fe-only dependent nitrogenases were labeled similarly, as V-nitrogenase, VFe-protein and Fe-only-nitrogenase, FeFe-protein, respectively. More recently, there has been a transition to labeling the nitrogenase proteins

Table 4.2: Commonly used nomenclature in the literature for the two-component systems of nitrogenase.

Nitrogenase species	Component 1 Catalytic unit	Component 2 Reductase unit
Mo-nitrogenase	MoFe protein NifDK *Av*I dinitrogenase	Fe protein NifH *Av*II dinitrogenase reductase
V-nitrogenase	VFe protein VnfDGK	Fe protein VnfH
Fe-nitrogenase	FeFe protein AnfDGK	Fe protein AnfH

as gene products for clarity, with *nif*, *vnf* and *anf* labels referring to the Mo-, V- and Fe-only nitrogenase systems, respectively. Mo-nitrogenase component 1 is designated as NifDK as the *nif*D and *nif*K genes encode for the subunits, and component 2 regulated by *nifH* is designated NifH. The component 1 proteins for the V- and Fe-only dependent nitrogenases require an additional subunit, encoded by *vnf*G and *anf*G, thus, component 1 in these systems is designated VnfDGK and AnfDGK, respectively. Similar to Mo-nitrogenase, component 2 from the alternative nitrogenases is designated VnfH and AnfH. In this chapter, we primarily use the gene product designation to refer to the specific protein.

4.2.1 Structural description of the catalytic component of nitrogenases

The Mo-nitrogenase from *A. vinelandii* NifDK is the best characterized nitrogenase system. The *nifD* and *K* genes encode for a heterotetrameric $\alpha_2\beta_2$ protein that is ~230 kDa in size (Figure 4.1A). NifDK houses two different complex metalloclusters in each αβ-dimer that are essential for ET and substrate turnover, designated as the P- and M-clusters (Figure 4.1B and C, respectively) [43–45]. The P-cluster is a [Fe_8S_7] cofactor that facilitates ET to the M-cluster during catalysis and is positioned at the α/β interface ~10 Å below the surface of the protein and is bound by three cysteine residues from each subunit ($Cys^{\alpha62}$, $Cys^{\alpha88}$, $Cys^{\alpha154}$, $Cys^{\beta70}$, $Cys^{\beta95}$ and $Cys^{\beta153}$). The cluster appears as two [Fe_4S_3] cubane cluster units that share a common vertex, a μ_6-sulfide (Figure 4.1B). The M-cluster is a [$MoFe_7S_9C$-(*R*)-homocitrate] cluster that is responsible for substrate reduction and this cofactor is buried ~10 Å below the surface of the protein in the α subunit ~14 Å from the P-cluster (Figure 4.1A). The cluster appears as two partial cubane units, [Fe_4S_3] and [$MoFe_3S_3$], with a μ_6-insterstitial

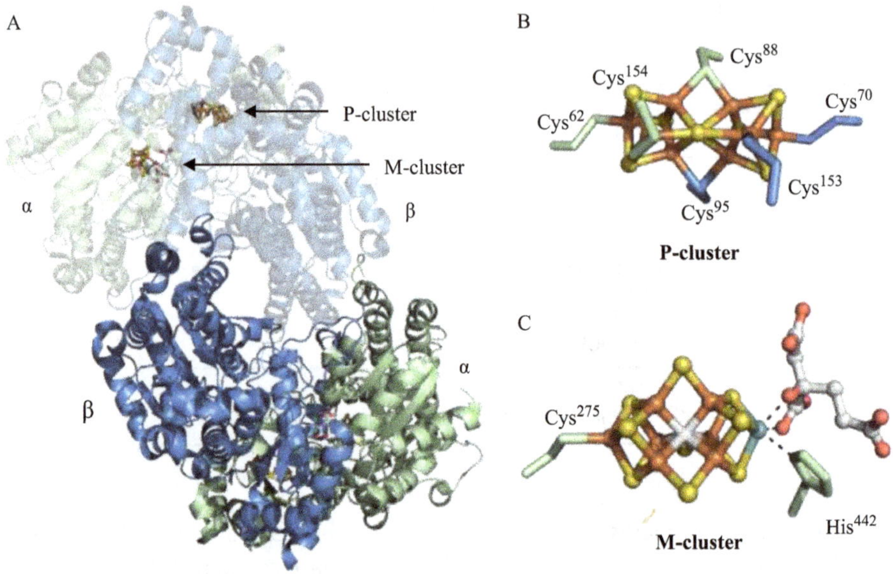

Figure 4.1: The crystal structure of NifDK from *A. vinelandii* at 1 Å resolution (PDB code: 3U7Q). A: Crystal structure cartoon of NifDK with the metallocofactors highlighted. B: A zoomed-in view of the P-cluster cofactor from NifDK. C: A zoomed-in view of the M-cluster cofactor from NifDK. The NifD subunit is represented in green and NifK subunit in blue. Atoms are colored as follows: Fe, rust; S, yellow; Mo, cyan; C, light gray; O, red. PyMOL was used to generate this figure [46].

carbide as a shared vertex, and additionally is bridged by three μ_2-sulfide ligands (Figure 4.1C) [34, 43–45]. The Mo atom is further coordinated to (*R*)-homocitrate by the 2-hydroxy and 2-carboxy groups of the organic acid, and the M-cluster is anchored to the protein by two amino acid residues: $Cys^{\alpha 275}$ at the Fe-capped end and $His^{\alpha 442}$ at the Mo-capped end (Figure 4.1C) [9]. The assembly of the P- and M-clusters will be discussed in more detail later.

The *A. vinelandii* V-nitrogenase catalytic component VnfDGK is encoded by the *vnfDGK* genes and forms a heterohexameric protein ($\alpha_2\beta_2\gamma_2$) ~244 kDa in size [14, 47, 48]. The exact function of the additional γ subunit is still unknown, but it has been proposed to be involved in the transfer of the V-cluster to the apo-enzyme based on sequence similarity to identified cofactor chaperones from *A. vinelandii* and *Klebsiella pneumoniae* [49, 50]. VnfDGK is homologous to NifDK but a recent crystal structure of VnfDGK from *A. vinelandii* has shed light on some key differences (Figure 4.2A) [48]. Like NifDK, VnfDGK houses two types of metalloclusters necessary for ET and substrate turnover, designated as the P*-cluster (Figure 4.2B) and V-cluster (Figure 4.2C). Prior to the recent crystal structure, the P*-cluster in VnfDGK and the P-cluster in NifDK were believed to have different structures, due in part to spectroscopic differences between the species [48, 51]. However, the VnfDGK structure reveals that the P*-cluster indeed appears as an $[Fe_8S_7]$ cluster

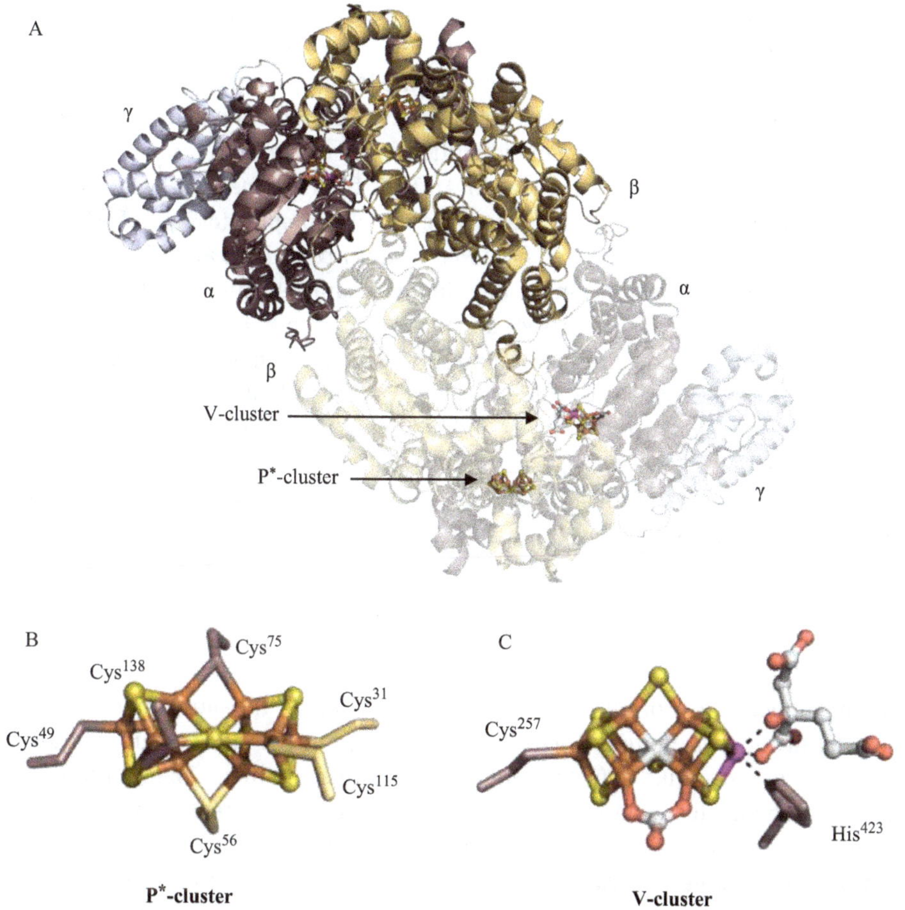

Figure 4.2: The crystal structure of VnfDGK from *A. vinelandii* at 1.35 Å resolution (PDB code: 5N6Y). A: Crystal structure cartoon of VnfDGK with the metallocofactors highlighted. B: A zoomed-in view of the P*-cluster cofactor from VnfDGK. C: A zoomed-in view of the V-cluster cofactor from VnfDGK. VnfD is represented in dark violet, VnfG in blue and VnfK in light orange. Atoms are colored as follows: Fe, rust; S, yellow; V, bright purple; C, light gray; O, red. PyMOL was used to generate this figure [46].

positioned between the VnfD and K subunits via six cysteine residues ($Cys^{\alpha49}$, $Cys^{\alpha75}$, $Cys^{\alpha138}$, $Cys^{\beta31}$, $Cys^{\beta56}$ and $Cys^{\beta115}$) and the overall cluster structure is remarkably similar to the P-cluster of NifDK (Figures 4.1B and 4.2B) [48]. The V-cluster is a [$VFe_7S_8C(CO_3)$(*R*-homocitrate)] cluster that is buried in the α-subunit ~15 Å from the P^*-cluster, and similarly to NifDK, is anchored to the protein by two residues, $Cys^{\alpha257}$ and $His^{\alpha423}$, on the Fe- and V-capped ends, respectively. The overall shape and structure of the V-cluster core is similar to that of M-cluster from NifDK, but with three differences: the V atom occupies the same place as the Mo center, including coordination by His and homocitrate ligands, the average V–Fe distance of 2.77 Å to the closest three Fe centers in the V-cluster is slightly longer than the analogous 2.69 Å Mo–Fe distance in the M-cluster and one of the μ_2-sulfide ligands is replaced by a CO_3^{2-} moiety [48]. There is currently no definitive function for or source of the carbonate ligand identified in the crystal structure.

The Fe-nitrogenase catalytic protein AnfDGK from *A. vinelandii* is encoded by the *anfDGK* genes, forms a heterohexameric protein complex ($\alpha_2\beta_2\gamma_2$) of ~250 kDa [14, 52] and has also been isolated from *Rhodobacter capsulatus* and *Rhodospirillum rubrum* [14, 53, 54]. Fe-nitrogenase has limited characterization compared to the other nitrogenase species discussed, and there is currently no crystal structure of the protein. However, spectroscopic analysis of AnfDGK from *R. capsulatus* by EPR, XAS and Mössbauer spectroscopies provided evidence that the protein contains clusters that are similar to the P/P*- and M/V-clusters of Mo- and V-nitrogenases [55, 56]. Based on this, it is proposed that the putative Fe-cluster of AnfDGK would have a similar structure to the V/M-clusters, but with Fe exclusively. Additional work is required to definitively address these issues.

4.2.1.1 EPR spectroscopy of the FeS clusters of NifDK

The metallocofactors of nitrogenase are arguably one of the most interesting aspects of the enzyme system. The M- and P-clusters housed in NifDK are unique to nitrogenase and unlike cofactors employed in other enzymes. The multinuclear transition metal clusters are capable of cycling through different oxidation states as electrons are transferred between them and understanding this electron flow can be a formidable undertaking. Fortunately, EPR spectroscopy lends itself well to the characterization of the nitrogenase clusters, providing an important and robust handle for following electronic changes in the system. The analogous clusters of the alternative nitrogenases can also be analyzed using EPR spectroscopy; however, there is far more extensive characterization of the Mo-dependent system, so this section will primarily focus on the Mo-dependent system. Described below are some of the important spectroscopic features from the EPR analysis of nitrogenase that may be useful for understanding metallocluster biosynthesis and reactivity.

As mentioned in the previous sections, NifDK from *A. vinelandii* is an $\alpha_2\beta_2$ heterotetramer with one M- and P-cluster pair per $\alpha\beta$-dimer; therefore, two separate catalytic sites are present in the tetramer. The NifDK-bound M-cluster displays a characteristic $S = 3/2$ EPR signal (Figure 4.3) with g-values at 4.7, 3.7 and 2.0 in the presence of excess dithionite [4]. This cluster state has been shown to undergo a reversible one-electron oxidation, which results in the disappearance of the $S = 3/2$ signal [4]. The M-cluster has also been successfully extracted from the protein, which involves carefully unfolding NifDK and releasing the intact cluster

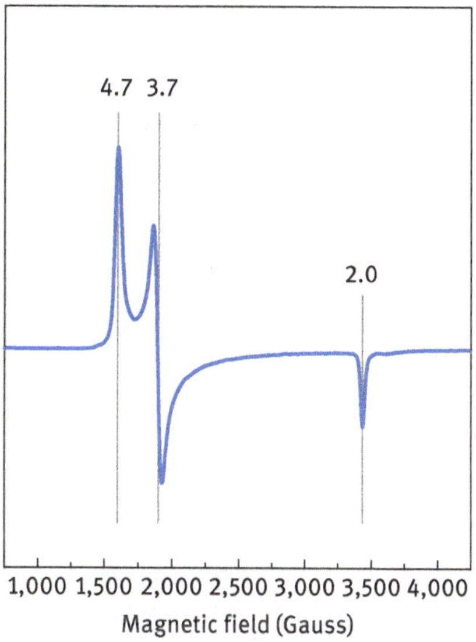

Figure 4.3: The characteristic $S = 3/2$ EPR signal of the M-cluster bound to NifDK. The g-values are observed at 4.7, 3.7 and 2.0 in the presence of excess dithionite, indicated by vertical lines. Conditions: four scans run at 10 K with a microwave power of 50 mW and microwave frequency of 9.62 GHz.

into organic solvents such as *N*-methylformamide and dimethylformamide [57]. The EPR signal of the extracted M-cluster displays similar, albeit broader, features than the protein-bound counterpart [57, 58]. This unique signal has been crucial for the study of M-cluster biosynthesis [9]. In one study, the appearance of the characteristic $S = 3/2$ signal on the scaffold protein from *A. vinelandii* NifEN indicated that M-cluster was fully synthesized outside of the nitrogenase active site on NifDK [23, 24]. More advanced EPR techniques, such as ENDOR, have also been used to characterize signals associated with M-cluster and identify intermediate species [59].

The P-cluster of NifDK primarily shuttles electrons from the reductase component, NifH, to the M-cluster, and three oxidation states of the P-cluster have been identified, designated as P^N, P^{1+} and P^{Ox} states, respectively [60–62]. In the presence of excess dithionite, the P-cluster is observed in the P^N all-ferrous diamagnetic state and this species does not exhibit an EPR signal. Addition of the oxidant indigo disulfonate (IDS), results in the two-electron oxidization of P-cluster to the P^{Ox} state, which exhibits an $S = 3$ or 4 spin state, reflected in an EPR signal with $g = 11.8$ (Figure 4.4A) [60–62]. Concomitant with the oxidation of the P^N (Figure 4.4B) to P^{Ox} (Figure 4.4C) state, the cluster attains a more open conformation, as observed by crystallography [43, 63, 64]. It is proposed that the cluster conformational change may allow for the two-electron reduction to occur; however, further studies are required to definitively address this issue. This change of the EPR signal with respect to the redox state and cluster conformation has also been used to study the stepwise synthesis of the P-cluster in NifDK [9].

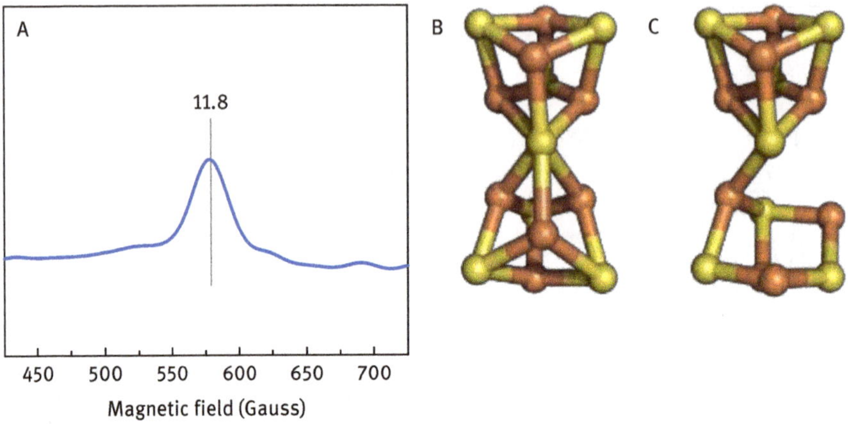

Figure 4.4: The characteristic EPR signal and structural changes of the P-cluster on NifDK. A: The $g = 11.8$ EPR signal of the P-cluster that reflects an $S = 3$ or 4 spin state. Conditions: 10 scans at 10 K with a microwave power of 50 mW, and a microwave frequency of 9.62 GHz. The crystal structure representations of the P^N (B) and P^{Ox} (C) states of the P-cluster (PDB code: 3U7Q). The cluster atoms are shown as a ball-and-stick model. Atoms are colored as follows: Fe, rust; S, yellow.

4.2.2 Structural description of the reductase component of nitrogenases

The *A. vinelandii* Mo-nitrogenase reductase component, NifH, is encoded by the *nifH* gene and forms a homodimer (γ_2) of approximately 60 kDa in size (Figure 4.5A and B). The protein contains a nucleotide-binding site in each subunit and a $[Fe_4S_4]$ cluster positioned at the subunit interface of the protein near the surface by the Cys^{97} and Cys^{132} residues from each subunit (Figure 4.5C) [65]. The FeS cluster of NifH can support three oxidations states, $[Fe_4S_4]^0$, $[Fe_4S_4]^{1+}$, and $[Fe_4S_4]^{2+}$, and the crystal structures of the protein in different states have been solved [66]. However, the structures of the $[Fe_4S_4]$ clusters do not greatly vary despite changes in the oxidation state and solution-state experiments that suggest otherwise [4, 66]. This difference is likely due either oxidation reactions that occur during the crystallization process, or to photoreduction of the metallocluster during data collection. NifH also binds ATP, one molecule to each subunit, in a Walker's motif A protein fold [67], found between residues 9 and 16 [65]. Several structures have been solved of NifH that have a nucleotide bound in this location (Table 4.3), and they show minor conformational changes between the nucleotide bound and unbound states [66]. However, a multitude of experimental results suggest that the structures of the protein and $[Fe_4S_4]$ cluster are dynamic in the presence of nucleotides [4], and these dynamics are better reflected in the crystal structures of the NifH:NifDK complex (*vide infra*).

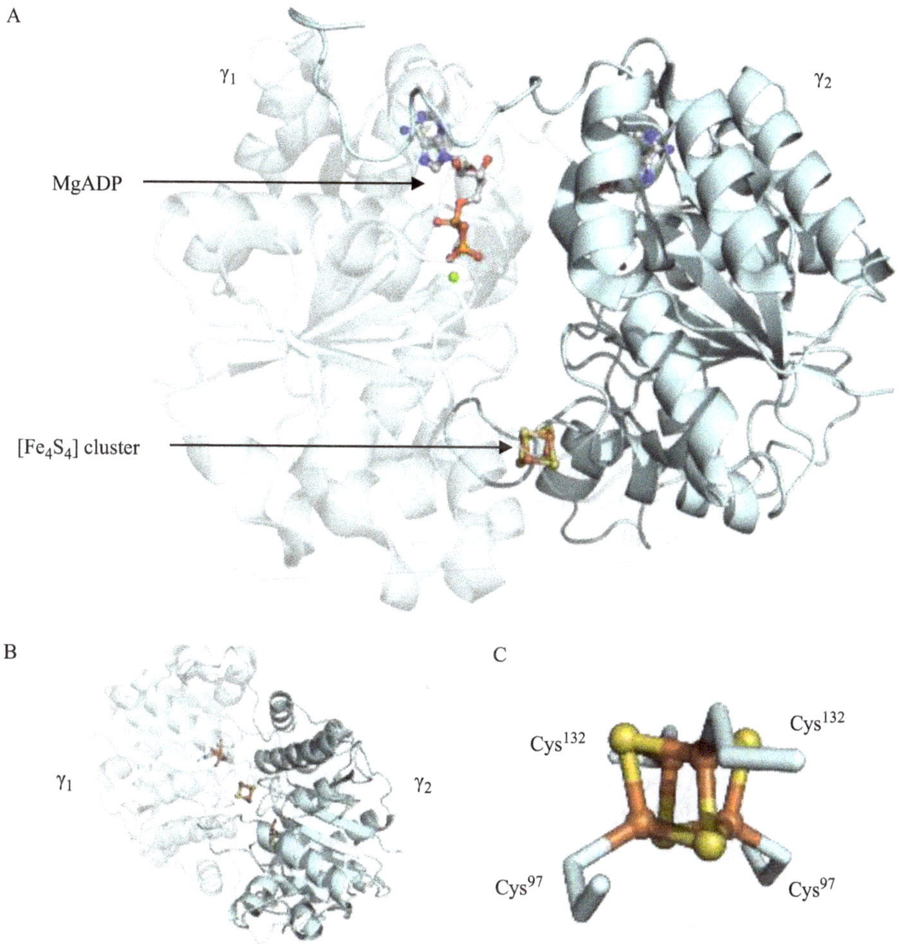

Figure 4.5: The crystal structure of NifH with ADP bound at 2.15 Å resolution (PDB code: 1FP6). A: Crystal structure cartoon of NifH with the metallocofactor and nucleotides highlighted. B: A 90° rotation of the NifH crystal structure, with a view down the twofold rotation axis. C: Ball-and-stick representation of the $[Fe_4S_4]$ cluster of NifH. Atoms are colored as follows: Fe, rust; S, yellow; C, white; N, blue; O, red; Mg, green. PyMOL was used to generate this figure [46].

Until recently, only crystal structures of the Mo-nitrogenase reductase protein were published, but spectroscopic characterization of the reductases from V- and Fe-nitrogenases, VnfH and AnfH, respectively, was known [14]. Analysis by EPR and XAS suggested that the $[Fe_4S_4]$ clusters in the alternative systems behave similarly to that of NifH, and sequence similarity between NifH, VnfH and AnfH suggested a similar overall protein structure [66]. In 2018, the Einsle group published a crystal structure of VnfH from *A. vinelandii* with ADP bound [68]. The protein is encoded by the *vnfH* gene and forms a homodimer (γ_2) of approximately 60 kDa in

Table 4.3: Select *Av*NifH crystal structures with corresponding PBD and resolutions.

Protein	PDB code	Resolution (Å)	Reference
NifH	1G5P	2.2	[69]
VnfH + MgADP	6Q93	2.2	[68]
NifH with $[Fe_4S_4]^0$ cluster	1G1M	2.25	[69]
NifH + MgADP	1FP6	2.15	[70]
ΔL127-NifH + MgATP	2C8V	2.5	[71]
ΔL127-NifH + NifDK + MgATP	1G21	3.0	[72]
NifH + NifDK + MgAMP · AlF_4^-	1N2C	3.0	[73]
NifH + NifDK + MgADP	2AFI	3.1	[74]

[a] At the time of writing, the PDB code was not yet released.

size, not dissimilar to the Mo-dependent system. The structure of VnfH, as predicted from spectroscopic and biochemical characterizations, is largely similar to the equivalent structure of NifH [68], but the complimentary structures of VnfH without nucleotide or bound in complex to VnfDGK have not yet been published.

4.2.2.1 Characterization of the $[Fe_4S_4]$ cluster of the Fe protein of nitrogenase

The Fe protein performs three known physiological roles in the nitrogenase system: (1) it mediates the ATP-dependent ET to the catalytic component, (2) it assists in the biosynthesis of the P-cluster on NifDK and (3) it serves as a molybdenum and homocitrate insertase for M-cluster maturation [27–29, 75]. Additionally, Hu and coworkers [76] showed that NifH could also facilitate the seemingly adventitious interconversion of CO_2 to CO. In all of these capacities, the $[Fe_4S_4]$ cluster of Fe protein plays a pivotal role for activity and is also sensitive to subtle changes in the reaction conditions [4, 66]. Substantial effort has gone into the characterization of this important FeS cluster both spectroscopically and biochemically; the results are summarized in the following sections.

4.2.2.1.1 EPR analysis of the Fe proteins

The $[Fe_4S_4]$ cluster of *A. vinelandii* Fe proteins can support three oxidation states: $[Fe_4S_4]^0$, $[Fe_4S_4]^{1+}$ and $[Fe_4S_4]^{2+}$, but only the $[Fe_4S_4]^{1+/2+}$ redox couple is thought to be physiologically relevant [4, 77]. The Fe protein, like most nitrogenase metallocofactors, is sensitive to O_2 and so is generally handled in the presence of the reductant and oxygen scavenger dithionite. Under these conditions, the $[Fe_4S_4]^{1+}$ state is readily accessible and exists as a mixture of $S = 1/2$ and $S = 3/2$ species (Figure 4.6B), with the rhombic $S = 1/2$ signal being the major component (*g*-values of 2.04, 1.93, 1.84) [78, 79]. However, chemical additives to the protein solution can

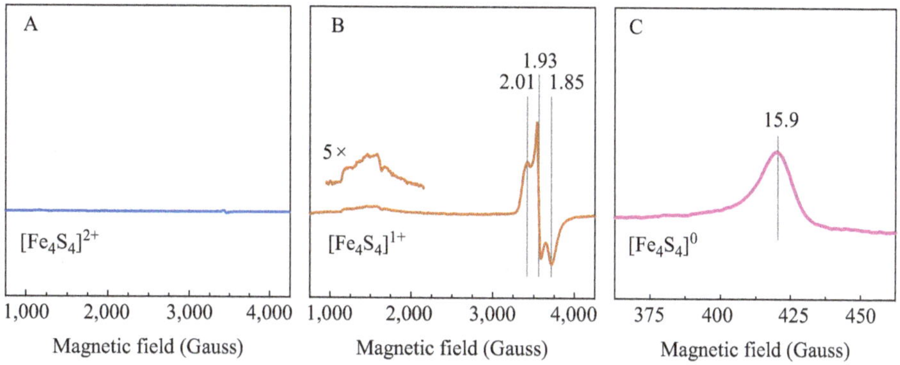

Figure 4.6: EPR spectra of NifH in the IDS-oxidized ($[Fe_4S_4]^{2+}$, A), dithionite-reduced ($[Fe_4S_4]^{1+}$, B), and Eu^{II}-reduced all-ferrous states ($[Fe_4S_4]^{0}$, C). The *g*-values are indicated by vertical lines. IDS = indigo disulfonate, Eu^{II} = europium (II) diethylenetriaminepentaacetic acid. Conditions: four scans run at 10 K with a microwave power of 50 mW, and a microwave frequency of 9.62 GHz.

change the ratio of the $S = 1/2$ and 3/2 species [80]. The cluster can be chemically oxidized to the $[Fe_4S_4]^{2+}$ state by the addition of IDS, resulting in a diamagnetic cluster with an overall $S = 0$ spin state, as determined by Mössbauer spectroscopy, which does not show a signal in the EPR spectrum (Figure 4.6A) [4]. The “all-ferrous” $[Fe_4S_4]^{0}$ state is generated by addition of a strong reductant, such as titanium (III) citrate or europium (II) diethylenetriaminepentaacetic acid (Eu^{II}-DTPA), and the EPR spectrum displays a $g = 15.9$ resonance in parallel mode, consistent with an $S = 4$ species (Figure 4.6C) [81–84]. The Fe proteins from the alternative nitrogenases, VnfH and AnfH, also exhibit similar EPR signals and cluster behavior as observed for NifH, and some of this information is summarized in Table 4.4 [79].

Table 4.4: EPR features of the reductase component of nitrogenase from *A. vinelandii.*

Protein	Oxidation state	Bound nucleotide	Spin state	g-Values			References
NifH	$[Fe_4S_4]^{2+}$	–	$S = 0$		–		[85]
	$[Fe_4S_4]^{1+}$	–	$S = 1/2, 3/2$	2.01	1.93	1.85	[80]
	$[Fe_4S_4]^{1+}$	ATP	$S = 1/2$	2.03	1.91		[86]
	$[Fe_4S_4]^{1+}$	ADP	$S = 1/2$	2.02	1.92		[87]
	$[Fe_4S_4]^{0+}$	–	$S = 4$		15.9		[84, 88]
VnfH	$[Fe_4S_4]^{2+}$	–	$S = 0$		–		[76]
	$[Fe_4S_4]^{1+}$	–	$S = 1/2, 3/2$	2.00	1.93	1.84	[79]
	$[Fe_4S_4]^{1+}$	ATP	$S = 1/2$	2.03	1.91		[76]
	$[Fe_4S_4]^{1+}$	ADP	$S = 1/2$	2.02	1.92		[89]
	$[Fe_4S_4]^{0+}$	–	$S = 4$		15.9		[76]

The $S = 1/2$ signals of the dithionite-reduced $[Fe_4S_4]^{1+}$ clusters of NifH and VnfH undergo a spectral change upon binding of nucleotides [4, 76]. The EPR signal for NifH changes from a rhombic (Figure 4.7, black) to an axial symmetry upon the binding of MgATP (Figure 4.7A, blue) or MgADP (Figure 4.7B, red). Similar effects are also observed for VnfH (not shown) [76]. This nucleotide-induced change in the EPR spectrum is consistent with the protein conformational changes that are observed for NifH in solution and will be described in more detail in Section 4.2.2.1.3. To date, detailed EPR analysis of AnfH with nucleotide binding has not been reported, however, due to the similarity of NifH and VnfH, AnfH would likely behave similarly.

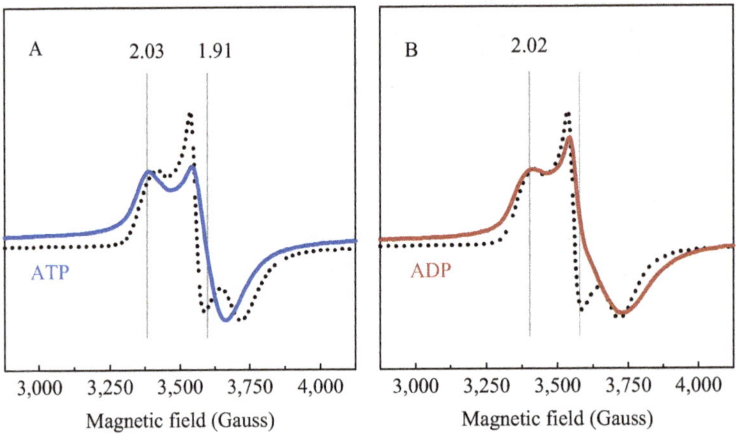

Figure 4.7: EPR spectra of NifH bound with nucleotides in the dithionite-reduced ($[Fe_4S_4]^{1+}$) state. A: EPR spectra of NifH in the presence (solid blue) and absence (dotted black) of ATP. B: EPR spectra of NifH in the presence (solid blue) and absence (dotted black) of ADP. The *g*-values of the nucleotide-bound spectra are indicated with vertical lines.

4.2.2.1.2 Reduction potentials of the Fe proteins

The $[Fe_4S_4]^{1+/2+}$ couple is proposed as the physiologically relevant redox event involved in ET by the Fe protein [66, 90]. The $[Fe_4S_4]^{1+}$ state can be accessed *in vitro* using excess dithionite, whereas flavodoxins and ferrodoxins drive the reduction *in vivo* [91–95]. To better understand the ET reactions facilitated by NifH, the midpoint reduction potential (E_m) was determined under a series of conditions using microcoulometry [96]. In this technique, a controlled potential is applied to a buffer solution in a glass electrochemical cell under anaerobic conditions, and then fully oxidized NifH ($[Fe_4S_4]^{2+}$) is introduced to the cell while the change in current was measured over time. This results in a measured E_m for the $[Fe_4S_4]^{1+/2+}$ couple of −290 mV versus the standard hydrogen electrode (SHE) (Figure 4.8A) [97]. The addition of either

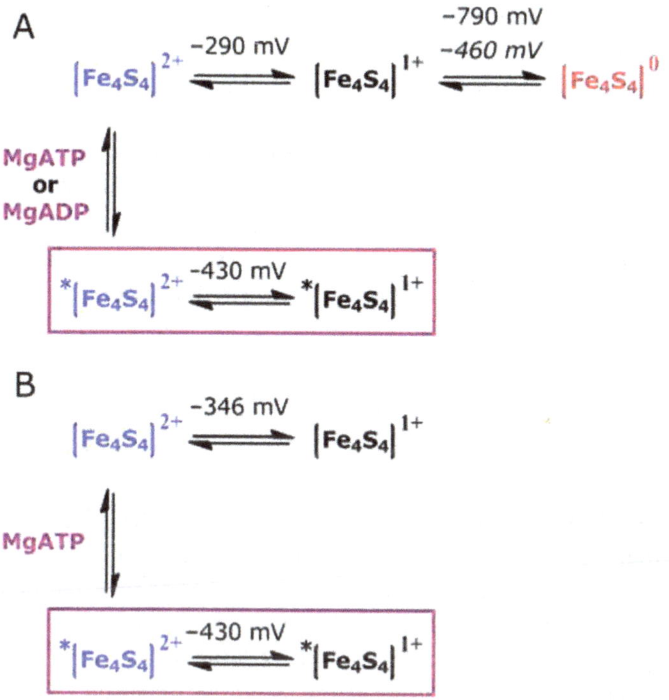

Figure 4.8: Summary of the midpoint potentials for the Fe proteins of nitrogenase from *A. vinelandii*. A: The midpoint potentials (E_m) for NifH with the oxidized, reduced and all-ferrous states shown in blue, black and pink, respectively. The E_m value for the $[Fe_4S_4]^{1+/0}$ couple reported for the $S = 4$ state is shown in plain text and the $S = 0$ state is shown in italics. The addition of nucleotide is depicted in the purple box. B: The E_m value for VnfH with the oxidized and reduced states shown in blue and black, respectively. Addition of MgATP is depicted in the purple box.

MgATP or MgADP to NifH results in a 140 mV shift to more negative potential (−430 mV) [97, 98]. The VnfH $[Fe_4S_4]^{1+/2+}$ couple (Figure 4.8B) was also measured but using a redox titration method that resulted in E_m = −346 mV versus SHE (NifH, E_m = −300 mV for the same method) and a shift to −430 mV upon addition of MgATP (NifH, E_m = −405 mV) [76]. The commensurate decrease in potential upon nucleotide binding makes the NifH a slightly stronger reductant, but this may not have a strong effect on catalysis, as the Fe proteins from different organisms (such as *K. pneumoniae* or *Clostridium pasteurianum*) can vary by as much as 100 mV and will still facilitate N_2 reduction even in the presence of the cross-catalytic component (e.g. *Av*NifH with *Kp*NifDK or vice versa) [4]. The midpoint potential of AnfH has not yet been reported.

For NifH, it was demonstrated that a "super-reduced" all-ferrous $[Fe_4S_4]^0$ state of the cluster was accessible, with a midpoint potential measured using microcoulometry of E_m = −460 mV versus the normal hydrogen electrode (NHE)

(Figure 4.8A) [88]. This species was reported by Watt and coworkers to be generated using reduced methyl viologen [88], or flavodoxin hydroquinone as redox agents and was proposed to have an $S = 0$ spin state [99]. Burgess and coworkers followed up on this finding, and using Ti(III) citrate as a strong reductant, they generated a NifH protein with an all-ferrous FeS cluster that was definitively assigned with an $S = 4$ spin state using extensive spectroscopic and computational characterization [66]. A midpoint potential of the $S = 4$ $[Fe_4S_4]^0$ NifH species was measured using spectroelectrochemical experiments and Cr (III)-EDTA (EDTA = ethylenediaminetetraacetic acid) as a redox mediator, and resulted in an E_m of −790 mV versus NHE (Figure 4.8) [100]. Subsequent work also found that the all-ferrous Fe protein could be rapidly generated using Eu (II) chelate complexes instead of the more sluggish Ti(III) citrate reagent [101].

4.2.2.1.3 Nucleotide-induced conformational changes of NifH

Calorimetric chelation studies were performed on NifH in the presence of nucleotides to assess the extent of structural change that occurred prior to the report of the first crystal structures of the protein [102–104]. These studies revealed that the $[Fe_4S_4]$ cluster was accessible, and the iron atoms were easily extracted from the Fe protein in the presence of chelating agents like 2,2′-bipyridine and bathophenanthrolinedisulfonate and MgATP. If MgADP was bound or no nucleotide was present, the rate of cluster degradation was much slower [102–104]. This led to the conclusion that the $[Fe_4S_4]$ cluster in NifH was more solvent exposed when MgATP was bound to the protein. Small-angle X-ray scattering (SAXS) studies, which are able to assess protein conformational changes without the need for crystalline material, also indicated that there was a more substantial protein rearrangement of NifH in the presence of MgATP compared to either MgADP or nucleotide-free samples, both of which were indistinguishable from each other [105]. Additional insights into nucleotide binding have come to light after crystal structures of NifH and NifH/NifDK were solved with different nucleotides bound.

When the noncomplexed NifH (protein, pale green; cluster, forest green) with and without MgADP (protein, light blue; cluster, sky blue) are overlaid (Figure 4.9A), there appears to be little difference in the overall structure of the proteins, consistent with the SAXS and chelation studies [102–105]. However, overlaying the MgADP-(protein, light blue; cluster, sky blue) [72] and MgATP-bound (pale yellow; cluster, bright orange) [74] NifH structures (Figure 4.9B) from the respective NifH/NifDK complexes reveals that the $[Fe_4S_4]$ cluster is ~3 Å closer to the protein surface in the MgATP structure than the MgADP structure [66]. It has been proposed that MgATP-bound NifH with a more surface-exposed cluster primes the protein to readily transfer electrons to NifDK, whereas the recession of the $[Fe_4S_4]$ cluster upon MgATP cleavage helps to prevent a backflow of electrons [4,106].

A

B

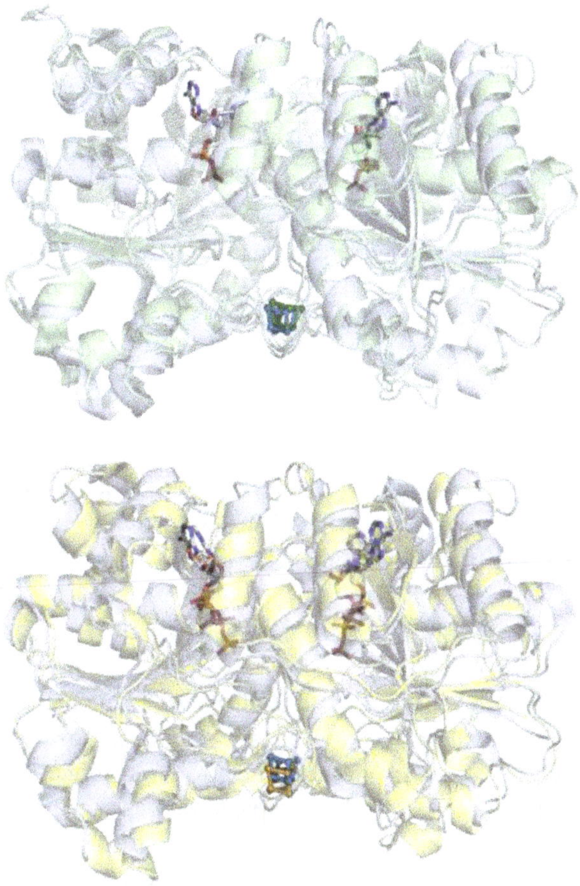

Figure 4.9: Structural comparison of nucleotide-free and nucleotide-bound NifH, both as free proteins and in complex with NifDK. A: The structures of the nucleotide-free (PDB code: 1G5P; protein, pale green; cluster, forest green) and the MgADP-bound (PDB code: 1FP6; protein, light blue; cluster, sky blue) free NifH are overlaid. B: The crystal structures of the MgADP-bound (PDB code: 2AFI; protein, light blue; cluster, sky blue) and MgATP-bound (PDB code: 1G21; protein, pale yellow; cluster, bright orange) NifH in complex with NifDK are overlaid (NifDK portion is not shown). Atoms are colored as follows: C, white; N, blue; O, red; Mg, green. PyMOL was used to generate this figure [46].

4.2.3 Structural description of the NifH:NifDK complex

While N_2 reduction is facilitated by NifDK at the M-cluster site, the reductase partner NifH is crucial for nitrogenase activity [4]. This raised questions about how the NifH and NifDK proteins bind to each other, as well as how these proteins interact during catalysis. Chemical cross-linking was one experimental approach that was used to understand which amino acid residues on NifH and NifDK came in contact with one another during the formation of a putative NifH:NifDK complex [107]. It was found that an isopeptide bond formed between the Glu^{112} of NifH and the Lys^{399} of the β-subunit of NifDK roughly indicating how these proteins could interact [107, 108]. Site-directed mutagenesis and posttranslational modifications were also used to modify residues of both NifH and NifDK to explore other potential points of interaction. Additional studies pushed the envelope further, expanding the strategies that were available to probe this protein–protein interaction. As mentioned previously, NifH binds nucleotides and subsequently undergoes conformational changes that affect the solvent access to the [Fe_4S_4] cluster depending on which nucleotide was bound to the protein [4, 66]. With the FeS cluster "exposed" in the presence of nucleotides, chelating agents are readily able to bind the iron atoms and degrade the cluster [102–104]. Taking advantage of this fact, NifH and NifDK were mixed together in solution in the presence of ATP, presumably forming a complex. Chelation agents were then added, which resulted in a greatly reduced rate of Fe chelation from NifH when NifDK was present, compared to NifH alone [102, 109]. This supported the notion that the [Fe_4S_4] cluster was protected in the presence of the catalytic component, implying the cluster-containing face of NifH docks to the surface of NifDK.

Similar experiments employing the use of NifH variants were also used to explore additional aspects of the NifH:NifDK complex. A set of experiments showed that K15Q and A157S NifH variants, proteins that could still bind MgATP but did not result in the associated MgATP-induced conformational change, were unable to compete with wild-type NifH in an activity assay despite being able to form normal cross-linked complexes with NifDK [110]. Conversely, NifH variants that form an MgATP-like or partial MgATP-like conformation in the absence of MgATP (e.g., D129E, ΔL127) bound to NifDK more tightly than nucleotide-free NifH [97, 111]. It was concluded from these experiments that an MgATP-induced conformational change was required to form an active NifH:NifDK complex [4]. Burgess and Lowe [4] further proposed that the structural change in nitrogenase evolved because the conformation of the proteins was a prerequisite to form an NifH:NifDK complex that optimally positioned the enzyme for MgATP hydrolysis and ET between metallocofactors.

The first NifH:NifDK complex structure was published by Rees and coworkers in 1997 (Figure 4.10, left) [73]. The structure shows the $\alpha_2\beta_2$ heterotetramer of NifDK, with one NifH docked to each αβ-dimer (Figure 4.10). This complex was "locked" in the ATP-bound conformation through the use of MgADP·AlF_4^-, a nonhydrolyzable ATP analog. In the complexed form, the [Fe_4S_4] cluster of NifH is positioned ~15 Å away from the P-cluster of NifDK, and the P-cluster is ~14 Å away from the M-cluster [43–45]. This protein complex places all of the FeS clusters in close proximity likely to better facilitate the interprotein ET that is required for substrate reduction. Electrons are transferred from the [Fe_4S_4] cluster of NifH to the P-cluster, and the P-cluster transfers electrons to the M-cluster; and this process occurs concomitantly with the hydrolysis of two molecules of MgATP by NifH for every electron transferred

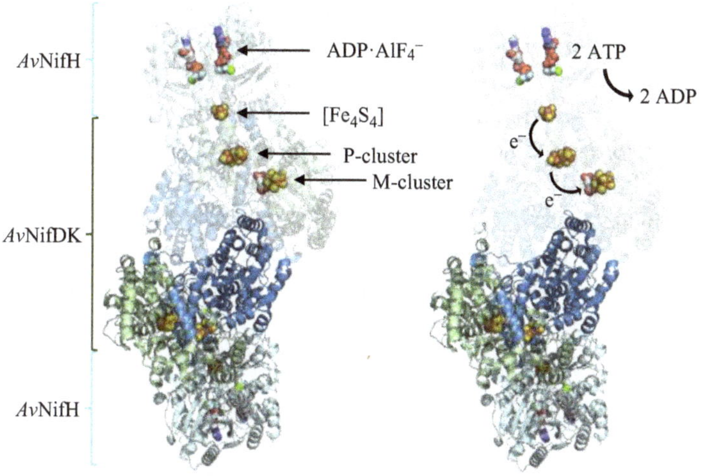

Figure 4.10: The cartoon crystal structure of the ADP·AlF_4^--stabilized NifH:NifDK complex (PDB code: 1N2C) highlighting the nucleotide and metalloclusters (left) as well as the components involved in electron transfer (right). NifH is represented in pale cyan, NifD in pale green and NifK in sky blue. Atoms are colored as follows: Fe, rust; S, yellow; Mo, cyan; C, white; N, blue; O, red; Mg, green; Al, gray; F, pale cyan. PyMOL was used to generate this figure [46].

(Figure 4.10, right) [65,112–116]. Additional insights into how the ET process occurs in nitrogenase will be described in further detail in the following section.

4.3 The biosynthesis of the metalloclusters of NifDK

The metalloclusters of nitrogenase are vital to the function of the enzyme, and unsurprisingly, the construction of these cofactors involves the involvement of and regulation by many other proteins in the *nif* operon [9]. The cluster biosynthesis can be generally split into two parts: an "ex situ" portion that describes the synthesis of the M-cluster, as it occurs on proteins other than NifDK, and an "in situ" portion that describes the P-cluster synthesis, as this takes place almost entirely on NifDK.

4.3.1 The ex situ biosynthesis of the M-cluster

The M-cluster of the nitrogenase active site (Figure 4.1C) is one of the most complex metalloclusters found in any known biological system. Efforts to synthetically replicate the cluster are challenging because of features such as the asymmetric Mo atom on one end of the cluster, the interstitial carbon atom, as well as the organic

R-homocitrate ligand. Despite these complications, a synthetic asymmetric mimic of the M-cluster was recently described, but this complex was not functional for N_2 reduction [117]. In parallel to synthetic approaches to M-cluster synthesis, the biosynthesis of the M-cluster has been a key focus of nitrogenase research that comes with its own set of challenges. The biosynthesis involves a cascade of proteins working in concert to produce the active cluster (Figure 4.11) [9]. There are multiple steps involved in M-cluster assembly, including the formation of [Fe_4S_4] clusters, the formation of an M-cluster precursor known as the L-cluster ([Fe_8S_9C]), Mo and homocitrate insertion into the L-cluster and the transfer of fully mature M-cluster to the active site of NifDK [9].

4.3.1.1 Formation of 2Fe and 4Fe clusters

Like other FeS cluster-containing nitrogenase proteins, M-cluster in *A. vinelandii* synthesis begins with the generation of smaller FeS clusters ([Fe_2S_2] and [Fe_4S_4]) by NifU and NifS (Figure 4.11) [16–21,120]. NifS is a pyridoxal-dependent cysteine desulfurase that catabolizes cysteine and functions to deliver a sulfur source to a scaffold protein NifU. The sulfur is then combined with iron atoms on NifU, and these components are processed into FeS clusters [16–21].

4.3.1.2 Formation of the L-cluster

NifB is a radical *S*-adenosyl-L-methionine (SAM)-dependent enzyme that plays a vital role as a scaffold for the assembly of the M-cluster. The importance of NifB was initially recognized in NifDK variants expressed in an *nifB*-deletion background (Δ*nifB* NifDK, or apo-NifDK), as these proteins lacked an M-cluster [121–124]. Subsequently, NifB was characterized as a ~ 55 kDa protein that supports three [Fe_4S_4] clusters, as well as it contains a canonical CXXXCXXC motif on the N-terminal end, characteristic of radical SAM-dependent enzymes [125]. One of the [Fe_4S_4] clusters on NifB reacts with SAM, while the other two clusters are the precursor to the M-cluster, called the K-cluster. Initial expression of NifB was not entirely fruitful; however, when NifB was fused to another scaffold protein NifEN, NifB was stabilized and biochemical characterization became feasible. This new protein was generated by fusing the 3′-end of the *nifEN* gene to the 5′-end of the *nifB* gene, and this protein was expressed in an *nifHDK*-deletion strain of *A. vinelandii* [126].

Recent work by Hu and coworkers [127] has shed light on the specific role that the three [Fe_4S_4] clusters play in L-cluster biosynthesis. A NifB protein from the organism *Methanosarcina acetivorans* heterologously expressed in *Escherichia coli* was used, and this NifB variant is stable, does not require fusion to another protein and is competent for M-cluster assembly with the other *A. vinelandii* proteins [118, 128].

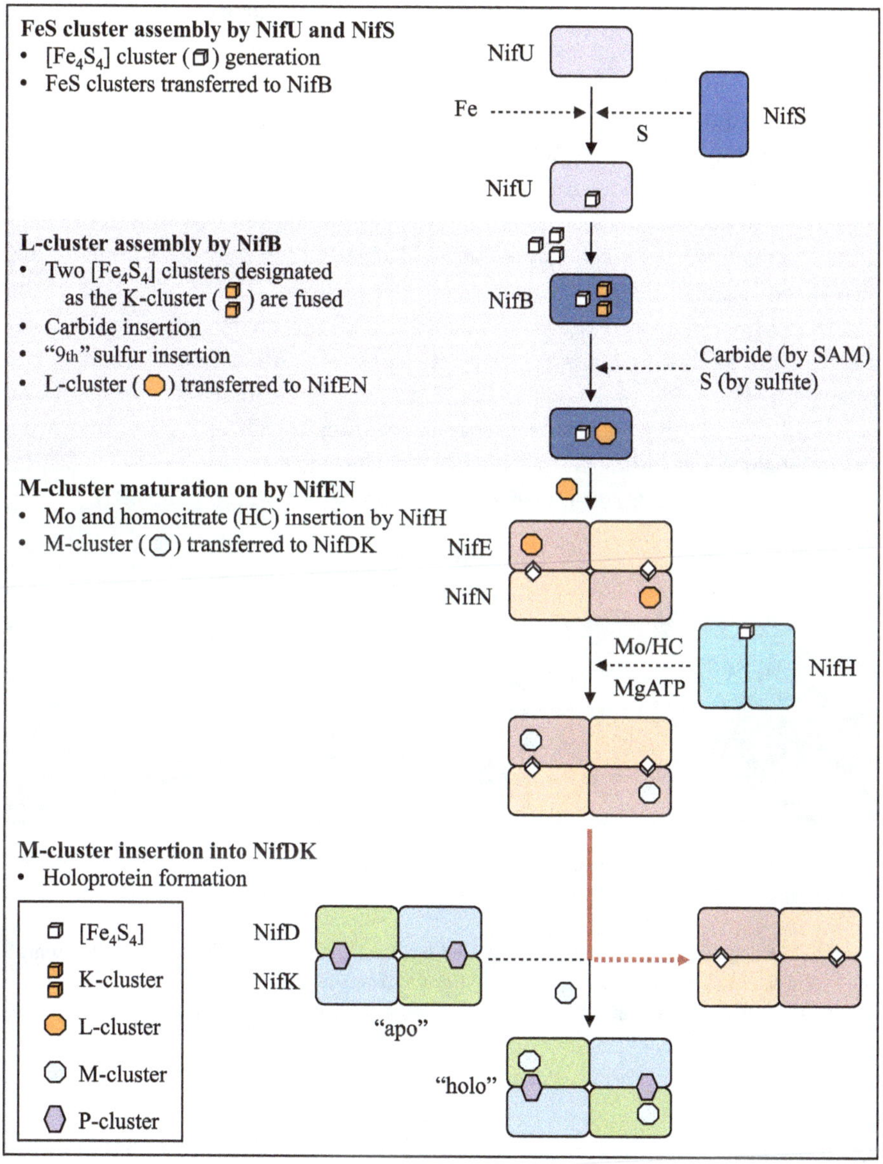

Figure 4.11: A schematic representation of the M-cluster biosynthetic pathway. In *A. vinelandii*, NifS delivers sulfur to NifU for the assembly of small FeS clusters. Three $[Fe_4S_4]$ clusters (K-cluster) synthesized on NifU are transferred to NifB. NifB facilitates the radical *S*-adenosyl methionine (SAM) cleavage reaction that converts the K-cluster into the L-cluster via carbide insertion and ninth sulfur insertion [118, 119]. The L-cluster subsequently transferred from NifB to NifEN for further processing. NifH acts as a Mo- and homocitrate insertase and interacts with the scaffold protein NifEN to convert the L-cluster into the M-cluster. The M-cluster is then transferred to apo-NifDK. NifS is represented in royal blue; NifU in light purple; NifB in dark blue; NifE in pink; NifN in orange; NifH in cyan; NifD in green; NifK in light blue; and the $[Fe_4S_4]$, K-, L-, P- and M-clusters are indicated on the legend.

A SAM molecule transfers a methyl to one of the $[Fe_4S_4]$ clusters of the K-cluster (Figure 4.11). Subsequently, a second SAM molecule interacts with the SAM-cleaving $[Fe_4S_4]$ cluster to form a 5′-deoxyadenosine radical (5′-dA·), an intermediate that is proposed to abstract a hydrogen atom from the methyl group bound to the K-cluster. This reaction results in a methylene radical bound to the K-cluster. Two additional hydrogen atoms are then lost, a structural rearrangement that presumably allows for the coupling of the two $[Fe_4S_4]$ clusters of the K-cluster concomitant with the generation of a carbide [22, 127]. The carbide-inserted cluster then receives a ninth sulfur, which forms a stable $[Fe_8S_9C]$ cluster called the L-cluster [118]. A structure of the L-cluster has been proposed and is based on spectroscopic characterization and comparison to the M-cluster; however, there are currently no crystal structures of the L-cluster [25, 129, 130]. Analogous to the M-cluster, the L-cluster is proposed to be two $[Fe_4S_3]$ partial cubanes that are bridged by three μ_2-sulfide ligands and a μ_6-insterstitial carbide coordinated in the central cavity (Figure 4.12) [25, 126, 129]. After L-cluster has been synthesized on NifB, it is transferred to NifEN (Figure 4.11).

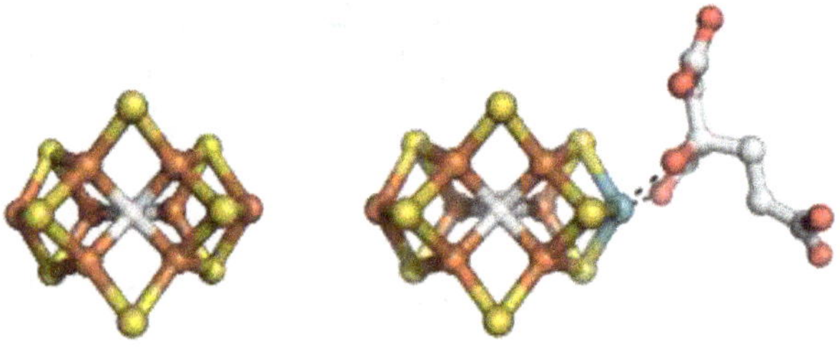

Figure 4.12: Structural comparison of the proposed L-cluster structure to the M-cluster. The atoms of the L- and M-clusters are shown as ball-and-stick models. The M-cluster coordinates were used from a crystal structure of NifDK (PDB code 3U7Q). Atoms are colored as follows: Fe, rust; S, yellow; Mo, cyan; C, light gray; O, red.

4.3.1.3 Mo and homocitrate insertion into the L-cluster

As mentioned previously, NifEN is a scaffold protein that is ~414 kDa in size and has a high degree of sequence homology to NifDK [131]. Like NifDK, NifEN also forms a $\alpha_2\beta_2$-heterotetramer (Figure 4.13) and houses two types of metalloclusters [25, 132]. The P-cluster analogous site on NifEN contains four cysteine residues ($Cys^{\alpha37}$, $Cys^{\alpha62}$, $Cys^{\alpha124}$, $Cys^{\beta44}$), compared to the six found in NifDK, and houses a $[Fe_4S_4]$ cluster called the O-cluster at the interface of the αβ-subunits. The specific L/M-cluster binding site on NifEN is still somewhat ambiguous, as the electron

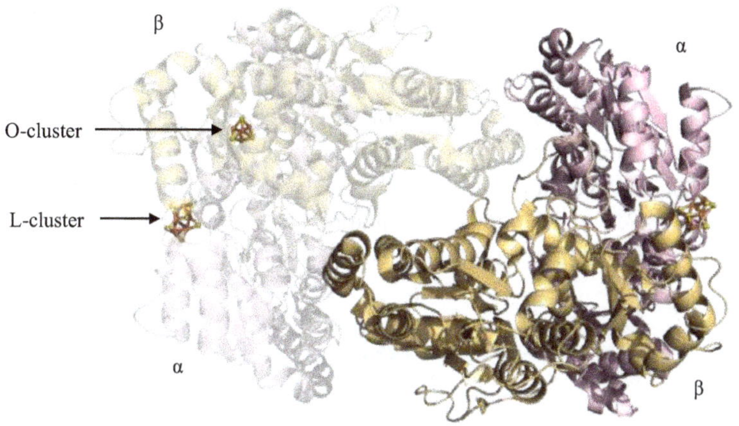

Figure 4.13: The cartoon representation of the crystal structure of NifEN from *A. vinelandii* at 2.4 Å resolution (PDB code: 3PDI). NifE is represented in light pink and NifN in light orange. Atoms are colored as follows: Fe, rust; S, yellow; Mo, cyan; O red. PyMOL was used to generate this figure [46].

density for the cluster in the crystal structure was not well defined [25]. However, spectroscopic and biochemical characterization supports the notion that NifEN receives the L-cluster after it has been assembled by NifB [9].

L-cluster conversion to M-cluster requires NifH and NifEN, emphasizing that NifH plays multiple roles in the nitrogenase system (Figure 4.11) [23, 24, 28]. This conversion takes place when NifEN loaded with L-cluster is incubated with NifH, MgATP, molybdate (MoO_4^{2-}) and *R*-homocitrate; a process that will be described with more detail in a subsequent section. The precise mechanism by which Mo and homocitrate insertion occurs is still unknown and requires additional study to fully characterize the reaction. Once the M-cluster is synthesized, the cluster can then be transferred to the active site on NifDK [23, 24, 133].

4.3.1.4 Transfer of the M-cluster from NifEN to the active site on NifDK

Biochemical analysis suggests that a conformational change in NifEN accompanies M-cluster maturation, and this change allows for the formation of a NifEN:NifDK complex and subsequent transfer of the M-cluster from NifEN to apo-NifDK (Figure 4.11) [133]. While there is currently no crystal structure of a NifEN:NifDK complex, comparison of the individual crystal structures paired with biochemical analysis helps to shed light on the process. The putative cluster insertion channel in apo-NifDK is rich in positively charged residues and may facilitate the transfer of the negatively charged M-cluster into NifDK [9]. Sequence alignment between NifEN and NifDK reveals that NifDK has unique residues around the M-cluster site that may allow the cluster to bind to the protein more tightly to NifDK than to NifEN [134]. It has therefore been postulated that M-cluster can be transferred from a lower affinity-binding site on NifEN to a higher affinity site on NifDK.

Once the cluster has been transferred to NifDK, there is an associated conformational change in NifDK that buries the cluster ~10 Å below the protein surface (Figures 4.1 and 4.11). The overall conformation of the protein becomes more compact as observed by SAXS, which provides information about the protein conformation without requiring a crystalline structure [135].

4.3.2 The in situ biosynthesis of the P-cluster

The P-cluster in NifDK from *A. vinelandii* (Figure 4.1B) is the metallocluster that supports nitrogen reduction catalysis through the transfer of electrons from the reductase partner NifH to the M-cluster. Unlike the M-cluster, the P-cluster has yet to be successfully extracted from the protein intact, likely due to rapid decomposition of the cluster once the bridging μ_2-cysteine ligands from the protein are removed [5, 9]. Synthetic efforts to synthesize the P-cluster have resulted in structural analogs that have been generated by the fusion of two cubane clusters, suggesting a possible mechanism in the biosynthetic pathway [136–138]. However, similar to the M-cluster assembly, the biosynthesis of the P-cluster is a multistep process involving several proteins, including an important role played by NifH.

4.3.2.1 Synthesis of the P-cluster precursor – P^P-cluster

P-cluster assembly is initiated by the synthesis of FeS clusters on NifU and NifS (Figure 4.14), the same as in the M-cluster biosynthesis [16–21]. Two $[Fe_4S_4]$ clusters are transferred from NifU to the P-cluster site of NifDK, and these clusters are referred to as the P-cluster precursor, the P^P-cluster. The P^P-cluster is observed in NifDK variants expressed in a *nifH*-deletion background (Δ*nifH* NifDK) (Figure 4.14) [27, 29, 139]. Fe K-edge XAS analysis of Δ*nifH* NifDK reveals that the P^P-cluster consists of a normal $[Fe_4S_4]$ cluster and an atypical $[Fe_4S_4]$ cluster that is either distorted by a bridging Cys ligand in place of a sulfide, or a coordinated light atom such as a N or O from Asp, Ser, His or adventitious water [27, 139]. This assignment is further supported by magnetic circular dichroism spectroscopy, which determined that one of the clusters exists as a $[Fe_4S_4]^{1+}$ cluster and the other as a diamagnetic $[Fe_4S_4]$-like cluster in the dithionite reduced state [140, 141].

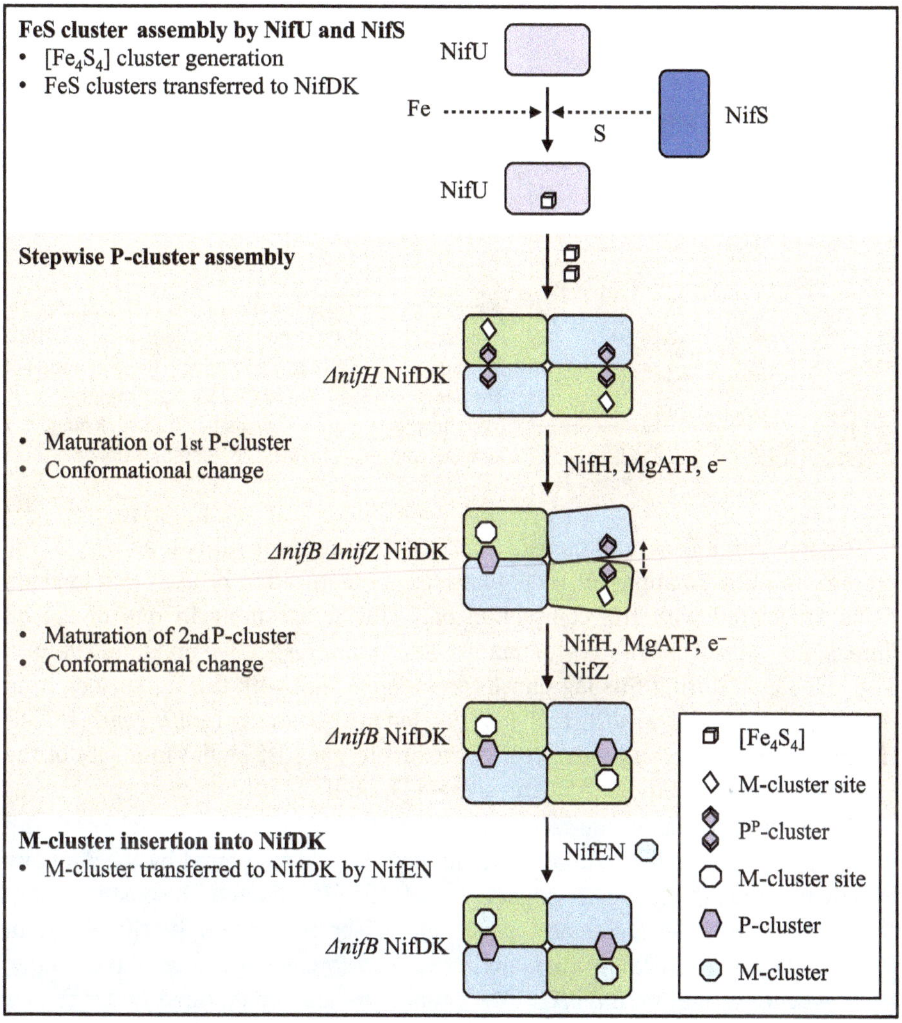

Figure 4.14: A schematic representation of the P-cluster biosynthetic pathway. NifH by itself is able to synthesize the first equivalent of P-cluster on NifDK, while the second P-cluster requires both NifH and NifZ. The P-cluster assembly results in a structural change to the cluster that stabilizes the protein and opens the active site to receive the M-cluster. NifD is represented in green; NifK in blue; and the $[Fe_4S_4]$, P- and M-clusters are indicated on the legend.

4.3.2.2 Conversion of the P^P-cluster to the P-cluster

P^P-cluster can be converted to P-cluster *in vitro* by incubating Δ*nifH* NifDK with NifH, MgATP, and dithionite (Figure 4.15) [27]. Cluster formation can be observed over time with EPR spectroscopy. As the reaction proceeds, the $S = 1/2$ signal associated with the P^P-cluster (g-values of 2.05, 1.93 and 1.90) decreases and the

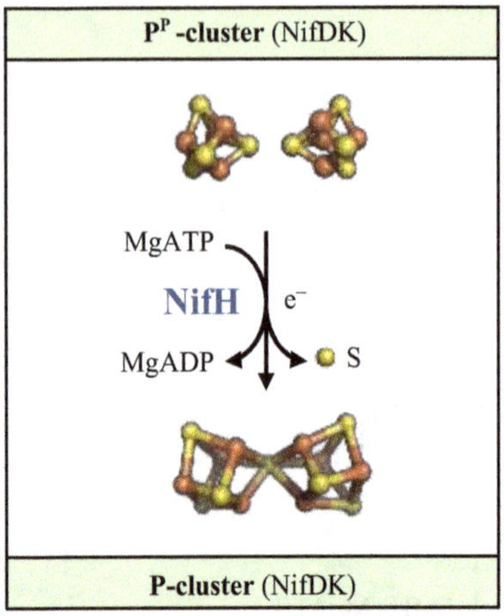

Figure 4.15: *In vitro* maturation of the first P-cluster requires Δ*nifH* NifDK, NifH, MgATP, and dithionite. NifZ has also been shown to play a role in the maturation of the second P-cluster on NifDK. Atoms of the metallocofactors are shown as ball-and-stick models and are colored as follows: Fe, rust; S, yellow.

P^{Ox}-cluster signal (g = 11.8) increases along with the reconstituted activity of the protein. There is an apparent "lag" phase in the maturation process that appears to be associated with the completion of P-cluster assembly in one of the αβ dimers, and this occurs at approximately 50% of the respective maximum values. The EPR signal during this lag phase corresponds well with the signal seen from the gene product of a *nifB*- and *nifZ*-deleted NifDK strain (Δ*nifB* Δ*nifZ* NifDK) (Figure 4.14) and is consistent with approximately a 50% signal intensity of the P^P-cluster signal and the P^{Ox}-cluster signal [33, 142].

P-cluster formation in the second αβ-dimer of Δ*nifH* NifDK can be achieved with longer incubation times (post-lag maturation), which is supported by the doubling of activities and P^{Ox}-cluster EPR signal [27, 30]. The EPR signal of the post-lag maturation Δ*nifH* NifDK variant coincides with that observed in the Δ*nifB* NifDK variant (Figure 4.14), which is known to contain two fully formed P-clusters. Interestingly, the P^P-cluster in the Δ*nifB* Δ*nifZ* NifDK variant can also be matured under similar conditions as the post-lag maturation of Δ*nifH* NifDK (Figure 4.15), but the addition of NifZ is required, suggesting that NifZ plays a role in P-cluster maturation [27, 30]. The activities and EPR signal of the matured Δ*nifB* Δ*nifZ* NifDK variant are consistent with those observed from the post-lag maturation of Δ*nifH* NifDK, likely indicating that the maturation process proceeds analogously. It is still uncertain the exact role that NifZ plays in P-cluster maturation, but the Δ*nifH* NifDK variant (which was expressed in an *nifZ*-intact system) did not require additional NifZ to mature the second P-cluster, whereas the Δ*nifB* Δ*nifZ* NifDK variant did. It has been proposed that NifZ may function in a chaperone-like role and possibly modifies certain residues in the second αβ-dimer to prepare it for cluster maturation [33, 30].

P-cluster biosynthesis prepares NifDK for M-cluster insertion. Prior to P-cluster biosynthesis, the M-cluster site occupies a closed conformation that does not allow for the insertion of M-cluster (Figure 4.14, Δ*nifH* NifDK). After conversion of the P^P-cluster to the P-cluster, NifDK undergoes a conformational change that opens the M-cluster site and allows for M-cluster insertion (Figure 4.14, Δ*nifB* NifDK) [143]. Once the M-cluster has been received, there is a subsequent conformational change that covers the P-cluster and protects it from solvent exposure (Figure 4.14, Δ*nifB* NifDK) producing functional nitrogenase [135].

4.4 The multiple functions of NifH

The reductase component from *A. vinelandii* plays a surprising number of important roles in the nitrogenase system. Most obviously, NifH facilitates ET from physiological sources to ultimately the M-cluster in the catalytic component NifDK. Along with this function, NifH also binds and cleaves the ATP equivalents that are required for substrate reduction, and as described in previous sections, the nucleotide binding and cleavage induces conformational changes that help to regulate the ET process. A role of NifH that is less obvious is in the involvement in the biosynthesis of the two important metallocofactors in NifDK, the P-cluster and M-cluster. Further still is that the Fe protein has been found to facilitate seemingly adventitious reactivity, interconverting CO_2 and CO [76], but this is beyond the scope of this chapter and will not be discussed in detail. This section covers the specific roles that NifH plays in nitrogenase with slightly more detail.

4.4.1 Nitrogenase cofactor maturation

As outlined in Section 4.3, nitrogenase cofactor biosynthesis is a multistep process that takes iron and sulfur and converts these components into multinuclear metalloclusters. These processes involve contributions from multiple proteins and cosubstrates such as SAM and MgATP. One factor that is not usually emphasized is that without NifH, not only would nitrogenase be unable to reduce N_2 to NH_3, functional P- and M-clusters could not be assembled.

The P-cluster maturation process has been explained in great detail in Section 4.3.2; however, there is a benefit to concisely summarize the specific role NifH plays here. When NifDK is expressed in a Δ*nifH* background, the P-cluster observed is incompletely formed and composed of two separate $[Fe_4S_4]$ units (called P^P-cluster) [27, 29, 139]. Each αβ-dimer of Δ*nifH* NifDK has one P^P-cluster and when NifH is incubated with the NifDK variant, reductant and MgATP *in vitro*, the P-cluster of one αβ-dimer is formed quickly, followed by the much slower formation of the P-cluster in the second αβ-dimer [27, 30]. However, NifH is not solely responsible for the fusion of the P-cluster precursor, as NifDK expressed in a Δ*nifH* Δ*nifZ* background will not have functional P-cluster without incubation of both NifH and a chaperone protein NifZ [33, 30]. Not only does NifH transfer electrons, it functions to help synthesize other ET cofactors.

M-cluster assembly (Section 4.3.1) is an arguably more complex multistep process than P-cluster assembly, which includes the formation of an iron-only [Fe_8S_9C] precursor called the L-cluster that is synthesized first on NifB and then is transferred to NifEN. With L-cluster loaded on NifEN, NifH is able to facilitate the insertion of molybdenum and homocitrate, converting the L- to the M-cluster (Figure 4.11). The gene products of an *A. vinelandii nifHDK*-deletion NifEN (Δ*nifHDK* NifEN) or *nifHDK*-deletion NifEN-B (Δ*nifHDK* NifEN-B) expression strains isolated with L-cluster are loaded in the proteins [26, 126, 130]. The L-cluster on these species is readily identified by a characteristic EPR at $g = 1.94$ in the IDS-oxidized state. *In vitro*, L-cluster maturation to M-cluster can occur directly on NifEN by, at minimum, incubating Δ*nifHDK* NifEN or Δ*nifHDK* NifEN-B with NifH, MgATP, dithionite, molybdate (MoO_4^{2-}), and *R*-homocitrate to form active M-cluster (Figure 4.16) [23, 24, 26, 28]. To monitor the conversion of L- to M-cluster, the L-cluster-specific signal at $g = 1.94$ was observed over the time course of the reaction. The decrease in intensity of the $g = 1.94$ resonance implied that the L-cluster was disappearing; however, a new signal with resonances at $g = 4.45$, 3.96 and 3.60 indicated that M-cluster was being generated [23, 24]. Fe and Mo K-edge XAS analyses also demonstrated that the M-cluster species on NifEN that resulted from incubation with NifH was nearly identical to that on wild-type NifDK, except for a slight perturbation of the Mo coordination environment [23, 24]. Importantly, it was observed that stepwise addition of Mo and homocitrate was insufficient for the conversion of the L- to the M-cluster, implying that their insertion may occur in one concerted step [133].

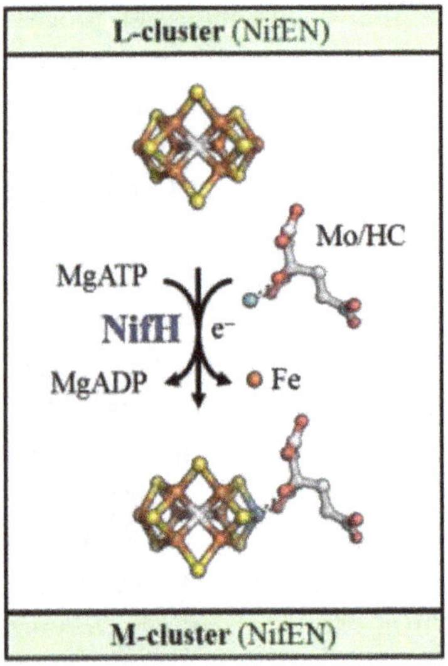

Figure 4.16: Summary of the *in vitro* maturation of the L- to M-cluster. The maturation of L-cluster requires, at minimum, NifEN, NifH, MgATP, dithionite, molybdate (MoO_4^{2-}) and *R*-homocitrate. NifH facilitates the insertion of Mo and *R*-homocitrate into the L-cluster to mature functional M-cluster. Atoms are colored as follows: Fe, rust; S, yellow; Mo, cyan; C, light gray; O, red.

After M-cluster maturation on NifEN, NifH can be re-isolated from the solution mixture to characterize any changes that may have occurred. Analysis of the repurified NifH showed that the protein had become loaded with Mo and homocitrate, and this loaded NifH could be used to insert Mo and homocitrate into NifEN-bound L-cluster [28]. Additionally, Mo K-edge XAS analysis was used to compare the loaded NifH with the MoO_4^{2-} starting material, and the results suggested that the Mo bound to NifH was in a different form than the starting material, likely a product of molybdate reduction [28]. The EPR spectrum of the Mo-bound NifH species features a characteristic $S = 1/2$ resonance but also has the addition of a small signal with *g*-values at 4.44, 4.05 and 3.96 that appears to coincide with Mo binding. One possibility for the Mo-binding location is in the position that is generally occupied by the γ-phosphate of ATP, as a partial molybdate occupancy was observed in this location of the first reported structure of NifH [65, 28]. Further studies are required to fully elucidate the mechanism of Mo and *R*-homocitrate insertion into the L-cluster, as well as the specific role that the Fe protein plays in these processes.

4.4.2 Electron transfer in nitrogenase

The ET process and subsequent substrate reduction of N_2 and substrate analogs by the catalytic component has been studied extensively over the past several decades [4, 90]. As eight electrons are required for the reduction of every equivalent of N_2, ET is a rather fundamental aspect of nitrogenase catalysis and the Fe protein is specifically responsible for this process. To understand the ET reactions in nitrogenase, methods such as stopped-flow and rapid freeze-quench techniques allow for the observation of spectroscopic changes on very fast timescales that would otherwise not be accessible.

The first step in what is referred to as the "Fe protein cycle" is the reduction of NifH to the $[Fe_4S_4]^{1+}$ state, followed by two equivalents of MgATP binding to the protein (Figure 4.17) [4, 90]. The reduction of the Fe protein can be facilitated *in vitro* by dithionite or likely *in vivo* by flavodoxins or ferredoxins [91–94]. The binding of ATP induces a conformational change that allows NifH interact with NifDK to make a NifH: NifDK complex (see Section 4.2.2.1.3). It is important to note that there are two equivalents of NifH that bind to NifDK in this complex, one NifH per αβ-dimer. Seefeldt and coworkers [144] have proposed that ATP-loaded NifH binding to NifDK results in a "conformation-gated" change in NifDK, which allows for an initial transfer of an electron from the all-ferrous P^N-cluster to the M-cluster. Once the P^N-cluster transfers an electron and converts to the P^{1+} state, the $[Fe_4S_4]^{1+}$ cluster of NifH rapidly transfers an electron to the P-cluster, generating the $[Fe_4S_4]^{2+}$ and P^N states on NifH and NifDK, respectively. This overall ET scheme is referred to as a "deficit-spending" model, as the P-cluster pushes an electron to the M-cluster before receiving exogenous electrons from NifH [90, 145]. It was initially unclear what relationship existed between ET and ATP hydrolysis in nitrogenase; whether these two events were separate or concerted

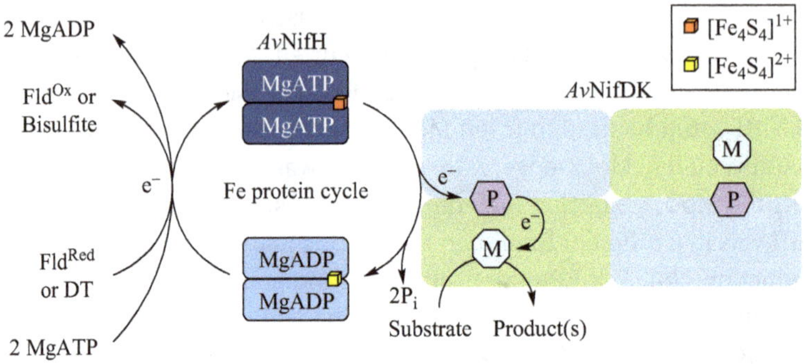

Figure 4.17: A depiction of MgATP-dependent electron transfer from NifH to NifDK. For electron transfer (ET) to occur, the Fe protein must be loaded with 2 equivalents of MgATP and reduced by either dithionite (*in vitro*) or flavodoxin (*in vivo*). A conformational change accompanies MgATP binding that allows for the Fe protein to interact with NifDK and form a NifH:NifDK complex. After binding, an electron is proposed to transfer from the P-cluster to the M-cluster, followed by ET from the [Fe_4S_4] cluster of NifH to the P-cluster. The MgATP nucleotides are then hydrolyzed and the phosphate dissociates from the Fe protein. Finally, the Fe protein dissociates from catalytic component, MgADP is released and the Fe protein is prepared for additional redox cycles. This process must be repeated until enough electrons have been transferred to completely reduce the substrate. Fld^{Ox}, oxidized flavodoxin; Fld^{Red}, reduced flavodoxin; and the P- and M-clusters are represented by hexagons and octagons, respectively.

[4]. Seefeldt and co-workers reported that ET precedes ATP hydrolysis in nitrogenase, supporting the notion that these two events are not concerted [146].

After ET to NifDK, the two ATP molecules on NifH are hydrolyzed to form two ADP and two P_i molecules, in a seemingly rapid reaction [90], but it is still unclear what specifically induces this hydrolysis. The next step, determined to be the rate-limiting step in the nitrogenase catalytic cycle, is the release of the two P_i from the complex, followed by the rapid release of the MgADP-bound NifH from NifDK [147]. It has been suggested that phosphate release from the NifH:NifDK complex serves to induce the conformational change observed in the MgADP-bound state of NifH, as described in the previous sections, which lowers the affinity of the protein for NifDK [90, 144]. Once freed, the $[Fe_4S_4]^{2+}$ cluster of the MgADP-bound NifH is then reduced, lowering the affinity of MgADP for binding NifH and allowing for the release of MgADP and subsequent replacement by two MgATP molecules completing the Fe protein cycle (Figure 4.17) [4, 90].

4.5 Nitrogenase substrate reduction mechanism

The multielectron reduction of the triple bond of N_2 has been the focus of a multipronged research effort over decades that incorporates spectroscopic, kinetic, biochemical and structural biology approaches. The two components of nitrogenase have been studied individually, providing useful information about spectroscopic and physical properties; however, the study of nitrogen reduction requires both components to function together which poses a significant challenge. Crystallization of proteins, including the NifH:NifDK complex, has shed light on structural aspects of nitrogenase catalysis, but crystals represent a solid-state snapshot of the protein and will not necessarily reflect the dynamics that may be present in the reaction mixture. Under physiological conditions, the steps in the nitrogenase reaction are fast, and require specialized techniques and methods to be able to observe intermediate steps, such as rapid freeze-quench, and stopped-flow methods [4, 90, 148]. Despite these apparent complications, many groups have contributed to the wealth of information that is available, so much so that covering it all in this chapter will not be possible. A simplified mechanistic model for N_2 reduction will be presented, primarily with studies involving the Mo-dependent nitrogenase because this system has been most extensively characterized. Though it is worth mentioning that insights into the mechanism of nitrogen fixation by the alternative V- and Fe-dependent nitrogenases have recently been published, further expanding the knowledge available to the field [52, 149].

4.5.1 The Lowe–Thorneley reaction model

The first comprehensive kinetic description of Mo-nitrogenase, including a mechanistic outline, involved the work of many research groups and resulted in the Lowe–Thorneley (LT) model (Figure 4.18) [150–153]. This model describes the kinetics of the chemical transformations that occur on NifDK during nitrogenase catalysis, and each identified state is designated by E_n notation, where n is the number of electrons and protons that has been accumulated on one αβ-dimer of NifDK. This formalism doesn't distinguish the location of the electrons, whether it be on the P-cluster or the M-cluster, just that the electron has been delivered to the system. These electrons are transferred by NifH (as described in Section 4.2) by the Fe protein cycle, so by analogy the LT model has also been referred to as the MoFe protein cycle [148].

The first state, E_0, represents the resting state of NifDK as found under dithionite reduced conditions and has the M-cluster in an $S = 3/2$ state [86]. Much of the earlier work on Mo-nitrogenase has involved the characterization of the resting state of the enzyme. The odd number of unpaired electrons in E_0 means that as electrons/protons are added, the even numbered states of the LT model ($n = 2, 4, 6, 8$) will have an odd number of electrons loaded on NifDK. This property potentially allows for a convenient spectroscopic handle to study these species using EPR techniques, including ENDOR and ESEEM, if they can be trapped or isolated [154]. Alternatively, in the odd numbered states ($n = 1, 3, 5, 7$) there will be an even

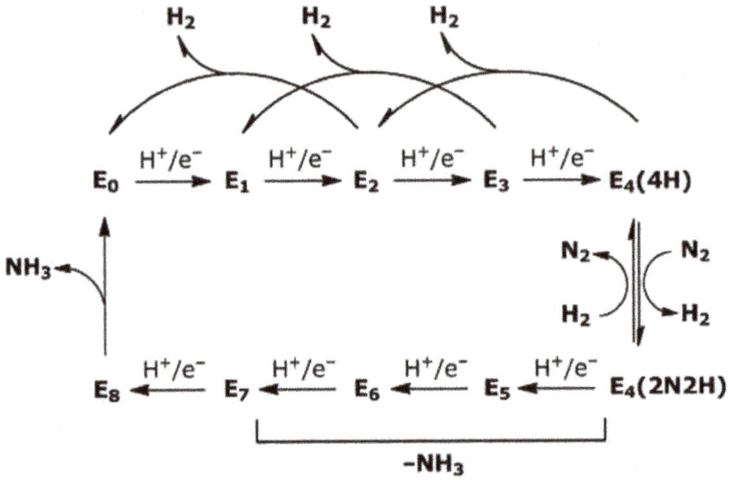

Figure 4.18: The modified LT kinetic scheme for N_2 reduction in Mo-nitrogenase. The E_n notation refers to the *n* number of proton and electron equivalents that have been loaded onto one αβ-dimer of NifDK. The E_0 state is designated as the resting state of the enzyme, and the release of NH_3 and H_2 during the cycle is indicated.

number of electrons, and this will either lead to an EPR-silent state (diamagnetic) or a state that will require advanced methods to probe EPR transitions [154]. The E_1 through E_3 states all precede the binding of N_2, which is proposed to occur in the E_4 or E_4(4H) state. Further addition of protons and electrons to the system through the E_5–E_8 states causes the scission of the N_2 bond, generating H_2 and 2 equivalents of NH_3, returning to the E_0 starting point (Figure 4.18). It should be noted that the E_2 and E_3 states are also able to generate H_2, converting back to the E_0 and E_1 states, respectively, which represents nonproductive hydrogen release that competes with N_2 reduction [148].

The E_4 state is inherently of interest within the LT scheme because it is a central point in the cycle that either releases H_2, reverting to the E_2 state, or can bind N_2 to the M-cluster moving the reaction cycle forward [148, 154]. When N_2 binds to the E_4 (4H), there is a concomitant loss of one equivalent of H_2 from NifDK, stoichiometrically leaving an E_5(2N2H) species. It is unclear why nitrogenase opts to produce H_2 as part of the N_2 reduction mechanism, as 25% of the ATP required for the reaction is funneled into this apparent byproduct. To understand the nature of the E_4 state, Hoffman et al. [154] have used freeze-quench spectroscopic experiments to trap E_4 as well as other intermediate states. A single-point mutant of NifDK has been identified in the E_4 state as being consistent with an M-cluster model with two μ_2-hydride (H^-) ligands and two protons (H^+), termed the "Janus intermediate" [155]. $^{1,2}H$, ^{95}Mo and ^{57}Fe ENDOR experiments demonstrated that the hydrides were each likely bridged between two of the core Fe atoms of the M-cluster, and did not involve the Mo center [156–158]. Subsequently, it was reported that spectroscopic features associated with the Janus intermediate were observed in the wild-type nitrogenase, suggesting that hydrides are involved in the mechanism of N_2 reduction [159].

4.5.2 N_2 hydrogenation pathways – moving beyond the E_4 state

The incorporation of hydride intermediates into the nitrogen fixation mechanism provides an elegant solution for introducing proton/electron equivalents to the system. However, it is unclear if additional hydride species are involved in the hydrogenation of the N_2, or even how the proton/electron equivalents are added to cleave the N_2 triple bond. There are two main pathways that have been proposed for the hydrogenation of N_2: the distal and alternating mechanisms (Figure 4.19). In both cases, N_2 is proposed to bind in an end-on mode to a metal center (M–N_2) of the M-cluster and eventually merge at the formation of a terminal amido species (M–NH_2). The final step of the mechanism is the delivery of one proton/electron equivalent to the terminal amido intermediate, releasing NH_3 and regenerating the E_0 state of NifDK. The two mechanisms generally differ in the specific sites of protonation of N_2, the intermediate species that are formed during the reaction and the E_n states that allow for NH_3 to be released.

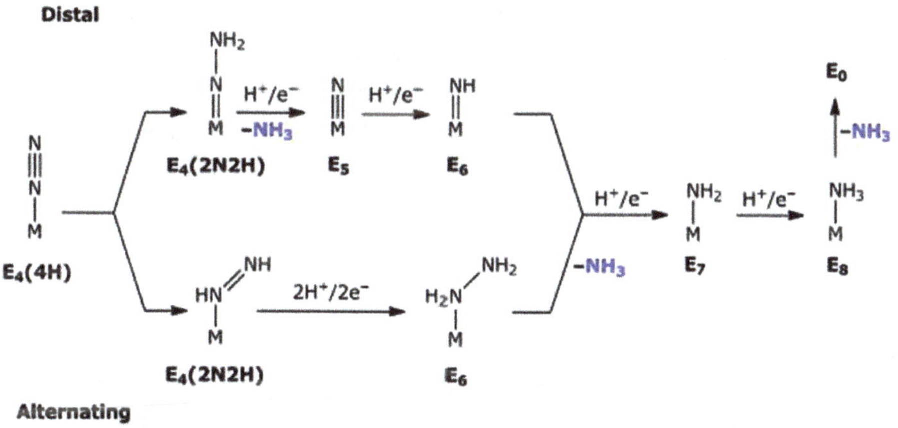

Figure 4.19: The mechanistic scheme presenting the distal (top) and alternating (bottom) pathways for N_2 hydrogenation by Mo-nitrogenase. The corresponding E_n states are indicated under each intermediate, and proposed release of NH_3 for each pathway is noted.

The development of the distal pathway has roots in the seminal work with inorganic Mo complexes by Chatt et al. [160–163] and Yandulov and Schrock [164, 165].

N_2 first binds to a metal center of the M-cluster in the $E_4(4H)$ state, and this causes two of the proton/electron equivalents loaded on NifDK to add to the distal site of N_2, forming a hydrazido ($M{=}N{-}NH_2$) species concomitant with release of H_2 (Figure 4.19). The next proton/electron equivalent (E_5) also adds to the distal site, forming a terminal metal nitride ($M{\equiv}N$) species and releasing NH_3. The subsequent two reduction and protonation steps (E_6, E_7) form metal imido ($M{=}NH$) and amido ($M{-}NH_2$) species, respectively, and the final proton/electron addition (E_8) generates and releases the second equivalent of NH_3.

While the distal pathway has been characterized for Mo complexes and proposed for nitrogenase, the alternating pathway of hydrogenation is derived from work with Fe complexes [166, 167]. As similarly described for the distal pathway, in the alternating mechanism N_2 binds to the $E_4(4H)$ state, but one proton/electron is added to each nitrogen atom, forming a diazene-bound species ($M{-}(H)N{=}NH$) with loss of H_2. The following two reduction/protonation steps (E_5 and E_6) form a hydrazine-bound species ($M{-}(H_2)N{-}NH_2$), alternating the nitrogen atom that is the site of hydrogen addition. Reduction and protonation in the next step (E_7) releases NH_3, generating an amido species that is liberated as the second equivalent of NH_3 in the following step (E_8).

There is no definitive consensus of the N_2 hydrogenation mechanism that is operative in nitrogenase, or if this reduction takes place at the Mo or Fe centers [154]. Generally, ammonia had been generated from N_2 using Mo complexes and this extensive and thorough body of work supported a Mo-centered distal pathway for mechanism in nitrogenase [154, 168]. However, there is growing support for Fe-centered N_2 reduction based on synthetic work with iron complexes [169–171]. In the biological system, hydrazine has been noted to be both a substrate for the production of ammonia and a product from acid- or base-quenching the protein under turnover conditions [4, 172]. These findings suggest the alternating mechanism for nitrogen reduction, and while the support continues to grow for this pathway, additional work and discussion is required to address this important issue.

4.6 Summary and outlook

Nitrogenase is an undeniably complex biological system that can cleave the strong triple bond of dinitrogen under ambient temperature and pressure. This is compared with energy-intensive industrial methods that require high temperatures, pressures, equipment and hydrogen gas. The two-component NifDK/NifH system uses unique metallocofactors to facilitate this reactivity that are not found elsewhere in Nature, and these cofactors have captivated research groups over many decades. Advances in spectroscopic characterization and crystallography have provided a robust understanding of the proteins, cofactors and some of their interactions, but there are still questions that remain to be answered. The Mo-nitrogenase mechanism has been studied extensively, but there is still no definitive proposal due to the complexity of targeting the M-cluster-bound intermediates during catalysis. Techniques like ENDOR and computations have shed light on trapped intermediate species under N_2 turnover conditions, pushing mechanistic studies forward. However, there is still much to be learned from the alternative nitrogenases, as these systems lack the characterization of the Mo-dependent system. V- and Fe-nitrogenases are also proposed to function similarly to Mo-nitrogenase, but further detailed kinetic and mechanistic studies are required to probe these proposals. V-nitrogenase as well as the nitrogenase Fe proteins are capable of seemingly adventitious reactivity such as the reduction of CO and CO_2 into longer chain hydrocarbon products, and there are efforts to modify the enzymes to understand and control this reactivity. Additionally, the transgenic expression of nitrogenase homologs from different organisms in *Escherichia coli* is another exciting front of nitrogenase research, with the goal of transferring the nitrogen fixation genes into agricultural crops that would not require fertilizers. The future of nitrogenase research is bright and will continue to capture the attention of investigators for years to come.

References

[1] Darwent, Bd. Bond dissociation energies in simple molecules, U.S. National Bureau of Standards; for sale by the Supt. of Docs., U.S. Govt. Print. Off., [Washington], 1970.

[2] Schlögl, R. Catalytic synthesis of ammonia—A "never-ending story"?. Angew Chem Int Ed 2003, 42, 2004–2008.

[3] Dybkjaer, I. Ammonia, catalysis and manufacture, Springer, Heidelberg, 1995.

[4] Burgess, BK., and Lowe, DJ. Mechanism of molybdenum nitrogenase. Chem Rev 1996, 96, 2983–3012.

[5] Howard, JB., and Rees, DC. Structural basis of biological nitrogen fixation. Chem Rev 1996, 96, 2965–2982.

[6] Navarro, RM., Peña, MA., and Fierro, JLG. Hydrogen production reactions from carbon feedstocks: fossil fuels and biomass. Chem Rev 2007, 107, 3952–3991.

[7] Palo, DR., Dagle, RA., and Holladay, JD. Methanol steam reforming for hydrogen production. Chem Rev 2007, 107, 3992–4021.

[8] Burk, D., Lineweaver, H., and Horner, CK. The specific influence of acidity on the mechanism of nitrogen fixation by *Azotobacter*. J Bacteriol 1934, 27, 325–340.

[9] Ribbe, MW., Hu, Y., Hodgson, KO., and Hedman, B. Biosynthesis of nitrogenase metalloclusters. Chem Rev 2014, 114, 4063–4080.

[10] Hardy, RWF., and Knight, E. ATP-dependent reduction of azide and HCN by N_2-fixing enzymes of *Azotobacter vinelandii* and *Clostridium pasteurianum*. Biochim Biophys Acta Enzymol 1967, 139, 69–90.

[11] Dilworth, MJ. Acetylene reduction by nitrogen-fixing preparations from *Clostridium pasteurianum*. Biochim Biophys Acta (BBA) – General Subjects 1966, 127, 285–294.

[12] Lee, CC., Hu, Y., and Ribbe, MW. Vanadium nitrogenase reduces CO. Science 2010, 329, 642.

[13] Seefeldt, LC., Rasche, ME., and Ensign, SA. Carbonyl sulfide and carbon dioxide as new substrates, and carbon disulfide as a new inhibitor, of nitrogenase. Biochemistry 1995, 34, 5382–5389.
[14] Eady, RR. Structure–function relationships of alternative nitrogenases. Chem Rev 1996, 96, 3013–3030.
[15] Joerger, RD., Bishop, PE., and Evans, HJ. Bacterial alternative nitrogen fixation systems. CRC Crit Rev Microbiol 1988, 16, 1–14.
[16] Dos Santos, PC., Johnson, DC., Ragle, BE., Unciuleac, M-C., and Dean, DR. Controlled expression of *nif* and *isc* iron-sulfur protein maturation components reveals target specificity and limited functional replacement between the two systems. J Bacteriol 2007, 189, 2854–2862.
[17] Smith, AD., Jameson, GNL., Dos Santos, PC. et al. NifS-mediated assembly of [4Fe–4S] clusters in the N- and C-terminal domains of the NifU scaffold protein. Biochemistry 2005, 44, 12955–12969.
[18] Yuvaniyama, P., Agar, JN., Cash, VL., Johnson, MK., and Dean, DR. NifS-directed assembly of a transient [2Fe-2S] cluster within the NifU protein. Proc Natl Acad Sci U S A 2000, 97, 599–604.
[19] Zheng, L., White, RH., Cash, VL., and Dean, DR. Mechanism for the desulfurization of L-cysteine catalyzed by the nifS gene product. Biochemistry 1994, 33, 4714–4720.
[20] Zheng, L., White, RH., Cash, VL., Jack, RF., and Dean, DR. Cysteine desulfurase activity indicates a role for NIFS in metallocluster biosynthesis. Proc Natl Acad Sci U S A 1993, 90, 2754–2758.
[21] Zheng, L., and Dean, DR. Catalytic formation of a nitrogenase iron-sulfur cluster. J Biol Chem 1994, 269, 18723–18726.
[22] Wiig, JA., Hu, Y., Lee, CC., and Ribbe, MW. Radical SAM-dependent carbon insertion into the nitrogenase M-cluster. Science 2012, 337, 1672–1675.
[23] Hu, Y., Corbett, MC., Fay, AW. et al. FeMo cofactor maturation on NifEN. Proc Natl Acad Sci U S A 2006, 103, 17119–17124.
[24] Yoshizawa, JM., Blank, MA., Fay, AW. et al. Optimization of FeMoco maturation on NifEN. J Am Chem Soc 2009, 131, 9321–9325.
[25] Kaiser, JT., Hu, Y., Wiig, JA., Rees, DC., and Ribbe, MW. Structure of precursor-bound NifEN: A nitrogenase FeMo cofactor maturase/insertase. Science 2011, 331, 91–94.
[26] Hu, Y., Fay, AW., and Ribbe, MW. Identification of a nitrogenase FeMo cofactor precursor on NifEN complex. Proc Natl Acad Sci U S A 2005, 102, 3236–3241.
[27] Lee, CC., Blank, MA., Fay, AW. et al. Stepwise formation of P-cluster in nitrogenase MoFe protein. Proc Natl Acad Sci U S A 2009, 106, 18474–18478.
[28] Hu, Y., Corbett, MC., Fay, AW. et al. Nitrogenase Fe protein: A molybdate/homocitrate insertase. Proc Natl Acad Sci U S A 2006, 103, 17125–17130.
[29] Ribbe, MW., Hu, Y., Guo, M., Schmid, B., and Burgess, BK. The FeMoco-deficient MoFe protein produced by a *nifH* deletion strain of *Azotobacter vinelandii* shows unusual P-cluster features. J Biol Chem 2002, 277, 23469–23476.
[30] Hu, Y., Fay, AW., Lee, CC., and Ribbe, MW. P-cluster maturation on nitrogenase MoFe protein. Proc Natl Acad Sci U S A 2007, 104, 10424–10429.
[31] Filler, WA., Kemp, RM., Ng, JC., Hawkes, TR., Dixon, RA., and Smith, BE. The nifH gene product is required for the synthesis or stability of the iron-molybdenum cofactor of nitrogenase from *Klebsiella pneumoniae*. Eur J Biochem 1986, 160, 371–377.
[32] Hernandez, JA., Curatti, L., Aznar, CP., Perova, Z., Britt, RD., and Rubio, LM. Metal trafficking for nitrogen fixation: NifQ donates molybdenum to NifEN/NifH for the biosynthesis of the nitrogenase FeMo-cofactor. Proc Natl Acad Sci U S A 2008, 105, 11679–11684.

[33] Hu, Y., Fay, AW., Dos Santos, PC., Naderi, F., and Ribbe, MW. Characterization of *Azotobacter vinelandii nifZ* deletion strains: indication of stepwise MoFe protein assembly. J Biol Chem 2004, 279, 54963–54971.
[34] Lancaster, KM., Roemelt, M., Ettenhuber, P. et al. X-ray emission spectroscopy evidences a central carbon in the nitrogenase iron-molybdenum cofactor. Science 2011, 334, 974–977.
[35] Palmer, G. Electron paramagnetic resonance of metalloproteins, Que L, Jr, ed, Physical methods in bioinorganic chemistry: spectroscopy and magnetism, University Science Books; 2000, Sausalito, CA.
[36] Chasteen, ND., and Snetsinger, PA. ESEEM and ENDOR spectroscopy, Que L, Jr, ed, Physical methods in bioinorganic chemistry: spectroscopy and magnetism, University Science Books; 2000, Sausalito, CA.
[37] Bencini, A., and Gatteschi, D. Electron paramagnetic resonance spectroscopy, Solomon EI, Lever ABP, eds, Inorganic electronic structure and spectroscopy, volume I: methodology, Wiley; 1999, New York, NY.
[38] Gütlich, P., and Ensling, J. Mössbauer spectroscopy, Solomon EI, Lever ABP, eds, Inorganic electronic structure and spectroscopy, volume I: methodology, Wiley; 1999, New York, NY.
[39] Münck, E. Aspects of ^{57}Fe Mössbauer spectroscopy, Que L, Jr, ed, Physical methods in bioinorganic chemistry: spectroscopy and magnetism, University Science Books, Sausalito, CA, 2000.
[40] Münck, E., Rhodes, H., Orme-Johnson, WH., Davis, LC., Brill, WJ., and Shah, VK. Nitrogenase. VIII. Mössbauer and EPR spectroscopy. the MoFe protein component from *Azotobacter vinelandii* OP. Biochimica et Biophysica Acta (BBA) – Protein Structure 1975, 400, 32–53.
[41] McLean, PA., Papaefthymiou, V., Orme-Johnson, WH., and Münck, E. Isotopic hybrids of nitrogenase. Mössbauer study of MoFe protein with selective ^{57}Fe enrichment of the P-cluster. J Biol Chem 1987, 262, 12900–12903.
[42] Yoo, SJ., Angove, HC., Papaefthymiou, V., Burgess, BK., and Münck, E. Mössbauer study of the MoFe protein of nitrogenase from *Azotobacter vinelandii* using selective ^{57}Fe enrichment of the M-centers. J Am Chem Soc 2000, 122, 4926–4936.
[43] Spatzal, T., Aksoyoglu, M., Zhang, L. et al. Evidence for interstitial carbon in nitrogenase FeMo cofactor. Science 2011, 334, 940.
[44] Einsle, O., Tezcan, FA., Andrade, SLA. et al. Nitrogenase MoFe-protein at 1.16 Å resolution: A central ligand in the FeMo-cofactor. Science 2002, 297, 1696–1700.
[45] Kirn, J., and Rees, DC. Crystallographic structure and functional implications of the nitrogenase molybdenum–iron protein from *Azotobacter vinelandii*. Nature 1992, 360, 553.
[46] Schrodinger, LLC. The PyMOL molecular graphics system, version 1.8. 2015.
[47] Hales, BJ., Case, EE., Morningstar, JE., Dzeda, MF., and Mauterer, LA. Isolation of a new vanadium-containing nitrogenase from *Azotobacter vinelandii*. Biochemistry 1986, 25, 7251–7255.
[48] Sippel, D., and Einsle, O. The structure of vanadium nitrogenase reveals an unusual bridging ligand. Nat Chem Biol 2017, 13, 956.
[49] Dyer, DH., Rubio, LM., Thoden, JB., Holden, HM., Ludden, PW., and Rayment, I. The three-dimensional structure of the core domain of Naf Y from *Azotobacter vinelandii* determined at 1.8-Å resolution. J Biol Chem 2003, 278, 32150–32156.
[50] Homer, MJ., Paustian, TD., Shah, VK., and Roberts, GP. The nifY product of *Klebsiella pneumoniae* is associated with apodinitrogenase and dissociates upon activation with the iron-molybdenum cofactor. J Bacteriol 1993, 175, 4907–4910.
[51] Lee, CC., Hu, Y., and Ribbe, MW. Unique features of the nitrogenase VFe protein from *Azotobacter vinelandii*. Proc Natl Acad Sci U S A 2009, 106, 9209–9214.

[52] Harris, DF., Lukoyanov, DA., Shaw, S. et al. Mechanism of N_2 reduction catalyzed by Fe-nitrogenase involves reductive elimination of H_2. Biochemistry 2018, 57, 701–710.
[53] Hu, Y., and Ribbe, MW. Nitrogenase and homologs. JBIC, J Biol Inorg Chem 2015, 20, 435–445.
[54] Mus, F., Alleman, AB., Pence, N., Seefeldt, LC., and Peters, JW. Exploring the alternatives of biological nitrogen fixation. Metallomics 2018, 10, 523–538.
[55] Siemann, S., Schneider, K., Dröttboom, M., and Müller, A. The Fe-only nitrogenase and the Mo nitrogenase from *Rhodobacter capsulatus*. Eur J Biochem 2002, 269, 1650–1661.
[56] Krahn, E., Weiss, B., Kröckel, M. et al. The Fe-only nitrogenase from *Rhodobacter capsulatus*: identification of the cofactor, an unusual, high-nuclearity iron-sulfur cluster, by Fe K-edge EXAFS and ^{57}Fe Mössbauer spectroscopy. JBIC, J Biol Inorg Chem 2002, 7, 37–45.
[57] Burgess, BK. The iron-molybdenum cofactor of nitrogenase. Chem Rev 1990, 90, 1377–1406.
[58] Fay, AW., Blank, MA., Lee, CC. et al. Characterization of isolated nitrogenase FeVco. J Am Chem Soc 2010, 132, 12612–12618.
[59] Cutsail, GE., Telser, J., and Hoffman, BM. Advanced paramagnetic resonance spectroscopies of iron–sulfur proteins: electron nuclear double resonance (ENDOR) and electron spin echo envelope modulation (ESEEM). Biochim Biophys Acta (BBA) – Molecular Cell Research 2015, 1853, 1370–1394.
[60] Pierik, AJ., Wassink, H., Haaker, H., and Hagen, WR. Redox properties and EPR spectroscopy of the P clusters of *Azotobacter vinelandii* MoFe protein. Eur J Biochem 1993, 212, 51–61.
[61] Surerus, KK., Hendrich, MP., Christie, PD., Rottgardt, D., Orme-Johnson, WH., and Munck, E. Mössbauer and integer-spin EPR of the oxidized P-clusters of nitrogenase: POX is a non-Kramers system with a nearly degenerate ground doublet. J Am Chem Soc 1992, 114, 8579–8590.
[62] Watt, GD., Burns, A., Lough, S., and Tennent, DL. Redox and spectroscopic properties of oxidized MoFe protein from *Azotobacter vinelandii*. Biochemistry 1980, 19, 4926–4932.
[63] Chan, M., Kim, J., and Rees, D. The nitrogenase FeMo-cofactor and P-cluster pair: 2.2 Å resolution structures. Science 1993, 260, 792–794.
[64] Peters, JW., Stowell, MHB., Soltis, SM., Finnegan, MG., Johnson, MK., and Rees, DC. Redox-dependent structural changes in the nitrogenase P-cluster. Biochemistry 1997, 36, 1181–1187.
[65] Georgiadis, M., Komiya, H., Chakrabarti, P., Woo, D., Kornuc, J., and Rees, D. Crystallographic structure of the nitrogenase iron protein from *Azotobacter vinelandii*. Science 1992, 257, 1653–1659.
[66] Jasniewski, A., Sickerman, N., Hu, Y., and Ribbe, M. The Fe protein: an unsung hero of nitrogenase. Inorganics 2018, 6, 25.
[67] Walker, JE., Saraste, M., Runswick, MJ., and Gay, NJ. Distantly related sequences in the alpha- and beta-subunits of ATP synthase, myosin, kinases and other ATP-requiring enzymes and a common nucleotide binding fold. EMBO J 1982, 1, 945–951.
[68] Rohde, M., Trncik, C., Sippel, D., Gerhardt, S., and Einsle, O. Crystal structure of VnfH, the iron protein component of vanadium nitrogenase. JBIC, J Biol Inorg Chem 2018.
[69] Strop, P., Takahara, PM., Chiu, H-J., Angove, HC., Burgess, BK., and Rees, DC. Crystal structure of the all-ferrous $[4Fe\text{-}4S]^0$ form of the nitrogenase iron protein from *Azotobacter vinelandii*. Biochemistry 2001, 40, 651–656.
[70] Jang, SB., Seefeldt, LC., and Peters, JW. Insights into nucleotide signal transduction in nitrogenase: structure of an iron protein with MgADP bound. Biochemistry 2000, 39, 14745–14752.
[71] Sen, S., Krishnakumar, A., McClead, J. et al. Insights into the role of nucleotide-dependent conformational change in nitrogenase catalysis: structural characterization of the

nitrogenase Fe protein Leu127 deletion variant with bound MgATP. J Inorg Biochem 2006, 100, 1041–1052.
[72] Chiu, H-J., Peters, JW., Lanzilotta, WN. et al. MgATP-bound and nucleotide-free structures of a nitrogenase protein complex between the Leu 127Δ-Fe-protein and the MoFe-protein. Biochemistry 2001, 40, 641–650.
[73] Schindelin, H., Kisker, C., Schlessman, JL., Howard, JB., and Rees, DC. Structure of ADP·AlF_4 – stabilized nitrogenase complex and its implications for signal transduction. Nature 1997, 387, 370.
[74] Tezcan, FA., Kaiser, JT., Mustafi, D., Walton, MY., Howard, JB., and Rees, DC. Nitrogenase complexes: multiple docking sites for a nucleotide switch protein. Science 2005, 309, 1377–1380.
[75] Rubio, LM., and Ludden, PW. Biosynthesis of the iron-molybdenum cofactor of nitrogenase. Annu Rev Microbiol 2008, 62, 93–111.
[76] Rebelein, JG., Stiebritz, MT., Lee, CC., and Hu, Y. Activation and reduction of carbon dioxide by nitrogenase iron proteins. Nat Chem Biol 2016, 13, 147–149.
[77] Rees, DC. Dinitrogen reduction by nitrogenase: if N_2 isn't broken, it can't be fixed: current opinion in structural biology. 3:921–928, Curr Opin Struct Biol 1993, 1993, 3, 921–928.
[78] Onate, YA., Finnegan, MG., Hales, BJ., and Johnson, MK. Variable temperature magnetic circular dichroism studies of reduced nitrogenase iron proteins and $[4Fe\text{-}4S]^+$ synthetic analog clusters. Biochim Biophys Acta, Protein Struct Mol 1993, 1164, 113–123.
[79] Blank, MA., Lee, CC., Hu, Y., Hodgson, KO., Hedman, B., and Ribbe, MW. Structural models of the $[Fe_4S_4]$ clusters of homologous nitrogenase Fe proteins. Inorg Chem 2011, 50, 7123–7128.
[80] Lindahl, PA., Day, EP., Kent, TA., Orme-Johnson, WH., and Münck, E. Mössbauer, EPR, and magnetization studies of the *Azotobacter vinelandii* Fe protein. evidence for a $[4Fe\text{-}4S]^{1+}$ cluster with spin S = 3/2. J Biol Chem 1985, 260, 11160–11173.
[81] Yoo, SJ., Angove, HC., Burgess, BK., Hendrich, MP., and Münck, E. Mössbauer and integer-spin EPR studies and spin-coupling analysis of the $[4Fe\text{-}4S]^0$ cluster of the Fe protein from *Azotobacter vinelandii* nitrogenase. J Am Chem Soc 1999, 121, 2534–2545.
[82] Chakrabarti, M., Deng, L., Holm, RH., Münck, E., and Bominaar, EL. Mössbauer, electron paramagnetic resonance, and theoretical studies of a carbene-based all-ferrous Fe_4S_4 cluster: electronic origin and structural identification of the unique spectroscopic site. Inorg Chem 2009, 48, 2735–2747.
[83] Chakrabarti, M., Münck, E., and Bominaar, EL. Density functional theory study of an all ferrous 4Fe-4S cluster. Inorg Chem 2011, 50, 4322–4326.
[84] Angove, HC., Yoo, SJ., Burgess, BK., and Münck, E. Mössbauer and EPR evidence for an all-ferrous Fe_4S_4 cluster with $S = 4$ in the Fe protein of nitrogenase. J Am Chem Soc 1997, 119, 8730–8731.
[85] Thorneley, RNF., and Ashby, GA. Oxidation of nitrogenase iron protein by dioxygen without inactivation could contribute to high respiration rates of *Azotobacter* species and facilitate nitrogen fixation in other aerobic environments. Biochem J 1989, 261, 181–187.
[86] Orme-Johnson, WH., Hamilton, WD., Jones, TL. et al. Electron paramagnetic resonance of nitrogenase and nitrogenase components from *Clostridium pasteurianum W5* and *Azotobacter vinelandii OP*. Proc Natl Acad Sci U S A 1972, 69, 3142–3145.
[87] Lindahl, PA., Gorelick, NJ., Münck, E., and Orme-Johnson, WH. EPR and Mössbauer studies of nucleotide-bound nitrogenase iron protein from *Azotobacter vinelandii*. J Biol Chem 1987, 262, 14945–14953.
[88] Watt, GD., and Reddy, KRN. Formation of an all ferrous Fe_4S_4 cluster in the iron protein component of *Azotobacter vinelandii* nitrogenase. J Inorg Biochem 1994, 53, 281–294.

[89] Hiller, CJ., Stiebritz, MT., Lee, CC., Liedtke, J., and Hu, Y. Tuning electron flux through nitrogenase with methanogen iron protein homologues. Chem Eur J 2017, 23, 16152–16156.

[90] Seefeldt, LC., Hoffman, BM., Peters, JW. et al. Energy transduction in nitrogenase. Acc Chem Res 2018.

[91] Yates, MG. Electron transport to nitrogenase in *Azotobacter chroococcum*: *Azotobacter* flavodoxin hydroquinone as an electron donor. FEBS Lett 1972, 27, 63–67.

[92] Bennett, LT., Jacobson, MR., and Dean, DR. Isolation, sequencing, and mutagenesis of the *nifF* gene encoding flavodoxin from *Azotobacter vinelandii*. J Biol Chem 1988, 263, 1364–1369.

[93] Thorneley, RN., and Deistung, J. Electron-transfer studies involving flavodoxin and a natural redox partner, the iron protein of nitrogenase. conformational constraints on protein-protein interactions and the kinetics of electron transfer within the protein complex. Biochem J 1988, 253, 587–595.

[94] Martin, AE., Burgess, BK., Iismaa, SE., Smartt, CT., Jacobson, MR., and Dean, DR. Construction and characterization of an *Azotobacter vinelandii* strain with mutations in the genes encoding flavodoxin and ferredoxin I. J Bacteriol 1989, 171, 3162–3167.

[95] Duyvis, MG., Wassink, H., and Haaker, H. Nitrogenase of *Azotobacter vinelandii*: Kinetic analysis of the Fe protein redox cycle. Biochemistry 1998, 37, 17345–17354.

[96] Watt, GD. An electrochemical method for measuring redox potentials of low potential proteins by microcoulometry at controlled potentials. Anal Biochem 1979, 99, 399–407.

[97] Lanzilotta, WN., Ryle, MJ., and Seefeldt, LC. Nucleotide hydrolysis and protein conformational changes in *Azotobacter vinelandii* nitrogenase iron protein: defining the function of aspartate 129. Biochemistry 1995, 34, 10713–10723.

[98] Sørlie, M., Seefeldt, LC., and Parker, VD. Use of stopped-flow spectrophotometry to establish midpoint potentials for redox proteins. Anal Biochem 2000, 287, 118–125.

[99] Lowery, TJ., Wilson, PE., Zhang, B. et al. Flavodoxin hydroquinone reduces *Azotobacter vinelandii* Fe protein to the all-ferrous redox state with a $S = 0$ spin state. Proc Natl Acad Sci U S A 2006, 103, 17131–17136.

[100] Guo, M., Sulc, F., Ribbe, MW., Farmer, PJ., and Burgess, BK. Direct assessment of the reduction potential of the $[4Fe–4S]^{1+/0}$ couple of the Fe protein from *Azotobacter vinelandii*. J Am Chem Soc 2002, 124, 12100–12101.

[101] Vincent, KA., Tilley, GJ., Quammie, NC. et al. Instantaneous, stoichiometric generation of powerfully reducing states of protein active sites using Eu(II) and polyaminocarboxylate ligands. Chem Commun 2003, 2590–2591.

[102] Walker, GA., and Mortenson, LE. Effect of magnesium adenosine 5′-triphosphate on the accessibility of the iron of *Clostridial* azoferredoxin, a component of nitrogenase. Biochemistry 1974, 13, 2382–2388.

[103] Ljones, T., and Burris, RH. Nitrogenase: the reaction between iron protein and bathophenanthrolinedisulfonate as a probe for interactions with MgATP. Biochemistry 1978, 17, 1866–1872.

[104] Deits, TL., and Howard, JB. Kinetics of MgATP-dependent iron chelation from the Fe-protein of the *Azotobacter vinelandii* nitrogenase complex. Evidence for two states. J Biol Chem 1989, 264, 6619–6628.

[105] Chen, L., Gavini, N., Tsuruta, H. et al. MgATP-induced conformational changes in the iron protein from *Azotobacter vinelandii*, as studied by small-angle X-ray scattering. J Biol Chem 1994, 269, 3290–3294.

[106] Tezcan, FA., Kaiser, JT., Howard, JB., and Rees, DC. Structural evidence for asymmetrical nucleotide interactions in nitrogenase. J Am Chem Soc 2015, 137, 146–149.

[107] Willing, AH., Georgiadis, MM., Rees, DC., and Howard, JB. Cross-linking of nitrogenase components. structure and activity of the covalent complex. J Biol Chem 1989, 264, 8499–8503.

[108] Willing, A., and Howard, JB. Cross-linking site in *Azotobacter vinelandii* complex. J Biol Chem 1990, 265, 6596–6599.

[109] Seefeldt, LC. Docking of nitrogenase iron-and molybdenum-iron proteins for electron transfer and MgATP hydrolysis: the role of arginine 140 and lysine 143 of the *Azotobacter vinelandii* iron protein. Protein Sci 1994, 3, 2073–2081.

[110] Gavini, N., and Burgess, BK. FeMo cofactor synthesis by a *nifH* mutant with altered MgATP reactivity. J Biol Chem 1992, 267, 21179–21186.

[111] Ryle, MJ., and Seefeldt, LC. Elucidation of a MgATP signal transduction pathway in the nitrogenase iron protein: formation of a conformation resembling the MgATP-bound state by protein engineering. Biochemistry 1996, 35, 4766–4775.

[112] Seefeldt, LC., Hoffman, BM., and Dean, DR. Electron transfer in nitrogenase catalysis. Curr Opin Chem Biol 2012, 16, 19–25.

[113] Thorneley, RNF., Lowe, DJ., Eady, RR., and Miller, RW. The coupling of electron transfer in nitrogenase to the hydrolysis of magnesium adenosine triphosphate. Biochem Soc Trans 1979, 7, 633–636.

[114] Hageman, RV., Orme-Johnson, WH., and Burris, RH. Role of magnesium adenosine 5′-triphosphate in the hydrogen evolution reaction catalyzed by nitrogenase from *Azotobacter vinelandii*. Biochemistry 1980, 19, 2333–2342.

[115] Yates, MG. The enzymology of molybdenum dependent nitrogen fixation. Stacey G, Burris RH, Evans HJ, eds, Biological nitrogen fixation, Chapman and Hall, New York, 1992, 685–735.

[116] Eady, RR., Lowe, DJ., and Thorneley, RNF. Nitrogenase of *Klebsiella pneumoniae*: A pre-steady state burst of ATP hydrolysis is coupled to electron transfer between the component proteins. FEBS Lett 1978, 95, 211–213.

[117] Tanifuji, K., Sickerman, N., Lee, CC. et al. Structure and reactivity of an asymmetric synthetic mimic of nitrogenase cofactor. Angew Chem 2016, 128, 15862–15865.

[118] Tanifuji, K., Lee, CC., Sickerman, NS. et al. Tracing the 'ninth sulfur' of the nitrogenase cofactor via a semi-synthetic approach. Nat Chem 2018, 10, 568–572.

[119] Wiig, JA., Lee, CC., Hu, Y., and Ribbe, MW. Tracing the interstitial carbide of the nitrogenase cofactor during substrate turnover. J Am Chem Soc 2013, 135, 4982–4983.

[120] Kennedy, C., and Dean, D. The nifU, nifS and nifV gene products are required for activity of all three nitrogenases of *Azotobacter vinelandii*. Mol Gen Genet MGG 1992, 231, 494–498.

[121] Christiansen, J., Goodwin, PJ., Lanzilotta, WN., Seefeldt, LC., and Dean, DR. Catalytic and biophysical properties of a nitrogenase apo-MoFe protein produced by a nifB-deletion mutant of *Azotobacter vinelandii*. Biochemistry 1998, 37, 12611–12623.

[122] Paustian, TD., Shah, VK., and Roberts, GP. Apodinitrogenase: purification, association with a 20-kilodalton protein, and activation by the iron-molybdenum cofactor in the absence of dinitrogenase reductase. Biochemistry 1990, 29, 3515–3522.

[123] Hawkes, TR., and Smith, BE. Purification and characterization of the inactive MoFe protein ($NifB^-Kp1$) of the nitrogenase from *nifB* mutants of *Klebsiella pneumoniae*. Biochem J 1983, 209, 43–50.

[124] Hawkes, TR., and Smith, BE. The inactive MoFe protein ($NifB^-Kp1$) of the nitrogenase from *nifB* mutants of *Klebsiella pneumoniae*. Its interaction with FeMo-cofactor and the properties of the active MoFe protein formed. Biochem J 1984, 223, 783–792.

[125] Schwarz, G., Mendel, RR., and Ribbe, MW. Molybdenum cofactors, enzymes and pathways. Nature 2009, 460, 839.

[126] Wiig, JA., Hu, Y., and Ribbe, MW. NifEN-B complex of *Azotobacter vinelandii* is fully functional in nitrogenase FeMo cofactor assembly. Proc Natl Acad Sci U S A 2011, 108, 8623–8627.

[127] Rettberg, LA., Wilcoxen, J., Lee, CC. et al. Probing the coordination and function of Fe_4S_4 modules in nitrogenase assembly protein NifB. Nat Commun 2018, 9, 2824.

[128] Fay, AW., Wiig, JA., Lee, CC., and Hu, Y. Identification and characterization of functional homologs of nitrogenase cofactor biosynthesis protein NifB from methanogens. Proc Natl Acad Sci U S A 2015, 112, 14829–14833.

[129] Fay, AW., Blank, MA., Lee, CC. et al. Spectroscopic characterization of the isolated iron–molybdenum cofactor (FeMoco) precursor from the protein NifEN. Angew Chem Int Ed 2011, 50, 7787–7790.

[130] Corbett, MC., Hu, Y., Fay, AW., Ribbe, MW., Hedman, B., and Hodgson, KO. Structural insights into a protein-bound iron-molybdenum cofactor precursor. Proc Natl Acad Sci U S A 2006, 103, 1238–1243.

[131] Brigle, KE., Weiss, MC., Newton, WE., and Dean, DR. Products of the iron-molybdenum cofactor-specific biosynthetic genes, *nifE* and *nifN*, are structurally homologous to the products of the nitrogenase molybdenum-iron protein genes, *nifD* and *nifK*. J Bacteriol 1987, 169, 1547–1553.

[132] Goodwin, PJ., Agar, JN., Roll, JT., Roberts, GP., Johnson, MK., and Dean, DR. The *Azotobacter vinelandii* NifEN complex contains two identical [4Fe-4S] Clusters. Biochemistry 1998, 37, 10420–10428.

[133] Fay, AW., Blank, MA., Yoshizawa, JM. et al. Formation of a homocitrate-free iron-molybdenum cluster on NifEN: Implications for the role of homocitrate in nitrogenase assembly. Dalton Trans 2010, 39, 3124–3130.

[134] Hu, Y., Fay, AW., Lee, CC., Yoshizawa, J., and Ribbe, MW. Assembly of nitrogenase MoFe protein. Biochemistry 2008, 47, 3973–3981.

[135] Corbett, MC., Hu, Y., Fay, AW. et al. Conformational differences between *Azotobacter vinelandii* nitrogenase MoFe proteins as studied by small-angle X-ray scattering. Biochemistry 2007, 46, 8066–8074.

[136] Ohki, Y., Tanifuji, K., Yamada, N., Cramer, RE., and Tatsumi, K. Formation of a Nitrogenase P-cluster $[Fe_8S_7]$ core via reductive fusion of Two All-Ferric $[Fe_4S_4]$ clusters. Chem – Asian J 2012, 7, 2222–2224.

[137] Zhang, Y., and Holm, RH. Structural conversions of molybdenum–iron–sulfur edge-bridged double cubanes and P^N-type clusters topologically related to the nitrogenase P-cluster. Inorg Chem 2004, 43, 674–682.

[138] Zhang, Y., Zuo, J-L., Zhou, H-C., and Holm, RH. Rearrangement of symmetrical dicubane clusters into topological analogues of the P cluster of nitrogenase: nature's choice?. J Am Chem Soc 2002, 124, 14292–14293.

[139] Corbett, MC., Hu, Y., Naderi, F., Ribbe, MW., Hedman, B., and Hodgson, KO. Comparison of iron-molybdenum cofactor-deficient nitrogenase MoFe proteins by X-ray absorption spectroscopy: implications for P-cluster biosynthesis. J Biol Chem 2004, 279, 28276–28282.

[140] Broach, RB., Rupnik, K., Hu, Y. et al. Variable-temperature, variable-field magnetic circular dichroism spectroscopic study of the metal clusters in the Δ*nifB* and Δ*nifH* MoFe proteins of nitrogenase from *Azotobacter vinelandii*. Biochemistry 2006, 45, 15039–15048.

[141] Rupnik, K., Lee, CC., Hu, Y., Ribbe, MW., and Hales, BJ. $[4Fe4S]^{2+}$ Clusters exhibit ground-state paramagnetism. J Am Chem Soc 2011, 133, 6871–6873.

[142] Cotton, MS., Rupnik, K., Broach, RB. et al. VTVH-MCD study of the Δ*nifB*Δ*nifZ* MoFe protein from *Azotobacter vinelandii*. J Am Chem Soc 2009, 131, 4558–4559.

[143] Schmid, B., Ribbe, MW., Einsle, O. et al. Structure of a cofactor-deficient nitrogenase MoFe protein. Science 2002, 296, 352–356.

[144] Danyal, K., Mayweather, D., Dean, DR., Seefeldt, LC., and Hoffman, BM. Conformational gating of electron transfer from the nitrogenase Fe protein to MoFe protein. J Am Chem Soc 2010, 132, 6894–6895.
[145] Danyal, K., Dean, DR., Hoffman, BM., and Seefeldt, LC. Electron transfer within nitrogenase: evidence for a deficit-spending mechanism. Biochemistry 2011, 50, 9255–9263.
[146] Duval, S., Danyal, K., Shaw, S. et al. Electron transfer precedes ATP hydrolysis during nitrogenase catalysis. Proc Natl Acad Sci U S A 2013, 110, 16414–16419.
[147] Yang, Z-Y., Ledbetter, R., Shaw, S. et al. Evidence that the P_i release event is the rate-limiting step in the nitrogenase catalytic cycle. Biochemistry 2016, 55, 3625–3635.
[148] Thorneley, RNF., and Lowe, DJ. Kinetics and mechanism of the nitrogenase enzyme system. Spiro TG, ed, Molybdenum enzymes, Wiley, New York, 1985, 221–284.
[149] Sippel, D., Rohde, M., Netzer, J. et al. A bound reaction intermediate sheds light on the mechanism of nitrogenase. Science 2018, 359, 1484–1489.
[150] Thorneley, RNF., and Lowe, DJ. The mechanism of *Klebsiella pneumoniae* nitrogenase action. Pre-steady-state kinetics of an enzyme-bound intermediate in N_2 reduction and of NH_3 formation. Biochem J 1984, 224, 887–894.
[151] Thorneley, RNF., and Lowe, DJ. The mechanism of *Klebsiella pneumoniae* nitrogenase action. Simulation of the dependences of H_2-evolution rate on component-protein concentration and ratio and sodium dithionite concentration. Biochem J 1984, 224, 903–909.
[152] Lowe, DJ., and Thorneley, RN. The mechanism of *Klebsiella pneumoniae* nitrogenase action. Pre-steady-state kinetics of H_2 formation. Biochem J 1984, 224, 877–886.
[153] Lowe, DJ., and Thorneley, RNF. The mechanism of *Klebsiella pneumoniae* nitrogenase action. The determination of rate constants required for the simulation of the kinetics of N_2 reduction and H_2 evolution. Biochem J 1984, 224, 895–901.
[154] Hoffman, BM., Lukoyanov, D., Yang, Z-Y., Dean, DR., and Seefeldt, LC. Mechanism of nitrogen fixation by nitrogenase: the next stage. Chem Rev 2014, 114, 4041–4062.
[155] Igarashi, RY., Laryukhin, M., Dos Santos, PC. et al. Trapping H^- bound to the nitrogenase FeMo-cofactor active site during H_2 evolution: characterization by ENDOR spectroscopy. J Am Chem Soc 2005, 127, 6231–6241.
[156] Kinney, RA., Saouma, CT., Peters, JC., and Hoffman, BM. Modeling the signatures of hydrides in metalloenzymes: ENDOR analysis of a Di-iron Fe(μ-NH)(μ-H)Fe core. J Am Chem Soc 2012, 134, 12637–12647.
[157] Lukoyanov, D., Yang, Z-Y., Dean, DR., Seefeldt, LC., and Hoffman, BM. Is Mo involved in hydride binding by the four-electron reduced (E4) intermediate of the nitrogenase mofe protein?. J Am Chem Soc 2010, 132, 2526–2527.
[158] Doan, PE., Telser, J., Barney, BM. et al. ^{57}Fe ENDOR spectroscopy and 'electron inventory' analysis of the nitrogenase E4 intermediate suggest the metal-ion core of FeMo-cofactor cycles through only one redox couple. J Am Chem Soc 2011, 133, 17329–17340.
[159] Lukoyanov, D., Khadka, N., Yang, Z-Y., Dean, DR., Seefeldt, LC., and Hoffman, BM. Reductive elimination of H_2 activates nitrogenase to reduce the N≡N triple bond: characterization of the E4(4H) janus intermediate in wild-type enzyme. J Am Chem Soc 2016, 138, 10674–10683.
[160] Chatt, J., Pearman, AJ., and Richards, RL. Diazenido (iminonitrosyl) (N_2H), Hydrazido(2-) (N_2H_2, and hydrazido(1-) (N_2H_3) ligands as intermediates in the reduction of ligating dinitrogen to ammonia. J Organomet Chem 1975, 101, C45–C7.
[161] Chatt, J., Pearman, AJ., and Richards, RL. The reduction of mono-coordinated molecular nitrogen to ammonia in a protic environment. Nature 1975, 253, 39.
[162] Chatt, J., Pearman, AJ., and Richards, RL. Relevance of oxygen ligands to reduction of ligating dinitrogen. Nature 1976, 259, 204.

[163] Chatt, J., Pearman, AJ., and Richards, RL. Conversion of dinitrogen in its molybdenum and tungsten complexes into ammonia and possible relevance to the nitrogenase reaction. J Chem Soc, Dalton Trans, 1977, 1852–1860.
[164] Yandulov, DV., and Schrock, RR. Reduction of dinitrogen to ammonia at a well-protected reaction site in a molybdenum triamidoamine complex. J Am Chem Soc 2002, 124, 6252–6253.
[165] Yandulov, DV., and Schrock, RR. Catalytic reduction of dinitrogen to ammonia at a single molybdenum center. Science 2003, 301, 76–78.
[166] Hinnemann, B., and Nørskov, JK. Catalysis by enzymes: the biological ammonia synthesis. Top Catal 2006, 37, 55–70.
[167] Tanabe, Y., and Nishibayashi, Y. Developing more sustainable processes for ammonia synthesis. Coord Chem Rev 2013, 257, 2551–2564.
[168] Schrock, RR. Catalytic reduction of dinitrogen to ammonia at a single molybdenum center. Acc Chem Res 2005, 38, 955–962.
[169] MacLeod, KC., and Holland, PL. Recent developments in the homogeneous reduction of dinitrogen by molybdenum and iron. Nat Chem 2013, 5, 559.
[170] Rodriguez, MM., Bill, E., Brennessel, WW., and Holland, PL. N_2 reduction and hydrogenation to ammonia by a molecular iron-potassium complex. Science 2011, 334, 780–783.
[171] Rittle, J., and Peters, JC. An Fe-N_2 complex that generates hydrazine and ammonia via Fe═NNH_2: demonstrating a hybrid distal-to-alternating pathway for N_2 reduction. J Am Chem Soc 2016, 138, 4243–4248.
[172] Thorneley, RNF., Eady, RR., and Lowe, DJ. Biological nitrogen fixation by way of an enzyme-bound dinitrogen-hydride intermediate. Nature 1978, 272, 557.

Xenia Engelmann, Teresa Corona and Kallol Ray

5 Oxidation of methane: methane monooxygenases

5.1 Introduction

The chemical activation of small molecules like H_2, O_2 or CH_4 has been at the center of scientific interest in biology, medicine, environment and catalysis in the last decades. These small molecules represent an almost inexhaustible reservoir of stored energy. Therefore, with respect to the continuous increase in energy demands, the controlled breakdown of these small molecules into their atomic components and conversion into alternative energy sources is of high social importance. By chemical means this breakdown is quite difficult, since these small molecules are thermodynamically inert and the selective cleavage of their bonds is often kinetically hindered. Methane, for example, the main constituent of natural gas, is one of the major energy sources on the Earth and can serve as a potential transitional energy source to replace the more polluting coal and oil until carbon-free energy sources become mature and deployed [1, 2]. However, methane is around 84 times more potent than carbon dioxide as a greenhouse gas [3] and human activities account for approximately two-thirds of the total methane emissions, including seepage from the exploitation of coal, oil and natural gas [4]. One way to tackle the problem would be the direct liquification and storage of methane, which is, however, too expensive. An alternate strategy would be to convert methane into liquid methanol; notably methanol represents a major carbon chemical feedstock and therefore a methanol economy is in general considered as one of the most promising alternative energy platforms that can be utilized to replace the fossil fuels in the future [5, 6]. Unfortunately, current strategies for the selective conversion of methane to methanol are neither economical nor sustainable [7]. At the moment, industrial methanol production is accomplished by steam reforming of methane (Reactions 5.1–5.2), and is a highly endothermic, energy-intensive and costly process, involving the complete dehydrogenation of methane [8, 9]. Notably, to date, no methods have been discovered that can convert methane to methanol directly from dioxygen in industrial scale (Reaction 5.3). This is due to the kinetic and thermodynamic barriers associated with methane activation. The C–H bonds of methane are extremely inert, with a dissociation energy of 104 kcal/mol. Furthermore, a spin-forbidden process occurs upon the direct reaction of triplet dioxygen with singlet methane to form singlet methanol. Finally, a substantial energy input (high

Xenia Engelmann, Teresa Corona, Kallol Ray, Department of Chemistry, Humboldt-Universität zu Berlin, Germany

https://doi.org/10.1515/9783110496574-005

temperatures, high pressure and high pK_a) is necessary to activate the conversion of methane to methanol:

$$CO\,(g) + 2H_2(g) \xrightarrow{Cu/ZnO/Al_2O_3} CH_3OH(g) \qquad \Delta H_{298} = -21.7\ \text{kcal/mol} \tag{5.1}$$

$$CH_4\,(g) + H_2O\,(g) \xrightarrow{Ni} CO\,(g) + 3H_2(g) \qquad \Delta H_{298} = 49.3\ \text{kcal/mol} \tag{5.2}$$

$$CH_4\,(g) + 1/2O_2\,(g) \rightarrow CH_3OH(I) \qquad \Delta H_{298} = -39.2\ \text{kcal/mol} \tag{5.3}$$

Due to the difficulty to achieve and control the activation of CH_4, the oxygenation of CH_4 under ambient conditions to CH_3OH using O_2 has been considered as one of the holy grails of organic chemistry [10]. Chemists draw inspiration from Nature for a steady improvement of the industrial CH_3OH production. Accordingly, much effort has been spent to understand the mechanistic details by which the CH_4 oxidation occurs in biology [11–14]. A class of bacteria known as methanotrophs employs enzymes called methane monooxygenases (MMOs), which utilize CH_4 as a sole carbon source under ambient conditions. These aerobic proteobacterias are able to hydroxylate the nonpolar and strong C–H bonds in CH_4 using O_2, protons and electrons under mild conditions to form CH_3OH (Reaction 5.4) as the first step of CH_4 metabolism [15]. However, their sophisticated enzyme machinery cannot be used and scaled up for industrial purpose, which is too expensive:

$$CH_4 + O_2 + NADH + H^+ \xrightarrow{MMO} CH_3OH + H_2O + NAD^+ \tag{5.4}$$

There are two forms of MMOs, which have been evolved to perform this reaction: one is membrane-bound and is a copper-containing particulate MMO (*p*MMO) (Figure 5.1B); and the other is water-soluble cytoplasmic and is an iron-containing soluble methane monooxygenase (sMMO) (Figure 5.1A). In both cases, the MMOs utilize transition metal centers to activate dioxygen in order to attack the strong C–H bonds of CH_4, but the structures, active sites, cofactor requirements and mechanisms of CH_4 activations are completely different in the two cases. Notably, compared to sMMO, *p*MMO is more efficient in oxidizing CH_4 and is the most efficient CH_4 oxidizer known to date. It is capable of activating CH_4 at a rate of one CH_4 molecule per second per enzyme, [16] and is preferably used in biology for methane activation; sMMO is operative only under copper-limiting conditions. The *p*MMO hydroxylase consists of the PmoB, PmoA and PmoC subunits. Unfortunately *p*MMO being a membrane-bound protein, it is extremely difficult to isolate and to purify it from the plasma membrane for biochemical and biophysical studies and, thus, studies on the *p*MMO are quite rare and a lot of issues are still under debate [17].

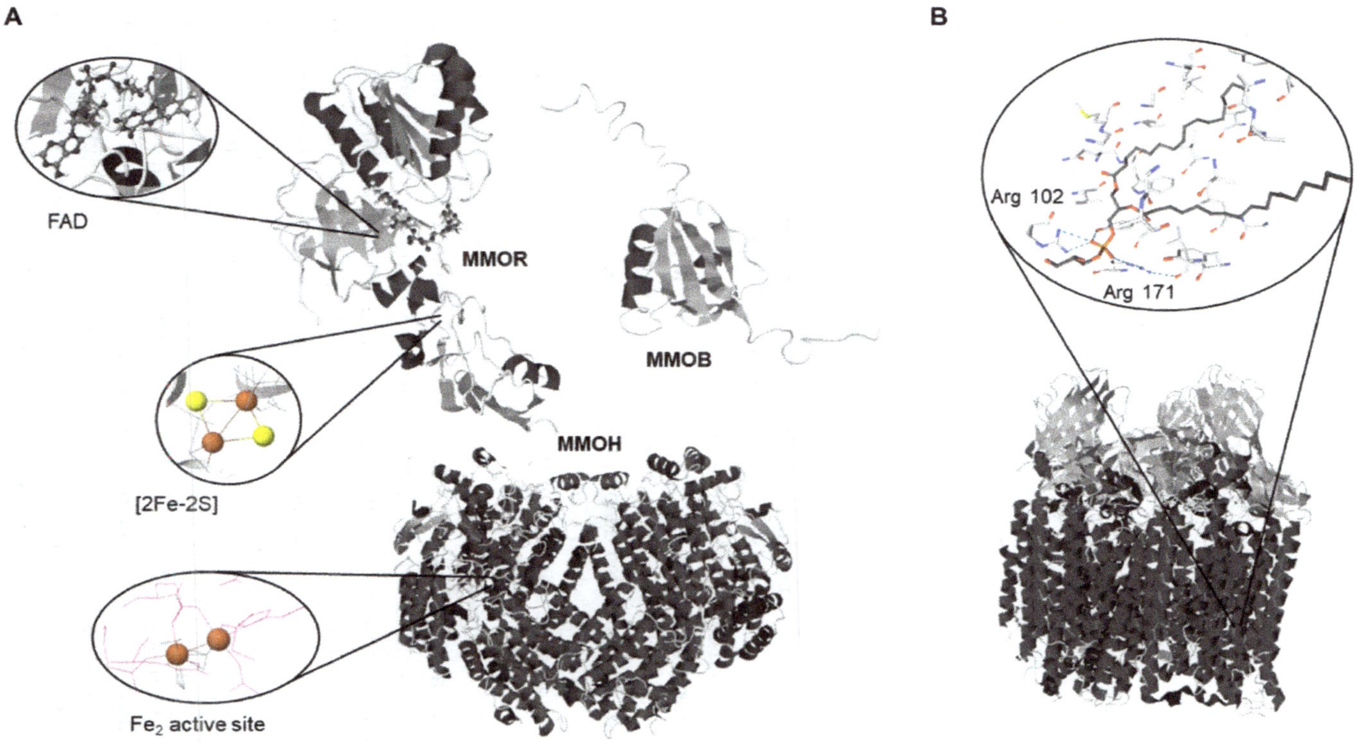

Figure 5.1: A) Overall architecture of sMMO from methylococcus capsulatus (Bath) consists of a hydroxylase (MMOH PDB reference 1MTY), an oxidoreductase (MMOR, 1TVC and 1JQ4) and a regulatory protein (MMOB, 1CKV). B) pMMO trimer (PDB, 4PHZ). Reproduced from [20] with permission from American Chemical Society.

For example, even the nature of the active site in *p*MMO is not unambiguous; it has been controversially discussed to contain a mono-, di-, trinuclear copper centers, as well as a dinuclear iron center [18, 19]. In contrast, the oxidative chemistry of *s*MMO has been investigated extensively for over 30 years and is better understood; this will be discussed in detail in this chapter.

5.2 Biological methane hydroxylation: soluble methane monooxygenase (*s*MMO)

5.2.1 Enzyme architecture, metal centers and active site structure

*s*MMO is a well-characterized member of the bacterial multicomponent monooxygenase (BMM) family as it is much more stable and is amenable to purification [21]. It employs three protein components for activity: a hydroxylase (MMOH), a reductase (MMOR) and a regulatory protein (MMOB) (Figure 5.1A). The 38 kDA reductase (MMOR) contains the FAD-binding and [2Fe-2S] ferredoxin domains that are involved in the two-electron transfer from NADH to the diiron center [22–24]. The 251 kDA hydroxylase component (MMOH) is an $\alpha_2\beta_2\gamma_2$ homodimer, consisting of three subunits. The diiron active site, which is a nonheme binuclear iron cluster, mediates the hydroxylation of CH_4 and is buried deep in the α subunit [25, 26]. Finally, the 16 kDA regulatory protein (MMOB), which plays a critical role in controlling the coupling of O_2 activation and substrate oxidation at the active center, has no cofactors and consists of seven β strands arranged in two β sheets and three α helices [27–29].

In MMOH there is only one type of metal center present. The first crystal structure of MMOH was achieved in 1993 [25]. Each αβγ protomer contains a nonheme diiron cluster that is able to activate dioxygen and mediate the hydroxylation chemistry [30]. This diiron center is housed within a four-helix bundle and has been crystallized in both the oxidized and reduced forms of MMOH. The oxidized (**Ox**; Figure 5.2) form represents the inactive state of the enzyme; only the reduced (**Red**; Figure 5.2) form reacts with dioxygen and results in the hydroxylation of CH_4. In the inactive **Ox**-form, one of the iron centers is coordinated by one histidine (His_{147}), one glutamate (Glu_{114}) and a solvent water molecule; the other iron center is coordinated by His_{246}, Glu_{243} and Glu_{209} (Figure 5.2). Two hydroxides and a Glu_{144} residue act as bridging ligands between the two iron centers. When reduction takes place, the Glu_{243} undergoes a carboxylate shift, thereby replacing the bridging hydroxide ions as a ligand to the diiron center (Figure 5.2). Note that the two iron centers move from being six coordinate in the inactive **Ox**-form to five coordinate in the **Red**-form; thus, the reduction process is accompanied by the lowering of the iron coordination number, which creates a dioxygen binding site at both the iron centers [31]. MMOH is only active in the presence of a protein cofactor, MMOB, which forms

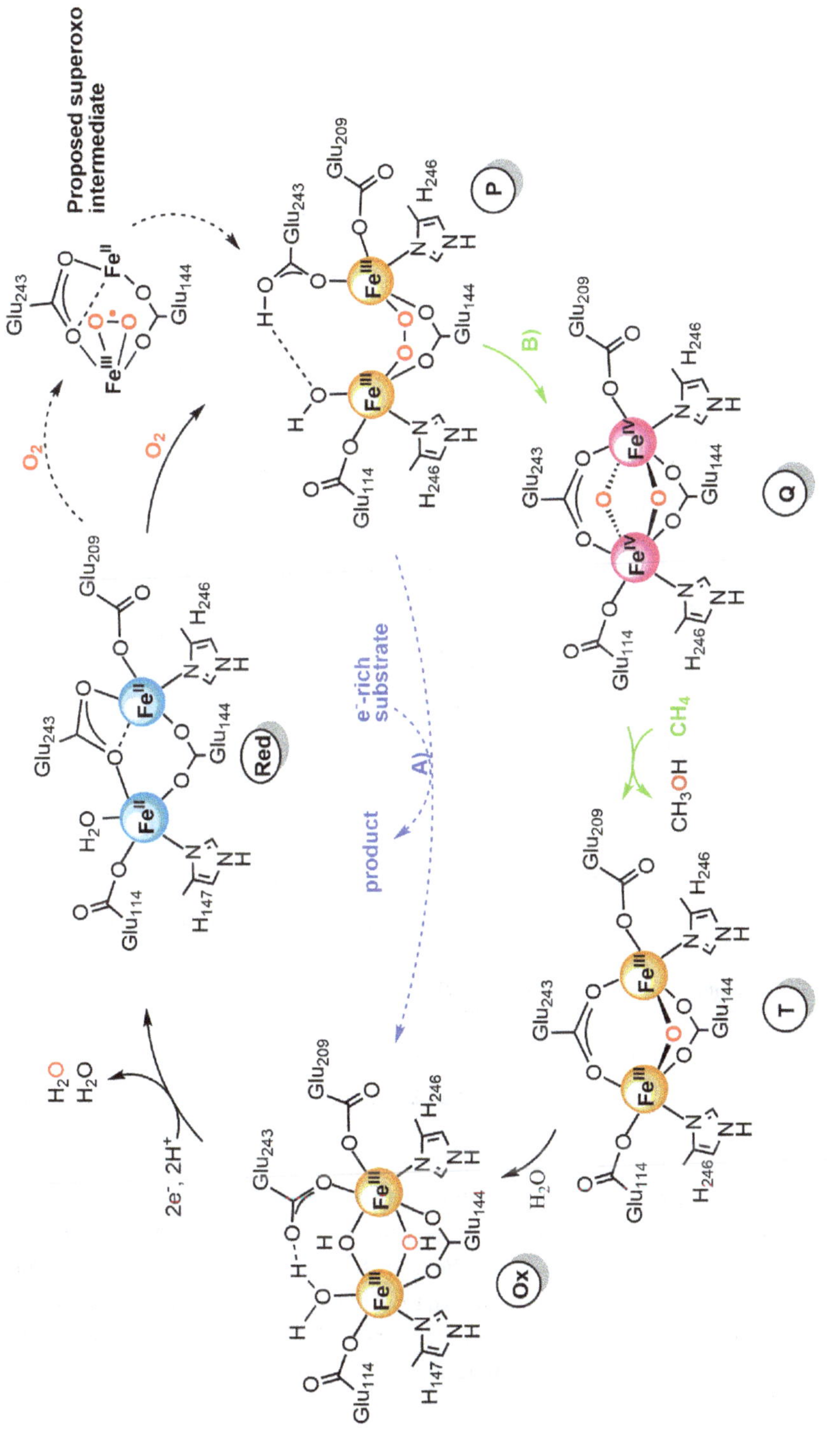

Figure 5.2: Schematic representation of the catalytic cycle of sMMO.

a specific complex with MMOH that indirectly affects the structure and reactivity of the diiron sites. In the oxidized MMOH, the two iron centers are present as an antiferromagnetically coupled diiron(III) center and in the reduced MMOH a weakly ferromagnetically coupled diiron(II) site exists [32–35]. To establish the electronic structure of the dinuclear iron center of the MMOH, a variety of spectroscopic studies, such as Mössbauer, extended X-ray absorption fine structure (EXAFS) and EPR, have been performed. Mössbauer spectra on the oxidized hydroxylase recorded in zero field at 4.2 K show an isomeric shift of δ = 0.50 mm/s and ΔE_Q = 1.07 mm/s, which are typical of high-spin iron(III) and are close to those of binuclear iron clusters such as uteroferrin or hemerythrin (Table 5.1). Spectra recorded in a 6.0 T applied field parallel to the γ-radiation reflect a diamagnetic environment, which underlines the assumption of a spin-coupled cluster with an even number of antiferromagnetically coupled iron(III) ions [32–35]. EPR studies on MMOH in the oxidized state reveals weak resonances at g = 4.3 (<5% of the total iron) and g_{av} = 1.85 (<5%), which correspond to signals arising from residual iron(III) and a one-electron reduced cluster, respectively; thus, the spectrum is in line with an essentially EPR silent native hydroxylase [32, 36, 37]. EXAFS analysis at the Fe-k-edge gives an Fe–Fe distance of 3.42 Å and an average first shell Fe–O/N distance of 2.04 Å [34]. Based on the absence of a short Fe–O distance in the EXAFS and the presence of a featureless UV/Vis absorption spectrum, it has been argued that no oxo-bridges are present between the iron(III) centers; instead a bridging hydroxo and one or two bridging carboxylates have been suggested (Figure 5.2) [34].

Table 5.1: Spectroscopic parameters of the oxidized and reduced MMOH.

	Mössbauer		EPR	EXAFS		Refs
	δ (mm/s)	ΔE_Q (mm/s)	g-Values	d(Fe–Fe) (Å)	Fe–O/N (Å)	
MMOH-ox	0.50	1.07	–	3.42	2.04	32–35
MMOH-red	1.06	3.01	15	–	2.15	32–35

Mössbauer studies at 80 K on the fully reduced MMOH reveal parameters of δ = 1.06 mm/s and ΔE_Q = 3.01 mm/s, which are typical of high-spin iron(II). In EPR a low-field resonance of g = 15 has been observed, which represents the majority of the total iron and is a characteristic of systems with integer electronic spin. The spectroscopic properties of the reduced MMOH have been interpreted in terms of the presence of two ferromagnetically coupled S = 2 high-spin iron(II) centers.

5.2.2 Catalytic mechanism

The details of how *s*MMO activates dioxygen have been extensively studied under single turnover conditions, resulting in the identification of transient oxygen intermediates, and establishment of their kinetic and spectroscopic properties [38–40]. At the initiation of the turnover of the enzyme, the high-spin diiron(III) center (**Ox**) in **MMOH-ox** is reduced by two electrons to high-spin diiron(II) state (**Red**), **MMOH-red** [41, 42]. After formation of **Red**, the diiron(II,II) center is activated by dioxygen in the presence of 2 equiv. of MMOB. Notably, both O_2 and CH_4 are carried to the active site through a hydrophobic cavity running through the protein structure [43]. The mechanism of the activation of dioxygen by **MMOH-red** has been intensively investigated in the last few decades. Either one or two electrons can be transferred from the diiron core to dioxygen forming the superoxo or peroxo species, respectively. Furthermore, the intermediates generated upon dioxygen activation can possess different geometrical structures; for example, O_2 can coordinate in both end-on or side-on binding-modes. In addition, the magnetic properties of the intermediates can be complicated by different possibilities for the spin coupling between the oxygen and the iron-spins.

The first intermediate resulting from dioxygen binding to **Red** is suggested by computational studies to be a mixed-valent diiron(II/III) superoxide species (Figure 5.2) [44, 45]. Even though an iron(II/III) superoxide species has never been directly observed in an enzyme in the reaction of a diiron(II) center with O_2, an oxygen kinetic isotope effect (KIE) of 1.016 has been determined for the oxidation of CH_3CN by MMOH, which correlates with a one-electron reduction of dioxygen (predicted value for a one-electron process: 1.033) [46].

The first spectroscopically detected intermediate in the catalytic cycle of MMOH is a peroxodiiron(III) species (**P**) (Figure 5.2) [47, 48]. Attempts to accumulate intermediate **P** in high yields were challenging due to their rapid rates of decay. However, their detection and characterization could be done under stopped-flow and/or rapid freeze quench conditions [47, 49, 50]. The Mössbauer spectrum of **P** gives parameters of δ = 0.66 mm/s and ΔE_Q = 1.51 mm/s, which reveal a diamagnetic center with near-identical iron(III) sites [49, 51]. The optical spectrum of **P** recorded by stopped-flow absorption spectroscopy exhibits visible absorption bands with $\lambda_{max} \approx$ 420 nm (ε = 3880 $M^{-1}cm^{-1}$) and a peroxo-to-iron charge-transfer transition at $\lambda_{max} \approx$ 720 nm (ε = 1,350 $M^{-1}cm^{-1}$) [47, 52]. Although, the O–O vibration energy in intermediate **P** has not been determined experimentally, a cis-μ-1-2-peroxodiiron(III) assignment for **P** has been made based on the resemblance of its Mössbauer and optical spectra to other enzymatic peroxo intermediates that have been characterized by resonance Raman and/or by X-ray crystallography in similar diiron enzymes [20, 39, 50–61].

Formation of **P** is pH-dependent and exhibits a kinetic solvent isotope effect in D_2O, indicating that proton transfer plays a role in its formation. Peroxo intermediates with a μ-1-2-peroxo binding mode have often been implied as an active intermediate responsible for a variety of C–H bond activation and oxygen-transfer reactions in biology [62]. The question whether intermediate **P** is the active species responsible for the oxidation of methane has therefore been intensively addressed. By transient kinetic studies through single-turnover stopped-flow methods, it has been demonstrated that in the presence of electron-rich substrates like ethyl vinyl ether, diethyl ether and propylene, [63, 64] the rate of the decay of an optical band at 720 nm corresponding to **P** is directly related to the increase in the substrate concentrations [47, 65]. In contrast, in the presence of CH_4, the rate of the overall CH_4 oxidation is driven by the conversion of the peroxo species into a new intermediate **Q** with a characteristic absorption feature at 420 nm. Therefore, two possible pathways can be generated depending on the nature of the entering substrates. For electron-rich substrates the MMOH can follow the sequence **Red** → **P** + Sub → product (Pathway A; Figure 5.2) and for harder to oxidize substrates like CH_4 a sequence of **Red** → **P** → **Q** + Sub → product is followed (Pathway B; Figure 5.2). Thus based on these studies, it has been concluded that although intermediate **P** is capable of performing hydroxylation of weak C–H bonds, it is not powerful enough to hydroxylate CH_4 and must be converted to a stronger oxidant **Q** in order to target the stronger C–H bonds of CH_4 [61].

Intermediate **Q** has also been trapped and characterized as a high-valent bis(μ-oxo)diiron(IV) species generated by a proton-dependent O–O bond homolysis of **P** [61]. It is the last intermediate that has been identified in the catalytic cycle of MMOH before methane is oxidized to methanol; accordingly, it is presently believed to be the key reactive intermediate responsible for methane hydroxylation (Figure 5.2). The structure of **Q** was elucidated via time-resolved resonance Raman spectroscopy, which displays a major Fe–O vibration mode at 690 cm^{-1}. Notably, this value is too low to be associated with a terminal Fe^{IV}=O core (expected Fe–O stretch in the 800 – 860 cm^{-1} range), which has also been implicated as a reactive intermediate in a number of mononuclear nonheme iron oxygenases [66–71]. However, it matches closely to the Fe–O vibration of 674 cm^{-1} observed in $[Fe^{IV}_2(\mu\text{-}O)_2(TPA^*)_2]^{4+}$ (TPA* = substituted tris(2-pyridylmethylamine)) [72, 73], which contains an Fe_2O_2 diamond core structure. Accordingly, a similar Fe_2O_2 diamond core structure is also suggested for **Q**, which is also supported by further spectroscopic investigation. Mössbauer studies revealed δ values of 0.14-0.21 mm/s and ΔE_Q of 0.55–0.68 mm/s in zero-field, consistent with the iron(IV) assignment for both the iron centers in **Q**. Furthermore, analysis of the Mössbauer spectrum in the presence of applied magnetic fields revealed a trend of "three large negative" A-tensors, which is typical of S = 2 iron(IV) centers; the two S = 2 iron centers are shown to be antiferromagnetically coupled and bridged together symmetrically by two oxygen atoms giving a diamagnetic ground state [74]. EXAFS data best fit with an Fe–Fe distance of 2.46 Å and an Fe–O distance of 1.78 Å, which additionally support the proposed diamond core structure of **Q**. Optical absorption bands at 420 nm ($\varepsilon \approx 7200\ M^{-1}cm^{-1}$) and 350 nm ($\varepsilon \approx 3600\ M^{-1}cm^{-1}$) have been identified for **Q**, which are assigned to oxo-to-iron charge transfer transitions [75].

Intermediate **Q** has been suggested to react with methane to insert a single oxygen atom derived from the bound O_2 via a still unknown mechanism. To investigate the C–H bond activation steps in sMMO, different approaches have been used to explore the various mechanistic possibilities. However, the measurement of kinetic isotope effects and computational analysis [76] have been performed to obtain insights into the C–H bond activation step [30]; the use of different substrate probes has been employed to differentiate between alkyl radical [77, 78], cationic [79–82] or concerted [83, 84] oxygen insertion mechanisms (Scheme 5.1) [30]. Two different kinds of probes have been used, namely, the chiral alkene and radical clock probes that will result in different rearranged (racemized, epimerized or ring-opened) products depending on the mechanism of the C–H bond activation reaction [30]. In case of a concerted mechanism, complete retention or inversion of the stereochemistry would be expected, while radical and cationic intermediates would lead to complete racemization products [30, 81, 84–96]. The results of these studies, however, proved to be not so straightforward to interpret; the mechanism was found to differ depending on the nature of the substrate probe used. Different computational studies using DFT methods to investigate the mechanism for the hydroxylation reaction favor the radical pathway for methane activation involving a hydrogen-atom abstraction process followed by an oxygen-rebound step [97–99]. The validity of this mechanism is also corroborated by the detection of methyl radicals in spin-trap experiments. Thus, despite all these experimental and computational results on the possible mechanism of the C–H bond activation, the results obtained so far are not straightforward and have been controversial discussed until now [30].

The reaction of **Q** with CH_4 is accompanied by a concomitant formation of a (μ-oxo)-bridged diiron(III) intermediate **T** (Figure 5.2). **T** has been characterized by time-resolved resonance Raman spectroscopy and shown to retain a single oxygen atom from dioxygen as a bridging ligand, while the other oxygen atom is incorporated into the product [72, 100]. In the near-ultraviolet region **T** exhibits a broad electronic absorption band; resonance enhancement with λ_{exc} = 315 nm revealed an Fe–O vibration at 556 cm^{-1} [9].

5.3 Chemical hydroxylation of methane

5.3.1 Chemical catalysts for the hydroxylation of methane

In contrast to the efficient conversion of methane to methanol in biology, which occurs at ambient conditions of pressure and temperature, chemical oxidation of CH_4 to CH_3OH has proved to be an extremely challenging transformation. Although thermodynamically favorable, this transformation is kinetically inhibited; it necessitates the use of expensive noble metals and/or harsh conditions (high temperature as well as

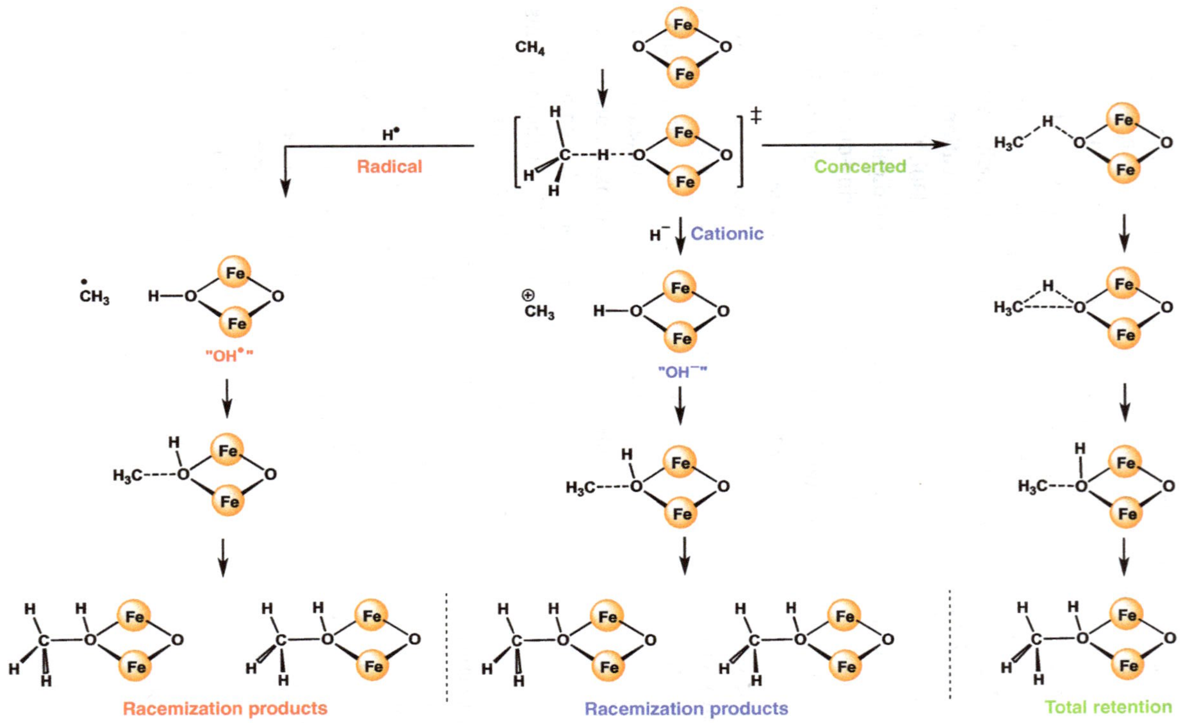

Scheme 5.1: Radical, cationic and concerted mechanism of the C–H bond activation of methane by intermediate **Q**. Adapted from [12] with permission from American Chemical Society.

high pressure) and often occurs with low selectivities and low product yields [101]. With regard to homogenous catalysts based on first-row transition metals, Chan et al. reported a bioinspired trinuclear copper cluster, being a model of *p*MMO for the hydroxylation of methane. The three copper atoms of the cluster are coordinated in 3,3′-(1,4-diazepane-1,4-diyl)bis[1-(4-ethylpiperazine-1-yl)propan-2-ol] ligand, capable to transform methane to methanol in the presence of both O_2 and H_2O_2 in an effective manner at room temperature and in acetonitrile [102]. According to the authors, the active species responsible for such conversion seems to be a mixed-valent $[Cu^{II}Cu^{II}(\mu\text{-}O)_2Cu^{III}(L)]^+$ center generated upon dioxygen activation at a tricopper(I) center. The low solubility of methane in acetonitrile as well as the side reaction of the oxidizing species $[Cu^{II}Cu^{II}(\mu\text{-}O)_2Cu^{III}(L)]^+$ with H_2O_2 instead of methane limitate the use of this catalyst [103].

Transition metal chloride salts like $FeCl_3$, $CoCl_2$, $RuCl_3$, $RhCl_3$, PdCl2, $OsCl_3$, $IrCl_3$, H_2PtCl_6, $CuCl_2$ and $HAuCl_4$ can also catalyze the partial oxidation of methane under homogeneous conditions using H_2O_2 as an oxidant at 90 ºC in water by a Fenton-type radical mechanism; however, the scope of the reaction is limited by the observed poor selectivities and low yields of the products (25–50% of product mixture containing methanol, formaldehyde, methyl hydroperoxide and formic acid). Out of all the metal salts, osmium(III) chloride showed the highest turnover frequency of 12 h^{-1} and UV/Vis spectroscopic measurements suggested an Os(IV) species as the active species responsible for CH_4 oxidation [104, 105]. Another example of a homogeneous methane-hydroxylation catalyst was reported by Shilov and coworkers. They were able to selectively convert CH_4 into CH_3OH using $PtCl_2$ as a catalyst in aqueous HCl at 100 ºC with $[PtCl_6]^{2-}$ acting as the active species responsible for oxidation [106, 107]. Afterward, Periana et al. modified the Pt(II) bipyridine complex, being able to oxidize methane in concentrated H_2SO_4 to afford methyl bisulfate in a 72% yield, which could then be further reacted with water to afford methanol [108]. Surprisingly, no other reports on the transition-metal-mediated conversion of methane-to-methanol under homogeneous conditions are known, although in recent years a large number of diiron, dicopper and tricopper biomimetic model complexes have been synthesized as biomimetic models of the active sites of MMOs.

As far as heterogeneous catalysis is concerned, Hutchings and coworkers showed that Au–Pd alloy nanoparticles supported on TiO_2 were capable to catalyze at 50 ºC and 30.5 bar the oxidation of methane by H_2O_2 in water providing, as main oxidation products, methanol and formic acid [109]. Further studies revealed that methanol can be produced in higher selectivity and yields (>80%) by employing Fe-containing zeolites as solid catalysts for methane oxidation in the presence of H_2O_2 (Scheme 5.2A) [105]. Based on UV/Vis, X-ray absorption near edge spectroscopy and DFT studies, the involvement of high-valent iron(IV)oxo cores in CH_4 oxidation reactions has been confirmed [110, 111]. Further, kinetic studies demonstrated that CH_3OOH is initially formed in the reaction, which is then subsequently converted to CH_3OH [112]. Fe-containing zeolites can also

A

B

$$(Fe^{II}) + N_2O \longrightarrow (Fe^{IV}=O) + N_2$$

$$2(Fe^{IV}=O) + \boxed{CH_4} \longrightarrow (Fe^{III}\text{-}OH) + (Fe^{III}\text{-}OCH_3)$$

$$(Fe^{III}\text{-}OCH_3) + H_2O \longrightarrow (Fe^{III}\text{-}OH) + \boxed{CH_3OH}$$

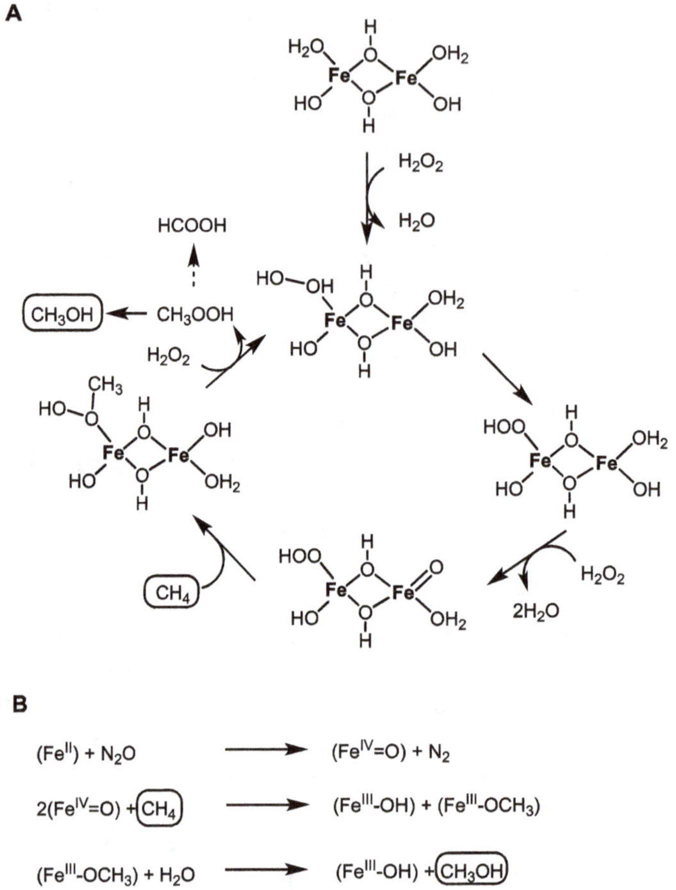

Scheme 5.2: A) Proposed catalytic cycle for the conversion of methane to oxygenates catalyzed with H_2O_2 by Fe-ZSM-5 or Fe-silicatie-1 in water. Adapted from [105] with permission from American Chemical Society. B) Proposed mechanism for the conversion of methane to methanol by Fe/ZSM-5 using N_2O as oxidant.

activate nitrous oxide to convert methane to methanol in a single turnover cycle at room temperature (Scheme 5.2B) [113–116]. The formation of CO and CO_2 upon heating confirmed the stoichiometry of the reaction, whereas the extraction with a mixture of acetonitrile and water afforded an excellent selectivity of 98%. Infrared spectroscopy showed the formation of methoxy groups, which could be hydrolyzed to methanol upon reaction with H_2O at 250 °C [117]. Copper-containing zeolites can also convert methane to methanol in the presence of dioxygen [118–121]. An intermediate with a characteristic UV/Vis absorption band at 440 nm ($\varepsilon \approx 22.700$ $M^{-1}cm^{-1}$) could be identified that disappeared in the presence of CH_4. In combination with EXAFS, resonance Raman, EPR and DFT studies, the active species responsible for methane oxidation has been assigned as a bis(μ-oxo)copper(II) species.

5.3.2 Model studies of the possible reactive intermediates

The challenging selective oxidation of C–H bonds under mild conditions has been successfully solved by nature using metalloenzymes. The metal, which constitutes less than 1% of the protein weight, is located in the protein-active site and it is essential for the enzyme function. During the last decades, many efforts have been dedicated to the elucidation of metalloprotein-active sites using spectroscopic techniques, crystallography, site-directed mutagenesis, mechanistic enzymology or theoretical calculations. An added benefit of knowing the metalloenzyme structure and function is its potential application in the design of model complexes. At the same time, these synthetic model systems constitute an attractive approach for gaining knowledge about the protein chemistry, providing mechanistic, structural and spectroscopic data of the active species involved in the natural enzymes (e.g., high-valent metal–oxygen species) [122]. Furthermore, although the enzymatic intermediates may have only a fleeting existence, the lifetimes of their synthetic analogs can be controlled by a systematic tuning of the electronics and the steric of the ancillary ligand systems. The proposed involvement of iron-peroxo, bis(μ-oxo)diiron(IV) species in biological methane oxidation reactions have made them attractive targets for biomimetic synthetic studies. Indeed, recent synthetic advances have led to the isolation and characterization of several well-characterized iron–dioxygen model complexes; detailed reactivity studies in conjunction with spectroscopy and theory have helped to understand how the steric and electronic properties of the iron centers modulate their reactivity. In this section, we will focus our attention on summarizing the progress achieved in the past decades on the development of model systems for intermediate **P** ((μ-1,2-peroxo)diiron (III) species) and intermediate **Q** (high-valent bis(μ-oxo)diiron(IV)). We will also show how the knowledge gained from these biomimetic studies has thrown light into the enzymatic processes.

5.3.2.1 (μ-1,2-Peroxo)diiron(III) intermediates: models of intermediate P

Most of the characterized synthetic peroxodiiron complexes provide useful points of comparison to the enzymatic peroxodiiron intermediates. In general, polydentate-supporting ligands with a combination of N-donors and O-donors are employed to mimic the enzyme environment (Figure 5.3). The employed N-donors are tertiary amines and pyridines, whereas the O-donors are carboxylate or alkoxide ligands. The N- and O-donors mimic the histidine or glutamate residues in the active site of sMMO. The generation of the peroxodiiron complexes has been achieved by bubbling O_2 or H_2O_2 into a solution of a diiron precursor. In most cases, low temperatures are needed to carry out the reactions to extend the lifetimes of the generated reactive intermediates and to allow for their spectroscopic characterization. Two different classes of intermediate **P** model compounds have been identified based on their

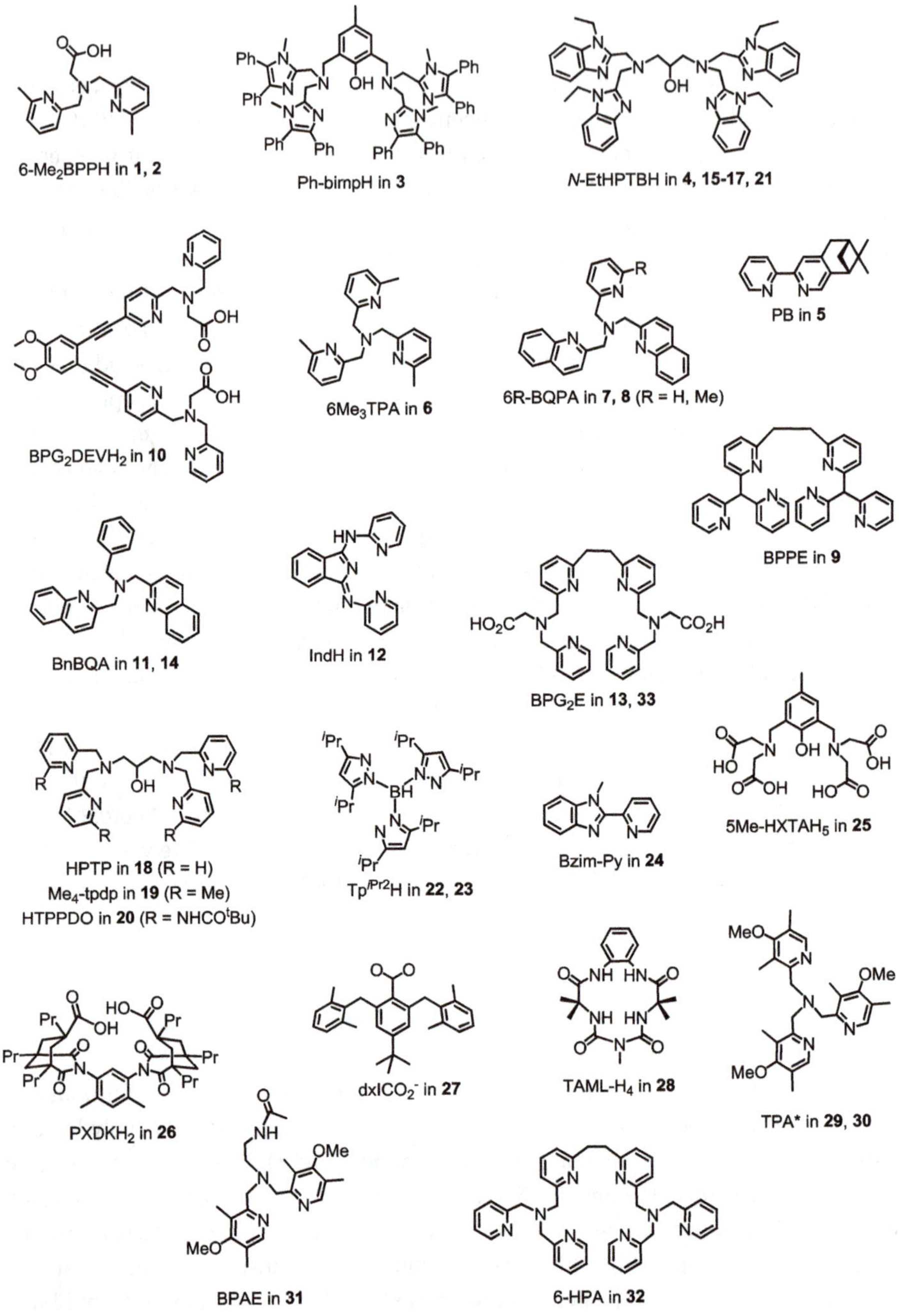

Figure 5.3: Structures of ligands used for the models of **P** and **Q** intermediates.

structural features: complexes with a cis(μ-1,2-peroxo)diiron(III) unit that have a single-atom bridge consisting of either an oxo, a hydroxo or an alkoxo ligand (Figure 5.4A) and cis(μ-1,2-peroxo)diiron(III) complexes containing only carboxylate bridges between the two iron centers (Figure 5.4B) [123]. The crystal structures of four (μ–OR)(μ-1,2-peroxo)diiron(III) type model complexes have been reported by the groups of Suzuki and Que, namely, the $[Fe^{III}_2(\mu\text{-}O)(O_2)(6\text{-}Me_2BPP)_2]^{2+}$, 6-$Me_2$BPP = bis (6-methyl-2-pyridylmethyl)-3-aminopropionate (**1**) [124], $[Fe^{III}_2(\mu\text{-}OH)(O_2)(6\text{-}Me_2BPP)_2]^+$ (**2**) [124, 125], $[Fe^{III}_2(Ph\text{-}bimp)(O_2)(OBz)]^{2+}$, Ph-bimp = 2,6-bis[bish2-(1-methyl-4,5-diphenylimidanzolyl)methylaminomethyl]-4-methylphenolate (**3**) [126] and $[Fe^{III}_2(O_2)(N\text{-}EtHPTB)(OPPh_3)_2]^{3+}$, *N*-EtHPTBH = tetrakis(2-benzimidazolylmethyl)-2-hydroxy-1,3-diaminpropane (**4**) [127] complexes (Figures 5.3 and 5.5). The nature of the single-atom (OR) bridge is shown to affect the Fe–O_{peroxo} and Fe–Fe distances, reflecting the basicity of the OR bridges. For example, the values for the Fe–O_{peroxo} bond length ranging from 2.07 Å to 2.11 Å is observed in **1** containing the highly basic bridging oxo ligand; in contrast a shorter Fe–O_{peroxo} distance of 1.86 Å to 1.90 Å have been observed for the hydroxo and alkoxo ligands (**2**, **3** and **4**). An opposite trend has been observed for the Fe–Fe distances, which is found to decrease with increasing basicity of the bridging ligand. For example, Fe–Fe distances of 3.171 Å, 3.395 Å, 3.328 Å, and 3.463 Å have been determined for **1**, **2**, **3** and **4**, respectively. The O–O distances are, however, independent of the nature of the bridging ligand and are found to be in the range of 1.40–1.43 Å. The (μ-1,2-peroxo)diiron complexes possess a five-membered ring, which is relatively flat (Figure 5.4A). In particular, the Fe–O–O–Fe dihedral angles of **1** and **4** are equal to 0° and of **2** and **3** are close to planarity (–14.5° and + 9.9°). A correlation between the spectroscopic parameters and the dihedral angle could not be observed so far. In contrast to the reactivity of intermediate **P**, which reacts with ethyl vinyl ether, diethyl ether and propylene, compounds **1** to **4** are inert toward C–H and O–H bonds. The lack of the reactivity of the model complexes compared to **P** is unclear, but further analysis on structural features, such as the peroxo binding geometry, may shed light on this issue.

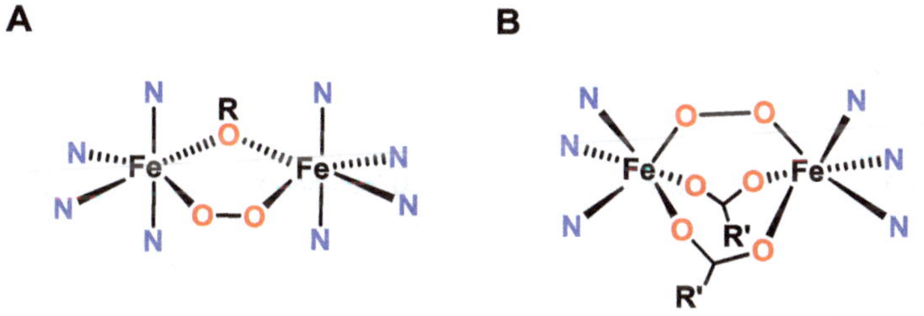

Figure 5.4: Two different classes of intermediate **P** model compounds. A) (μ-OR)(μ-1,2-peroxo)diiron (III) core with a bridging oxo, hydroxo or alkoxo unit. B) cis(μ-1,2-Peroxo)diiron(III) core without an alkoxo, hydroxo or oxo bridging unit. Reproduced from [40] with permission from American Chemical Society.

"Peroxodiiron(III) model complexes"

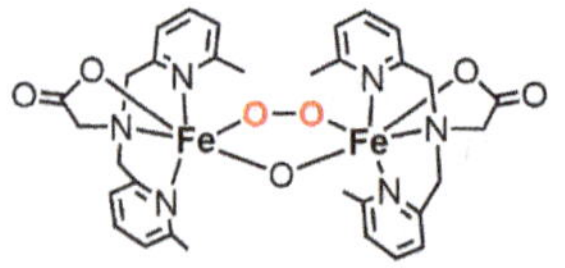

$[Fe^{III}_2(\mu\text{-}O)(O_2)(6\text{-}Me_2BPP)_2]^{2+}$ (**1**)

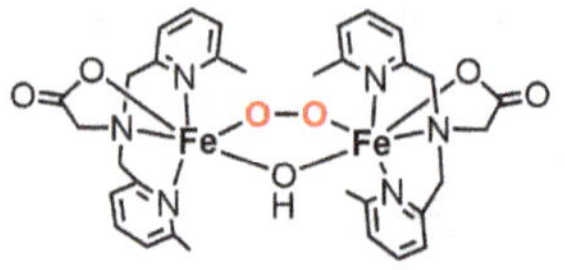

$[Fe^{III}_2(\mu\text{-}OH)(O_2)(6\text{-}Me_2BPP)_2]^{2+}$ (**2**)

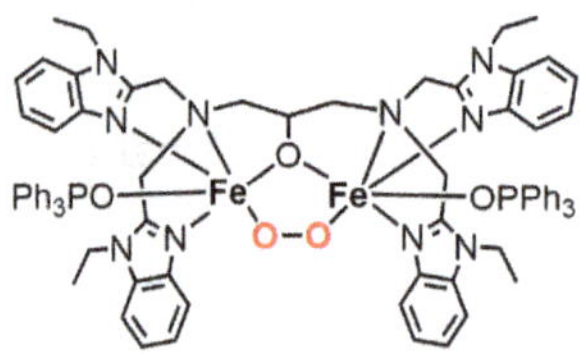

$[Fe^{III}_2(O_2)(N\text{-}EtHPTB)(OPPh_3)_2]^{3+}$ (**4**)

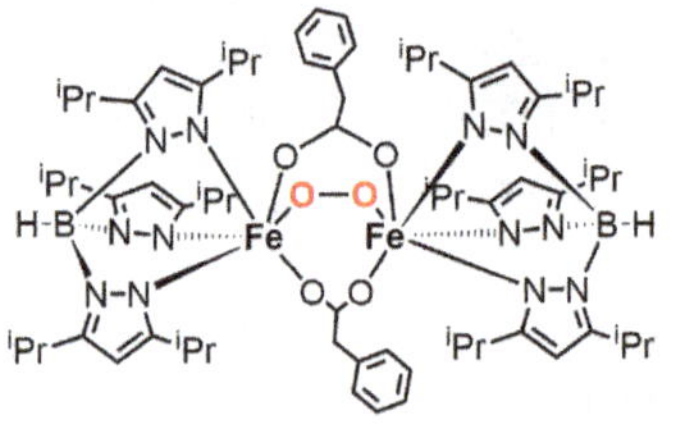

$[Fe^{III}_2(O_2)(Tp^{iPr2})_2(PPA)_2]$ (**22**)

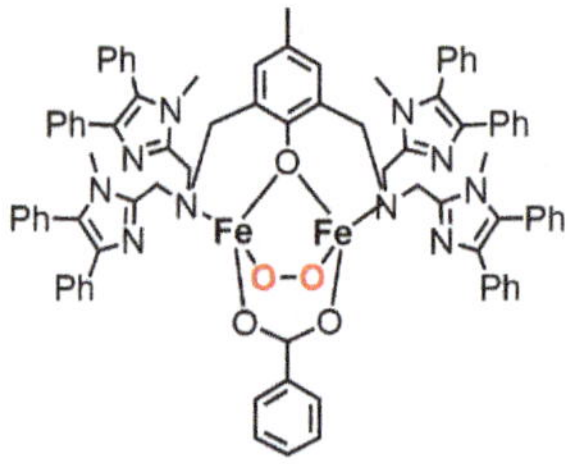

$[Fe^{III}_2(Ph\text{-}bimp)(O_2)(OBz)]^{2+}$ (**3**)

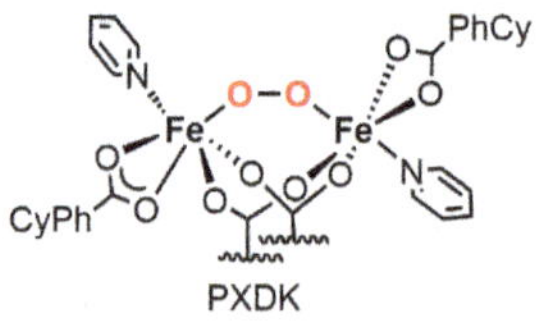

$[Fe^{III}_2(O_2)(PXDK)(O_2CPhCy)_2(py)_2]$ (**26**)

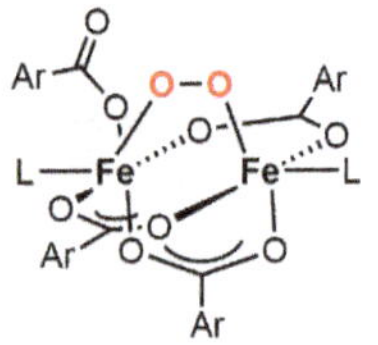

$[Fe^{III}_2(dxlCO_2)_4(O_2)(Py)_2]$ (**27**)

"Bis(μ-oxo)diiron(IV) model complexes"

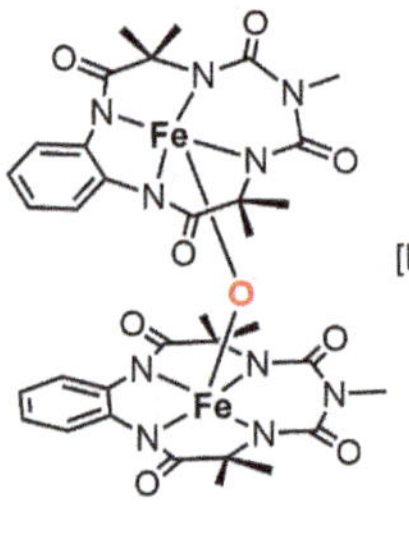

$[Fe^{IV}_2(\mu\text{-}O)(TAML)_2]^{2-}$ (**28**)

Figure 5.5: Structurally characterized model complexes for **P** and **Q** intermediates.

In addition to the four crystallized model complexes, several other (μ-1,2-peroxo)diiron complexes (**5**–**21**) (Table 5.2) involving oxo, hydroxo and alkoxo bridging ligands have been obtained in solution and characterized by a variety of spectroscopic methods like EXAFS, Mössbauer and UV/Vis absorption spectroscopy. The spectroscopic properties of the complexes are summarized in Table 5.2. The Fe–Fe and Fe–O distances, as obtained from EXAFS analysis, are again found to be controlled by the nature of the bridging ligand. The Mössbauer isomer shifts (δ) are all within the range of 0.48 – 0.65 mm/s, which is close to what could be observed for the biological intermediate **P** (δ = 0.66 mm/s) and are typical of high-spin (S = 5/2) iron(III) centers. Furthermore observance of quadrupole doublets in zero-field Mössbauer spectroscopy for all the complexes indicates that the iron ions are antiferromagnetically coupled to afford an S = 0 ground spin state. In resonance Raman spectroscopy, the oxygen-sensitive ν(O–O) stretches are observed between 816–928 cm^{-1} with a shift of −33 to −53 cm^{-1} in the ^{18}O-labeled species, thereby supporting the peroxo assignment of the intermediates. Furthermore, optical spectroscopic features ranging from 450 – 730 nm ($\mathcal{E} \approx$ 1000 – 3,000 $M^{-1}cm^{-1}$) have been observed, which can be assigned to the oxo-to-iron(III) and peroxo-to-iron(III) LMCT transitions. In the case of (μ-oxo)(μ-1,2-peroxo)diiron species, a double humped absorption band, which corresponds to the oxo-to-iron(III) LMCT transition for the higher-energy band (450 – 505 nm) and to peroxo-to-iron(III) LMCT transition for the lower-energy band (546 – 690 nm), has been detected [128].

In addition to the cis(μ-1,2-peroxo)diiron model complexes with a single-atom bridge that have been discussed so far, spectroscopic properties of the peroxo complexes containing only carboxylate bridges between the two iron centers have also been investigated (Figure 5.4B). Noteworthy to mention are the complexes $[Fe^{III}_2(O_2)(Tp^{iPr2})_2(R)_2]$, Tp^{iPr2} = tris(3,5-diisoporpyl-1-pyrazol)borate, R = PPA, PPA = phenylacetic acid (**22**) [48]; OBz (**23**) [65], which were first characterized by Kitajima et al. [129] and then crystallized by Kim and Lippard [48] (Figures 5.3 and 5.5). The dioxygen in **22** binds in a *cis* mode with an Fe–O–O–Fe dihedral angle of 52.9°, and average O–O, Fe–O_{peroxo} and Fe–Fe distances of 1.408 Å, 1.885 Å and 4.004 Å, respectively. Notably, in comparison to the group of peroxo complexes with an additional single-atom bridge, the Fe–Fe distance and the dihedral angle in **22** are larger. Nevertheless, the Mössbauer parameters (δ = 0.66 mm/s, ΔE_Q = 1.40 mm/s) reflect very closely the values determined for the **P** intermediate of sMMOH; accordingly, complex **22** is considered as a reasonable model for the O_2 adducts of MMOH. The optical features of **22**, **23** and $[Fe^{III}_2(O_2)(Bzim\text{-}Py)_4(MeCN)]^{4+}$, (Bzim-Py = 2-(2′-pyridyl)-*N*-methylbenzimidazole (**24**) [130], Figure 5.3) also reproduce the low-energy feature observed in the absorption spectrum of intermediate **P** in sMMOH. The absence of highly basic single-atom bridges in these compounds results in a significant red shift of the peroxo LMCT transition to ~700 nm.

Table 5.2: Spectroscopic parameters of peroxodiiron(III) and high-valent bis(μ-oxo)diiron(IV) cores in chemistry and biology.

			d(Fe–Fe)	Mössbauer		Raman		optical		Refs
			(Å)	δ (mm/s)	ΔE_Q (mm/s)	ν(O–O) (cm^{-1})	$\Delta^{18}O_2$ (cm^{-1})	λ_{max} (nm)	ε ($M^{-1}cm^{-1}$)	
		MMOH-P		0.66	1.51			420 720	3880 1350	[47, 51, 52]
(μ-oxo)(μ-1,2-peroxo) diiron species	1	$[Fe^{III}_2(\mu\text{-}O)(O_2)(6\text{-}Me_2BPP)_2]^{2+}$	3.171	0.50	1.46	847	−33	450 577	1000 1500	[124]
	5	$[Fe^{III}_2(O_2)(\mu\text{-}O)(PB)_4]^{2+}$	–	0.49	−0.62	868	−48	680	2000	[134, 135]
	6	$[Fe^{III}_2(O)(O_2)(6\text{-}Me_3TPA)_2]^{2+}$	3.14	0.54	1.68	847	−44	494 648	1100 1200	[136, 137]
	7	$[Fe^{III}_2(\mu\text{-}O)(O_2)(BQPA)_2]^{2+}$	3.13	–	–	844	−44	480 620	1000 1000	[136]
	8	$[Fe^{III}_2(\mu\text{-}O)(O_2)(6Me\text{-}BQPA)_2]^{2+}$	3.15	–	–	853	−45	495 640	1200 1300	[136]
	9	$[Fe^{III}_2(\mu\text{-}O)(O_2)(OAc)(BPPE)]^{+}$	3.04	0.53	1.67	816	−45	505 595	1500 1400	[136, 138]
	10	$[Fe^{III}_2(\mu\text{-}O)(O_2)(BPG_2DEV)]$	–	0.58	0.58	845 819	−48 −47	490	1500	[139]
	11	$[Fe^{III}_2(\mu\text{-}O)(O_2)(BnBQA)_2]^{2+}$	3.16	0.55	1.43	854	−47	505 650	1250 1300	[140]
	12	$[Fe^{III}_2(\mu\text{-}O)(O_2)(IndH)_2]^{2+}$	3.13	–	–	874	−38	690	1500	[129]
	13	$[Fe^{III}_2(\mu\text{-}O)(O_2)(BPG_2E)]$	–	0.48	1.66	835	−51	452 546	1420 1300	[141]

(μ-OH or *μ*-OR)(μ-1,2-peroxo) diiron species	**2**	$[Fe^{III}_2(\mu\text{-OH})(O_2)(6\text{-Me}_2BPP)_2]^+$ (2)	3.395	0.50	1.31	908	−47	644	3000	[124]
	3	$[Fe^{III}_2(\text{Ph-bimp})(O_2)(OBz)]^{2+}$ (3)	3.328	0.58 0.65	0.74 1.70	–	–	500–700	~1700	[126]
	4	$[Fe^{III}_2(O_2)(N\text{-Et-HPTB})(OPPh_3)2]^{3+}$ (4)	3.463							[127]
	14	$[Fe^{III}_2(O_2)(OH)(BnBQA)_2]^{3+}$	3.46	0.57 0.56	1.35 0.96	928	−53	730	2400	[140]
	15	$[Fe^{III}_2(O_2)(N\text{-EtHPTB})(\eta 1\text{-}O_2PPh_2)(MeCN)]^{2+}$	3.47	0.53	−1.03	897	−49	621	1800	[142]
	16	$[Fe^{III}_2(O_2)(O_2PPh_2)(N\text{-EtHPTB})]^{2+}$	3.25	0.56	−1.26	849	−42	678	2100	[142]
	17	$[Fe^{III}_2(O_2)(\mu\text{-}1,3\text{-}O_2AsMe_2)(N\text{-EtHPTB})]^{2+}$	3.27	–	–	845	−49	632	2100	[142]
	18	$[Fe^{III}_2(HPTP)(O_2)(OBz)]^{2+}$	–	–	–	885	−51	572	2060	[143]
	19	$[Fe^{III}_2(Me_4\text{-tpdp})(O_2)(OBz)]^{2+}$	–	–	–	905	−47	616	2000	[144]
	20	$[Fe^{III}_2(HTTPDO)(O_2)(OBz)]^{2+}$	–	–	–	887 873	−48 −48	610	1700	[145]
	21	$[Fe^{III}_2(O_2)(O_2CPh)(N\text{-EtHPTB})]^{2+}$	–	–	–	900	−50	500	1500	[142]

(continued)

Table 5.2 (continued)

			d(Fe–Fe) (Å)	Mössbauer δ (mm/s)	Mössbauer ΔE_Q (mm/s)	Raman ν(O–O) (cm^{-1})	Raman $\Delta^{18}O_2$ (cm^{-1})	optical λ_{max} (nm)	optical ε ($M^{-1}cm^{-1}$)	Refs
(μ-1,2-peroxo) diiron species without an additional bridge	**22**	$[Fe^{III}_2(O_2)(Tp^{iPr2})_2(PPA)_2]$	4.004	0.66	1.40	888	−46	694	2650	[48]
	23	$[Fe^{III}_2(O_2)(Tp^{iPr2})_2(OBz)_2]$	–	–	–	876	−48	682	3450	[65, 129]
	24	$[Fe^{III}_2(O_2)(Bzim\text{-}Py)_4(MeCN)]^{4+}$	–	–	–	876	−50	685	1400	[130]
	25	$[Fe^{III}_2(O_2)(5Me\text{-}HXTA)(OAc)]^{2+}$	–	–	–	884	–	470	1700	[131]
	26	$[Fe^{III}_2(O_2)(PXDK)(O_2CPhCy)_2(py)_2]$	–	0.47 0.63	0.88 1.20	861	−50	580	1200	[132]
	27	$[Fe^{III}_2(dxlCO_2)_4(O_2)(Py)_2]$	–	0.65 0.52	1.27 0.71	822	−43	500	1000	[133]
		MMOH-Q	2.46	0.21 0.14	0.68 0.55	690 556	−36 −23	420 350	7200 3600	[39, 49, 50, 52, 74, 75]
bis-(μ-oxo) diiron(IV) species	**28**	$[Fe^{IV}_2(\mu\text{-}O)(TAML)_2]^{2-}$ (28)	3.35	−0.07	3.3	–	–	–	–	[146]
	29	$[Fe^{IV}_2(\mu\text{-}O)_2(TPA^*)_2]^{3+}$ (29)	2.73	−0.04	2.09	674	−30	485 875	9800 2200	[73]
	30a		–	0.14	0.52	–	–			[147]
	30b		–	−0.02 0.14	−1.17 −0.82	–	–			[147]
	31	$[Fe^{IV}_2O(BPAE)_2]^{4+}$	3.08	−0.05	2.14	–	–	–	–	[74]
	32	$[Fe^{III}_2(6\text{-}HPA)(O)_2(O)]^{2+}$	–	0.13	0.44	820	−43	–	–	[148, 149]
	33	$[Fe^{III}_2(O_2)(\mu\text{-}O)(BPG_2E)]$	–	0.20	0.40	–	–	–	–	[141, 149]

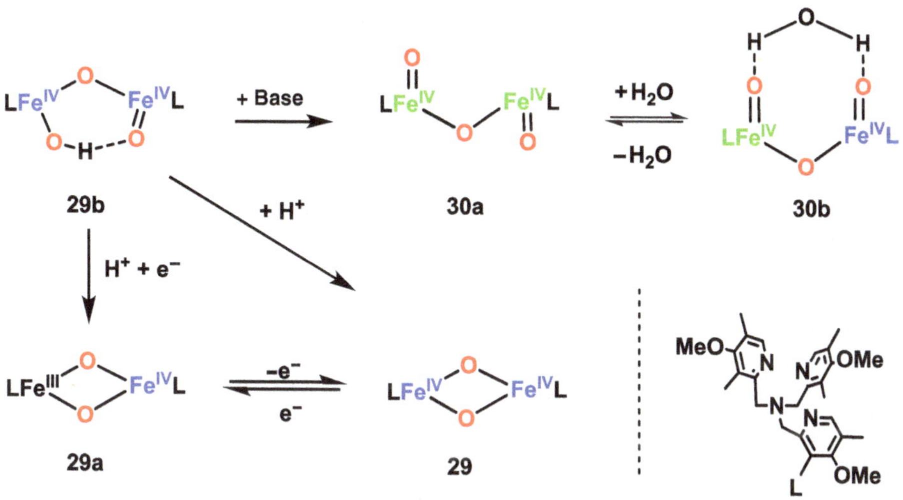

Scheme 5.3: Schematic representation of the diamond core compounds **29, 29a, 29b** and related open-core compound **30** with exchange couplings in **30a** and **30b**. For **30b** the two Fe^{IV}=O bonds adopt an approximately perpendicular orientation, which is stabilized by the interaction of **30b** with a water molecule. Blue indicates an Fe^{IV}=O unit with an $S = 1$ center and green indicates an $S = 2$ center. Adapted from [40] with permission from American Chemical Society.

Based on the predominance of the carboxylate-based ligands at the active site of sMMO, efforts were also made to generate the peroxodiiron complexes in carboxylate-rich ligand environments. Accordingly, a few peroxodiiron complexes with three carboxylate-rich ligands have been synthesized: $[Fe^{III}_2(O_2)(5Me\text{-}HXTA)(OAc)]^{2+}$, 5Me-HXTAH$_5$ = *N,N'*-(2-hydroxy-5-methyl-1,3-xylylene)bis(*N*-carboxymethylglycine) **(25)** [131], $[Fe^{III}_2(O_2)(PXDK)(O_2CPhCy)_2(py)_2]$, PXDK = an analog of the m-xylylenediamine bis(Kemp's triacidimide) (H2XDK) ligand; HO$_2$CPhCy = 1-phenylcyclohexanecarboxylic acid **(26)** [132] and $[Fe^{III}_2(dxlCO_2)_4(O_2)(Py)_2]$, dxlCO$_2^-$ = 2,6-bis[(2,6-dimethylphenyl) methyl]-4-*tert*-butylbenzoate **(27)** [133] (Figures 5.3 and 5.5). Complexes **26** and **27** employ sterically bulky carboxylate ligands, which are able to provide at least two carboxylates that can bridge a diiron site (Figure 5.5). LeCloux and coworkers [132] were successful in assembling diiron complexes that have a ligand arrangement close to that found for the reductive MMOH-active site, including the incorporation of a μ-1,1-carboxylate bridge. Upon O_2 bubbling through a solution of the diiron(II) complex, an adduct **26** with a broad absorption feature at 580 nm and ν(O–O) of 861 cm^{-1} was generated. Mössbauer spectroscopy suggests an asymmetric diiron center with an isomeric shift of 0.47 mm/s and 0.63 mm/s. Unfortunately, the binding mode of the peroxo in **26** is still unknown. Another interesting complex is **27**, [133] where the two iron(III) centers are bridged via four bidentate carboxylate units to form a paddlewheel-like complex (Figure 5.5). Complex **27** exhibits a broad band

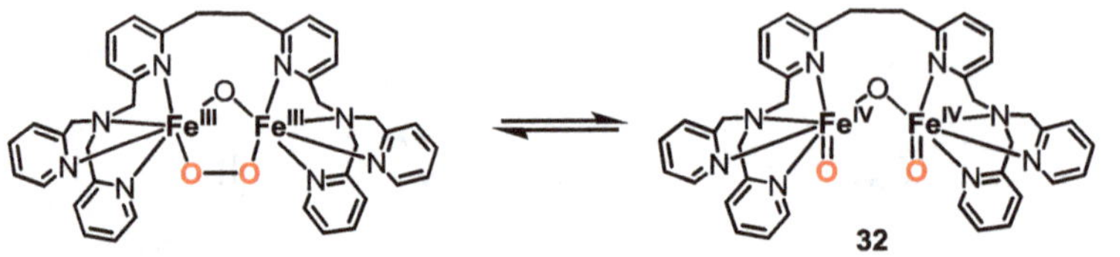

32

Scheme 5.4: Proposed reversible conversion between (μ-oxo)(μ-1,2-peroxo)diiron(III) species and the corresponding high-valent bis(μ-oxo)diiron(IV) species for **32** with the temperature.

centered at ~ 500 nm in the UV/Vis absorption spectrum. Furthermore, similar to **26**, complex **27** also possess an asymmetric dimeric center with isomeric shifts of 0.52 mm/s and 0.65 mm/s and a relatively low ν(O–O) of 822 cm^{-1}.

In the absence of a crystal structure of **P**, efforts were made to compare its spectroscopic properties with that of the well-defined model complexes (Table 5.2), in order to obtain structural insights for **P**. However, the results are contradictory. While model diironperoxo complexes with an additional μ-oxo bridge could successfully reproduce the short Fe–Fe distance and the "two-hump" feature in the absorption spectrum of **P** (Table 5.2), the carboxylate bridged complex **22** in the absence of any μ-oxo bridge better reproduces the Mössbauer parameters and the lower-energy (at 720 nm) absorption feature of **P**. Thus, although the presence of μ-hydroxo or μ-alkoxo bridged diiron complexes in **P** can be ruled out, whether the cis(μ-1,2-peroxo)diiron(III) core in **P** contains an additional μ-oxo bridge is presently not clear.

5.3.2.2 Bis(μ-oxo)diiron(IV) intermediates: models of intermediate Q

One challenge for the bioinorganic chemists is the generation of a biomimetic model of intermediate **Q**, a high-valent bis(μ-oxo)diiron(IV) species, that is responsible for the oxidation of methane to methanol. Efforts to generate the synthetic high-valent bis(μ-oxo)diiron(IV) complexes involved the reaction of a precursor diiron(II) or diiron(III) complexes with H_2O_2. To date, only a few synthetic models for intermediate **Q** have been generated (Table 5.2); of them only $[Fe^{IV}_2(\mu\text{-O})(TAML)_2]^{2-}$, tetra-anionic macrocyclic ligand (TAML; **28**) [146] (Figure 5.5), has been crystallographically characterized. The Fe–μ-O_{oxo} distance of 1.7284 Å, an Fe–Fe distance of 3.35 Å and Mössbauer parameters with an isomeric shift of −0.07 mm/s and quadrupole splitting of 3.3 mm/s are consistent with the iron(IV) assignment of the iron centers in **28**. Notably, in $[Fe^{IV}_2(\mu\text{-O})(TAML)_2]^{2-}$ the Fe–Fe distance at 3.35 Å is significantly larger than that in intermediate **Q** (Fe–Fe distance of 2.46 Å), which can be attributed to the presence of an additional oxygen bridge in **Q**. Although $[Fe^{IV}_2(\mu\text{-O})(TAML)_2]^{2-}$ can oxidize PPh_3 to $Ph_3P{=}O$ and alcohols to aldehydes, it is not capable of performing C–H bond hydroxylation reactions.

A structural model complex for **Q** with a diamond core Fe_2O_2 structure was not known in the literature, until very recently. By modification of the *tris*(2-pyridylmethyl) amine (TPA) ligand by introducing electron donating substituents on the 3–, 4– and 5-positions of each pyridine, Que and coworkers were successful to synthesize the most relevant model complex for intermediate **Q**, $[Fe^{IV}_2(\mu\text{-}O)_2(TPA^*)_2]^{4+}$ (**29**) [150–152]. EXAFS analysis of **29** revealed an Fe–μ-O_{oxo} distance of 1.77 Å and an Fe–Fe distance of 2.73 Å that are similar to intermediate **Q** (Fe–O of 1.78 Å and Fe–Fe of 2.46 Å). Complex **29** can be generated in two different ways. Either by one-electron electrochemical oxidation of the mixed-valent $[Fe^{III}Fe^{IV}(\mu\text{-}O)_2(TPA^*)_2]^{3+}$ (**29a**) precursor or by acidification of $[Fe^{IV}_2(\mu\text{-}O)(O)(OH)(TPA^*)_2]^{3+}$ (**29b**) (Scheme 5.3). [153–156] Although complex **29b** is the product of the reaction of H_2O_2 with the diiron precursor $[(H_2O)(TPA^*)Fe^{III}–O–Fe^{III}(TPA^*)\text{-}(OH)]^{3+}$, mixed valent **29a** with a diamond core structure can be generated by a proton-coupled electron transfer to the open $(O)Fe^{IV}\text{-}O\text{-}Fe^{IV}\text{-}OH$ core of **29b** (Scheme 5.3). The addition of base to **29b**, however, results in another open-core complex **30** involving a diiron(IV) center. In frozen solutions of **30**, two different isomeric forms **30a** and **30b** can be identified; based on parallel Mössbauer and EPR studies [147], **30a** has been assigned to antiferromagnetically coupled $S = 2$ $Fe^{IV}{=}O$ sites, whereas **30b** to ferromagnetically coupled $S = 1$ and $S = 2$ Fe^{IV} centers. Notably, the Mössbauer parameters of **30a**, with an isomeric shift of 0.14 mm/s and a quadrupole splitting of 0.52 mm/s, reproduce very closely the values obtained for intermediate **Q**. Furthermore, comparative reactivity studies have established that the presence of a terminal $S = 2$ Fe(IV)=O core is prerequisite for achieving the high C–H bond hydroxylating ability. Accordingly it is suggested that the $[Fe_2(\mu\text{-}O)_2]$ diamond core of **Q** as deduced from EXAFS analysis is actually the resting state of the enzyme [49]; it is activated for methane oxidation by isomerization to a more reactive ring-opened form with a terminal $Fe^{IV}{=}O$ unit [157–159].

Further model complexes for **Q** were reported: $[Fe^{IV}_2O(BPAE)_2]^{4+}$, BPAE = *N,N*-bis(3′,5′-dimethyl-4′-methoxypyridyl-2′-methyl)-*N′*-acetyl-1,2-diaminoethane (**31**) [74], could be generated by one-electron oxidation of $[Fe^{III}Fe^{IV}O(BPAE)_2]^{3+}$, which itself was generated by the electrochemical oxidation of its precursor $[Fe^{III}_2O(BPAE)_2]^{2+}$ (Figure 5.3). EXAFS studies showed an Fe–μ-O_{oxo} distance of 1.71 Å and an Fe–Fe distance of 3.08 Å. [52, 74] Later, Kodera et al. reported new provocative results by employing ligands that are essentially dimeric versions of the tripodal tetradentate ligand (TPA) such as 6-HPA = 1,2-bis[2-{bis(2-pyridylmethyl)amino-methyl}-6-pyridyl]ethane [148] and BPG_2E = 1,2-bis[2-(*N*-2-pyridyl-methyl-*N*-glycinylmethyl)-6-pyridyl]ethane [141] (Figure 5.3). The reaction of the (μ-oxo) diiron complex precursor with H_2O_2 results in the formation of the corresponding (μ-oxo)(μ-1,2-peroxo)diiron species, which efficiently catalyze the oxidation of alkanes and alkenes at 25 °C [148, 160]. Interestingly, Kodera showed a temperature-dependent reversible O–O bond cleavage between the (μ-oxo)(μ-1,2-peroxo) diiron(III) species and the corresponding high-valent bis(μ-oxo)diiron(IV) species by using Mössbauer spectroscopy (Scheme 5.4). Two quadrupole doublets are observed; one corresponding to the (μ-oxo)(μ-1,2-peroxo)diiron(III) species (δ = 0.35 mm/s and ΔE_Q = 1.64 mm/s) for $[Fe^{III}_2(6\text{-HPA})(O_2)(O)]^{2+}$ (**32**) [149] and δ = 0.48 mm/s and ΔE_Q = 1.66 mm/s for $[Fe^{III}_2(O_2)(\mu\text{-}O)(BPG_2E)]$ (**33**)) and the other corresponding to the S = 2 bis(μ-oxo)diiron(IV) species (δ = 0.13 mm/s and ΔE_Q = 0.44 mm/s for **32** and δ = 0.2 mm/s and ΔE_Q = 0.4 mm/s for **33**). For both the complexes the (μ-oxo)(μ-1,2-peroxo)diiron(III) species isomer is favored at 23 K (ratio of 60:40 for 6-HPA and 90:10 for BPG_2E). In contrast, the high-valent bis(μ-oxo)diiron(IV) species is favored at 295 K (ratios of 15:85 for 6-HPA and 45:55 for BPG_2E). The study of Kodera et al. presented the first spectroscopic observation of dioxygen activation via reversible O–O bond scission of a peroxo-diiron(III) to a high-spin S = 2 oxodiiron(IV), and therefore serves as a functional model in the dioxygen activation of sMMO for the conversion of **P** to **Q**.

5.4 Conclusion

Employing reactive complexes of abundant metals for synthesis, catalysis and energy supply is of great current interest. Selective functionalization of unactivated C–H bonds in organic compounds, for example, is a highly attractive strategy in organic synthesis, and the oxidation of methane is considered "holy grails" in synthetic chemistry. A range of metalloenyzmes achieve the challenging task of hydroxylating methane in biology by activating dioxygen and using cheap and abundant first-row transition metals, like iron and copper. Such reactions are carried out under ambient conditions with high efficiency and high stereospecificity. Detailed mechanistic studies on the biological methane activation process lend credence to the participation of high-valent bis(μ-oxo)diiron(IV) diamond cores as the active species responsible for the hydroxylation of methane. Few model metal-oxo complexes involving the diamond core structure have now been synthesized and they show intriguing reactivities, which in turn have provided vital insights into the modeled enzymatic reactions. Among the most significant conclusion of these studies is the increased reactivity of the linear $[(O)Fe^{IV}\text{–}O\text{–}Fe^{III}(OH_2)]^{2\,+}$ model complex, as compared with the ring-like $[Fe^{IV}_2(\mu\text{-}O)_2]^{2+}$ core, that provides evidence for a comparable, more ring-opened form of **Q** with a terminal Fe^{IV}=O unit as the active species in the reactivity of sMMO.

There are nevertheless still some gaps in our present understanding of the sMMO chemistry and its extension to potential technological applications. Catalysts based on cheap first-row transition metals that can hydroxylate methane with comparable efficiency as in MMOs are not known. Furthermore, the reactions exhibited by the model complexes of intermediate **Q** containing an Fe_2O_2 core are mostly non-catalytic, and an Fe_2O_2-mediated methane oxidation reaction is yet to be reported. The low reactivity of the model complexes can be explained by the inability of synthetic chemists to exactly reproduce the biological ligands and protein environments. Although O-donor ligands are ubiquitous in biology, most of the model compounds are based on N-rich ligands. In addition, the factors that control the O–O bond lysis in the peroxo intermediate **P** that would lead to the formation of the catalytically relevant intermediate **Q** is also not well understood. Thus, new and innovative synthetic strategies are needed to generate superoxidized metal centers in ligand environments that better resemble the active site of the metalloenzymes. These goals may eventually lead to the development of cheap and efficient bioinspired/biomimetic catalysts for methane oxidation that will help to influence the energy landscape of our society.

References

[1] Kerr, R. A. Natural gas from shale bursts onto the scene. Science 2010, 328, 1624–1626.
[2] Girod, B., van Vuuren, D. P., and Hertwich, E. G. Climate policy through changing consumption choices: options and obstacles for reducing greenhouse gas emissions. Glob Environ Chang 2014, 25, 5–15.
[3] Bibler, C. J., Marshall, J. S., and Pilcher, R. C. Status of worldwide coal mine methane emissions and use. Int J Coal Geol 1998, 35, 283–310.
[4] Nisbet, E. G., Dlugokencky, E. J., and Bousquet, P. Methane on the Rise-Again. Science 2014, 343, 493–495.
[5] Olah, G. A. Beyond Oil and Gas: the methanol economy. Angew Chem Int Ed 2005, 44, 2636–2639.
[6] Olah, G. A. Towards oil independence through renewable methanol chemistry. Angew Chem Int Ed 2013, 52, 104–107.
[7] Arakawa, H., Aresta, M., Armor, J. N., Barteau, M. A., Beckman, E. J., Bell, A. T., Bercaw, J. E., Creutz, C., Dinjus, E., Dixon, D. A., Domen, K., DuBois, D. L., Eckert, J., Fujita, E., Gibson, D. H., Goddard, W. A., Goodman, D.W., Keller, J., Kubas, G. J., Kung, H. H., Lyons, J. E., Manzer, L. E., Marks, T. J., Morokuma, K., Nicholas, K. M., Periana, R., Que, L., Rostrup-Nielson, J., Sachtler, W.M.H., Schmidt, L. D.,Sen, A., Somorjai,G.A., Stair, P. C., Stults, B. R., and Tumas, W. Catalysis research of relevance to carbon management: progress, challenges, and opportunities. Chem Rev 2001, 101, 953–996.
[8] Tang, P., Zhu, Q. J., Wu, Z. X., and Ma, D. Methane activation: the past and future. Energy Environ Sci 2014, 7, 2580–2591.
[9] Wang, V. C.-C., Maji, S., Chen, P. P.-Y., Lee, H. K., Yu, S. S.-F., and Chan, S. I. Alkane oxidation: methane monooxygenases, related enzymes, and their biomimetics. Chem Rev 2017, 117, 8574-8621.
[10] Bard, A. J., Whitesides, G. M., Zare, R. N., and Mclafferty, F. W. Holy grails in chemistry. Acc Chem Res 1995, 28, 91–91.
[11] Sazinsky, M. H., and Lippard, S. J. Correlating structure with function in bacterial multicomponent monooxygenases and related diiron proteins. Acc Chem Res 2006, 39, 558–566.
[12] Merkx, M., Kopp, D. A., Sazinsky, M. H., Blazyk, J. L., Müller, J., and Lippard, S. J. Dioxygen activation and methane hydroxylation by soluble methane monooxygenase: a tale of two irons and three proteins. Angew Chem Int Ed 2001, 40, 2782–2807.

[13] Kovaleva, E. G., Neibergall, M. B., Chakrabarty S., and Lipscomb, J. D. Finding intermediates in the O_2 activation pathways of non-heme iron oxygenases. Acc Chem Res 2007, 40, 475–483.
[14] Murray, L. J., and Lippard, S. J. Substrate trafficking and dioxygen activation in bacterial multicomponent monooxygenases. Acc Chem Res 2007, 40, 466–474.
[15] Hanson, R. S., and Hanson, T. E. Methanotrophic bacteria. Microbiol Rev 1996, 60, 439–471.
[16] Semrau, J. D., DiSpirito, A. A., and Yoon, S. Methanotrophs and copper. FEMS Microbiol Rev 2010, 34, 496–531.
[17] Chan, S. I., and Yu, S. S. F. Controlled oxidation of hydrocarbons by the membrane-bound methane monooxygenase: the case for a tricopper cluster. Acc Chem Res 2008, 41, 969–979.
[18] Culpepper, M. A., and Rosenzweig, A. C. Architecture and active site of particulate methane monooxygenase. Crit Rev Biochem Mol Biol 2012, 47, 483–492.
[19] Cao, L., Caldararu, O., Rosenzweig, A. C., and Ryde, U. Quantum refinement does not support dinuclear copper sites in crystal structures of particulate methane monooxygenase. Angew Chem Int Ed 2018, 57, 162 –166.
[20] Friedle, S., Reisner, E., and Lippard, S. J. Current challenges of modeling diiron enzyme active sites for dioxygen activation by biomimetic synthetic complexes. Chem Soc Rev 2010, 39, 2768–2779.
[21] Tinberg, C. E., and Lippard, S. J. Dioxygen activation in soluble methane monooxygenase. Acc Chem Res 2011, 44, 280–288.
[22] Lund, J., and Dalton, H. Further characterization of the FAD and Fe_2S_2 redox centers of component C, the NADH: acceptor reductase of the soluble methane monooxygenase of methylococcus capsulatus (Bath). Eur J Biochem 1985, 147, 291–296.
[23] Liu, Y., Nesheim, J. C., Paulsen, K. E., Stankovich, M. T., and Lipscomb, J. D. Roles of the methane monooxygenase reductase component in the regulation of catalysis. Biochemistry 1997, 36, 5223–5233.
[24] Kopp, D. A., Gassner, G. T., Blazyk, J. L., and Lippard, S. J. Electron-transfer reactions of the reductase component of soluble methane monooxygenase from methylococcus capsulatus (Bath). Biochemistry 2001, 40, 14932–14941.
[25] Elango, N., Radhakrishnan, R., Froland, W. A., Wallar, B. J., Earhart, C. A., Lipscomb, J. D., and Ohlendorf, D. H. Crystal structure of the hydroxylase component of methane monooxygenase from methylosinus trichosporium Ob3b. Protein Sci 1997, 6, 556–568.
[26] Rosenzweig, A. C., Frederick, C. A., Lippard, S. J., and Nordlund, P. Crystal structure of a bacterial nonheme iron hydroxylase that catalyzes the biological oxidation of methane. Nature 1993, 366, 537–543.
[27] Wallar, B. J., and Lipscomb, J. D. Dioxygen activation by enzymes containing binuclear Non-heme iron clusters. Chem Rev 1996, 96, 2625–2657.
[28] Gassner, G. T., and Lippard, S. J. Component interactions in the soluble methane monooxygenase system from methylococcus capsulatus (Bath). Biochemistry 1999, 38, 12768–12785.
[29] Liu, Y., Nesheim, J. C., Lee, S. K., and Lipscomb, J. D. Gating effects of component B on oxygen activation by the methane monooxygenase hydroxylase component. J Biol Chem 1995, 270, 24662–24665.
[30] Baik, M.-H., Newcomb, M., Friesner, R. A., and Lippard, S. J. Mechanistic studies on the hydroxylation of methane by methane monooxygenase. Chem Rev 2003, 103, 2385–2419.
[31] Whittington, D. A., and Lippard, S. J. Crystal structures of soluble methane monooxygenase hydroxylase from methylococcus capsulatus (Bath) demonstrating geometrical variability at the dinuclear iron active site. J Am Chem Soc 2001, 123, 827–838.

[32] Fox, B. G., Surerus, K. K., Münck, E., and Lipscomb, J. D. Evidence for a μ-oxo-bridged binuclear iron cluster in the hydroxylase component of methane monooxygenase. J Biol Chem 1988, 263, 10553–10556.

[33] Hendrich, M. P., Münck, E., Fox, B. G., and Lipscomb, J. D. Integer-spin EPR studies of the fully reduced methane monooxygenase hydroxylase component. J Am Chem Soc 1990, 112, 5861–5865.

[34] DeWitt, J. G., Bentsen, J. G., Rosenzweig, A. C., Hedman, B., Green, J., Pilkington, S., Papaefthymiou, G. C., Dalton, H., Hodgson, K. O., and Lippard, S. J. X-ray absorption, mössbauer, and EPR studies of the dinuclear iron center in the hydroxylase component of methane monooxygenase. J Am Chem Soc 1991, 113, 9219–9235.

[35] Fox, B. G., Hendrich, M. P., Surerus, K. K., Andersson, K. K., Froland, W. A., Lipscomb, J. D., and Münck, E. Mössbauer, EPR, and ENDOR studies of the hydroxylase and reductase components of methane monooxygenase from methylosinus trichosporium OB3b. J Am Chem Soc 1993, 115, 3688–3701.

[36] Prince, R. C., George, G. N., Savas, J. C., Cramer, S. P., and Patel, R. N. Spectroscopic properties of the hydroxylase of methane monooxygenase. Biochim Biophys Acta 1988, 952, 220-229.

[37] Woodland,. P., Pa Til, D. S.¡., Cammack, R., and Dalton, H. ESR studies of protein A of the soluble methane monooxygenase from methylococcus capsulatus (Bath). Biochim Biophys Acta 1986, 873, 237-242.

[38] Sazinsky, M. H., and Lippard, S. J. Methane monooxygenase: functionalizing methane at iron and copper. Met Ions Life Sci 2015, 15, 205–256.

[39] Lee, S.-K., Nesheim, J. C., and Lipscomb, J. D. Transient intermediates of the methane monooxygenase catalytic cycle. J Biol Chem 1993, 268, 21569–21577.

[40] Jasnewski, A. J., and Que, Jr. L. Dioxygen activation by nonheme diiron enzymes: diverse dioxygen adducts, high-valent intermediates, and related model complexes. Chem Rev 2018, 118, 2554-2592.

[41] Lee, S. J., McCormick, M. S., Lippard, S. J., and Cho, U. S. control of substrate access to the active site in methane monooxygenase. Nature 2013, 494, 380–384.

[42] Wang, W. X., Iacob, R. E., Luoh, R. P., Engen, J. R., and Lippard, S. J. Electron transfer control in soluble methane monooxygenase. J Am Chem Soc 2014, 136, 9754–9762.

[43] Sazinsky, M. H., and Lippard, S. J. Product bound structures of the soluble methane monooxygenase hydroxylase from methylococcus capsulatus (Bath): Protein Motion in the A-Subunit. J Am Chem Soc 2005, 127, 5814–5825.

[44] Gherman, B. F., Maik, M.H., Lippard, S. J., and Friesner, R. A. Dioxygen activation in methane monooxygenase: a theoretical study. J Am Chem Soc 2004, 126, 2978–2990.

[45] Rinaldo, D., Philipp, D. M., Lippard, S. J., and Friesner, R. A. Intermediates in dioxygen activation by methane monooxygenase: a QM/MM study. J Am Chem Soc 2007, 129, 3135–3147.

[46] Stahl, S. S., Francisco, W. A., Merkx, M., Klinman, J. P., and Lippard, S. J. Oxygen Kinetic Isotope Effects in Soluble Methane Monooxygenase. J Biol Chem 2001, 276, 4549–4553.

[47] Valentine, A. M., Stahl, S. S., and Lippard, S. J. Mechanistic studies of the reaction of reduced methane monooxygenase hydroxylase with dioxygen and substrates. J Am Chem Soc 1999, 121, 3876–3887.

[48] Kim, K., and Lippard, S. J. Structure and mössbauer spectrum of a (μ-1,2-Peroxo)bis(μ-carboxylato)diiron(III) model for the peroxo intermediate in the methane monooxygenase hydroxylase reaction cycle. J Am Chem Soc 1996, 118, 4914–4915.

[49] Shu, L. J., Nesheim, J. C., Kauffmann, K., Münck, E., Lipscomb, J. D., and Que, L. An ($Fe_2^{IV}O_2$) diamond core structure for the key intermediate q of methane monooxygenase. Science 1997, 275, 515–518.
[50] Liu, K. E., Valentine, A. M., Wang, D. L., Huynh, B. H., Edmondson, D. E., Salifoglou, A., and Lippard, S. J. kinetic and spectroscopic characterization of intermediates and component interactions in reactions of methane monooxygenase from methylococcus capsulatus (Bath). J Am Chem Soc 1995, 117, 10174–10185.
[51] Liu, K. E., Wang, D., Huynh, B. H., Edmondson, D. E., Salifoglou, A., and Lippard, S. J. spectroscopic detection of intermediates in the reaction of dioxygen with the reduced methane monooxygenase hydroxylase from methylococcus capsulatus (Bath). J Am Chem Soc 1994, 116, 7465–7466.
[52] Tinberg, C. E., and Lippard, S. J. revisiting the mechanism of dioxygen activation in soluble methane monooxygenase from M. capsulatus (bath): evidence for a multi-step, proton dependent reaction pathway. Biochemistry 2009, 48, 12145–12158.
[53] Moënne-Loccoz, P., Baldwin, J., Ley, B. A., Loehr, T. M., and Bollinger, J. M., Jr. O_2 activation by non-heme diiron proteins: identification of a symmetric μ-1,2-peroxide in a mutant of ribonucleotide reductase. Biochemistry 1998, 37, 14659–14663.
[54] Skulan, A. J., Brunold, T. C., Baldwin, J., Saleh, L., Bollinger, J. M., Jr., and Solomon, E. I. nature of the peroxo intermediate of the W48F/D84E ribonucleotide reductase variant: implications for O_2 activation by binuclear non-heme iron enzymes. J Am Chem Soc 2004, 126, 8842–8855.
[55] Moënne-Loccoz, P., Krebs, C., Herlihy, K., Edmondson, D. E., Theil, E. C., Huynh, B. H., and Loehr, T. M. the ferroxidase reaction of ferritin reveals a diferric μ-1,2 bridging peroxide intermediate in common with other O_2-activating non-heme diiron proteins. Biochemistry 1999, 38, 5290–5295.
[56] Broadwater, J. A., Ai, J., Loehr, T. M., Sanders-Loehr, J., and Fox, B. G. Peroxodiferric intermediate of stearoyl-acyl carrier protein Δ^9 desaturase: oxidase reactivity during single turnover and implications for the mechanism of desaturation. Biochemistry 1998, 37, 14664–14671.
[57] Vu, V. V., Emerson, J. P., Martinho, M., Kim, Y. S., Münck, E., Park, M. H., and Que, L., Jr. human deoxyhypusine hydroxylase, an enzyme involved in regulating cell growth, activates O_2 with a nonheme diiron center. Proc Natl Acad Sci U.S.A 2009, 106, 14814–14819.
[58] Bailey, L. J., and Fox, B. G. Crystallographic and catalytic studies of the peroxide-shunt reaction in a diiron hydroxylase. Biochemistry 2009, 48, 8932–8939.
[59] Tshuva, E. Y., and Lippard, S. J. Synthetic models for non-heme carboxylate-bridged diiron metalloproteins: strategies and tactics. Chem Rev 2004, 104, 987–1012.
[60] Han, W. G., and Noodleman, L. Structural model studies for the peroxo intermediate P and the reaction pathway from P–>Q of methane monooxygenase using broken-symmetry density functional calculations. Inorg Chem 2008, 47, 2975–2986.
[61] Lee, S.-K., and Lipscomb, J. D. Oxygen activation catalyzed by methane monooxygenase hydroxylase component: proton delivery during the O-O bond cleavage steps. Biochemistry 1999, 38, 4423–4432.
[62] Sazinsky, M. H., and Lippard, S. J. In Sustaining Life on Planet Earth: Metalloenzymes Mastering Dioxygen and Other Chewy Gases, Metal Ions in Life Sciences, Kroneck, P. M. H., Torres, M. E. S., Eds.; Springer, Cham, 2015, 205-256.
[63] Beauvais, L. G., and Lippard, S. J. Reactions of the peroxo intermediate of soluble methane monooxygenase hydroxylase with ethers. J Am Chem Soc 2005, 127, 7370–7378.

[64] Tinberg, C. E., and Lippard, S. J. Oxidation reactions performed by soluble methane monooxygenase hydroxylase intermediates hperoxo and Q proceed by distinct mechanisms. Biochemistry 2010, 49, 7902–7912.
[65] Brunold, T. C., Tamura, N., Kitajima, N., Moro-Oka, Y., and Solomon, E. I. Spectroscopic study of [Fe2(O2)(Obz)$_2${Hb(Pz′)$_3$}$_2$]: nature of the μ-1,2 peroxide-Fe(III) Bond and its possible relevance to O_2 activation by non-heme iron enzymes. J Am Chem Soc 1998, 120, 5674–5690.
[66] McDonald, A. R., and Que, L. High-valent nonheme iron-oxo complexes: synthesis, structure, and spectroscopy. Coord Chem Rev 2013, 257, 414–428.
[67] Klein, J. E. M. N., and Que, L., Jr. Biomimetic High-Valent Mononuclear Nonheme Iron-Oxo Chemistry, In Encyclopedia of Inorganic and Bioinorganic Chemistry, Scott, R. A., Ed.; John Wiley, Chichester, U.K., 2016, DOI: 10.1002/9781119951438.eibc2344.
[68] Engelmann, X., Monte-Pérez, I., and Ray, K. Oxidation reactions with bioinspired mononuclear non-heme metal-oxo complexes. Angew Chem Int Ed 2016, 55, 7632-7649.
[69] Heims, F., Pfaff, F. F., and Ray, K. Terminal oxo and imido transition metal complexes of groups 9-11. Eur J Inorg Chem 2013, 22-23, 3784-3807.
[70] Hong, S., Lee, Y.-M., Ray, K., and Nam, W. Dioxygen activation chemistry by synthetic mononuclear nonheme iron, copper and chromium complexes. Coord Chem Rev 2016, 334, 25-42.
[71] Pfaff, F. F., Wang, B., Ray, K., and Nam, W. Status of reactive non-heme metal-oxygen intermediates in chemical and enzymatic reactions. J Am Chem Soc 2014, 136, 13942.
[72] Banerjee, R., Proshlyakov, Y., Lipscomb, J. D., and Proshlyakov, D.A. Structure of the key species in the enzymatic oxidation of methane to methanol. Nature 2015, 518, 431–434.
[73] Xue, G., Wang, D., De Hont, R., Fiedler, A. T., Shan, X., Münck, E., and Que, L. A synthetic precedent for the [Fe^{IV}_2(μ-O)$_2$] diamond core proposed for methane monooxygenase intermediate Q. Proc Natl Acad Sci U S A 2007, 104, 20713–20718.
[74] Wang, D., Farquhar, E. R., Stubna, A., Münck, E., and Que, Jr., L. A diiron(IV) complex that cleaves strong C–H and O–H bonds. Nat Chem 2009, 1, 145–150.
[75] Lee, S.-K., Fox, B. G., Froland, W. A., Lipscomb, J. D., and Münck, E. A transient intermediate of the methane monooxygenase catalytic cycle containing an $Fe^{IV}Fe^{IV}$ cluster. J Am Chem Soc 1993, 115, 6450–6451.
[76] Huang, S.-P., Shiota, Y., and Yoshizawa, K. DFT study of the mechanism for methane hydroxylation by soluble methane monooxygenase (sMMO): effects of oxidation state, spin state, and coordination number. Dalton Trans 2013, 42, 1011–1023.
[77] Deighton, N., Podmore, I. D., Symons, M. C., Wilkins, P. C., and Dalton, H. Substrate radical intermediates are involved in the soluble methane monooxygenase catalysed oxidations of methane, methanol and acetonitrile. J Chem Soc Chem Commun 1991, 16, 1086–1088.
[78] Brazeau, B. J., Austin, R. N., Tarr, C., Groves, J. T., and Lipscomb, J. D. Intermediate Q from soluble methane monooxygenase hydroxylates the mechanistic substrate probe norcarane: evidence for a stepwise reaction. J Am Chem Soc 2001, 123, 11831–11837.
[79] Wilkins, P. C., Dalton, H., Podmore, I. D., Deighton, N., and Symons, M. C. R. Biological methane activation involves the intermediacy of carbon-centered radicals. Eur J Biochem 1992, 210, 67–72.
[80] Dalton, H., Wilkins, P. C., Deighton, N., Podmore, I. D., and Symons, M. C. R. Electron paramagnetic resonance studies of the mechanism of substrate oxidation by methane monooxygenase. Faraday Discuss 1992, 93, 163–171.
[81] Jin, Y., and Lipscomb, J. D. Probing the mechanism of C–H activation: oxidation of methylcubane by soluble methane monooxygenase from methylosinus trichosporium OB3b. Biochemistry 1999, 38, 6178–6186.

[82] Liu, A., Jin, Y., Zhang, J., Brazeau, B. J., and Lipscomb, J. D. Substrate radical intermediates in soluble methane monooxygenase. Biochem Biophys Res Commun 2005, 338, 254–261.

[83] Liu, K. E., Johnson, C. C., Newcomb, M., and Lippard, S. J. Radical clock substrate probes and kinetic isotope effect studies of the hydroxylation of hydrocarbons by methane monooxygenase. J Am Chem Soc 1993, 115, 939–947.

[84] Valentine, A. M., LeTadic-Biadatti, M.- H., Toy, P. H., Newcomb, M., and Lippard, S. J. Oxidation of ultrafast radical clock substrate probes by the soluble methane monooxygenase from methylococcus capsulatus (Bath). J Biol Chem 1999, 274, 10771–10776.

[85] Sears, T. J., Johnson, P. M., Jin, P., and Oatis, S. Infrared laser transient absorption spectroscopy of the ethyl radical. J Chem Phys 1996, 104, 781-792.

[86] Gherman, B. F., Dunietz, B. D., Whittington, D. A., Lippard, S. J., and Friesner, R. A. Activation of the C–H bond of methane by intermediate Q of methane monooxygenase: a theoretical study. J Am Chem Soc 2001, 123, 3836–3837.

[87] Priestley, N. D., Floss, H. G., Froland, W. A., Lipscomb, J. D., Williams, P. G., and Morimoto, H. Cryptic stereospecificity of methane monooxygenase. J Am Chem Soc, 1992, 114, 7561–7562.

[88] Valentine, A. M., Wilkinson, B., Liu, K. E., Komar-Panicucci, S., Priestley, N. D., Williams, P. G., Morimoto, H., Floss, H. G., and Lippard, S. J. Tritiated chiral alkanes as substrates for soluble methane monooxygenase from methylococcus capsulatus (Bath): probes for the mechanism of hydroxylation. J Am Chem Soc, 1997, 119, 1818–1827.

[89] McMurry, T. J., and Groves, J. T. In Cytochrome P-450 Structure, Mechanism, and Biochemistry, Ortiz de Montellano,P. R., Ed.; Plenum Publishing Corp.: ,New York, 1986, 1-28.

[90] Ortiz de Montellano, P. R. In Cytochrome P-450 Structure, Mechanism, and Biochemistry, Ortiz de Montellano, P. R, Ed.; Plenum Publishing Corp, New York, 1986, 217-271.

[91] Mansuy, D., and Battioni, P. In Activation and Functionalization of Alkanes, Hill, C. L., Ed.; Wiley, New York, 1989, Chapter VI.

[92] Guengerich, F. P. In Biological Oxidation Systems, Reddy, C. C., Hamilton, G. A., Madyastha, K. M., Eds.; Academic Press, San Diego, 1990, 1, 51-67.

[93] Jin, Y., and Lipscomb, J. D. Mechanistic insights into C–H activation from radical clock chemistry: oxidation of substituted methylcyclopropanes catalyzed by soluble methane monooxygenase from Methylosinus trichosporium OB3b. Biochim Biophys Acta 2000, 1543, 47-59

[94] Ruzicka, F., Huang, D. S., Donnelly, M. I., and Frey, P. A. Methane monooxygenase catalyzed oxygenation of 1,1-dimethylcyclopropane. Evidence for radical and carbocationic intermediates. Biochemistry 1990, 29, 1696–1700.

[95] Choi, S.-Y., Eaton, P. E., Hollenberg, P. F., Liu, K. E., Lippard, S. J., Newcomb, M., Putt, D. A., Upadhyaya, S. P., and Xiong, Y. Regiochemical variations in reactions of methylcubane with tert-butoxyl radical, cytochrome P-450 enzymes, and a methane monooxygenase system. J Am Chem Soc 1996, 118, 6547–6555.

[96] Choi, S.-Y., Eaton, P. E., Kopp, D. A., Lippard, S. J., Newcomb, M., and Shen, R. Cationic species can be produced in soluble methane monooxygenase-catalyzed hydroxylation reactions; radical intermediates are not formed. J Am Chem Soc 1999, 121, 12198–12199.

[97] Siegbahn, P. E. M., and Crabtree, R. H. mechanism of C–H activation by diiron methane monooxygenases: quantum chemical studies. J Am Chem Soc 1997, 119, 3103–3113.

[98] Siegbahn, P. E. M. Theoretical model studies of the iron dimer complex of MMO and RNR. Inorg Chem 1999, 38, 2880–2889.

[99] Dunietz, B. D., Beachy, M. D., Cao, Y., Whittington, D. A., Lippard, S. J., and Friesner, R. A. Large scale ab initio quantum chemical calculation of the intermediates in the soluble methane monooxygenase catalytic cycle. J Am Chem Soc 2000, 122, 2828–2839.
[100] Nesheim, J. C., and Lipscomb, J. D. Large kinetic isotope effects in methane oxidation catalyzed by methane monooxygenase: evidence for C–H bond cleavage in a reaction cycle intermediate. Biochemistry, 1996, 35, 10240-10247.
[101] Guo, Z., Liu, B., Zhang, Q., Deng, W., Wang, Y., and Yan, Y. Recent advances in heterogeneous selective oxidation catalysis for sustainable chemistry. Chem Soc Rev 2014, 43, 3480-3524.
[102] Chan, S. I., Lu, Y.-J., Nagababu, P., Maji, S., Hung, M.-C., Lee, M. M., Hsu, I. J., Minh, P. D., Lai, J. C. H., Ng, K. Y., Ramalingam, S., Yu, S. S. F., and Chan, M. K. Efficient oxidation of methane to methanol by dioxygen mediated by tricopper clusters. Angew Chem, Int Ed 2013, 52, 3731-3735.
[103] Liu, C.-C., Mou, C.-Y., Yu, S. S. F., and Chan, S. I. Heterogeneous formulation of the tricopper complex for efficient catalytic conversion of methane into methanol at ambient temperature and pressure. Energy Environ Sci 2016, 9, 1361-1374.
[104] Yuan, Q., Deng, W., Zhang, Q., and Wang, Y. Osmium-catalyzed selective oxidations of methane and ethane with hydrogen peroxide in aqueous medium. Adv Synth Catal 2007, 349, 1199-1209.
[105] Hammond, C., Forde, M. M., Ab Rahim, M. H., Thetford, A., He, Q., Jenkins, R. L., Dimitratos, N., Lopez-Sanchez, J. A., Dummer, N. F., Murphy, D. M., Carley, A. F., Taylor, S. H., Willock, D. J., Stangland, E. E., Kang, J., Hagen, H., Kiely, C. J., and Hutchings, G. J. Angew Chem, Int Ed 2012, 51, 5129-5133.
[106] Shilov, A. E., and Shulpin, G. B. Activation of C-H bonds by metal complexes. Chem Rev 1997, 97, 2879–2932.
[107] Labinger, J. A., and Bercaw, J. E. Mechanistic studies on the shilov system: a retrospective. J Organomet Chem 2015, 793, 47–53.
[108] Periana, R. A., Taube, D. J., Gamble, S., Taube, H., Satoh, T., and Fujii, H. Platinum catalysts for the high-yield oxidation of methane to a methanol derivative. Science 1998, 280, 560–564.
[109] Rahim, M. H. A., Forde, M. M., Jenkins, R. L., Hammond, C., He, Q., Dimitratos, N., Lopez-Sanchez, J. A., Carley, A. F., Taylor, S. H., Willock, D. J., Murphy, D. M., Kiely, C. J., and Hutchings, G. J. Oxidation of methane to methanol with hydrogen peroxide using supported gold-palladium alloy nanoparticles. Angew Chem, Int Ed 2013, 52, 1280–1284.
[110] Hammond, C., Jenkins, R. L., Dimitratos, N., Lopez-Sanchez, J. A., Rahim, M. H. A., Forde, M. M., Thetford, A., Murphy, D. M., Hagen, H., Stangland, E. E., Moulijn, M., Taylor, J. S. H., Willock, D. J., and Hutchings, G. J. Catalytic and mechanistic insights of the low-temperature selective oxidation of methane over Cu-promoted Fe-ZSM-5. Chem–Eur J 2012, 18, 15735–15745.
[111] Hammond, C., Dimitratos, N., Jenkins, R. L., Lopez-Sanchez, J. A., Kondrat, S. A., Hasbi Ab Rahim, M., Forde, M. M., Thetford, A., Taylor, S. H., Hagen, H., Stangland, E. E., Kang, J. H., Moulijn, J. M., and Willock, D. J. Hutchings, G. J. ACS Catal 2013, 3, 689-699.
[112] Hammond, C., Dimitratos, N., Lopez-Sanchez, J. A., Jenkins, R. L., Whiting, G., Kondrat, S. A., Ab Rahim, M. H., Forde, M. M., Thetford, A., Hagen, H., Stangland, E. E., Moulijn, J. M., Taylor, S. H., Willock, D. J., and Hutchings, G. J. ACS Catal 2013, 3, 1835-1844.
[113] Liu, H. F.; Liu, R. S.; Liew, K. Y.; Johnson, R. E.; Lunsford, J. H. Partial oxidation of methane by nitrous oxide over molybdenum on silica. J Am Chem Soc 1984, 106, 4117–4121.
[114] Sobolev, V. I., Dubkov, K. A., Panna, O. V., and Panov, G. I. Selective oxidation of methane to methanol on a FeZSM-5 surface. Catal Today 1995, 24, 251–252.

[115] Dubkov, K. A., Sobolev, V. I., Talsi, E. P., Rodkin, M. A., Watkins, N. H., Shteinman, A. A., and Panov, G. I. Kinetic isotope effects and mechanism of biomimetic oxidation of methane and benzene on FeZSM-5 zeolite. J Mol Catal A: Chem 1997, 123, 155–161.
[116] Starokon, E. V., Parfenov, M. V., Pirutko, L. V., Abornev S. I., and Panov, G. I. Room-temperature oxidation of methane by α-oxygen and extraction of products from the FeZSM-5 surface. J Phys Chem A 2011, 115, 2155–2161.
[117] Strong, P. J., Xie, S., and Clarke, W. P. Environ Sci Technol 2015, 49, 4001-4018.
[118] Smeets, P. J., Groothaert, M. H., and Schoonheydt, R. A. Cu based zeolites: A UV-vis study of the active site in the selective methane oxidation at low temperatures. Catal Today 2005, 110, 303–309.
[119] Groothaert, M. H., Smeets, P. J., Sels, B. F., Jacobs, P. A., and Schoonheydt, R. A. Selective oxidation of methane by the Bis(μ-oxo)dicopper core stabilized on ZSM-5 and mordenite zeolites. J Am Chem Soc 2005, 127, 1394–1395.
[120] Beznis, N. V., van Laak, A. N. C., Weckhuysen, B. M., and Bitter, J. H. Oxidation of methane to methanol and formaldehyde over Co–ZSM-5 molecular sieves: tuning the reactivity and selectivity by alkaline and acid treatments of the zeolite ZSM-5 agglomerates. Micropor Mesopor Mat 2011, 138, 176-183.
[121] Beznis, N. V., Weckhuysen, B. M., and Bitter, J. H. Partial oxidation of methane over Co-ZSM -5: tuning the oxygenate selectivity by altering the preparation route. Catal Lett 2009, 136, 52–56.
[122] Kraatz, H. B., and Metzler-Nolte, N. Concepts and Models in Bioinorganic Chemistry, Wiley-VCH, Weinheim, 2006.
[123] Siewert, I., and Limberg, C. Low-molecular-weight analogues of the soluble methane monooxygenase (sMMO): from the structural mimicking of resting states and intermediates to functional models. Chem–Eur J 2009, 15, 10316–10328.
[124] Zhang, X., Furutachi, H., Fujinami, S., Nagatomo, S., Maeda, Y., Watanabe, Y., Kitagawa, T., and Suzuki, M. Structural and spectroscopic characterization of (μ-Hydroxo or μ-Oxo)(μ-Peroxo)-Diiron(III) complexes: models for peroxo intermediates of non-heme diiron proteins. J Am Chem Soc 2005, 127, 826–827.
[125] Shan, X., and Que, L., Jr. Intermediates in the oxygenation of a nonheme diiron(II) complex, including the first evidence for a bound superoxo species. Proc Natl Acad Sci U S A 2005, 102, 5340–5345.
[126] Ookubo, T., Sugimoto, H., Nagayama, T., Masuda, H., Sato, T., Tanaka, K., Maeda, Y., Okawa, H., Hayashi, Y., Uehara, A., and Suzuki, M. *cis*-μ-1,2-peroxo diiron complex: structure and reversible oxygenation. J Am Chem Soc 1996, 118, 701–702.
[127] Dong, Y., Yan, S., Young, V. G., Jr., and Que, L., Jr. Crystal structure analysis of a synthetic nonheme diiron-O_2 adduct: insight into oxygen activation. Angew Chem Int Ed Engl 1996, 35, 618–620.
[128] Pap, J. S., Cranswick, M. A., Balogh-Hergovich, E., Barath, G., Giorgi, M., Rohde, G. T., Kaizer, J., Speier, G., and Que, L., Jr. An Iron(II)[1,3-Bis(2′-Pyridylimino)Isoindoline] complex as a catalyst for substrate oxidation with H_2O_2 - evidence for a transient peroxidodiiron(III) species. Eur J Inorg Chem 2013, 22-23, 3858–3866.
[129] Kitajima, N., Tamura, N., Amagai, H., Fukui, H., Moro-oka, Y., Mizutani, Y., Kitagawa, T., Mathur, R., and Heerwegh, K. Monomeric carboxylate ferrous complexes as models for the dioxygen binding sites in non-heme iron proteins. the reversible formation and characterization of μ-peroxo diferric complexes. J Am Chem Soc 1994, 116, 9071–9085.
[130] Pap, J. S., Draksharapu, A., Giorgi, M., Browne, W. R., Kaizer, J., and Speier, G. Stabilisation of M-peroxido-bridged Fe(III) intermediates with non-symmetric bidentate *N*-donor ligands. Chem Commun 2014, 50, 1326–1329.

[131] Murch, B. P., Bradley, F. C., and Que, L., Jr. A binuclear iron peroxide complex capable of olefin epoxidation. J Am Chem Soc 1986, 108, 5027–5028.
[132] LeCloux, D. D., Barrios, A. M., Mizoguchi, T. J., and Lippard, S. J. Modeling the diiron centers of non-heme iron enzymes. preparation of sterically hindered diiron(II) tetracarboxylate complexes and their reactions with dioxygen. J Am Chem Soc 1998, 120, 9001–9014.
[133] Chavez, F. A., Ho, R. Y. N., Pink, M., Young, V. G., Jr., Kryatov, S. V., Rybak-Akimova, E. V., Andres, H., Münck, E., Que, L., Jr., and Tolman, W. B. Unusual peroxo intermediates in the reaction of dioxygen with carboxylate-bridged diiron(II,II) paddlewheel complexes. Angew Chem Int Ed 2002, 41, 149–152.
[134] Hummel, H., Mekmouche, Y., Duboc-Toia, C., Ho, R. Y. N., Que, L., Jr., Schunemann, V., Thomas, F., Trautwein, A. X., Lebrun, C., Fontecave, M., and Menage, S. A diferric peroxo complex with an unprecedented spin configuration: an $S = 2$ system arising from an $S = 5/2$, 1/2 pair. Angew Chem Int Ed 2002, 41, 617–620.
[135] Duboc-Toia, C., Ménage, S., Ho, R. Y. N., Que, L., Lambeaux, C., and Fontecave, M. Enantioselective sulfoxidation as a probe for a metal-based mechanism in H_2O_2-dependent oxidations catalyzed by a diiron complex. Inorg Chem 1999, 38, 1261–1268.
[136] Fiedler, A. T., Shan, X., Mehn, M. P., Kaizer, J., Torelli, S., Frisch, J. R., Kodera, M., and Que, L., Jr. spectroscopic and computational studies of (μ-Oxo)(μ-1,2-Peroxo)diiron(III) complexes of relevance to nonheme diiron oxygenase intermediates. J Phys Chem A 2008, 112, 13037–13044.
[137] Dong, Y., Zang, Y., Kauffmann, K., Shu, L., Wilkinson, E. C., Münck, E., and Que, L., Jr. Models for nonheme diiron enzymes. assembly of a high-valent $Fe2(\mu\text{-}O)_2$ diamond core from its peroxo precursor. J Am Chem Soc 1997, 119, 12683–12684.
[138] Kodera, M., Taniike, Y., Itoh, M., Tanahashi, Y., Shimakoshi, H., Kano, K., Hirota, S., Iijima, S., Ohba, M., and Okawa, H. Synthesis, Characterization, and activation of thermally stable μ-1,2-peroxodiiron(III) complex. Inorg Chem 2001, 40, 4821–4822.
[139] Friedle, S., Kodanko, J. J., Morys, A. J., Hayashi, T., Moënne-Loccoz, P., and Lippard, S. J. Modeling the syn disposition of nitrogen donors in non-heme diiron enzymes. synthesis, characterization,and hydrogen peroxide reactivity of diiron(III) complexes with the Syn N-donor ligand H_2BPG_2dev. J Am Chem Soc 2009, 131, 14508–14520.
[140] Cranswick, M. A., Meier, K. K., Shan, X., Stubna, A., Kaizer, J., Mehn, M. P., Münck, E., and Que, L. Jr. Protonation of a peroxodiiron- (III) complex and conversion to a diiron(III/IV) intermediate: implications for proton-assisted O-O bond cleavage in nonheme diiron enzymes. Inorg Chem 2012, 51, 10417–10426.
[141] Kodera, M., Tsuji, T., Yasunaga, T., Kawahara, Y., Hirano, T., Hitomi, Y., Nomura, T., Ogura, T., Kobayashi, Y., Sajith, P. K., Shiota, Y., and Yoshizawa, K. Roles of carboxylate donors in O-O bond scission of peroxodiiron(III) to high-spin oxodiiron(IV) with a new carboxylate-containing dinucleating ligand. Chem Sci 2014, 5, 2282–2292.
[142] Frisch, J. R., Vu, V. V., Martinho, M., Münck, E., and Que, L., Jr. Characterization of two distinct adducts in the reaction of a nonheme diiron(II) complex with O_2. Inorg Chem 2009, 48, 8325–8336.
[143] Dong, Y., Menage, S., Brennan, B. A., Elgren, T. E., Jang, H. G., Pearce, L. L., and Que, L., Jr. Dioxygen binding to diferrous centers. models for diiron-oxo proteins. J Am Chem Soc 1993, 115, 1851–1859.
[144] Hayashi, Y., Kayatani, T., Sugimoto, H., Suzuki, M., Inomata, K., Uehara, A., Mizutani, Y., Kitagawa, T., and Maeda, Y. Synthesis, characterization, and reversible oxygenation of μ-alkoxo diiron(II) complexes with the dinucleating ligand *N,N,N′,N′*-TETRakis{(6- Methyl-2-Pyridyl)Methyl}-1,3-Diamino-Propan-2-Olate. J Am Chem Soc 1995, 117, 11220–11229.

[145] Arii, H., Nagatomo, S., Kitagawa, T., Miwa, T., Jitsukawa, K., Einaga, H., and Masuda, H. A novel diiron complex as a functional model for hemerythrin. J Inorg Biochem 2000, 82, 153–162.
[146] Ghosh, A., Tiago de Oliveira, F., Yano, T., Nishioka, T., Beach, E. S., Kinoshita, I., Münck, E., Ryabov, A. D., Horwitz, C. P., and Collins, T. J. Catalytically active μ-oxodiiron(IV) oxidants from iron(III) and dioxygen. J Am Chem Soc 2005, 127, 2505–2513.
[147] Stoian, S. A., Xue, G., Bominaar, E. L., Que, L., and Münck, E. Spectroscopic and theoretical investigation of a complex with an [$O{=}Fe^{IV}{-}O{-}Fe^{IV}{=}O$] core related to methane monooxygenase intermediate Q. J Am Chem Soc 2014, 136, 1545–1558.
[148] Kodera, M., Kawahara, Y., Hitomi, Y., Nomura, T., Ogura, T., and Kobayashi, Y. Reversible O–O bond scission of peroxodiiron(III) to high-spin oxodiiron(IV) in dioxygen activation of a diiron center with a Bis-Tpa dinucleating ligand as a soluble methane monooxygenase model. J Am Chem Soc 2012, 134, 13236–13239.
[149] Kodera, M., Ishiga, S., Tsuji, T., Sakurai, K., Hitomi, Y., Shiota, Y., Sajith, P. K., Yoshizawa, K., Mieda, K., and Ogura, T. Formation and high reactivity of the anti-dioxo form of high-spin μ-oxodioxodiiron(IV) as the active species that cleaves strong C–H bonds. Chem - Eur J 2016, 22, 5924–5936.
[150] Dong, Y., Fujii, H., Hendrich, M. P., Leising, R. A., Pan, G., Randall, C. R., Wilkinson, E. C., Zang, Y., Que, L., Jr., Fox, B. G., Kauffmann, K., and Münck, E. A high-valent nonheme iron intermediate. structure and properties of [$Fe_2(\mu\text{-}O)_2(5\text{-Me-TPA})_2](ClO_4)_3$. J Am Chem Soc 1995, 117, 2778–2792.
[151] Dong, Y., Que, L., Jr., Kauffmann, K., and Münck, E. An exchange-coupled complex with localized high-spin FeIV and FeIII sites of relevance to cluster X of escherichia coli ribonucleotide reductase. J Am Chem Soc 1995, 117, 11377–11378.
[152] Hsu, H.-F., Dong, Y., Shu, L., Young, V. G., Jr., and Que, L., Jr. Crystal structure of a synthetic high-valent complex with an $Fe_2(\mu\text{-}O)_2$ diamond core. implications for the core structures of methane monooxygenase intermediate Q and ribonucleotide reductase intermediate X. J Am Chem Soc 1999, 121, 5230–5237.
[153] Xue, G., Fiedler, A. T., Martinho, M., Münck, E., and Que, L. Insights into the P-to-Q conversion in the catalytic cycle of methane monooxygenase from a synthetic model system. Proc Natl Acad Sci U S A 2008, 105, 20615–20620.
[154] Xue, G., De Hont, R., Münck, E., and Que, L. Million-fold activation of the [$Fe_2(\mu\text{-}O)_2$] diamond core for C–H bond cleavage. Nat Chem 2010, 2, 400–405.
[155] Xue, G., Pokutsa, A., and Que, L. Substrate-triggered activation of a synthetic [$Fe_2(\mu\text{-}O)_2$] diamond core for C–H bond cleavage. J Am Chem Soc 2011, 133, 16657–16667.
[156] Xue, G.; Geng, C.; Ye, S.; Fiedler, A. T.; Neese, F.; Que, L. Hydrogen-bonding effects on the reactivity of [$X{-}Fe^{III}{-}O{-}Fe^{IV}{=}O$] (X = OH, F) complexes toward C–H bond cleavage. Inorg Chem 2013, 52, 3976–3984.
[157] Hirao, H., Kumar, D., Que, L., and Shaik, S. Two-state reactivity in alkane hydroxylation by non-heme iron–oxo complexes. J Am Chem Soc 2006, 128, 8590–8606.
[158] Shaik, S., Hajime Hirao, H., and Kumar, D. Reactivity of high-valent iron–oxo species in enzymes and synthetic reagents: a tale of many states. Acc Chem Res 2007, 40, 532–542.
[159] Bernasoni, L., Louwerse M. J., and Baerends, E. J. The role of equatorial and axial ligands in promoting the activity of non-heme oxidoiron(IV) catalysts in alkane hydroxylation. Eur J Inorg Chem 2007, 3023-3033.
[160] Kodera, M., Itoh, M., Kano, K., Funabiki, T., and Reglier, M. A. Diiron center stabilized by a Bis-TPA ligand as a model of soluble methane monooxygenase: predominant alkene epoxidation with H_2O_2. Angew Chem Int Ed 2005, 44, 7104–7106.

Part II: Organometallic Enzyme Reactions

Bernhard Kräutler

6 Organometallic B_{12}-derivatives in life processes

6.1 Introduction

The discovery of deep red vitamin B_{12} (cyanocobalamin, **CNCbl**) and its characterization as an exceptionally complex cobalt corrin have provided first insights into the possible biological roles of this vitamin [1–4]. When, in the 1960s, coenzyme B_{12} was identified as the organometallic vitamin B_{12} analogue 5′-deoxyadenosylcobalamin (**AdoCbl**) [3] the importance of organometallic life processes became apparent, as well [5, 6].

Vitamin B_{12} (**CNCbl**) is the most important commercially available vitamin form of the naturally occurring B_{12}-derivatives. Besides, **CNCbl** is the most complex member of the natural tetrapyrroles [7]. Its total synthesis by Eschenmoser [8] and Woodward [9] has been an absolute highlight in synthesis. Strikingly, **CNCbl** itself has no proper physiological function in humans [10]. Physiologically directly relevant B_{12}-derivatives are, however, the organometallic analogues, coenzyme B_{12} (**AdoCbl**) and methylcobalamin (**MeCbl**). By engaging the reactivity of their (Co–C)-bond they help to catalyze exceptional enzyme processes [6, 11–13]. The B_{12}-derivatives represent the physiologically most broadly relevant organometallic cofactors. They are required, not only by humans [14–16], but by a broad range of organisms [11, 17, 18]. Among these, only specific "primitive" microorganisms have the capacity to also synthesize B_{12} and other natural corrinoids [19]. All other B_{12}-dependent organisms rely on the supply of B_{12}-derivatives as their vitamins [14]. Hence, their metabolism depends not only on the functioning catalysis by the indispensable B_{12}-dependent enzymes [11–13, 20], but also on the adequate uptake and cellular import of B_{12}-derivatives [21] and on the intracellular transformation of these to relevant B_{12}-cofactors [10]. Indeed, important neuropathological and various other physiological effects of B_{12} in humans are puzzling or still need to be identified [16, 22], so that B_{12} has been called a "moonlighting" vitamin [23].

Dedication: Dedicated to Professor Helmut Sigel, on the occasion of his 80th birthday

Bernhard Kräutler, Institute of Organic Chemistry and Centre of Molecular Biosciences, University of Innsbruck, Innsbruck, Austria

https://doi.org/10.1515/9783110496574-006

6.2 Structures of organometallic B$_{12}$-derivatives

6.2.1 "Complete" and "incomplete" corrinoids

Vitamin B$_{12}$ (**CNCbl**) and coenzyme B$_{12}$ (**AdoCbl**) are "complete" corrinoids, in which a pseudonucleotide function is attached via the f-side chain to the corrin moiety [4]. In contrast, adenosyl-cobyrate, the cobalt-adenosylated cobyric acid (Cby) moiety of **AdoCbl**, is a biosynthetically important "incomplete" corrinoid [19, 24]. Adenosyl-co-binamide (**AdoCbi**) is, likewise, an "incomplete" corrin moiety of **AdoCbl**, in which only the (R)-isopropanol-amine linker section of the nucleotide "loop" extends from its f-side chain (see Figure 6.1). The natural "complete" and "incomplete" corrinoids are frequently classified as B$_{12}$-derivatives (or "B$_{12}$"), collectively.

CN,H$_2$OCby: R^1=NH$_2$, R^2=OH,
 L$_\alpha$=LCN$^-$, L$_\beta$=H$_2$O
AdoCby: R^1=NH$_2$, R^2=OH,
 L$_\alpha$=LH$_2$O, L$_\beta$=5'-deoxy-5'-adenosyl
AdoCbi: R^1=NH$_2$, R^2=NHCH$_2$CH(CH$_3$)OH
 L$_\alpha$=LH$_2$O, L$_\beta$=5'-deoxy-5'-adenosyl
(CN)$_2$cobester: R^1=R^2=OCH$_3$, L$_\alpha$=L$_\beta$=CN$^-$
Co(II)cobester: R^1=R^2=OCH$_3$, L$_\beta$=ClO$_4^-$

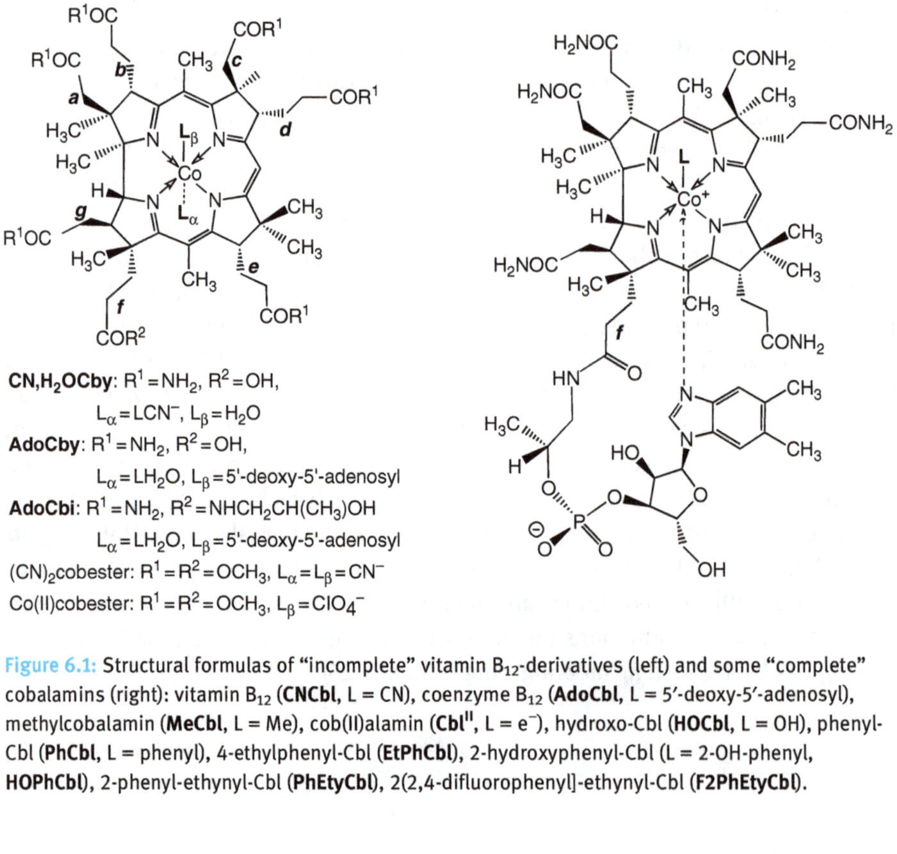

Figure 6.1: Structural formulas of "incomplete" vitamin B$_{12}$-derivatives (left) and some "complete" cobalamins (right): vitamin B$_{12}$ (**CNCbl**, L = CN), coenzyme B$_{12}$ (**AdoCbl**, L = 5'-deoxy-5'-adenosyl), methylcobalamin (**MeCbl**, L = Me), cob(II)alamin (**CblII**, L = e$^-$), hydroxo-Cbl (**HOCbl**, L = OH), phenyl-Cbl (**PhCbl**, L = phenyl), 4-ethylphenyl-Cbl (**EtPhCbl**), 2-hydroxyphenyl-Cbl (L = 2-OH-phenyl, **HOPhCbl**), 2-phenyl-ethynyl-Cbl (**PhEtyCbl**), 2(2,4-difluorophenyl]-ethynyl-Cbl (**F2PhEtyCbl**).

The crystalline "incomplete" corrinoid cyano-aquo-cobyric acid (**CN,H$_2$OCby**) played important roles, historically, as its X-ray investigation led to the first correct characterization of the corrin ligand (see [3, 25]) and it also represented the direct corrinoid target of the total synthesis of vitamin B_{12} [8, 9]. The lipophilic dicyano-Co(III)-heptamethyl-cobyrinate ("cobester") was prepared as a B_{12}-model compound [26], and its crystal structure was analyzed [27]. The crystal structure of the Co(II)-heptamethyl-cobyrinate perchlorate ("cob(II)ester") provided first detailed insights into the structure of a paramagnetic Co(II)-corrin [28], revealing its Co(II)-center as five-coordinate, as expected [5].

The "complete" corrinoids are conjugates of the natural cobyrates with unusual, B_{12}-typical α-pseudonucleotide functions, such as 5′, 6′-dimethylbenzimidazole (DMB), which coordinates the cobalt ion from the "lower" axial (or α) side in the Cbls [4, 11, 29]. In **CNCbl**, a cyanide ligand is bound at the "upper" axial coordination site (or β-face) [30]. Hence, Cbls and other "complete" corrinoids have a unique 3D geometry and they adopt a topologically chiral architecture when their α-pseudo-nucleotide function is cobalt-coordinated in an intramolecular fashion [3, 25, 31–33].

In various microorganisms, purinyl-cobamides occur as a second important class of "complete" corrinoids [34, 35], such as pseudovitamin B_{12} (a 7′-adeninyl-cobamide), factor A (a 7′-[2-methyl]adeninylcobamide) or purinyl-cobamide, in which adenine, 2-methyl-adenine or purine, respectively, replace the DMB nucleotide function of the Cbls (see Figure 6.2) [36–39]. As models for the "base-off/His-on" forms of enzyme-bound Cbls, the "complete" imidazolylcobamides were prepared, in which an imidazole substitutes for the DMB-base of the Cbls, as in

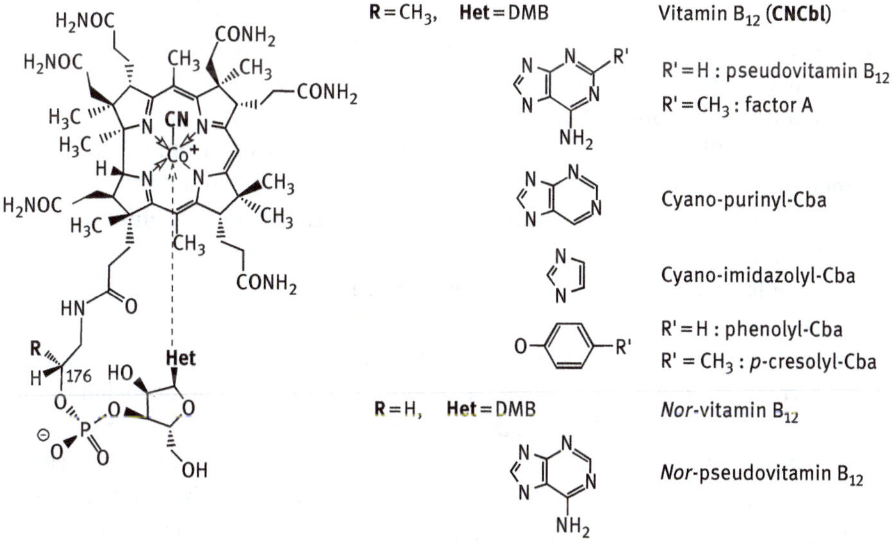

Figure 6.2: Structural formulas of natural "complete" corrinoids: cobamides (**R** = CH_3) and 176-*nor*-cobamides (**R** = H). Note: the phenol-type pseudonucleotides of phenolyl-Cba/*p*-cresolyl-Cba are noncoordinating.

Co_β-cyano-imidazolylcobamide [40]. *Nor*-pseudovitamin B_{12} (Co_β-cyano-7′-adeninyl-176-norcobamide) is a natural "complete" B_{12}-derivative lacking the methyl group at C176 of the cobamide (Cba) moiety [41]. This modification destabilizes the B_{12}-"base-on" forms [42], as was discovered and shown to be relevant in some B_{12}-dependent dehalogenases [41, 43, 44] (see below). In phenolyl- and *p*-cresolyl-cobamides, the noncoordinating aglycons phenol or 4-methylphenol, respectively, are present in their nucleotide moiety [45].

Crystallographic studies revealed both axial bonds of coenzyme B_{12} (**AdoCbl**) to be relatively long [(Co_β–C) (2.030 Å) and (Co_α–N) (2.237 Å)] [3, 31]. The crystal structure of methylcobalamin (**MeCbl**), the "second" biologically important organocobalamin, showed two axial bonds shorter than the ones in **AdoCbl** [46]. Crystal structures of a range of other Cbls and other "complete" corrinoids have been solved [11, 31, 32], one of them is the "inorganic" B_{12}-derivative aquocobalamin perchlorate (**H_2OCbl^+** ClO_4^-), which showed a remarkably short axial (Co_α–N)-bond (1.925 Å) [32, 47].

Cob(II)alamin (**Cbl^{II}**) is the product of (Co–C)-bond homolysis of coenzyme B_{12} (**AdoCbl**), and occurs during the catalytic cycle of coenzyme B_{12}-dependent enzymes. Surprisingly, the comparison of the crystal structures of the oxygen-sensitive **Cbl^{II}** [48] and of **AdoCbl** [3, 31] has revealed very similar corrin moieties [48]. These observations suggested that, in **AdoCbl**-dependent enzymes, the enigmatic protein-induced activation of the bound **AdoCbl** toward homolysis of its (Co–C)-bond could not largely come about by way of a conformational distortion of the corrin ring [49], but by strong binding of the separated homolysis fragments of **AdoCbl**, assisting the protein-induced separation of these two components [48]. In the "stretched" homologue of coenzyme B_{12}, "homocoenzyme B_{12}" (Co_β-(5′-deoxy-5′-adenosylmethyl)-Cbl), the distance between the cobalt center and C5′ of the organometallic group was increased to 2.99 Å [50]. This is roughly the same distance as the one found between the corrin-bound cobalt center and C5′ in one of the two "activated" forms of **AdoCbl** in the crystal structure of glutamate mutase [51]. Hence, "homocoenzyme B_{12}" was suggested to function as a covalent structural mimic of the hypothetical enzyme-bound "activated" state of the B_{12}-cofactor [50].

In single crystals, **Cbl^{II}** bound molecular oxygen reversibly at low temperature, giving a complex described as superoxo-cob(III)alamin [52]. Vinyl-cobalamin and *cis*-chlorovinyl-cobalamin [53] were the first organo-cobalamins with sp^2-hybridized carbon ligands studied by crystallography. The crystal structures of the recently prepared phenylcobalamins [54, 55] and of the newly available phenylalkynylcobalamins [56–59], with aromatic sp^2- or sp-carbons bound to cobalt, revealed particularly short – and apparently strong – (Co–C)-bonds.

6.2.2 The "base-on/base-off" switch of "complete corrinoids"

The DMB-base of Cbls can either be cobalt-coordinated ("base-on") or may be de-coordinated, generating the "base-off" form (see Figure 6.3) [11, 33]. Cbls and related "complete" corrinoids are, therefore, natural "molecular switches" that exist in two states with different reactivity, relevant in biological organometallic reactions [11, 33]. The particular structure (with the exceptional α-configuration) of the nucleotide moiety of Cbls and of other "complete" corrinoids allows for a stable intramolecular coordination of the heterocyclic base to the "lower" α-axial coordination site of the corrin-bound cobalt center [3, 25, 31, 32]. The DMB-nucleotide function of Cbls undergoes intramolecular cobalt coordination with negligible build-up of strain [26, 42]. A thermodynamic effect of the DMB coordination may, therefore, stabilize the "base-on" form of (organo)-Cbls significantly [29, 30, 60]. The coordinating DMB-nucleotide function also exerts a kinetic effect and steers the face selectivity at the corrin-bound cobalt center [37]. By coordinating to the "lower" face, it may direct alkylation (and other ligation) reactions (in cobalamins) to the "upper" (or β-face).

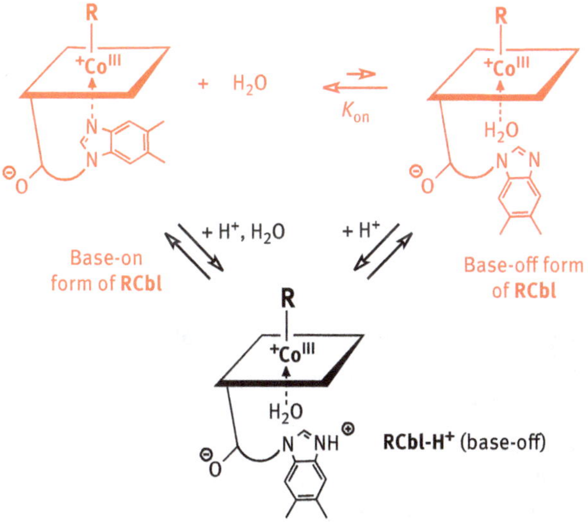

Figure 6.3: Organocobalamins act as "molecular switches" [33]. Their DMB-base is cobalt-coordinated in the "base-on" form, or de-coordinated in the less stable "base-off" form (for **R** = CH_3 (**MeCbl**): K_{on} = 93 at 25 °C [29]). Protonation of the DMB-base of **RCbl** furnishes the stable protonated "base-off" form **RCbl-H**$^+$.

A complete "base-on" to "base-off" switch results from protonation of the nucleotide base and de-coordination from the corrin-bound cobalt-ion. The associated acidity of the protonated "base-off" form **RCbl-H**$^+$ (as expressed by its pK_a) reflects, quantitatively, the strength of the intramolecular DMB coordination. The (protonated) "base-off" forms of **AdoCbl** (pK_a[**AdoCbl-H**]$^+$ = 3.67, i.e., K_{on}(**AdoCbl**) = 73 at 25 °C) and of **MeCbl** (pK_a[**MeCbl-H**]$^+$ = 2.9, i.e., =K_{on}(**MeCbl**) = 93 at 25 °C) [29, 33] are more readily accessible and less acidic than those of **CNCbl** (pK_a[**CNCbl-H**]$^+$ = 0.10) [29] and alkynyl-Cbls (pK_a[**F2PhEtyCbl-H**]$^+$ = 0.75) [57], indicating strong resistance of the "base-on" forms of the latter two against the proton-assisted de-coordination of DMB.

The existence of two forms ("base-off" or "base-on") also implies, correspondingly, a significant restructuring of the "complete" corrinoids. Their "base-on" or "base-off" forms represent isomers that may (or may not) be structured correctly for recognition by specific macromolecular B_{12}-binding partners. The complete structure of Cbls is, clearly, an effective determinant for their selective, and tight, binding by B_{12}-binding proteins [61]. This is crucial, for example, for discriminating Cbls from various other natural B_{12}-derivatives by the B_{12}-uptake and transport system of mammals, which recognizes and binds its Cbl load in the "base-on" form [21, 62]. Indeed, a protein environment may be prepared to bind a Cbl in its more stable "base-on" form [63] and to even switch the bound B_{12}-cofactors from "base-off" to "base-on" [61, 64] or, else, from "base-on" to "base-off" [65, 66] (see below).

6.3 Basic reactions and redox chemistry of organometallic B_{12}-derivatives

6.3.1 Cleavage and formation of the (Co–C)-bond of organometallic B_{12}-derivatives

Cleavage and formation of the (Co–C)-bond are basic processes of organometallic B_{12}-cofactors in solution and are key steps in the reactions catalyzed by B_{12}-dependent enzymes [11–13, 20, 29, 37, 49, 67, 68].

Cleavage and formation of the (Co–C)-bond of organo-Cbls have been observed in water in four important reaction modes that involve one or two of the three basic oxidation levels of the corrin-bound cobalt center [29, 37, 60]:

(i) The homolytic mode – essential for the cofactor role of coenzyme B_{12} (**AdoCbl**):

$$\text{5'-adenosyl-Co(III)-corrin} \rightleftarrows \text{Co(II)-corrin} + \text{5'-adenosyl radical:}$$

This reaction results in the cleavage or formation of the axial bond at the β-face of the cobalt center. In a formal sense, this reaction mode represents a one-electron reduction/oxidation of the corrin-bound cobalt center [11].

(ii) The nucleophile-induced heterolytic mode – typical of the reactivity of **MeCbl**:

$$\text{methyl-Co(III)-corrin} + \text{nucleophile} \rightleftarrows \text{Co(I)-corrin} + \text{methylated nucleophile}$$

Formally, this reaction mode involves a two-electron reduction/oxidation of the corrin-bound cobalt center and the cleavage or formation of both axial bonds [11].

(iii) The radical abstraction mode involving alkyl-corrinoids, such as **MeCbl**

methyl-Co(III)-corrin + radical ⟶ Co(II)-corrin + methylated radical

Formally, this reaction mode represents a one-electron ("inner sphere") reduction of the corrin-bound cobalt center. It is a thermodynamically highly favorable reaction involving cleavage of a weak (Co–C)-bond and formation of a strong (C–C)-bond [69].

(iv) The electrophile-induced mode of heterolysis of the (Co–C)-bond

alkyl-Co(III)-corrin + electrophile ⇌ Co(III)-corrin + alkylated electrophile

In this reaction mode, the oxidation state of the corrin-bound Co(III)-center remains unchanged; the cobalt-bound alkyl group is replaced by another ligand (in general, water or other solvent molecules).

6.3.1.1 The homolytic mode – the "program" of coenzyme B_{12}

Coenzyme B_{12} (**AdoCbl**) undergoes selective thermal homolysis of its organometallic bond readily (see Figure 6.4), and it has been considered a "reversibly functioning source of a free radical" [49]. The homolytic mode of cleavage of the (Co–C)-bond of **AdoCbl** is also particularly important for its role as a cofactor in radical enzymes (see below). Using detailed kinetic analyses of the thermal decomposition of **AdoCbl** in aqueous or glycerol solutions, the homolytic (Co–C)-bond dissociation energy (BDE) of **AdoCbl** was calculated to amount to about 31 kcal/mol [49, 70, 71]. Significant cage effects and the presence of both "base-on" and "base-off" forms

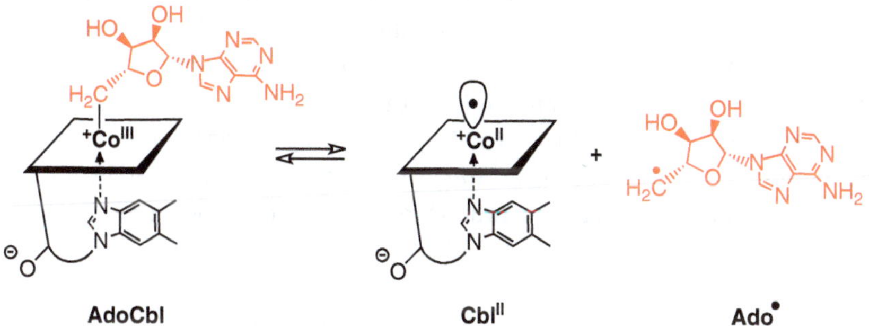

Figure 6.4: The (Co–C)-bond of coenzyme B_{12} (**AdoCbl**) cleaves by thermally reversible homolysis, furnishing the "radical trap" cob(II)alamin (**Cbl^{II}**) and the 5′-desoxy-5′adenosyl radical (**Ado$^{\bullet}$**), reversibly.

hampered the quantitative determination of the homolytic (Co–C)-BDE of **AdoCbl** [71]. The higher homolytic (Co–C)-BDE of **MeCbl** of about 37 kcal/mol has been determined in a similar way [72]. In mass spectrometric experiments, the homolytic gas-phase (Co–C)-BDEs of the "incomplete" organocorrinoids adenosylcobinamide (**AdoCbi**) and methylcobinamide (**MeCbi**) were determined as 41.5 and 44.6 kcal/mol, respectively [73].

The nucleotide-coordinated "base-on" forms of some sterically demanding organocobalamins decomposed considerably faster than their (protonated) "base-off" forms or the related "incomplete" organocobinamides [74]. Therefore, the intramolecular coordination of the nucleotide was associated with a weakening of the (Co–C)-bond of organometallic B_{12}-derivatives [49, 74]. However, the nucleotide coordination in **AdoCbl** activates the homolysis of its (Co–C)-bond by about 0.7 kcal/mol only, as derived on the basis of available thermodynamic data concerning the stabilization by DMB coordination of **AdoCbl** and of cob(II)alamin (**Cbl^{II}**) [60, 75]. Likewise, the methyl group transfer equilibrium between **MeCbl**/cob(II)inamide (**Cbi^{II}**) and methylcobinamide (**MeCbi**)/**Cbl^{II}** even indicated a slight increase by about 0.3 kcal/mol of the homolytic (Co–C)-BDE of **MeCbl**, as a result of the intramolecular DMB coordination [60].

Organometallic B_{12}-derivatives have long been known to be sensitive to daylight [76, 77]. Indeed, absorption of visible light induces very effective homolytic cleavage of the (Co–C)-bond of **AdoCbl** and **MeCbl**, as well as most of other organocorrinoids [78]. Hence, by the use of visible irradiation, organocorrinoids are a convenient source for organic radicals [79]. Interestingly, (phenyl)alkynyl-Cbls are a striking exception, and **PhEtyCbl** is inert to (Co–C)-bond cleavage upon absorption of visible light [80].

Organocobalamins are also accessible via reaction between Co(II)corrins and organic radicals. The pentacoordinated low-spin Co(II)-ion of **Cbl^{II}** is the radicaloid center of a highly efficient, persistent "radical trap" [48]: **Cbl^{II}** reacts with alkyl radicals with negligible restructuring of the cobalt corrin moiety, furnishing **AdoCbl**, and other organo-Cbls, by direct formation of the (Co–C)-bond in the "reverse homolytic" (or radical recombination) mode [48]. The structural features of **Cbl^{II}** also help to rationalize its remarkably fast reaction with alkyl radicals (such as the 5′-deoxy-5′-adenosyl radical), as well as the diastereospecificity for the reaction at the β-face [11, 81]. **Cbl^{II}** even traps the very short-lived acetyl radicals efficiently, providing an effective synthetic route to acetyl-cobalamin [82] by "slaving-in" radicals [83].

6.3.1.2 Formation and induced cleavage of the (Co–C)-bond of methyl corrinoids

Typical alkylating agents react efficiently via bimolecular nucleophilic substitution (S_N2) reaction with the highly nucleophilic Co(I)-corrins [84, 85]. This mechanism, probably, represents the best established mode of formation of the (Co–C)-bond (see Figure 6.5) and is particularly relevant in typical enzyme-catalyzed methyl transfer reactions, as well as in the biosynthesis of adenosyl-corrinoids [12, 68, 86, 87]. In a similar way, attack of nucleophiles on methyl-corrinoids may induce effective heterolytic cleavage of the (Co–C)-bond. Indeed, the S_N2 path of formation and cleavage of (Co–C)-bonds is, biologically, broadly relevant and takes place without the intermediary existence of free methyl or alkyl cations.

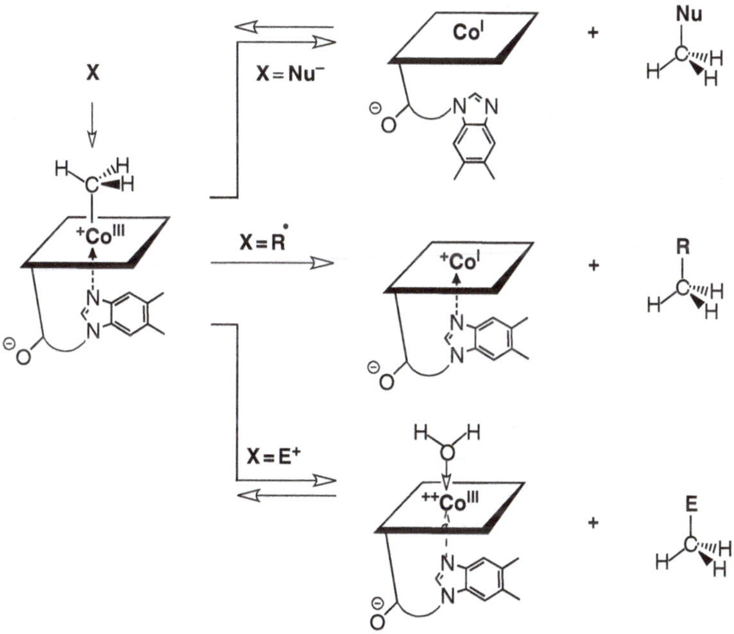

Figure 6.5: Important modes of induced cleavage and formation of the (Co-C)-bond of the methyl corrinoid **MeCbl**. Abstraction of the cobalt-bound methyl group of **MeCbl** may be induced by nucleophiles (X = **Nu**⁻), radicals (X = **R**·) or electrophiles (X = **E**⁺).

Alkylation of cob(I)alamin (**Cbl**$^{I-}$) normally proceeds via the "classical" S_N2 mechanism, where the Co(I)-corrin **Cbl**$^{I-}$ acts as a strong nucleophile [68, 84, 87, 88]. The high nucleophilicity of the corrin-bound Co(I)-center is an inherent property of Co(I)-corrins and is virtually independent of the presence, or absence, of a (cobalt-coordinating) nucleotide base. Both "complete" and "incomplete" Co(I)-corrins react with similar rates and strong preference for their β-face. The latter is, therefore, the more nucleophilic of the two diastereofaces of the corrin-bound Co(I)-center [37, 88]. In certain cases, however, alkylation with Co(I)-corrins occurs via a two-step one-electron transfer path, where Co(I)-corrins act as strong one-electron reducing agents. This mechanism involves Co(II)-corrin and radical intermediates, resulting in aberrant (kinetic) face selectivity in the formation of the (Co–C)-bonds [55, 75, 88].

The nucleophile-induced demethylation and heterolysis of the (Co–C)-bond of methyl-Co(III)-corrins is the, biologically, important basis for methylation by **MeCbl**. It furnishes a strongly reducing Co(I)-corrin [68], and represents, formally, a reductive *trans*-elimination at cobalt [37]. Strong nucleophiles, such as thiolates, demethylate the "incomplete" **MeCbi**$^+$-ion approximately 1,000 times faster than **MeCbl** [89], reflecting the strong stabilization by the coordinated DMB-nucleotide in **MeCbl**, which amounts to about 4 kcal/mol [37, 60]. This effect is of relevance also for enzymatic methyl-group transfer reactions involving protein-bound methyl-Co(III)-corrins, where the coordinated histidine ligand plays a significant role [68, 90].

A recently recognized further mode of cleavage of the (Co–C)-bond of **MeCbl** involves the abstraction at the cobalt-bound methyl group by a radical (see Figure 6.5) [11, 69]. This type of substitution reaction has been observed in the reaction of a malonyl-methyl-radical with **MeCbl** and may allow for a second biological role of methyl-corrinoids [69]. Related radical abstraction mechanisms of the formation of (C–C)-bond have been deduced with other organo-corrinoids [91].

Indeed, the abstraction of the cobalt-bound methyl group of **MeCbl** by an alkyl radical is thermodynamically very favorable. It is also surprisingly effective kinetically. An enormous number of unusual biological (C–C)-bond-forming reactions [92] and biosynthetic methylations at inactivated carbon centers have recently been reported and have been proposed to proceed by B_{12}- and *S*-adenosylmethionine (SAM)-dependent radical enzymes (so-called class B radical SAM enzymes) [93–95]. Hence, the abstraction of the methyl group of **MeCbl** by radicals [69] is now considered a likely step in the biochemical mechanism of the "class B" radical SAM methyl transferases (see below).

In a formally related radical abstraction reaction, the cobalt-bound methyl group of methyl corrinoids, such as **MeCbl**, is rapidly abstracted by Co(II)corrinoids, such as **Cbi**^{II}, giving, for example, **MeCbi** and **Cbl**^{II} [60]. This type of reaction does not involve free methyl radicals and, under appropriate conditions (aprotic solvents), it is (even) insensitive to the presence of molecular oxygen [96].

The electrophile-induced dealkylation of the cobalt-bound methyl group of **MeCbl** by polarizable metal ions, such as Hg^{2+} ions, is an environmentally important (possibly nonenzymatic) path to the poisonous $Hg^{II}–CH_3$ cation [97, 98] (see Figure 6.5). The coordination of the DMB nucleotide modifies the reactivity of the cobalt center by enhancing the ease of abstraction by electrophiles, in both a kinetic and thermodynamic sense [75, 98]. The (Co–C)-bond of alkyl-Co(III)-corrins is remarkably inert against proteolytic cleavage under physiological conditions, and acid-induced heterolytic cleavage of the (Co–C)-bond of **MeCbl** is unknown. However, the proteolysis of the (Co–C)-bond of **AdoCbl** occurs slowly at low pH in aqueous solution [99]. Likewise, (phenyl)-alkynyl-Cbls are slowly cleaved in acidic medium by proton-induced heterolysis of the (Co–C)-bond, resulting in the formation of (substituted phenyl)ethyne and aquo-Cbl [56].

6.3.2 Redox chemistry of B_{12}-derivatives

Under physiological conditions, vitamin B_{12}-derivatives exist as Co(III)-, Co(II)- or Co(I)-corrins, each oxidation state possessing different coordination properties and, correspondingly, differing structure and reactivity [5, 11, 29, 37]. Electrochemistry has been used for determining the crucial redox potentials in solution [100], for controlled electrosynthesis of organometallic B_{12}-derivatives [101], for generation of reduced forms of protein-bound B_{12}-derivatives [102] and of electrode-bound B_{12}-derivatives for analytical applications [103].

Axial coordination to the corrin-bound cobalt center depends on the formal oxidation state of the cobalt ion, and, as a rule, the number of axial ligands decreases with the cobalt oxidation state [11, 32, 100]. In the thermodynamically predominating forms of cobalt-corrins, two axial ligands are bound at the diamagnetic Co(III) center. The paramagnetic (low-spin) Co(II) has one axial ligand bound, while axial ligands are not bound, or only very weakly, at the diamagnetic Co(I)-center (see Figure 6.6). Electron transfer reactions involving B_{12}-derivatives are, therefore, accompanied by a change in the number of axial ligands, which considerably influences the thermodynamic and kinetic features of the redox processes of cobalt corrins (reviewed in [100, 101]).

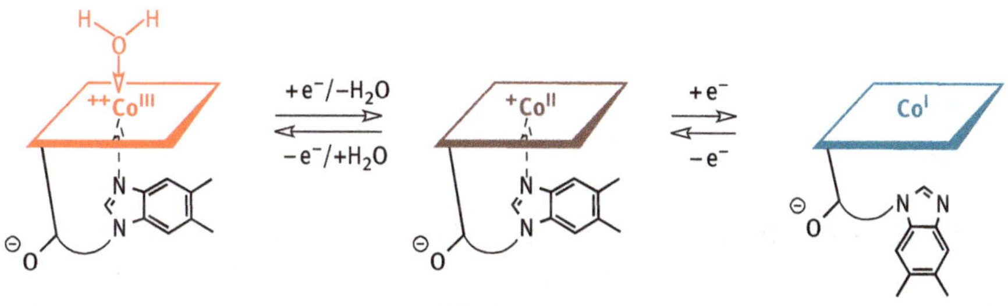

6-Coordinate Co(III)-center aquocobalamin (**H_2OCbl^+**) | 5-Coordinate Co(II)-center cob(II)alamin (**Cbl^{II}**) | 4-Coordinate Co(I)-center cob(I)alamin (**Cbl^{I-}**)

Figure 6.6: Reversible one-electron reduction of the red Co(III)-corrin **H_2OCbl^+** converts it to the brown Co(II)-corrin **Cbl^{II}**, which may be reduced reversibly at more negative potential to the green Co(I)-corrin **Cbl^{I-}**. The reduction steps are accompanied by de-coordination (loss) of one axial ligand (oxidation steps, in reverse, by the coordination of axial ligands).

The electrochemistry of organometallic studies of B_{12}-derivatives is generally complicated due to a rapid and irreversible loss of the organic ligand upon reduction [100,

104]. Organo-corrinoids are, hence, labile to strong one-electron reducing agents, and the (Co–C)-bond of **MeCbl** is weakened strongly by the reduction [72, 100, 105]. However, the standard potential of the typical Co(III)/Co(II)-redox pair of organometallic B_{12}-derivatives is strikingly more negative than that of $\mathbf{Cbl^{II}}$/$\mathbf{Cbl^{I-}}$ and out of the reach of biological reductants [100]. Thus, further reduction of organometallic Co(III)-corrins does, typically, not occur at the potentials needed for electrogeneration of the highly nucleophilic Co(I)-corrins. As a consequence, the selective electrochemical production of Co(I)-corrins in the presence of suitable alkylating agents is an efficient and selective means for preparing alkyl-corrinoids [101] and other complex organo-corrinoids [33, 50, 106, 107]. However, organo-corrinoids with electron-withdrawing substituents in their cobalt-bound organic ligand are reduced more easily than **MeCbl** [108, 109], often rendering it difficult to prepare such organo-Cbls by alkylation of the strongly reducing $\mathbf{Cbl^{I-}}$ as a nucleophilic intermediate [110].

Methyl- and adenosyl-corrinoids, such as **MeCbl** and **AdoCbl**, are often observable in functioning enzymes [11], where such "complete" corrinoids may be bound in their characteristic "base-on" [63, 111], "base-off/His-on" [65, 66, 112] or "base-off" forms [113]. The important protein-bound cofactor form $\mathbf{Cbl^{II}}$ has also been characterized by crystallography, eventually (see, e.g., [114, 115]). However, for reasons of its thermodynamic instability and inaccessibility under typical physiological conditions, the $\mathbf{Cbl^{I-}}$ form exists but fleetingly and has only rarely been observed in enzyme reactions [12, 87, 116].

6.4 Enzymatic reactions with organometallic B_{12}-cofactors

6.4.1 Enzymes dependent on coenzyme B_{12} and related adenosylcobamides

In the classical **AdoCbl**-dependent enzymes, 5′-deoxy-5′-adenosyl radical (**Ado·**) is the actual reactive agent, which originates from homolysis of the (Co–C)-bond of **AdoCbl**. The structurally sophisticated **AdoCbl** functions as a reversible source for the **Ado**-radical and acts as a "pre-catalyst" (or catalyst precursor) [11, 117]. On the other hand, the question has been a matter of a controversial discussion, whether $\mathbf{Cbl^{II}}$ (the Co(II)-corrin fragment from **AdoCbl** homolysis) is a mere "spectator" in the rearrangement step of the isomerase processes, or takes part as a participating "conductor" [118, 119]. Without a doubt, the critical isomerization reactions rely mainly on the reactivity of bound organic radicals, which are formed (directly or indirectly) by abstraction of an H-atom from the substrate by **Ado**.

Ten B_{12}-dependent isomerases and a B_{12}-dependent ribonucleotide reductase are known, which use **AdoCbl** (or an AdoCba) as cofactors of radical enzymes

[11, 20, 67, 111, 120]. These isomerases are five carbon skeleton mutases named methylmalonyl-CoA mutase (MCM) [120], ethylmalonyl-CoA-mutase [121], glutamate mutase [51], methylene glutarate mutase [13] and isobutyryl-CoA-mutase [122], the two isomerases diol dehydratase and glycerol dehydratase [111], ethanolamine ammonia lyase [111], and the two aminomutases ornithine-4,5-aminomutase and D-lysine/L-β-lysine-5,6-aminomutase [20, 123]. The **AdoCbl**-dependent enzymes are unevenly distributed in living organisms. MCM (see Figure 6.7) is the only **AdoCbl**-dependent enzyme required for a functioning metabolism of humans and other mammals [12].

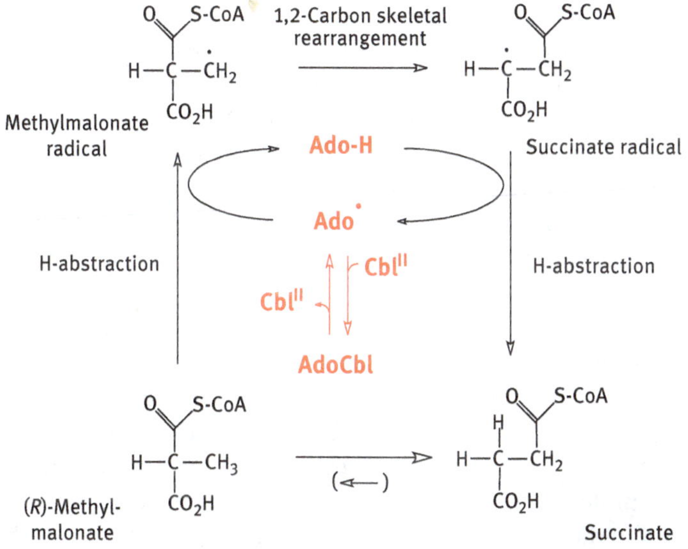

Figure 6.7: Methylmalonyl-CoA mutase (MCM) is relevant in human metabolism, where this carbon skeleton mutase converts (*R*)-methylmalonyl-CoA to succinyl-CoA by a radical process. It is induced by (Co–C)-homolysis of enzyme-bound **AdoCbl**, with simultaneous formation of **Cbl^{II}** and of the 5′-deoxy-5′-adenosyl radical (**Ado·**). The proposed mechanism involves H-atom abstraction by **Ado·**, furnishing the 2-methyl-malon-2′-yl-CoA radical (top left). This enzyme-bound radical undergoes carbon skeleton rearrangement to the succin-3′-yl-CoA radical (top, right), which abstracts an H-atom from 5′-deoxyadenosine (**Ado-H**), providing succinyl-CoA and regenerating **Ado·**, ready for recombination with **Cbl^{II}**. The sequence of steps of this reversible enzyme reaction is only depicted in the figure in the discussed "forward" direction.

The light-sensitive organometallic **AdoCbl** has also found use in nature in amazing photoregulatory roles in bacteria. As discovered recently, exceptional ("nonclassical") **AdoCbl**-dependent DNA-binding enzymes repurpose the efficient light-induced cleavage of the (Co–C)-bond of **AdoCbl** in some microorganisms, in order to achieve very effective photoregulation of gene expression (see below) [76, 124–126].

6.4.1.1 Adenosylcobamide-dependent isomerases

The adenosylcobamide-dependent enzymes make use of the protein-activated homolysis of the weak (Co–C)-bond of **AdoCbl**, furnishing **Cbl^{II}** and the 5′-deoxy-5′-adenosyl radical (**Ado·**) [117]. The primary radical **Ado·** is tightly protein-bound and abstracts an H-atom from its substrate with high selectivity, inducing further exceptional enzymatic processes [13]. Indeed, adenosylcobamide-dependent enzymes perform chemical transformations that are difficult to achieve by typical "organic reactions." With the exception of the enzymatic ribonucleotide reduction [127], the results of adenosylcobamide-catalyzed enzymatic processes correspond to isomerization reactions with vicinal exchange of a hydrogen atom and of a migrating group with a variety of heavier atom centers.

The homolysis of the (Co–C)-bond of the protein-bound organometallic "radical-starter" **AdoCbl** needs to be activated, and accelerated, by a factor of about 10^{12} to come up with the observed reaction rates of the **AdoCbl**-dependent enzymes [49, 70]. The deduced dramatic destabilization of the bound organometallic cofactor toward homolysis of the (Co–C)-bond and its mechanism are intriguing, and still much discussed, facets of the **AdoCbl**-dependent enzymes [48, 51, 111, 120, 128, 129]. Covalent restructuring of the bound cofactor (except for the formation of the "base-off/His-on"-form in the carbon skeleton mutases) is not indicated. In addition, protein and solvent molecules can only weakly stabilize a radical center [130]. Activation of the (Co–C)-bond toward homolysis may come about largely from a protein- and substrate-assisted separation of the largely non-strained homolysis fragments, a 5′-deoxy-5′-adenosyl radical and **Cbl^{II}** (in either a "base-off/His-on" or "base-on" form), and through strong binding of the separated fragments by the protein [48, 50, 128, 131]. The existence in some of these enzymes of a binding interface (e.g., of an "adenosine-binding pocket"), which implies a mode of strained binding of the organometallic moiety, helps to support this picture [50, 112].

Adenosylcobamide-dependent enzymes come in two structural classes with respect to the bound B_{12}. In one of them, the B_{12}-cofactor is bound "base-on," as found for example, in diol-dehydratases and in B_{12}-dependent ribonucleotide reductase [111, 127, 132]. However, as discovered in MCM, the B_{12}-cofactor is bound in a "base-off" (and "His-on") form in the carbon skeleton mutases [66]. Fixed placement of the corrin moiety at the interfaces of the B_{12}-binding and substrate-binding/activating domains of MCM appears to be of high significance. The "regulatory His–Asp–Ser triad," provided by one protein domain (or subunit), may not be involved in any proton transfer steps in the mutases and, probably, conserves its structure largely during enzymatic turnover. Indeed, "electronic effects" of the axial *trans* ligand on the (Co–C)-bond homolysis in **AdoCbl** and in **MeCbl** are less important [37]. The proper substrate-to-product rearrangement steps of **AdoCbl**-dependent enzymatic rearrangements are accomplished by tightly protein-bound radicals that are controlled in their reaction space in the second protein domain (or subunit) [11, 117]. The major functions of the protein part of the enzyme concern not only the assistance in its proper reactions (by activation of protein-bound **AdoCbl**) but also the reversible generation of the radical intermediates and the protection of its protein environment from nonspecific radical chemistry, a role classified as "negative catalysis" [130].

6.4.1.2 Coenzyme B_{12} as light receptor in the photoregulation of gene expression

Coenzyme B_{12} (**AdoCbl**) is a notoriously light-sensitive compound. This particular photochemical property of **AdoCbl** is used in a remarkable type of photoregulation of gene expression, a complete twist from the known biological roles of adenosylcobamides in radical enzymes to the function of a light-sensing cofactor [124, 133, 134]. In the bacterium *Myxococcus xanthus*, the **AdoCbl**-based photoreceptor CarH regulates biosynthesis of the photoprotecting carotenoids. CarH is a representative of an apparently abundant class of B_{12}-based bacterial photoregulators [135]. How the organometallic B_{12}-cofactor **AdoCbl** could be repurposed to play a broadly relevant light-sensing gene-regulatory role has been largely clarified in very elegant crystallographic, mutational and mechanistic studies [125, 136, 137]. These investigations have shown that **AdoCbl** is not directly involved in DNA binding [125]. The mechanism of gene regulation by CarH relies on the modulation of the structure of the protein through interaction with the organometallic B_{12}-cofactor. Under low-light conditions, holo-CarH forms a dimer-of-dimer-type tetramer. In this state, CarH, with intact **AdoCbl** bound, binds to the promoter region of biosynthesis genes with high affinity, which are coding for carotenoid photoprotectors, inhibiting their transcription. Upon photolytic cleavage of the (Co–C)-bond of **AdoCbl**, the adenosyl group is lost as unreactive 4′,5′-anhydroadenosine [136]. A conformational change of the protein leads to the disintegration of the protein tetramer, which can no longer bind strongly to DNA [125]. It is still a puzzle how the protein moiety "reprograms" the path of the light-triggered cleavage of the (Co–C)-bond to the here observed unique ("heterolytic") mode, thus repurposing **AdoCbl** perfectly for effective photoregulation [76, 138]. Another **AdoCbl**-dependent photoregulator, AerR, has surfaced in the photosynthetic bacterium *Rhodobacter capsulatus* [139]. The B_{12}-binding protein AerR operates by a scheme related to the one in CarH [125], but interacting with CrtJ, the regulator of genes coding for tetrapyrrole biosynthesis [139]. Several other bacteria also appear to use related B_{12}-dependent photoregulators [126].

6.4.2 Enzymes catalyzing B_{12}-dependent methyl group transfer

B_{12}-dependent methyltransferases are widespread in nature and play important biosynthetic roles in a broad variety of organisms. Two basic biochemical mechanisms of methyl group transfer are now established [87, 93]. B_{12}-dependent methionine synthases use nucleophile-induced heterolytic methyl group transfer steps and occur in many organisms, including humans [87]. Related B_{12}-dependent methyltransferases operate broadly in the basic one-carbon metabolism of many anaerobic microorganisms, most notably in methanogens [140] and in acetogens [141] (in an organometallic pathway of CO_2 fixation) [102, 142–144], as well as (in other anaerobic microorganisms) in acetic acid catabolism to methane and CO_2 [145, 146]. The activities of the Co(I)-forms of the B_{12}-cofactors as nucleophiles and of the methyl-Co(III)-forms as methylating agents provide the basis for the catalysis of these enzymatic methyl group transfer reactions [11, 29, 68]. Alternatively, bifunctional B_{12}- and SAM-dependent radical methyl transferases (now classified as "class B" radical SAM methyl transferases) were also recognized recently as an important group of biosynthetic methyl group transferases [93–95, 147], as also discussed below.

6.4.2.1 B_{12}-dependent methionine synthase

B_{12}-dependent methionine synthase (MetH) is a widespread enzyme and, probably, the most extensively studied B_{12}-dependent methyltransferase [12, 87, 148]. The enzyme MetH from *Escherichia coli* has been a particularly useful representative of the B_{12}-dependent methionine synthases [87, 148]. MetH is a basic model also for other methyl transferases that operate via a nucleophile-induced methyl group transfer [142]. Methyl group transfer, catalyzed by MetH, follows a two-step ping-pong mechanism. In a first step, the protein-bound **MeCbl** is de-methylated by activated homocysteine, furnishing protein-bound $\mathbf{Cbl^{I-}}$ and methionine. In the second step, activated *N*-methyl-tetrahydrofolate remethylates the protein-bound $\mathbf{Cbl^{I-}}$, generating tetrahydrofolate and regenerating the protein-bound **MeCbl** (see Figure 6.8) [142].

The two steps proceed with an overall retention of configuration at the methyl group carbon, consistent with two S_N2-type nucleophilic displacement steps, each occurring with inversion of configuration [68, 149]. These two methyl group transfer steps are considered to be subject to the strict geometric control of S_N2 reactions, that is, to require in-line arrangements of the incoming nucleophile, of the transferred CH_3 group and of the leaving group. The methyl group transfer catalyzed by MetH involves (nucleophile-induced) heterolytic cleavage and formation of the (Co–CH_3)-bond, and represents, in a formal sense only (!), a methyl "cation" transfer. However, in such S_N2 processes free methyl cations, or radicals, are excluded as intermediates.

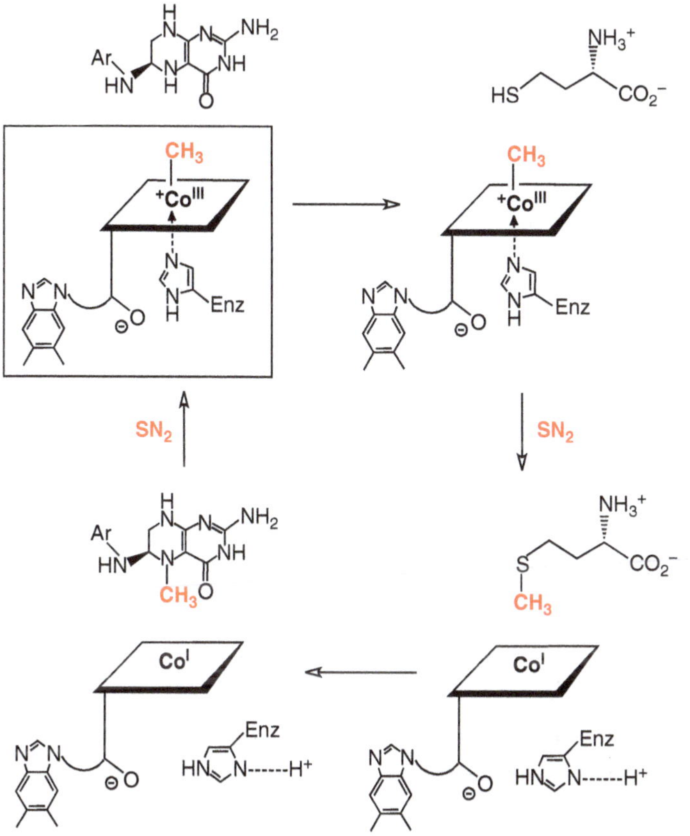

Figure 6.8: Methionine synthase (MetH) catalyzes the Cbl-dependent formation of methionine from homocysteine and demethylation of N^5-methyltetrahydrofolate to tetrahydrofolate, involving the protein-bound Cbls **MeCbl** in a "base-off/His-on"-state and **Cbl**$^{I-}$ (the imidazole ring symbolizes His759 of MetH). The heterolytic methyl group transfer occurs in a ping-pong mechanism via two nucleophilic substitution (S_N2) steps. The resting state of MetH with **MeCbl**-bound "base-off/His-on" is highlighted on the top left.

During turnover, striking structural changes accompany the transitions of the enzyme MetH between its state with (tetracoordinate) **Cbl**$^{I-}$ bound and the one with (hexacoordinate) "base-off/His-on" **MeCbl** (see Figure 6.8). The protein environment plays a crucial role in controlling substrate positions, as well as in providing access to the catalytic center [142].

The X-ray crystal analysis of the B_{12}-binding domain of MetH provided the first insight into the three-dimensional structure of a B_{12}-dependent enzyme [65, 150]. A true eye-opener of this work was the finding that the cobalt-coordinating DMB-nucleotide tail of the protein-bound organometallic cofactor **MeCbl** was displaced at cobalt by the histidine of a conserved His–Asp–Ser triad [65, 150]. Consequently, in MetH, the corrinoid cofactor is bound in a "base-off/His-on" mode. In various other B_{12}-dependent methyltransferases, the methyl-Co(III)-corrinoid cofactor has been observed in a "base-off/His-on" binding mode, or even in a "base-off" form (i.e., without DMB- or His-coordination) [151].

The axial bond of the histidine of the His–Asp–Ser triad to the protein-bound **MeCbl** in MetH helps to position the corrinoid cofactor for methyl group transfer [150, 152]. A thermodynamic role of the histidine coordination in the methyl transfer reactions of MetH has also been discussed [75, 87, 90, 150]. Indeed, a significant thermodynamic *trans*-effect of the DMB coordination in **MeCbl** on heterolytic methyl group transfer reactions has been observed in aqueous solution [37, 60, 75]. The coordinating DMB-ligand stabilized **MeCbl** and opposed nucleophilic abstraction of the methyl group by about 4 kcal/mol [60]. The His–Asp–Ser triad may, furthermore, represent a "relay" for H^+-uptake/release accompanying the enzymatic methylation/demethylation cycles [152, 153]. It may, thus, fine-tune the bound corrinoid for enzyme catalysis: weakening of the axial (Co–N)-bond activates both the methyl group of **MeCbl** for abstraction by a nucleophile and adventitiously formed **Cbl^{II}** for re-reduction to protein-bound **Cbl^{I-}**.

6.4.2.2 B_{12}- and SAM-dependent radical methyl transferases

A puzzling biochemical discovery, made in the late 1980s, concerned the incorporation of intact methionine-derived methyl groups in the course of the biosynthetic methylation at saturated carbon positions of some antibiotics [154]. This scenario was incompatible with a nucleophilic mechanism like the one of MetH, suggesting a new biological path for methylation. The eventual identification of a class of abundant enzymes with protein signatures of "radical" SAM enzymes [155] was consistent with the broad biosynthetic involvement of radicals [156, 157]. The thermodynamically very favorable direct abstraction of the cobalt-bound methyl group of **MeCbl** by a carbon radical was conceived as a first model radical methylation reaction [69]. Indeed, the primary (malonate-derived) radical underwent efficient and, kinetically, very effective methylation by **MeCbl**, suggesting methylation of radicals to be an important "second" biological role of methyl-corrinoids [69].

Evidence for a large subclass of B_{12}-dependent "radical" SAM enzymes has been obtained in the meantime, many (but not all) of which appear to be B_{12}-dependent methyltransferases [93, 95, 158, 159]. Specific methylation of radicals by **MeCbl** is now considered an accepted mechanism for the methyltransferase Fom3 in the course of the biosynthesis of fosfomycin (see Figure 6.9) [95].

Fom3 methylates C2 of the fosfomycin precursor 2-hydroxy-ethylphosphonate stereoselectively. Overall, this methylation reaction occurs with loss of H_R to give (*S*)-

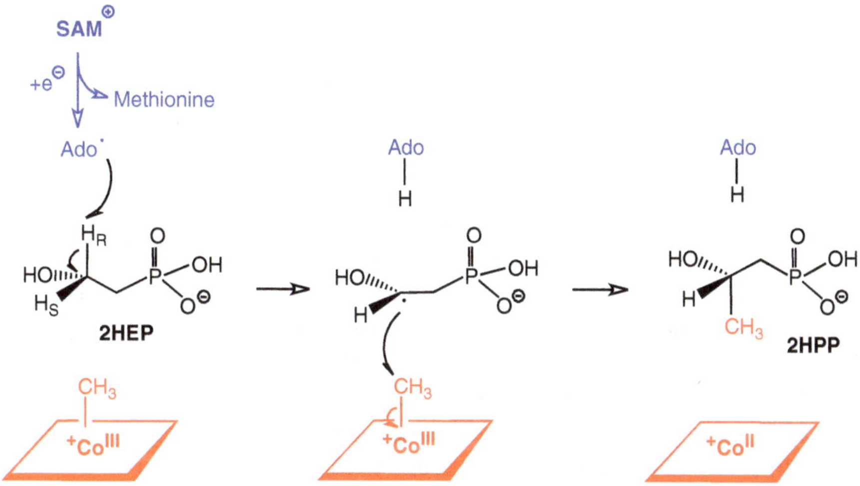

Figure 6.9: The **MeCbl**- and radical SAM-dependent methyl transferase Fom3 catalyzes the methylation of 2-hydroxyethylphosphonate (2HEP) at C2, which generates (*S*)-2-hydroxypropylphosphonate (2HPP).

2-hydroxypropyl-phosphonate and is, thus, indicated to involve stereochemical inversion at carbon-2 of the phosphonate. Fom3 was shown to consist of a B_{12}-binding domain and a "radical" SAM domain [95, 160]. A coupled H-atom abstraction/methylation process was proposed for the formation of (*S*)-2-hydroxypropyl-phosphonate. This involves abstraction of an H-atom (H_R) from hydroxyethyl-2-phosphonate by **Ado·**, produced in the radical SAM domain, and generation of the short-lived hydroxyethyl-2-phosphonate radical. Rapid, stereoselective methylation of the latter radical by **MeCbl** in the B_{12}-binding domain would lead to (*S*)-2-hydroxypropyl-phosphonate (see Figure 6.9) [11, 95]. The observed stereochemical inversion (at C-2 of hydroxyethyl-2-phosphonate) is consistent with a preorganized enzyme, in which the substrate (hydroxyethyl-2-phosphonate) is sandwiched between the radical-generating SAM and the methylating B_{12}-domain that presents **MeCbl** from its β-face. The biosynthesis of fosfomycin and of some other natural products (see [93, 95, 161]) is now indicated to depend upon striking radical methylation processes in which **MeCbl** and other methyl corrinoids would serve as methylating reagents. However, in other cases, the class B radical SAM methylases appear to methylate their substrates by yet less established pathways [93]. A biochemical curiosity, at present, is the B_{12}- and SAM-dependent radical enzyme that catalyzes the (furane to oxetane) ring contraction step in the biosynthesis of the bacterial antibiotic Oxetanocin A, in which a function for the B_{12}-cofactor is not yet implicated [162].

6.4.3 B_{12}-dependent dehalogenases in anaerobes

The ability of some anaerobic microorganisms, using reduced corrinoids as cofactors to dehalogenate haloalkanes and haloaromatics reductively, is a globally relevant biological feature [163, 164]. A variety of environmentally relevant dehalogenation reactions have been described, such as those of chloromethane [165], of chloroethenes [43], of trichloroethane [166], of hexachlorocyclohexane [167] and of chlorinated phenols [12, 168]. The anaerobic bacterium *Sulfurospirillum multivorans* dechlorinates tetrachloroethene by its B_{12}-dependent tetrachloroethene-reductive dehalogenase [43]. This membrane-bound dehalogenase uses a reduced form of an unusual "complete" corrinoid cofactor, isolated as *nor*-pseudovitamin B_{12} [41], in order to reduce tetrachloroethene to trichloroethene (first) and to *cis*-dichloroethene [43] (see Figure 6.10). The *nor*-cobamide from *Dehalospirillum multivorans* was about 50 times more active than Cbls in an in vitro reduction of trichloroacetate [169], and addition of DMB into the growth medium, inducing formations of *nor*-cobalamins (*nor*-Cbls), actually reduced bacterial growth [44]. In aqueous solution, *nor*-cobamides (and *nor*-Cbls) show a lesser preference than Cbls, to assemble to B_{12} "base-on" forms [42], enabling redox reactions with *nor*-cobamides to occur at potentials less negative than those of the corresponding Cbl redox couples [170].

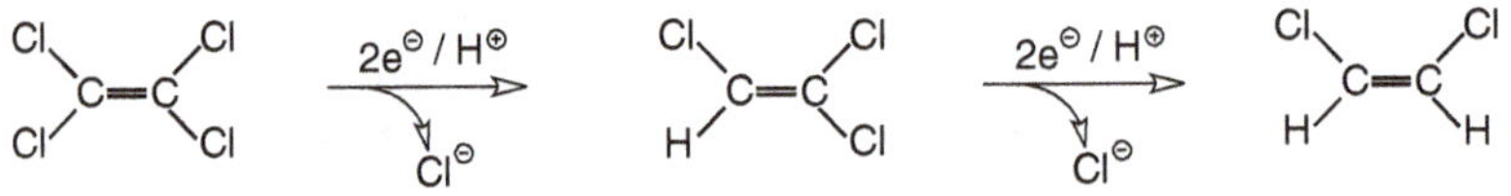

Figure 6.10: Tetrachloroethene reductive dehalogenase of *Sulfurospirillum multivorans* reduces tetrachloroethene to trichloroethene (first), and to *cis*-dichloroethene (subsequently), by *nor*-B_{12}-mediated electron transfer reactions.

Recent X-ray analytical studies have revealed the structures of the tetrachloroethene reductive dehalogenase of *S. multivorans* [169] with the "base-off" Co(II)-form of *nor*-pseudovitamin B_{12} bound [43], as well as of a dehalogenase associated with respiratory chlorophenol dechlorination, with the "base-off" form of **Cbl^{II}** bound [171]. In the case of PCE of *S. multivorans*, the Co(I)-form of the enzyme is proposed to reduce the haloalkene via a long-distance electron transfer [172], whereas, in the latter situation, the structural data suggest an "inner-sphere" reduction by **Cbl^{I}** [75] of the halophenol via a formal halogen atom transfer [171]. Formation of organometallic intermediates, as considered earlier [53, 168, 173], appears unlikely in these two enzymes. Interestingly, the B_{12}-binding region of both dehalogenases is structured similarly as in the B_{12}-processing enzyme CblC [57, 113] (see below).

6.5 Organometallic B_{12}-derivatives as ligands in B_{12}-binding proteins and nucleic acids

Corrinoids, such as the Cbls, are synthesized exclusively by some "primitive" microorganisms. Humans, and other omnivorous vertebrates, have evolved a highly complex uptake mechanism to efficiently absorb and transport Cbls to each cell, and to further process the Cbls delivered there [10, 62]. The question, how dietary Cbls are transported to cells in humans, has been thoroughly studied over the years: the process is accomplished by the three Cbl-binding proteins, haptocorrin (HC), intrinsic factor (IF) and transcobalamin (TC), which operate in largely successive steps, as well as by corresponding membrane-bound receptors for the B_{12}-protein complexes (see below) [21, 62].

The metabolism of archaea of many bacteria and algae also depends on B_{12}-derivatives, such as Cbls, or may take advantage of them, and the growth of some bacteria [174] and algae [35] is boosted strongly by external supply with corrinoids. The Gram-negative bacterium *E. coli* can synthesize its needed organometallic Cbls from "incomplete" cobinamides. *E. coli* has been used as a model organism for studying bacterial B_{12}-import and biosynthesis. It acquires corrinoids available externally via its complex "B-twelve-uptake" (Btu) system [174, 175]. The Btu system accomplishes the import of corrinoids, and its regulation, by engaging several proteins, first of all, the outer membrane transporter BtuB [174, 175].

6.5.1 B_{12}-binding proteins in uptake and transport in mammals

In humans, freely available dietary corrinoids are first bound by HC, which is secreted in saliva. However, the major part of these corrinoids is bound in food components until liberated by the combined action of low pH and proteolysis by pepsin in the stomach [62]. A second transport protein, IF, produced in the stomach, traps the free Cbls (also those eventually released from HC) in the duodenum after neutralization of the gastric juice. IF strongly selects for "base-on" Cbls [61] and transports them to the terminal ileum. There, the IF–Cbl complex docks to cubam ("cubilin-amnionless," the receptor complex at the brush border membrane of enterocytes) [176], and undergoes receptor-mediated endocytosis. Internalized Cbl-loaded IF is degraded by lysosomal enzymes, and the released Cbl emerges into the portal circulation, where it is trapped by TC [21]. Cells absorb TC–Cbl complexes from plasma by the specific endocytic TC receptor [177]. The Cbls are then released in the cells, for eventual further processing to the organometallic B_{12}-cofactors [10].

Hence, the three Cbl-binding proteins, HC, IF and TC, are crucial for transport of the dietary Cbls. The protein moieties of IF, HC and TC have molecular weights of about 45 kDa and share high sequence identity, especially near the cobalamin-binding pocket (IF and HC are heavily glycosylated, in contrast to TC) [62]. The proposed two domain architectures of these B_{12}-transporters [178] and the "base-on" nature of the bound Cbls [61] were verified in crystal structure analyses of the three Cbl-binding proteins [179–181]. The larger N-terminal "α-domain" spans ~300 residues and consists of 12 α-helices. The smaller C-terminal "β-domain" is composed of β-strands mainly, providing a good first binding interface for "base-on" Cbls [62, 182, 183]. Cbl-binding induces strong tightening and mutual structuring of the protein domains in the holoprotein (Figure 6.11).

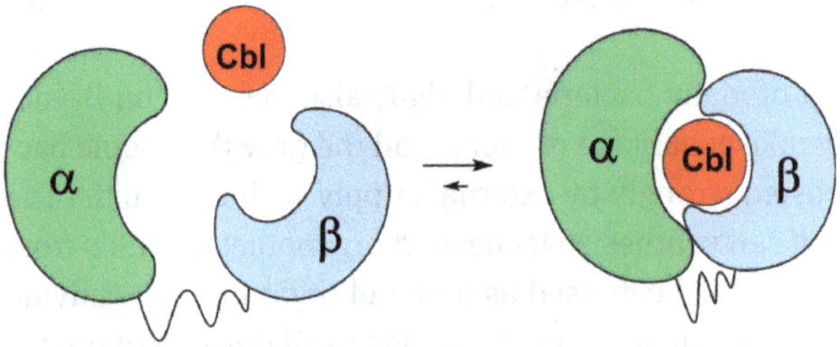

Figure 6.11: Qualitative model of the human Cbl-binding proteins, haptocorrin, intrinsic factor and transcobalamin, and highlighting the induced restructuring from Cbl-binding.

X-ray crystallographic studies of all three holo-proteins showed Cbls-bound "base-on" at the interface of their two structured domains connected by a linker (see Figure 6.11). The biostructural studies also indicated significant tolerance for binding of organometallic Cbls with differing groups attached at the "upper" axial position, or with 5′-OH of the nucleotide ribose unit linked with further (bulky) substituents. The human B_{12}-transporters show considerable promiscuity in binding suitably structured Cbls. Hence, a variety of Cbl-conjugates [184] (and "antivitamins B_{12}" [185], see below) have been designed and were tested for their capacity to function as vectors (as "Trojan horses") for the cellular transport of diagnostic and therapeutic cargo.

In contrast, the human B_{12}-transporters discriminate Cbls very effectively from the various "incomplete" analogues and from other natural "complete" cobamides, having another nucleotide moiety than the Cbls [62]. IF and TC, in particular, exhibit a high specificity for Cbls, which are bound in their "base-on" forms (discriminating against other corrinoids, such as the abundant purinyl-Cbas). Strikingly, upon binding to these B_{12}-transporters, the "base-off" forms of some natural purinyl-Cbas are restructured into their "base-on" forms [61].

6.5.2 B_{12}-binding oligonucleotides – B_{12}-aptamers and B_{12}-riboswitches

6.5.2.1 B_{12}-aptamers

Aptamers are natural, or artificial, short pieces of RNA that bind natural products and other ligands effectively and selectively [186]. In the course of explorations of the artificial evolution of RNA aptamers, the selection of aptamers specific for vitamin B_{12} (**CNCbl**) was also tested with the help of in vitro selection techniques. This furnished a 35nt RNA construct, eventually, that selectively bound **CNCbl** with a K_D of 90 nM [187]. The crystal structure of this B_{12}-aptamer represented the first three-dimensional high-resolution structure of a B_{12}–RNA complex [188]. Interestingly, this in vitro evolved B_{12}-aptamer did not fold around the bound **CNCbl**, but formed a stable framework of secondary structure elements for docking of the Cbl. **CNCbl** contacts the binding pocket of the RNA with the β-axial side, whereas the DMB-nucleotide loop of the bound **CNCbl** is exposed to the solvent [189].

6.5.2.2 B_{12}-riboswitches

Riboswitches are natural RNA sequences within the 5′-untranslated region of mRNAs, which regulate gene expression (transcription or translation) upon binding of a specific ligand [190–192]. The discovery of riboswitches at the beginning of this century is linked to puzzling and surprising ways of genetic control by coenzyme B_{12} (**AdoCbl**) [193]. Breaker and coworkers provided first evidence for direct binding of **AdoCbl** to the *E. coli btuB* riboswitch (K_D roughly 90 nM^{-1}) [194] and inducing a structural reorganization of the RNA [195]. The *E. coli btuB* B_{12}-riboswitch regulates the expression of the outer membrane protein BtuB, which is essential for the uptake of Cbls by *E. coli* cells [174]. Related B_{12}-riboswitches occur in a broad variety of B_{12}-dependent microorganisms [193]. Besides the *E. coli btuB* B_{12}-riboswitch [195, 196], various other riboswitches sensing "complete" corrinoids and other important cofactors and metabolites were identified [192, 197, 198]. The crystal structures of three B_{12}-riboswitches are now known: The **AdoCbl**-binding riboswitch from *Symbiobacterium thermophilum* revealed the interaction between the "base-on"-bound B_{12}-cofactor and the large ligand-binding pocket around it, presenting elements for recognition of the Cbl and Ado moieties [199]. Two more B_{12}-riboswitches were studied by crystallography [200], one of which had aquocobalamin (**H_2OCbl^+**) bound (it is bound better than **AdoCbl**, in this case), as was rationalized by the specific structure of the Cbl/RNA interface.

Interestingly, the *E. coli btuB* riboswitch is still poorly characterized. Like other riboswitches [190, 192], B_{12}-riboswitches consist of an aptamer domain that selectively recognizes the B_{12}-ligand, and an expression platform that undergoes a structural switch upon B_{12} binding (to the aptamer region) [192]. To accommodate **AdoCbl**, the largest known cofactor, the aptamer domain of the *E. coli btuB* B_{12}-riboswitch forms a highly complex binding pocket comprising ~200 nucleotides, one of the largest in size compared to other riboswitches. The specific relevance of basic structural features of **AdoCbl** for binding to the *E. coli btuB* B_{12}-riboswitch has been tested. **CNCbl** binding was studied as a Cbl lacking the "upper" β-axial adenosyl ligand of **AdoCbl**, as well as **AdoCbi**, which lacks the "lower" α-coordinated DMB nucleotide of **AdoCbl**, and adenosyl factor A (**Ado-FA**), an organometallic purinyl-Cba that prefers to exist in its "base-off" form in aqueous solution [194]. The *E. coli btuB* B_{12}-riboswitch bound all of these **AdoCbl** analogues, which induced essentially the same structural reorganization of the RNA as from **AdoCbl** binding, although much higher concentrations of some of these analogues were required to bind, and to induce correct folding. Hence, the common corrin moiety of all of these four B_{12}-derivatives appears to be the determinant for correct folding of the aptamer domain and for triggering the conformational switch [194]. Remarkably, this result with the *E. coli btuB* B_{12}-riboswitch correlates with the corrinoid specificity of the outer membrane transporter BtuB, which binds and transports also "incomplete" corrinoids, such as cobinamides [175]. In conclusion, B_{12}-riboswitches are not inherently selective for binding **AdoCbl**, but may be targeted for corrinoids that are physiologically more crucial for the respective organism [200].

6.5.3 Covalent B_{12}-oligonucleotide conjugates

6.5.3.1 Reversed B_{12}-riboswitches

The discovery of the important interaction between nucleic acids and B_{12} in B_{12}-riboswitches has fueled basic ideas concerning evolution of life in an "RNA-world," including the "early" existence and relevance of important cofactors, such as the corrinoids [201, 202]. Indeed, the interaction of an RNA environment and a B_{12}-ligand could induce structural adaptations in the bound corrinoid, and "activate" it into a catalytically more effective cofactor in an organometallic ribozyme [33, 107] (or a DNAzyme [203] in a DNA-based environment). How organometallic B_{12}-derivatives behave in a nucleic acid environment has, therefore, become an intriguing question. To address this question, several covalent model RNA- and DNA-conjugates of the organometallic Cbls **MeCbl** and **AdoCbl** were studied (Figure 6.12) [33, 107, 203].

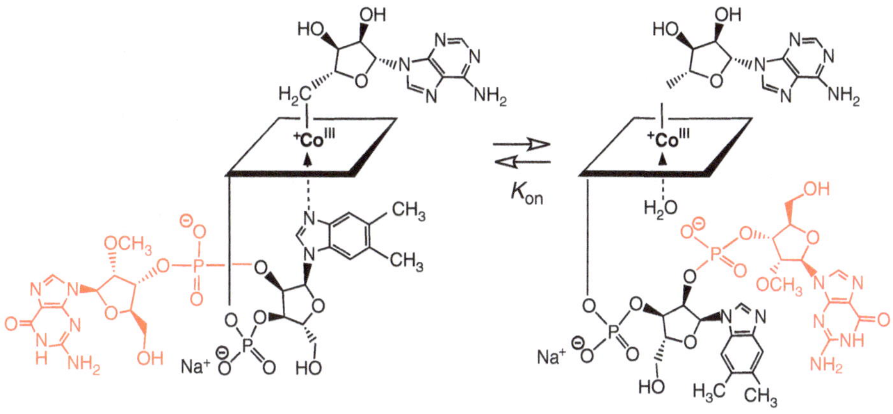

Figure 6.12: A guanosyl unit (marked red), attached at the ribose O2′ in the conjugate **Ado-G$_R$Cbl**, helps shifting the "base-on/base-off" equilibrium toward "base-off".

A significant destabilization of the "base-on" form could be achieved, for example, in **Ado-G$_R$Cbl**, an artificial covalent O2′-methylguanosyl conjugate of **AdoCbl** (see Figure 6.12). The equilibrium between its "base-on" form and the "activated" DMB de-coordinated "base-off" form, K_{on} = 1.3 at 25 °C, was shifted about 60-fold toward "base-off," when compared with **AdoCbl** (K_{on}(**AdoCbl**) = 73) [107] . In relation to the B_{12}-riboswitches, in such (simple) organometallic B_{12}-RNA or B_{12}-DNA conjugates, the roles of the nucleic acid and of the B_{12}-moieties are reversed. They have, hence, been classified as retro-riboswitches or reversed riboswitches [11, 33].

Clearly, organometallic Cbls, and other natural "complete" corrinoids, may be induced by a basic artificial or naturally evolved nucleic acid environment to switch from their "base-on" forms toward "base-off," resulting in activation in organometallic ribozymes [33, 107] or DNAzymes [203], similar to the situation observed in B_{12}-dependent enzymes having a Cbl-bound "base-off" [11, 115].

6.5.3.2 Organometallic B_{12}–DNA conjugates as DNA vectors

The proteins of the Cbl-uptake system in humans and in other mammals have an unmatched affinity for Cbls, but are rather promiscuous with respect to some conjugations of the Cbls. Hence, conjugates of Cbls with biologically interesting loads have been developed, and their biomedical diagnostic and therapeutic applications have been a subject of great interest [184, 204–206]. By carrying toxins as covalent "cargo," Cbl conjugates may function as "Trojan horses" for the transport of specific cell poisons into the organs of laboratory animals [184, 207, 208].

A particularly interesting application of this transport capacity of Cbls would concern the cellular delivery of short pieces of DNA or RNA, a subject of immense interest [209–214]. Indeed, the organometallic conjugates of single-stranded DNA and Cbl, **25nt-DNA-Cbl** and two of its homologues, bound quickly and effectively to TC [106, 215]. The crystal structure of the complex of essential parts of the TC-receptor and holo-TC [177] suggested that the nucleotide moiety of such Cbl–DNA conjugates would not interfere with receptor binding (and eventual import into the cells). All of this implies the possible capacity of Cbl–DNA for transporting single-stranded DNA into cells (see Figure 6.13). Once the Cbl–DNA conjugates are delivered into the cells, their nucleotide moieties may remain bound to their Cbl carrier, or could be set free by B_{12}-processing enzymes such as CblC (see below).

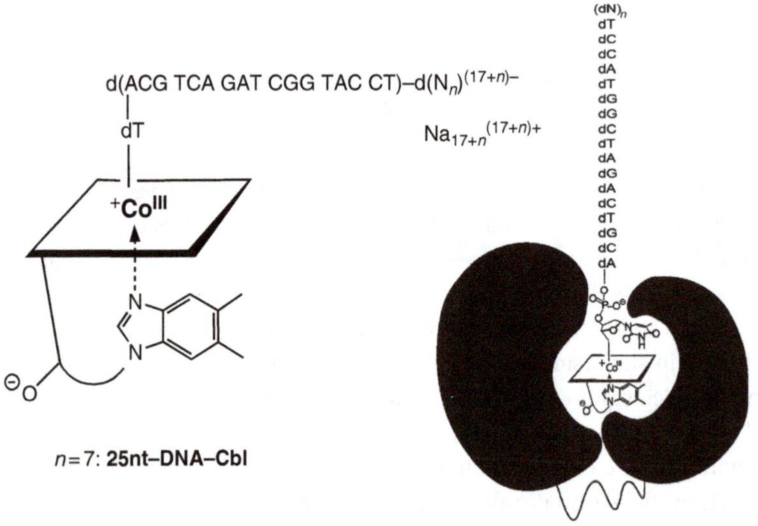

$n=7$: **25nt-DNA-Cbl**

Figure 6.13: Covalent organometallic DNA–Cbl conjugates, such as **25nt-DNA-Cbl** ($n = 7$), bind the human B_{12}-binding protein transcobalamin (TC) with high rates and affinity.

6.6 B_{12}-processing enzymes

In some people, Cbl deficiency prevails despite their adequate supply with **CNCbl** [216]. Indeed, vitamin B_{12} (**CNCbl**), the most common form of B_{12}, has no direct physiological functions in humans and needs to be transformed inside healthy human cells into the organometallic cofactors methylcobalamin (**MeCbl**) and coenzyme B_{12} (**AdoCbl**) [10]. Eight genes, involved in the diagnosed aberrant Cbl metabolism, have been identified in the human genome [217]. Two of them turned out to be responsible for the intracellular processing of Cbls. One of them, named *cblC*, encodes for the Cbl-processing enzyme CblC [218], the other one (*cblB*) for the **AdoCbl**-synthesizing adenosyltransferase ACA [219].

The *cblC*-type Cbl disorders caused metabolic accumulation of methylmalonate and of homocysteine. The genetic locus responsible for this disorder was named the MMACHC gene (for methylmalonic aciduria type C and homocystinuria) [218]. The translational product of this gene, the protein CblC, "tailors" different cobalamins to **Cbl^{II}**, which is processed further within the cell to the physiologically active cofactors **MeCbl** and **AdoCbl** [220]. Depending on the Cbl substrate, CblC employs two different mechanisms to produce **Cbl^{II}** (Figure 6.14) [221]: When **CNCbl** is bound to CblC, it catalyzes the reductive decyanation by NADPH via flavin cofactors [222]. Alternative binding of alkylcobalamins (such as **MeCbl**) initiates the removal of the upper ligand via nucleophilic substitution by glutathione (GSH) [223]. The crystal structure of human CblC in complex with **MeCbl** (but lacking GSH and the flavin) allowed for detailed insights into the three-dimensional structure of this unique 26 kDa protein [113]. The corrinoid substrate **MeCbl** was bound "base-off" in a large cavity of the protein with the nucleotide tail buried in the N-terminal domain. The bound **MeCbl** was, thus, activated by having a five-coordinate cobalt center. The X-ray crystal structure of the complex of CblC with GSH and the Cbl-inhibiting (see below) "antivitamin B_{12}" difluorophenylethynylcobalamin (**F2PhEtyCbl**) revealed the crucial further structuring of CblC and binding of GSH near the corrin [57].

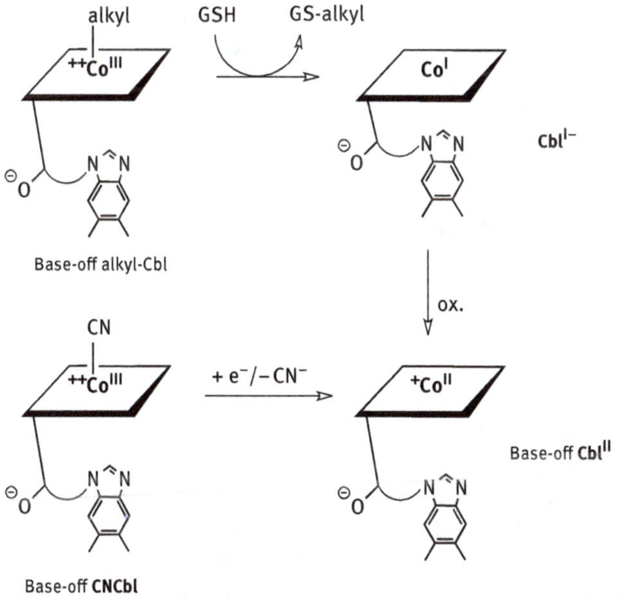

Figure 6.14: The dual-performance enzyme CblC uses two mechanisms for "tailoring" Cbls. The common product **Cbl^{II}** is formed in "base-off" form either by reductive decyanation of (base-off) **CNCbl** or by nucleophilic substitution of (base-off) alkylcobalamins (such as **MeCbl**) by glutathione (GSH), followed by oxidation of the directly resulting **Cbl^{I-}**.

The *cblB*-type disorders are due to a nonfunctional enzyme ATP:Cbl^{I} adenosyltransferase (ACA), responsible for the biosynthesis of **AdoCbl** from the precursor corrinoid **Cbl^{II}** [10, 224]. The enzyme-catalyzed adenosyl transfer is based on the intermediate formation of the nucleophilic **Cbl^{I-}**, which attacks the 5′-carbon of correctly bound ATP (see Figure 6.15). As the reduction of "base-on" **Cbl^{II}** to **Cbl^{I-}** requires a reduction potential beyond the capacities of in vivo reducing agents, ACA activates **Cbl^{II}** for reduction by binding it in a four-coordinate "base-off" form. This was shown by crystallographic snapshots of the ACA structure with **Cbl^{II}** and ATP and the partially formed reaction product **AdoCbl** [224], bound in the active site. The remarkable enzyme-catalyzed adenosyl transfer makes use of a customized protein environment around the Cbl and the triphosphate moiety of ATP that facilitates not only the reduction to **Cbl^{I-}**, but lowers the energy barrier for adenosylation (so far, the direct adenosylation of **Cbl^{I-}** with ATP has not been successful).

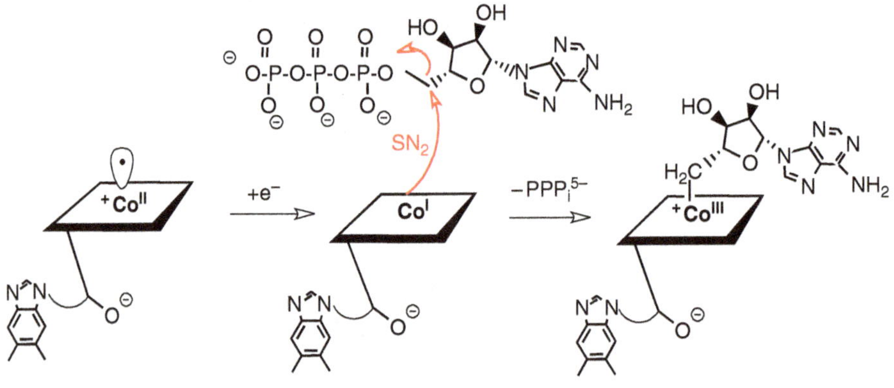

Figure 6.15: The mitochondrial human ATP:**Cbl^{I-}** adenosyltransferase (ACA) generates coenzyme B_{12} (**AdoCbl**) from **Cbl^{II}**.

6.7 Antivitamins B_{12}

Organo-Cbls that resist tailoring by CblC to **Cbl^{II}** cannot furnish **AdoCbl** and **MeCbl** in the cell, the functional cofactors of the enzymes MCM and MetH. Indeed, "functional" Cbl deficiency is induced in animals by the supply of such metabolically inert Cbls [225], which have been classified as "antivitamins B_{12}" [185, 226]. So far, two organometallic classes of Cbls have been shown to be inert to tailoring by CblC and, hence, to represent potential "antivitamins B_{12}": (i) aryl-Cbls, which have a substitution and reduction inert organometallic bond, such as 4-ethylphenylcobalamin (**EtPhCbl**) [54] or **PhCbl** [55], and (ii) hydrolysis-resistant alkynyl-Cbls such as 2-phenyl-ethynylcobalamin (**PhEtyCbl**) [56] and 2-(2,4-difluorophenyl)-ethynyl-cobalamin (**F2PhEtyCbl**) (see Figure 6.16) [57].

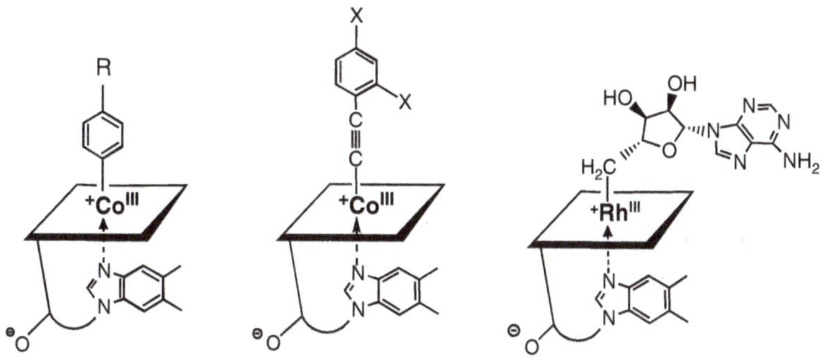

Figure 6.16: Symbolic formulae of some "antivitamins B_{12}." Left and center: the arylcobalamins **EtPhCbl** (R = Et) and **PhCbl** (R = H) and the phenylalkynyl-cobalamins **F2PhEtyCbl** (X = F) and **PhEtyCbl** (X = H) are organo-Cbls with strong (Co-C_{sp2})- and (Co-C_{sp})-bonds, respectively. Right: **AdoRhbl**, the Rh-analogue of **AdoCbl**, is an inhibitor of AdoCbl-dependent enzymes.

The antivitamins B_{12} **EtPhCbl** and **PhEtyCbl** bound strongly to the three important human B_{12}-transporting proteins IF, TC and HC [62], suggesting these inert Cbls to be imported and to induce functional B_{12} deficiency, when administered orally [185]. Indeed, the intravenous application of **EtPhCbl** induced functional Cbl deficiency in laboratory animals [225]. Likewise, when B_{12}-dependent bacteria are supplied with antivitamins B_{12} resistant to the metabolism of these microorganisms, these metabolically inert B_{12}-derivatives may function as useful antibiotics [185, 226, 227].

Cbl analogues, in which the B_{12}-specific cobalt-center is replaced by the homologous rhodium-ions, that is, rhodibalamins (Rhbls), may resemble the corresponding Cbls by their structure, whereas their organometallic reactivity may differ considerably. Such suitably structured Rhbls would also have the features of "antivitamins B_{12}" [185]. Indeed, the crystal structure of adenosyl-rhodibalamin (**AdoRhbl**), the organometallic Rh-analogue of **AdoCbl**, is very similar to the one of **AdoCbl**. **AdoRhbl** inhibited most effectively a bacterial **AdoCbl**-dependent diol-dehydratase, as well as the growth of *Salmonella enterica* [228].

6.8 Conclusions

The dependence of many forms of life on organometallic B_{12}-derivatives became apparent when the organometallic nature of coenzyme B_{12} was discovered in the Hodgkin labs [3]. In the various organisms, organometallic B_{12}-cofactors help to drive metabolism and to control gene expression, and are attached to B_{12}-binding macromolecules, either proteins [229] or RNA [230]. Obviously, nature makes remarkable and multiple uses of organometallic B_{12}-derivatives [11]. The organometallic corrinoids have a unique capacity for biological catalysis and for control of life processes. Better insights into organisms from most kingdoms of life assign previously unrecognized biological roles to the complex and sparse corrinoids [34, 231], which may, hence, also constitute a versatile basis for new biomedical applications [93, 106, 134, 228, 232, 233]. Indeed, the still puzzling effects of Cbl deficiency in human and in mammalian physiology [16, 22, 23] call for new, and better, diagnostic tools in B_{12} biomedicine. In this respect, "antivitamins B_{12}" [185, 226] and suitably structured organometallic DNA or RNA conjugates of Cbl may, possibly, find use in more advanced studies. B_{12}-derivatives have also a remarkable potential for the development of novel applications based on "purely" chemical (organometallic and analytical) research [184]. Doubtlessly, future biostructural, chemical, biological, biochemical and biomedical studies with organometallic B_{12}-cofactors [11, 134, 221, 229] will uncover further fascinating roles of organometallic B_{12}-derivatives in the living nature.

Acknowledgments: The author would like to thank his past and present coworkers and collaborators for their great experimental and intellectual input in his research on B_{12}, and the Austrian Science Foundation (FWF) for generous support over many years.

List of Abbreviations

ACA	ATP:**Cbl**$^{I-}$ adenosyltransferase
Ado	5′-deoxy-5′-adenosyl (group or radical)
AdoCbi	adenosyl-cobinamide
AdoCbl	coenzyme B_{12} (adenosylcobalamin)
AdoRhbl	adenosylrhodibalamin
BDE	bond dissociation energy
BtuB	outer membrane B_{12}-transporter (in *E. coli*)
Cba	cob(III)amide
Cbi	cob(III)inamide
Cbi^{II}	cob(II)inamide
Cbl	cob(III)alamin (DMB-cob(III)amide)
Cbl^{II}	cob(II)alamin
Cbl^{I-}	cob(I)alamin (anion)
Cby	cob(III)yric acid
CNCbl	vitamin B_{12} (cyanocob(III)alamin)
CN,H_2O-Cby	cyano, aquo-cobyrate
DMB	5,6-dimethylbenzimidazole
E. coli	*Escherichia coli*
EtPhCbl	4-ethylphenylcobalamin
F2PhEtyCbl	2(2,4-difluorophenyl)-ethynylcobalamin
2HEP	2-hydroxyethylphosphonate

H_2OCbl^+	aquocobalamin (cation)
HOCbl	hydroxocobalamin
2HPP	2-hydroxypropylphosphonate
MCM	methylmalonyl-CoA mutase
MeCbl	methylcobalamin
$MeCbi^+$	methylcobinamide
MetH	B_{12}-dependent methionine synthase
NADH	nicotine-adenine-dinucleotide hydride
PhEtyCbl	2-phenylethynylcobalamin
SAM	*S*-adenosyl-methionine

References

[1] Rickes, EL., Brink, NG., Koniuszy, FR. Wood, TR., and Folkers, K.. Crystalline vitamin B_{12}. Science 1948, 107, 396–397.
[2] Smith, EL., and Parker, LFJ. Purification of anti-pernicious anaemia factor. Biochem J 1948, 43, R8-R9.
[3] Hodgkin, DC. X-ray analysis of complicated molecules. Science 1965, 150, 979–988.
[4] Friedrich, W. Vitamins, Walter de Gruyter, Berlin, 1988.
[5] Pratt, JM. Inorganic chemistry of vitamin B_{12}, Academic Press, New York, 1972.
[6] Dolphin D. ed, B_{12}, Vol. I and II, John Wiley & Sons, New York, Chichester, 1982.
[7] Battersby, AR. How nature builds the pigments of life – the conquest of vitamin B_{12}. Science 1994, 264, 1551–1557.
[8] Eschenmoser, A., and Wintner, CE. Natural product synthesis and vitamin-B_{12}. Science 1977, 196, 1410–1426.
[9] Woodward, RB. The total synthesis of vitamin B_{12}. Pure Appl Chem 1973, 33, 145–177.
[10] Banerjee, R., Gherasim, C., and Padovani, D. The Tinker, Tailor, Soldier in Intracellular B_{12} Trafficking. Curr Opin Chem Biol 2009, 13, 484–491.
[11] Kräutler, B., and Puffer, B. Vitamin B_{12}-derivatives: organometallic catalysts, cofactors and ligands of bio-macromolecules, Kadish KM, Smit, KM, Guilard R, eds, Handbook of Porphyrin Science, World Scientific, Singapore, 2012, Vol. 6, 133–265.
[12] Banerjee, R., and Ragsdale, SW. The many faces of vitamin B_{12}: catalysis by cobalamin-dependent enzymes. Ann Rev Biochem 2003, 72, 209–247.
[13] Buckel, W., and Golding, BT. Radical enzymes, Chatgilialoglu C, Studer A., eds, Encyclopedia of Radicals in Chemistry, Biology and Materials, John Wiley & Sons, 2012.
[14] Obeid, R. Vitamin B_{12} – Advances and Insights, CRC Press, Taylor & Francis Group, 2017.
[15] Green, R., and Miller, JW. Vitamin B_{12}, Zempleni J, Suttie JW, Gregory JF, Stover PJ., eds, Handbook of Vitamins, CRC Press, Boca Raton, USA, 2014, 447–489.
[16] Green, R., Allen, LH., Bjorke-Monsen, AL., Brito, A., Gueant, JL. et al. Vitamin B12 deficiency. Nat Rev Dis Primers 2017, 3, 17040.
[17] Kräutler, B, Arigoni D, Golding BT eds, Vitamin B_{12} and B_{12}-Proteins, John Wiley VCH, Weinheim, 1998.
[18] Banerjee R ed., Chemistry and Biochemistry of B_{12}, John Wiley & Sons, New York, Chichester, 1999.
[19] Warren, MJ., Raux, E., Schubert, HL., and Escalante-Semerena, JC. The biosynthesis of adenosylcobalamin (Vitamin B_{12}). Nat Prod Rep 2002, 19, 390–412.

[20] Frey, PA., Hegeman, AD., and Reed, GH. Free radical mechanisms in enzymology. Chem Rev 2006, 106, 3302–3316.
[21] Nielsen, MJ., Rasmussen, MR., Andersen, CBF., Nexo, E., and Moestrup, SK. Vitamin B_{12} transport from food to the body's cells-a sophisticated, multistep pathway. Nat Rev Gastroenterol Hepatol 2012, 9, 345–354.
[22] Chan, W., Almasieh, M., Catrinescu, MM., and Levin, LA. Cobalamin-associated superoxide scavenging in neuronal cells is a potential mechanism for vitamin B-12-deprivation optic neuropathy. Am J Pathol 2018, 188, 160–172.
[23] Scalabrino, G. The multi-faceted basis of vitamin B_{12} (Cobalamin) neurotrophism in adult central nervous system: lessons learned from its deficiency. Prog Neurobiol 2009, 88, 203–220.
[24] Widner, FJ., Gstrein, F., and Kräutler, B. Partial synthesis of coenzyme B-12 from cobyric acid. Helv Chim Acta 2017, 100, e1700170.
[25] Glusker, JP. X-Ray Crystallography of B_{12} and Cobaloximes, Dolphin D, ed, B_{12}, John Wiley & Sons, New York, 1982, Vol. I, 23–106.
[26] Eschenmoser, A. Vitamin-B_{12} – experiments concerning the origin of its molecular-structure. Angew Chem Int Ed 1988, 27, 5–39.
[27] Fischli, A., and Daly, JJ. Cob(I)alamin as catalyst. 8. Cob(I)alamin and heptamethyl cob(I) yrinate during the reduction of α,β-unsaturated carbonyl derivatives. Helv Chim Acta 1980, 63, 1628–1643.
[28] Kräutler, B., Keller, W., Hughes, M., Caderas, C., and Kratky, C. A crystalline Cobalt(II) corrinate derived from Vitamin B_{12}: preparation and X-ray crystal structure. J Chem Soc Chem Comm, 1987, 1678–1680.
[29] Brown, KL. Chemistry and enzymology of vitamin B_{12}. Chem Rev 2005, 105, 2075–2149.
[30] Kräutler, B. B_{12}-Nomenclature and a suggested atom-numbering, Kräutler B, Arigoni D, Golding BT, eds, Vitamin B_{12} and B_{12}-Proteins, Wiley VCH, Weinheim, 1998, 517–521.
[31] Randaccio, L., Geremia, S., Nardin, G., and Würges, J. X-ray structural chemistry of cobalamins. Coord Chem Rev 2006, 250, 1332–1350.
[32] Kratky, C., and Kräutler, B. Molecular Structure of B_{12} cofactors and other B_{12} derivatives, Banerjee R., ed., Chemistry and Biochemistry of B_{12}, John Wiley & Sons, New York, Chichester, 1999, 9–41.
[33] Gschösser, S., Gruber, K., Kratky, C., Eichmüller, C., and Kräutler, B. B_{12}-retro-riboswitches: Constitutional switching of B_{12} coenzymes induced by nucleotides. Angew Chem Int Ed 2005, 44, 2284–2288.
[34] Helliwell, KE., Lawrence, AD., Holzer, A., Kudahl, UJ., Sasso, S., Kräutler, B., Scanlan, D J., Warren, M J., and Smith, AG. Cyanobacteria and eukaryotic algae use different chemical variants of vitamin B_{12}. Curr Biol 2016, 26, 999–1008.
[35] Croft, MT., Warren, MJ., and Smith, AG. Algae need their vitamins. Eukaryot Cell 2006, 5, 1175–1183.
[36] Hoffmann, B., Oberhuber, M., Stupperich, E., Bothe, H., Buckel, W., Konrat, R., and Kräutler, B. Native corrinoids from *Clostridium cochlearium* are adeninylcobamides: spectroscopic analysis and identification of pseudovitamin B_{12} and factor A. J Bacteriol 2000, 182, 4773–4782.
[37] Kräutler, B. Organometallic chemistry of B_{12}-coenzymes, Sigel A, Sigel H, Sigel RKO, eds., Metal-Ions in Life Sciences, RSC Publishing, Cambridge, UK, 2009, Vol. 6, 1–51.
[38] Taga, ME., and Walker, GC. Pseudo-B_{12} joins the cofactor family. J Bacteriol 2008, 190, 1157–1159.
[39] Yan, J., Bi, M., Bourdon, AK., Farmer, AT., Wang, PH. et al. Purinyl-cobamide is a native prosthetic group of reductive dehalogenases. Nat Chem Biol 2018, 14, 8–14.
[40] Kräutler, B., Konrat, R., Stupperich, E., Färber, G., Gruber, K., and Kratky, C. Direct evidence for the conformational deformation of the corrin ring by the nucleotide base in vitamin-B_{12} –

synthesis and solution spectroscopic and crystal-structure analysis of $Co_β$-Cyano-Imidazolyl-Cobamide. Inorg Chem 1994, 33, 4128–4139.
[41] Kräutler, B., Fieber, W., Ostermann, S., Fasching, M. et al. The cofactor of tetrachloroethene reductive dehalogenase of *Dehalospirillum multivorans* is norpseudo-B_{12}, a new type of a natural corrinoid. Helv Chim Acta 2003, 86, 3698–3716.
[42] Butler, PA., Eber, MO., Lyskowski, A., Gruber, K., Kratky, C., and Kräutler, B. Vitamin B_{12} – a methyl group without a job?. Angew Chem Int Ed 2006, 45, 989–993.
[43] Bommer, M., Kunze, C., Fesseler, J., Schubert, T., Diekert, G., and Dobbek, H. Structural basis for organohalide respiration. Science 2014, 346, 455–458.
[44] Keller, S., Ruetz, M., Kunze, C., Kräutler, B., Diekert, G., and Schubert, T. Exogenous 5,6-dimethylbenzimidazole caused production of a non-functional tetrachloroethene reductive dehalogenase in *Sulfurospirillum multivorans*. Environm Microbiol 2014, 16, 3361–3369.
[45] Stupperich, E., Eisinger, HJ., and Kräutler, B. Diversity of corrinoids in acetogenic bacteria: P-cresolylcobamide from *Sporomusa ovata*, 5-methoxy-6-methylbenzimidazolylcobamide from *Clostridium formicoaceticum* and vitamin-B_{12} from *Acetobacterium woodii*. Eur J Biochem 1988, 172, 459–464.
[46] Rossi, M., Glusker, JP., Randaccio, L., Summers, MF., Toscano, PJ., and Marzilli, LG. The structure of a B_{12} coenzyme – methylcobalamin studies by X-ray and NMR methods. J Am Chem Soc 1985, 107, 1729–1738.
[47] Kratky, C., Färber, G., Gruber, K. et al. Accurate structural data demystify B_{12} – high-resolution solid-state structure of aquocobalamin perchlorate and structure-analysis of the aquocobalamin ion in solution. J Am Chem Soc 1995, 117, 4654–4670.
[48] Kräutler, B., Keller, W., and Kratky, C. Coenzyme B_{12}-chemistry: the crystal and molecular structure of cob(ii)alamin. J Am Chem Soc 1989, 111, 8936–8938.
[49] Halpern, J. Mechanisms of coenzyme B_{12}-dependent rearrangements. Science 1985, 227, 869–875.
[50] Gschösser, S., Hannak, RB., Konrat, R., Gruber, K., Mikl, C., Kratky, C., and Kräutler, B. Homocoenzyme B_{12} and bishomocoenzyme B_{12}, covalent structural mimics for homolyzed, enzyme-bound coenzyme B_{12}. Chem Eur J 2005, 11, 81–93.
[51] Gruber, K., and Kratky, C. Coenzyme B_{12} dependent glutamate mutase. Curr Op Chem Biol 2002, 6, 598–603.
[52] Hohenester, E., Kratky, C., and Kräutler, B. Low-temperature crystal-structure of superoxocobalamin obtained by solid-state oxygenation of the B_{12} derivative cob(ii)alamin. J Am Chem Soc 1991, 113, 4523–4530.
[53] McCauley, KM., Pratt, DA., Wilson, SR., Shey, J., Burkey, TJ., and van der Donk, WA. Properties and reactivity of chlorovinylcobalamin and vinylcobalamin and their implications for vitamin B_{12}-catalyzed reductive dechlorination of chlorinated alkenes. J Am Chem Soc 2005, 127, 1126–1136.
[54] Ruetz, M., Gherasim, C., Fedosov, SN., Gruber, K., Banerjee, R., and Kräutler, B. Radical synthesis opens access to organometallic aryl-cobaltcorrins – 4-ethylphenyl-cobalamin, a potential "antivitamin B_{12}". Angew Chem Int Ed 2013, 52, 2606–2610.
[55] Brenig, C., Ruetz, M., Kieninger, C., Wurst, K., and Krautler, B. Alpha- and beta-diastereoisomers of phenylcobalamin from cobalt-arylation with diphenyliodonium chloride. Chem Euro J 2017, 23, 9726–9731.
[56] Ruetz, M., Salchner, R., Wurst, K., Fedosov, S., and Kräutler, B. Phenylethynylcobalamin: a light-stable and thermolysis-resistant organometallic vitamin B_{12} derivative prepared by radical. Angew Chem Int Ed 2013, 52, 11406–11409.
[57] Ruetz, M., Shanmuganathan, A., Gherasim, C., Karasik, A., Salchner, R., Kieninger, C., Wurst, K., Banerjee, R., Koutmos, M., and Kräutler, B. Antivitamin B-12 inhibition of the human B-12-

processing enzyme CblC: crystal structure of an inactive ternary complex with glutathione as the cosubstrate. Angew Chem Int Ed 2017, 56, 7387–7392.
[58] Chrominski, M., Lewalska, A., and Gryko, D. Reduction-free synthesis of stable acetylide cobalamins. Chem Comm 2013, 49, 11406–11408.
[59] Chrominski, M., Lewalska, A., Karczewski, M., and Gryko, D. Vitamin B_{12} derivatives for orthogonal functionalization. J Org Chem 2014, 79, 7532–7542.
[60] Kräutler, B. Thermodynamic trans-effects of the nucleotide base in the B_{12} coenzymes. Helv Chim Acta 1987, 70, 1268–1278.
[61] Fedosov, SN., Fedosova, NU., Kräutler, B., Nexo, E., and Petersen, TE. Mechanisms of discrimination between cobalamins and their natural analogues during their binding to the specific B_{12}-transporting proteins. Biochemistry 2007, 46, 6446–6458.
[62] Fedosov, SN. Physiological and molecular aspects of cobalamin transport, Stanger O, ed, Water Soluble Vitamins, Springer, 2012, 347–368.
[63] Shibata, N., Masuda, J., Tobimatsu, T. et al. A new mode of B_{12} binding and the direct participation of a potassium ion in enzyme catalysis: X-ray structure of diol dehydratase. Structure 1999, 7, 997–1008.
[64] Hannak, RB., Konrat, R., Schüler, W., Kräutler, B., Auditor, MTM., and Hilvert, D. An antibody that reconstitutes the "base-on" form of B_{12} coenzymes. Angew Chem Int Ed 2002, 41, 3613–3616.
[65] Drennan, CL., Huang, S., Drummond, JT., Matthews, RG., and Ludwig, ML. How a protein binds B_{12} – a 3.0-angstrom X-ray structure of B_{12}-binding domains of methionine synthase. Science 1994, 266, 1669–1674.
[66] Ludwig, ML., and Evans, PR. X-Ray crystallography of B_{12} Enzymes: methylmalonyl-CoA mutase and methionine synthase, Banerjee R, ed, Chemistry and Biochemistry of B_{12}, John Wiley & Sons, New York, Chichester, 1999, 595–632.
[67] Marsh, ENG., and Drennan, CL. Adenosylcobalamin-dependent isomerases: new insights into structure and mechanism. Curr Opin Chem Biol 2001, 5, 499–505.
[68] Matthews, RG. Cobalamin-dependent methyltransferases. Acc Chem Res 2001, 34, 681–689.
[69] Mosimann, H., and Kräutler, B. Methylcorrinoids methylate radicals – their second biological mode of action?. Angew Chem Int Ed 2000, 39, 393–200.
[70] Finke, RG., and Hay, BP. Thermolysis of adenosylcobalamin – a product, kinetic, and Co-C5′ bond-dissociation energy study. Inorg Chem 1984, 23, 3041–3043.
[71] Finke, RG. Coenzyme B_{12}-based chemical precedent for Co-C bond homolysis and other key elementary step, Kräutler B, Arigoni D, Golding BT, eds, Vitamin B_{12} and B_{12}-Proteins, Wiley-VCH, Weinheim, 1998, 383–402.
[72] Martin, BD., and Finke, RG. Methylcobalamins full-strength vs half-strength cobalt-carbon sigma-bonds and bond-dissociation enthalpies – a-greater-than-10(15) Co-CH_3 homolysis rate enhancement following one-antibonding-electron reduction of methylcobalamin. J Am Chem Soc 1992, 114, 585–592.
[73] Kobylianskii, I., Widner, F., Kräutler, B., and Chen, P. Co–C bond energies in adenosylcobinamide and methylcobinamide in the gas phase and in silico. J Am Chem Soc 2013, 135, 13648–13651.
[74] Grate, JH., and Schrauzer, GN. Chemistry of cobalamins and related compounds. 48. sterically induced, spontaneous dealkylation of secondary alkylcobalamins due to axial base coordination and conformational-changes of the corrin ligand. J Am Chem Soc 1979, 101, 4601–4611.
[75] Kräutler, B. B_{12} coenzymes, the central theme, Kräutler B, Arigoni D, Golding BT, eds, Vitamin B_{12} and B_{12} Proteins, Wiley-VCH, Weinheim, 1998, 3–43.
[76] Jones, AR. The photochemistry and photobiology of vitamin B12. Photochem Photobiol Sci 2017, 16, 820–834.

[77] Rury, AS., Wiley, TE., and Sension, RJ. Energy cascades, excited state dynamics, and photochemistry in cob(III)alamins and ferric porphyrins. Acc Chem Res 2015, 48, 860–867.
[78] Sension, RJ., Harris, DA., and Cole, AG. Time-resolved spectroscopic studies of B-12 coenzymes: Comparison of the influence of solvent on the primary photolysis mechanism and geminate recombination of methyl-, ethyl-, n-propyl-, and 5 '-deoxyadenosylcobalamin. J Phys Chem B 2005, 109, 21954–21962.
[79] Hisaeda, Y., Tahara, K., Shimakoshi, H., and Masuko, T. Bioinspired catalytic reactions with vitamin B_{12} derivative and photosensitizers. Pure Appl Chem 2013, 85, 1415–1426.
[80] Miller, NA., Wiley, TE., Spears, KG., Ruetz, M., Kieninger, C., Kräutler, B., and Sension, RJ. Toward the design of photoresponsive conditional antivitamins B_{12}: a transient absorption study of an arylcobalamin and an alkynylcobalamin. J Am Chem Soc 2016, 138, 14250–14256.
[81] Endicott, JF., and Netzel, TL. Early events and transient chemistry in the photohomolysis of alkylcobalamins. J Am Chem Soc 1979, 101, 4000–4002.
[82] Kräutler, B. Acetyl-cobalamin from photoinduced carbonylation of methyl-cobalamin. Helv Chim Acta 1984, 67, 1053–1059.
[83] Fischer, H. Unusual selectivities of radical reactions by internal suppression of fast modes. J Am Chem Soc 1986, 108, 3925–3927.
[84] Schrauzer, GN., Deutsch, E., and Windgassen, RJ. The nucleophilicity of vitamin B_{12s}. J Am Chem Soc 1968, 90(9), 2441–2442.
[85] Haglund, J., Magnusson, A-L., Ehrenberg, L., and Törnqvist, M. Introduction of Cob(I)alamin as an analytical tool: application to reaction-kinetic studies of oxiranes. Toxicol Environ Chem 2003, 85, 81–94.
[86] Stich, TA., Buan, NR., Escalante-Semerena, JC., and Brunold, TC. Spectroscopic and computational studies of the ATP: Corrinoid adenosyltransferase (CobA) from Salmonella enterica: Insights into the mechanism of adenosylcobalamin biosynthesis. J Am Chem Soc 2005, 127, 8710–8719.
[87] Matthews, RG., Koutmos, M., and Datta, S. Cobalamin-dependent and cobamide-dependent methyltransferases. Curr Op Struct Biol 2008, 18, 658–666.
[88] Kräutler, B., and Caderas, C. Complementary diastereoselective cobalt methylations of the vitamin-B_{12} derivative cobester. Helv Chim Acta 1984, 67, 1891–1896.
[89] Hogenkamp, HPC., Bratt, GT., and Sun, S. Methyl transfer from methylcobalamin to thiols – a reinvestigation. Biochemistry 1985, 24, 6428–6432.
[90] Dorweiler, JS., Matthews, RG., and Finke, RG. Providing a chemical basis toward understanding the histidine base-on motif of methylcobalamin-dependent methionine synthase: an improved purification of methylcobinamide, plus thermodynamic studies of methylcobinamide binding exogenous imidazole and pyridine bases. Inorg Chem 2002, 41, 6217–6224.
[91] Kräutler, B., Dérer, T., Liu, PL., Mühlecker, W., Puchberger, M., Kratky, C., and Gruber, K. Oligomethylene-bridged vitamin-B_{12} dimers. Angew Chem Int Ed 1995, 34, 84–86.
[92] Galliker, PK., Gräther, O., Rümmler, M., Fitz, W., and Arigoni, D. New structural and biosynthetic aspects of the unusual core lipids from Archaebacteria, Kräutler B, Arigoni D, Golding BT, eds, Vitamin B_{12} and B_{12}-Proteins, Wiley-VCH, Weinheim, 1998, 447–458.
[93] Benjdia, A., Balty, C., and Berteau, O. Radical SAM enzymes in the biosynthesis of ribosomally synthesized and post-translationally modified peptides (RiPPs). Front Chem 2017, 5.
[94] Fujimori, DG. Radical SAM-mediated methylation reactions. Curr Opin Chem Biol 2013, 17, 597–604.
[95] Zhang, Q., van der Donk, W., and Liu, W. Radical-mediated enzymatic methylation: a tale of two SAMS. Acc Chem Res 2012, 45, 555.

[96] Kräutler, B., Hughes, M., and Caderas, C. Thermal methyl-group transfer between methylcobalt(III) corrinates and cobalt(II) corrinates – equilibration experiments with heptamethyl cobyrinates and cobalamins. Helv Chim Acta 1986, 69, 1571–1575.
[97] Craig PJ, Glockling F eds, The Biological Alkylation of Heavy Elements, Royal Society of Chemistry, London, 1988.
[98] Desimone, RE., Penley, MW., Charbonn., L. et al. Kinetics and mechanism of cobalamin-dependent methyl and ethyl transfer to mercuric ion. Biochim Biophys Acta 1973, 304, 851–863.
[99] Jensen, MP., and Halpern, J. Dealkylation of coenzyme B_{12} and related organocobalamins: ligand structural effects on rates and mechanisms of hydrolysis. J Am Chem Soc 1999, 121, 2181–2192.
[100] Lexa, D., and Savéant, JM. The electrochemistry of vitamin B_{12}. Acc Chem Res 1983, 16, 235–243.
[101] Kräutler, B. Electrochemistry and organometallic electrochemical synthesis, Banerjee R, ed, Chemistry and Biochemistry of B_{12}, John Wiley, New York, 1999, 315–339.
[102] Ragsdale, SW. Enzymology of the acetyl-CoA pathway of CO_2 fixation. Crit Rev Biochem Mol Biol 1991, 26, 261–300.
[103] Hisaeda, Y., and Shimakoshi, H. Bioinspired catalysts with B_{12} enzyme functions, Kadish KM, Smith, K. M., Guilard, R., ed, Handbook of Porphyrin Science, 2010, 313–364. World Scientific Publishing Co. Pte. Ltd.
[104] Birke, RL., Huang, QD., Spataru, T., and Gosser, DKj. Electroreduction of a series of alkylcobalamins: mechanism of stepwise reductive cleavage of the Co–C bond. J Am Chem Soc 2006, 128, 1922–1936.
[105] Scheffold, R., Rytz, G., Walder, L., Orlinski, R., and Chilmonczyk, Z. Formation of (C-C)Bonds catalyzed by vitamin-B12. Pure Appl Chem 1983, 55, 1791–1797.
[106] Hunger, M., Mutti, E., Rieder, A., Enders, B., Nexo, E., and Kräutler, B. Organometallic B_{12}-DNA-conjugate: synthesis, structure analysis and studies of binding to human B_{12}-transporter proteins. Chem Eur J 2014, 20, 13103–13107.
[107] Gschösser, S., and Kräutler, B. B_{12}-retro-riboswitches: guanosyl-induced constitutional switching of B_{12}-coenzymes. Chem Eur J 2008, 14, 3605–3619.
[108] Tinembart, O., Walder, L., and Scheffold, R. Reductive cleavage of the Co,C-bond of [(methoxycarbonyl)methyl]cobalamin. Ber Bunsen Ges Phys Chem Chem Phys 1988, 92, 1225–1231.
[109] Puchberger, M., Konrat, R., Kräutler, B., Wagner, U., and Kratky, C. Reduction-labile organo-cob(III)alamins via cob(II)alamin: Efficient synthesis and solution and crystal structures of [(methoxycarbonyl)methyl]cob(III)alamin. Helv Chim Acta 2003, 86, 1453–1466.
[110] Zhou, DL., Tinembart, O., Scheffold, R., and Walder, L. Influence of the axial alkyl ligand on the reduction potential of alkylcob(III)alamins and alkylcob(III)yrinates. Helv Chim Acta 1990, 73, 2225–2241.
[111] Toraya, T. Cobalamin-dependent dehydratases and a deaminase: radical catalysis and reactivating chaperones. Arch Biochem Biophys 2014, 544, 40–57.
[112] Gruber, K., Reitzer, R., and Kratky, C. Radical shuttling in a protein: Ribose pseudorotation controls alkyl-radical transfer in the coenzyme B_{12} dependent enzyme glutamate mutase. Angew Chem Int Ed 2001, 40, 3377–3380.
[113] Koutmos, M., Gherasim, C., Smith, JL., and Banerjee, R. Structural basis of multifunctionality in a vitamin B_{12}-processing enzyme. J Biol Chem 2011, 286, 29780–29787.
[114] Bauer, CB., Fonseca, MV., Holden, HM. et al. Three-dimensional structure of ATP: corrinoid adenosyltransferase from *Salmonella typhimurium* in its free state, complexed with MgATP, or complexed with hydroxycobalamin and MgATP. Biochemistry 2001, 40, 361–374.

[115] Moore, TC., Newmister, SA., Rayment, I., and Escalante-Semerena, JC. Structural insights into the mechanism of four-coordinate Cob(II)alamin formation in the active site of the salmonella enterica ATP:Co(I)rrinoid adenosyltransferase enzyme: critical role of residues Phe91 and Trp93. Biochem 2012, 51, 9647–9657.
[116] Li, Z., Lesniak, NA., and Banerjee, R. Unusual aerobic stabilization of Cob(I)alamin by a B-12-trafficking protein allows chemoenzymatic synthesis of organocobalamins. J Am Chem Soc 2014, 136, 16108–16111.
[117] Marsh, ENG., Patterson, DP., and Adenosyl Radical:, Li L. Reagent and catalyst in enzyme reactions. ChemBioChem 2010, 11, 604–621.
[118] Buckel, W., Kratky, C., and Golding, BT. Stabilization of methylene radicals by Cob(II)alamin in coenzyme B_{12} dependent mutases. Chem Eur J 2006, 12, 352–362.
[119] Buckel, W., Friedrich, P., and Golding, BT. Hydrogen bonds guide the short-lived 5′-deoxyadenosyl radical to the place of action. Angew Chem Int Ed 2012, 51, 9974–9976.
[120] Banerjee, R. Radical carbon skeleton rearrangements: catalysis by coenzyme B_{12}-dependent mutases. Chem Rev 2003, 103, 2083–2094.
[121] Erb, TJ., Retey, J., Fuchs, G., and Alber, BE. Ethylmalonyl-CoA mutase from Rhodobacter sphaeroides defines a new subclade of coenzyme B-12-dependent acyl-CoA mutases. J Biol Chem 2008, 283, 32283–32293.
[122] Jost, M., Born, DA., Cracan, V., Banerjee, R., and Drennan, CL. Structural basis for substrate specificity in adenosylcobalamin-dependent isobutyryl-CoA mutase and related Acyl-CoA mutases. J Biol Chem 2015, 290, 26882–26898.
[123] Berkovitch, F., Behshad, E., Tang, K-H., Enns, EA., Frey, PA., and Drennan, CL. A locking mechanism preventing radical damage in the absence of substrate, as revealed by the X-ray structure of lysine 5,6-aminomutase. Proc Natl Acad Sci USA 2004, 101, 15870–15875.
[124] Ortiz-Guerrero, JM., Polanco, MC., Murillo, FJ., Elias-Arnanz, M., and Padmanabhan, S. Light-dependent gene regulation by a coenzyme B_{12}-based photoreceptor. Proc Natl Acad Sci USA 2011, 108, 7565–7570.
[125] Jost, M., Fernandez-Zapata, J., Polanco, MC. et al. Structural basis for gene regulation by a B12-dependent photoreceptor. Nature 2015, 526, 536–541.
[126] Romine, MF., Rodionov, DA., Maezato, Y. et al. Elucidation of roles for vitamin B-12 in regulation of folate, ubiquinone, and methionine metabolism. Proc Natl Acad Sci USA 2017, 114, E1205-E14.
[127] Stubbe, J. Ribonucleotide reductases: the link between an RNA and a DNA world?. Curr Opin Struct Biol 2000, 10, 731–736.
[128] Robertson, WD., Wang, M., and Warncke, K. Characterization of protein contributions to cobalt-carbon bond cleavage catalysis in adenosylcobalamin-dependent ethanolamine ammonia-lyase by using photolysis in the ternary complex. J Am Chem Soc 2011, 133, 6968–6977.
[129] Roman-Melendez, GD., von Glehn, P., Harvey, JN., Mulholland, AJ., and Marsh, ENG. Role of active site residues in promoting cobalt-carbon bond homolysis in adenosylcobalamin-dependent mutases revealed through experiment and computation. Biochemistry 2014, 53, 169–177.
[130] Rétey, J. Enzymatic-reaction selectivity by negative catalysis or how do enzymes deal with highly reactive intermediates. Angew Chem Int Ed 1990, 29, 355–361.
[131] Fukuoka, M., Nakanishi, Y., Hannak, RB., Kräutler, B., and Toraya, T. Homoadenosylcobalamins as probes for exploring the active sites of coenzyme B_{12}-dependent diol dehydratase and ethanolamine ammonia-lyase. FEBS J 2005, 272, 4787–4796.

[132] Sintchak, MD., Arjara, G., Kellogg, BA., Stubbe, J., and Drennan, CL. The crystal structure of class II ribonucleotide reductase reveals how an allosterically regulated monomer mimics a dimer. Nat Struct Biol 2002, 9, 293–300.
[133] Pérez-Marín, MC., Padmanabhan, S., Polanco, MC., Murillo, FJ., and Elías-Arnanz, M. Vitamin B_{12} partners the CarH repressor to downregulate a photoinducible promoter in *Myxococcus xanthus*. Mol Microbiol 2008, 67, 804–819.
[134] Bridwell-Rabb, J., and Drennan, CL. Vitamin B-12 in the spotlight again. Curr Opin Chem Biol 2017, 37, 63–70.
[135] Elias-Arnanz, M., Padmanabhan, S., and Murillo, FJ. Light-dependent gene regulation in nonphototrophic bacteria. Curr Opin Microbiol 2011, 14, 128–135.
[136] Jost, M., Simpson, JH., and Drennan, CL. The transcription factor CarH safeguards use of adenosylcobalamin as a light sensor by altering the photolysis products. Biochemistry 2015, 54, 3231–3234.
[137] Kutta, RJ., Hardman, SJO., Johannissen, LO. et al. The photochemical mechanism of a B_{12}-dependent photoreceptor protein. Nat Comm 2015, 6.
[138] Gruber, K., and Kräutler, B. Coenzyme B_{12} repurposed for photoregulation of gene expression. Angew Chem Int Ed 2016, 55, 5638–5640.
[139] Cheng, Z., Li, KR., Hammad, LA., Karty, JA., and Bauer, CE. Vitamin B_{12} regulates photosystem gene expression via the CrtJ antirepressor AerR in Rhodobacter capsulatus. Mol Microbiol 2014, 91, 649–664.
[140] Sauer, K., and Thauer, RK. The role of corrinoids in methanogenesis, Banerjee R, ed, Chemistry and Biochemistry of B_{12}, New York, Chichester, John Wiley & Sons, 1999, 655–679.
[141] Stupperich, E. Corrinoid-dependent mechanism of acetogenesis from methanol. Acetogenesis 1994, 180–194.
[142] Matthews, RG. Cobalamin- and corrinoid-dependent enzymes, Sigel A, Sigel H, Sigel RKO, eds., Metal-Ions in Life Sciences, RSC Publishing, Cambridge, 2009, Vol. 6, 53–114.
[143] Can, M., Armstrong, FA., and Ragsdale, SW. Structure, function, and mechanism of the nickel metalloenzymes, CO dehydrogenase, and acetyl-CoA synthase. Chem Rev 2014, 114, 4149–4174.
[144] Appel, AM., Bercaw, JE., Bocarsly, AB. et al. Frontiers, opportunities, and challenges in biochemical and chemical catalysis of CO2 fixation. Chem Rev 2013, 113, 6621–6658.
[145] Thauer, RK., Mollerzinkhan, D., and Spormann, AM. Biochemistry of acetate catabolism in anaerobic chemotropic bacteria. Annu Rev Microbiol 1989, 43, 43–67.
[146] Ragsdale, SW. Metals and their scaffolds to promote difficult enzymatic reactions. Chem Rev 2006, 106, 3317–3337.
[147] Chan, KKJ., Thompson, S., and O'Hagan, D. The mechanisms of radical SAM/cobalamin methylations: an evolving working hypothesis. ChemBioChem 2013, 14, 675–677.
[148] Ludwig, ML., and Matthews, RG. B_{12}-dependent methionine synthase: a structure that adapts to catalyze multiple methyl transfer reactions. ACS Symposium Series: Structures and Mechanisms, from Ashes to enzymes 2002:186–201.
[149] Zydowsky, TM., Courtney, LF., Frasca, V. et al. Stereochemical analysis of the methyl transfer catalyzed by cobalamin-dependent methionine synthase from *Escherichia coli B*. J Am Chem Soc 1986, 108, 3152–3153.
[150] Ludwig, ML., Drennan, CL., and Matthews, RG. The reactivity of B_{12} cofactors: the proteins make a difference. Structure 1996, 4, 505–512.
[151] Svetlitchnaia, T., Svetlitchnyi, V., Meyer, O., and Dobbek, H. Structural insights into methyl transfer reactions of a corrinoid iron-sulfur protein involved in acetyl-CoA synthesis. Proc Natl Acad Sci USA 2006, 103, 14331–14336.

[152] Bandarian, V., Ludwig, ML., and Matthews, RG. Factors modulating conformational equilibria in large modular proteins: a case study with cobalamin-dependent methionine synthase. Proc Natl Acad Sci USA 2003, 100, 8156–8163.

[153] Kräutler, B., and Kratky, C. Vitamin B_{12}: the haze clears. Angew Chem Int Ed 1996, 35, 167–170.

[154] Zhou, P., Ohagan, D., Mocek, U. et al. Biosynthesis of the antibiotic thiostrepton – methylation of tryptophan in the formation of the quinaldic acid moiety by transfer of the methionine methyl-group with net retention of configuration. J Am Chem Soc 1989, 111, 7274–7276.

[155] Sofia, HJ., Chen, G., Hetzler, BG., Reyes-Spindola, JF., and Miller, NE. Radical SAM, a novel protein superfamily linking unresolved steps in familiar biosynthetic pathways with radical mechanisms: functional characterization using new analysis and information visualization methods. Nucl Acids Res 2001, 29, 1097–1106.

[156] Frey, PA., and Magnusson, OT. *S*-Adenosylmethionine: A wolf in sheep's clothing, or a rich man's adenosylcobalamin?. Chem Rev 2003, 103, 2129–2148.

[157] Frey, PA., and Hegeman, AD. Enzymatic Reaction Mechanisms, Oxford University Press, New York, 2007.

[158] Akiva, E., Brown, S., Almonacid, DE. et al. The structure–function linkage database. Nucl Acids Res 2014, 42, D521-D30.

[159] Freeman, MF., Helf, MJ., Bhushan, A., Morinaka, BI., and Piel, J. Seven enzymes create extraordinary molecular complexity in an uncultivated bacterium. Nat Chem 2017, 9, 387–395.

[160] Woodyer, RD., Li, G., Zhao, H. and van der Donk, WA.. New insight into the mechanism of methyl transfer during the biosynthesis of fosfomycin. Chem Commun 2007, 4, 359–361.

[161] Zhang, Q., and Liu, W. Complex biotransformations catalyzed by radical S-Adenosylmethionine enzymes. J Biol Chem 2011, 286, 30245–30252.

[162] Bridwell-Rabb, J., Zhong, A., Sun, HG., Drennan, CL., and Liu, HW. A B_{12}-dependent radical SAM enzyme involved in oxetanocin A biosynthesis. Nature 2017, 544, 322–326.

[163] Jugder, BE., Ertan, H., Lee, M., Manefield, M., and Marquis, CP. Reductive dehalogenases come of age in biological destruction of organohalides. Trends Biotechnol 2015, 33, 595–610.

[164] Smidt, H., and de Vos, WM. Anaerobic microbial dehalogenation. Ann Rev Microbiol 2004, 58, 43–73.

[165] Studer, A., Stupperich, E., Vuilleumier, S., and Leisinger, T. Chloromethane:tetrahydrofolate methyl transfer by two proteins from *Methylobacterium chloromethanicum* strain CM4. Eur J Biochem 2001, 268, 2931–2938.

[166] Sun, BL., Griffin, BM., Ayala-del-Rio, HL., Hashsham, SA., and Tiedje, JM. Microbial dehalorespiration with 1,1,1-trichloroethane. Science 2002, 298, 1023–1025.

[167] Raina, V., Rentsch, D., Geiger, T. et al. New metabolites in the degradation of α- and γ-hexachlorocyclohexane (HCH): Pentachlorocyclohexenes are hydroxylated to cyclohexenols and cyclohexenediols by the haloalkane dehalogenase LinB from *Sphingobium indicum* B90A. J Agric Food Chem 2008, 56, 6594–6603.

[168] Krasotkina, J., Walters, T., Maruya, KA., and Ragsdale, SW. Characterization of the B_{12}- and iron-sulfur-containing reductive dehalogenase from *Desulfitobacterium chlororespirans*. J Biol Chem 2001, 276, 40991–40997.

[169] Neumann, A., Siebert, A., Trescher, T., Reinhardt, S., Wohlfarth, G., and Diekert, G. Tetrachloroethene reductive dehalogenase of *Dehalospirillum multivorans*: substrate specificity of the native enzyme and its corrinoid cofactor. Arch Microbiol 2002, 177, 420–426.

[170] Diekert, G., Gugova, D., Limoges, B., Robert, M., and Saveant, J-M. Electroenzymatic reactions. Investigation of a reductive dehalogenase by means of electrogenerated redox cosubstrates. J Am Chem Soc 2005, 127, 13583–13588.

[171] Payne, KAP., Quezada, CP., Fisher, K. et al. Reductive dehalogenase structure suggests a mechanism for B_{12}-dependent dehalogenation. Nature 2015, 517, 513–516.
[172] Kunze, C., Bommer, M., Hagen, WR. et al. Cobamide-mediated enzymatic reductive dehalogenation via long-range electron transfer. Nat Commun 2017, 8.
[173] Holliger, C., Wohlfarth, G., and Diekert, G. Reductive dechlorination in the energy metabolism of anaerobic bacteria. FEMS Microbiol Rev 1999, 22, 383–398.
[174] Bradbeer, C. Cobalamin Transport in Bacteria, Banerjee R, ed, Chemistry and Biochemistry of B_{12}, New York, John Wiley & Sons, 1999, 489–506.
[175] Kenley, JS., Leighton, M., and Bradbeer, C. Transport of vitamin B_{12} in *Escherichia coli* – corrinoid specificity of outer membrane receptor. J Biol Chem 1978, 253, 1341–1346.
[176] Andersen, CBF., Madsen, M., Storm, T., Moestrup, SK., and Andersen, GR. Structural basis for receptor recognition of vitamin-B_{12}-intrinsic factor complexes. Nature 2010, 464, 445–448.
[177] Alam, A., Woo, J-S., Schmitz, J. et al. Structural basis of transcobalamin recognition by human CD320 receptor. Nat Commun 2016, 7, 12100.
[178] Fedosov, SN., Fedosova, NU., Berglund, L., Moestrup, SK., Nexo, E., and Petersen, TE. Assembly of the intrinsic factor domains and oligomerization of the protein in the presence of cobalamin. Biochem 2004, 43, 15095–15102.
[179] Würges, J., Garau, G., Geremia, S., Fedosov, SN., Petersen, TE., and Randacco, L. Structural basis for mammalian vitamin B_{12} transport by transcobalamin. Proc Natl Acad Sci USA 2006, 103, 4386–4391.
[180] Mathews, FS., Gordon, MM., Chen, Z. et al. Crystal structure of human intrinsic factor: cobalamin complex at 2.6 Å resolution. Proc Natl Acad Sci USA 2007, 104, 17311–17316.
[181] Furger, E., Frei, DC., Schibli, R., Fischer, E., and Prota, AE. Structural basis for universal corrinoid recognition by the cobalamin transport protein haptocorrin. J Biol Chem 2013, 288, 25466–25476.
[182] Wuerges, J., Geremia, S., and Randaccio, L. Structural study on ligand specificity of human vitamin B_{12} transporters. Biochem J 2007, 403, 431–440.
[183] Bloch, JS., Ruetz, M., Krautler, B., and Locher, KP. Structure of the human transcobalamin beta domain in four distinct states. PloS one 2017, 12, e0184932.
[184] Zelder, F., and Alberto, R. Vitamin B12 Derivatives for spectroanalytical and medicinal applications, Kadish KM, Smith, K. M., Guilard, R., ed, Handbook of Porphyrin Science, World Scientific, 2012, 84–132.
[185] Kräutler, B. Antivitamins B_{12} – a structure- and reactivity-based concept. Chem Eur J 2015, 21, 11280–11287.
[186] Famulok, M. Oligonucleotide aptamers that recognize small molecules. Curr Opin Struct Biol 1999, 9, 324–329.
[187] Lorsch, JR., and Szostak, JW. *In vitro* selection of RNA aptamers specific for cyanocobalamin. Biochemistry 1994, 33, 973–982.
[188] Sussman, D., Nix, JC., and Wilson, C. The structural basis for molecular recognition by the vitamin B_{12} RNA aptamer. Nat Struct Biol 2000, 7, 53–57.
[189] Sussman, D., and Wilson, C. A water channel in the core of the vitamin B_{12} RNA aptamer. Structure 2000, 8, 719–727.
[190] Breaker, RR. Complex riboswitches. Science 2008, 319, 1795–1797.
[191] Deigan, KE., and Ferre-D'Amare, AR. Riboswitches: discovery of drugs that target bacterial gene-regulatory RNAs. Acc Chem Res 2011, 44, 1329–1338.
[192] Serganov, A., and Nudler, E. A decade of riboswitches. Cell 2013, 152, 17–24.
[193] Nahvi, A., Barrick, JE., and Breaker, RR. Coenzyme B_{12} riboswitches are widespread genetic control elements in prokaryotes. Nucl Acids Res 2004, 32, 143–150.

[194] Gallo, S., Oberhuber, M., Sigel, RKO., and Kräutler, B. The corrin moiety of coenzyme B_{12} is the determinant for switching the btuB riboswitch of E.*coli*. ChemBioChem 2008, 9, 1408–1414.
[195] Nahvi, A., Sudarsan, N., Ebert, MS., Zou, X., Brown, KL., and Breaker, RR. Genetic control by a metabolite binding mRNA. Chem Biol 2002, 9, 1043–1049.
[196] Mandal, M., and Breaker, RR. Gene regulation by riboswitches. Nat Rev Mol Cell Biol 2004, 5, 451–463.
[197] Breaker, RR. Prospects for riboswitch discovery and analysis. Mol Cell 2011, 43, 867–879.
[198] Nudler, E., and Mironov, AS. The riboswitch control of bacterial metabolism. Trends Biochem Sci 2004, 29, 11–17.
[199] Peselis, A., and Serganov, A. Structural insights into ligand binding and gene expression control by an adenosylcobalamin riboswitch. Nat Struct Mol Biol 2012, 19, 1182–1184.
[200] Johnson, JE., Reyes, FE., Polaski, JT., and Batey, RT. B_{12} Cofactors directly stabilize an mRNA regulatory switch. Nature 2012, 492, 133–137.
[201] Joyce, GF. The antiquity of RNA-based evolution. Nature 2002, 418, 214–221.
[202] Eschenmoser, A. Etiology of potentially primordial biomolecular structures: from vitamin B_{12} to the nucleic acids and an inquiry into the chemistry of life's origin – a retrospective. Angew Chem Int Ed 2011, 50, 12412–12472.
[203] Fasching, M., Perschinka, H., Eichmüller, C., Gschösser, S., and Kräutler, B. Enhancing the methyl-donor activity of methylcobalamin by covalent attachment of DNA. Chem Biodiv 2005, 2, 178–197.
[204] Clardy, SM., Allis, DG., Fairchild, TJ., and Doyle, RP. Vitamin B_{12} in drug delivery: breaking through the barriers to a B_{12} bioconjugate pharmaceutical. Expert Opin Drug Deliv 2011, 8, 1–14.
[205] Ruiz-Sanchez, P., Konig, C., Ferrari, S., and Alberto, R. Vitamin B_{12} as a carrier for targeted platinum delivery: in vitro cytotoxicity and mechanistic studies. J Biol Inorg Chem 2011, 16, 33–44.
[206] Hogenkamp, HPC., Collins, DA., Live, D., Benson, LM., and Naylor, S. Synthesis and characterization of nido-carborane-cobalamin conjugates. Nucl Med Biol 2000, 27, 89–92.
[207] Bagnato, JD., Eilers, AL., Horton, RA., and Grissom, CB. Synthesis and characterization of a cobalamin-colchicine conjugate as a novel tumor-targeted cytotoxin. J Org Chem 2004, 69, 8987–8996.
[208] Hogenkamp, HPC., Collins, DA., Grissom, CB., and West, FG. Diagnostic and therapeutic analogues of cobalamin, Banerjee R, ed, Chemistry and Biochemistry of B_{12}, New York, John Wiley & Sons, 1999, 385–410.
[209] Wagner, E. Tumorspezifischer transfer von anti-microRNA zur Krebstherapie – pHLIP ist der Schlüssel. Angew Chem 2015, 127, 5918–5920.
[210] Wang, Z., Wang, Z., Liu, D. et al. Biomimetic RNA-silencing nanocomplexes: overcoming multidrug resistance in cancer cells. Angew Chem 2014, 126, 2028–2032.
[211] Shi, S., Lu, H., Gong, T., Zhang, Z., and Sun, X. Systemic delivery of microRNA-34a for cancer stem cell therapy. Angew Chem 2013, 125, 3993–3997.
[212] Garzon, R., Marcucci, G., and Croce, CM. Targeting microRNAs in cancer: rationale, strategies and challenges. Nat Rev Drug Discovery 2010, 9, 775–789.
[213] Castanotto, D., and Rossi, JJ. The promises and pitfalls of RNA-interference-based therapeutics. Nature 2009, 457, 426–433.
[214] Watts, JK., and Corey, DR. Silencing disease genes in the laboratory and the clinic. J Pathol 2012, 226, 365–379.
[215] Mutti, E., Hunger, M., Fedosov, S., Nexo, E., and Krautler, B. Organometallic DNA-B-12 conjugates as potential oligonucleotide vectors: synthesis and structural and binding studies with human cobalamin-transport proteins. ChemBioChem 2017, 18, 2280–2291.

[216] Quadros, EV. Advances in the understanding of cobalamin assimilation and metabolism. Br J Haematol 2009, 148, 195–204.
[217] Watkins, D., and Rosenblatt, DS. Inborn errors of cobalamin absorption and metabolism, Am J Med Genet C 2011, 157C, 33–44.
[218] Lerner-Ellis, JP., Tirone, JC., Pawelek, PD. et al. Identification of the gene responsible for methylmalonic aciduria and homocystinuria, cblC type. Nat Genet 2006, 38, 93–100.
[219] Dobson, CM., Wai, T., Leclerc, D. et al. Identification of the gene responsible for the cblB complementation group of vitamin B12-dependent methylmalonic aciduria. Human Mol Genet 2002, 11, 3361–3369.
[220] Banerjee, R. B_{12} trafficking in mammals: a case for coenzyme escort service. ACS Chem Biol 2006, 1, 149–159.
[221] Gherasim, C., Lofgren, M., and Banerjee, R. Navigating the B_{12} road: assimilation, delivery, and disorders of cobalamin. J Biol Chem 2013, 288, 13186–13193.
[222] Kim, J., Gherasim, C., and Banerjee, R. Decyanation of vitamin B_{12} by a trafficking chaperone. Proc Natl Acad Sci USA 2008, 105, 14551–14554.
[223] Hannibal, L., Kim, J., Brasch, NE. et al. Processing of alkylcobalamins in mammalian cells: A role for the MMACHC (cblC) gene product. Mol Genet Metab 2009, 97, 260–266.
[224] St Maurice, MS., Mera, P., Park, K., Brunold, TC., Escalante-Semerena, JC., and Rayment, I. Structural characterization of a human-type corrinoid adenosyltransferase confirms that coenzyme B_{12} is synthesized through a four-coordinate intermediate. Biochemistry 2008, 47, 5755–5766.
[225] Mutti, E., Ruetz, M., Birn, H., Kräutler, B., and Nexo, E. 4-ethylphenyl-cobalamin impairs tissue uptake of vitamin B_{12} and causes vitamin B_{12} deficiency in mice. Plos One 2013, 8, e75312.
[226] Zelder, F., Sonnay, M., and Prieto, L. Antivitamins for medicinal applications. ChemBioChem 2015, 16, 1264–1278.
[227] Guzzo, MB., Nguyen, HT., Pham, TH. et al. Methylfolate trap promotes bacterial thymineless death by sulfa drugs. Plos Pathog 2016, 12, e1005949.
[228] Widner, FJ., Lawrence, AD., Deery, E., Heldt, D., Frank, S., Gruber, K., Wurst, K., Warren, MJ., and Kräutler, B. Total synthesis, structure, and biological activity of adenosylrhodibalamin, the non-natural rhodium homologue of coenzyme B_{12}. Angew Chem Int Ed 2016, 55, 11281–11286.
[229] Gruber, K., Puffer, B., and Kräutler, B. Vitamin B_{12}-derivatives – enzyme cofactors and ligands of proteins and nucleic acids. Chem Soc Rev 2011, 40, 4346–4363.
[230] Winkler, WC., and Breaker, RR. Regulation of bacterial gene expression by riboswitches. Ann Rev Microbiol 2005, 59, 487–517.
[231] Croft, MT., Lawrence, AD., Raux-Deery, E., Warren, MJ., and Smith, AG. Algae acquire vitamin B_{12} through a symbiotic relationship with bacteria. Nature 2005, 438, 90–93.
[232] Rownicki, M., Wojciechowska, M., Wierzba, AJ. et al. Vitamin B-12 as a carrier of peptide nucleic acid (PNA) into bacterial cells. Sci Rep 2017, 7.
[233] Prieto, L., Rossier, J., Derszniak, K. et al. Modified biovectors for the tuneable activation of anti-platelet carbon monoxide release. Chem Comm 2017, 53, 4.

Paul A. Lindahl

7 Acetyl-coenzyme A synthase: a beautiful metalloenzyme

7.1 The big picture

Organisms can be categorized according to what they eat. *Heterotrophs* exclusively eat food derived from other living things. Humans, even vegans, are heterotrophs. *Autotrophs*, in contrast, live on matter with nonbiological origins – on inorganic chemicals like CO, CO_2, H_2, H_2S and N_2. The enzyme to be discussed in this chapter, *acetyl-coenzyme A synthase* (ACS), is found in certain autotrophs and in some heterotrophs that can also grow autotrophically [1, 2]. ACS along with carbon monoxide dehydrogenase (CODH/ACS) allows these organisms to grow autotrophically. Viewed anthropomorphically, heterotrophs are like thieves and scavengers who repurpose stolen or discarded items for their own needs and throw out what they don't want. Autotrophs are like survivalists who live off the land, making everything they need with their own hands. They live simply and efficiently, and generate little waste. Viewed from this perspective, autotrophs deserve our respect and admiration – they coexist peacefully with all living things on the planet and are hard-core environmentalists besides.

How do they do it? Many metabolic processes occurring within autotrophs are actually no different from those in heterotrophs. Rather, autotrophs convert inorganic nutrient molecules into metabolites that are central to the metabolism they share with heterotrophs. Acetyl-coenzyme A (coenzyme A is abbreviated CoA or CoASH) is one such metabolite. Once formed, acetyl-CoA can be used by both autotrophs and heterotrophs to help generate macromolecules required for life, including proteins, nucleic acids, lipids and other metabolites. ACS catalyzes the synthesis of the acetyl group of acetyl-CoA from CO and a methyl group, both of which ultimately originate from CO_2. Both carbon moieties bind to a particular nickel in the enzyme (called Ni_p) that is the heart of the so-called *A-cluster* active site. CoA is a carrier molecule of nucleotide origin that is regenerated when the acetyl group is transferred for use in downstream metabolic processes.

Paul A. Lindahl, Department of Chemistry, Texas A&M University, TX, USA

https://doi.org/10.1515/9783110496574-007

7.2 Organisms harboring ACSs

These enzymes are found in a wide distribution of anaerobic lineages within both archaeal and bacterial domains [3]. A group of organisms that contain ACSs are *obligate chemolithoautotrophs*; these are strict anaerobes that grow exclusively on CO_2 and H_2 as a source of carbon and reducing equivalents, respectively. Indeed, ACS is exquisitely O_2-sensitive and it can be inactivated with stoichiometric amounts of O_2. In practical terms, this means that all experiments involving ACS must be performed anaerobically (a major technical challenge). Certain obligate chemolithoautotrophs generate methane from CO_2 and H_2; examples include *Methanocaldococcus jannaschii*, *Methanothermobacter thermoautotrophicus* and *Methanopyrus kandleri* [4]. These organisms are thermophiles whose optimal growth temperature hovers around 100 °C. Despite being rugged individualists, the genomes of these organisms, which encode all of the tools they need for survival, are remarkably small. For this reason, *M. jannaschii* was selected to be the first organism to have its genome sequenced [5].

7.3 Connections to the origin of life

Early ideas regarding the origin of life implied that the first living cell was a heterotroph [6]. Writing in the 1870s Darwin suggested that life originated in "some warm little pond" teeming with organic compounds that spontaneously organized into the first cell. A few decades later, Haldane formally proposed the same – that life began in a rich "primordial soup" or broth. These ideas gained traction in 1953 when Miller and Urey demonstrated that amino acids and other biologically relevant organic molecules could be generated from aqueous solutions of simple inorganic molecules that were exposed to electrical discharges and heat. Such heterotrophic reasoning developed into the currently popular *RNA World* scenario according to which a self-replicating strand of the most complicated molecule on the planet formed spontaneously from a pool of organic molecules [7]. Accordingly, this was followed by the evolution of the genetic code and related proteins, all using molecules present in the primordial soup. In this scenario, metabolism evolved *last* – only after the metabolites in the rich broth were consumed and the heterotrophic organisms were forced to work for a living.

In 1988, Gunter Wächtershäuser proposed exactly the opposite – that life began under autotrophic conditions (a thin broth), namely on a mineral surface of iron sulfide (perhaps also containing nickel) [8, 9]. According to what has become known as the *Iron–Sulfur World* scenario, the first step leading to life was the reduction of CO_2 on this surface by a chemical reductant (e.g., H_2S) to generate organic functional units such as acetyl groups (Figure 7.1, left panel). This is essentially the same reaction as catalyzed by ACSs, with a Ni–Fe–S surface acting as the primordial precursor of the A-cluster active site of the modern enzyme. These ancient events were

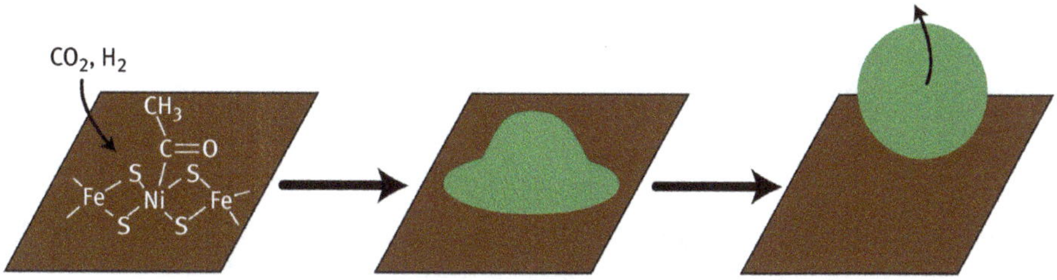

Figure 7.1: Steps in the origin of life, as proposed by the iron–sulfur world scenario. Left panel: The initial step was the reduction of CO_2 on a mineral surface of iron sulfur (and nickel) near a hydrothermal vent. A variety of reduced carbon species formed on this surface, including activated acetyl groups similar to the formation of acetyl-CoA on the Ni_p site of the A-cluster of extant CODH/ACS enzymes. Middle panel: The metabolism of the surface organism evolved to generate macromolecules, including a hemispherical membrane "droplet" covering the surface cell. Right panel: Once transcription, translation and DNA replication had evolved, the membrane pinched off of the surface and the first cell was born.

proposed to have occurred near hydrothermal ocean vents when the young Earth was hot but cooling. Activated organic groups on the surface eventually organized into an autocatalytic "surface metabolist" in which the citric acid cycle operated in the reductive direction, reducing CO_2 to organic molecules on that surface. Through the process of waste conversion [10], some of these molecules, which originally served no function, eventually contributed to the growing and evolving autocatalytic system. Once that happened, the new molecule and associated reaction became part of the surface metabolist. According to the Fe/S world model, metabolism evolved *first*, whereas proteins, nucleic acids and the genetic code came later [11]. A critical event in this process occurred when the surface metabolist becomes covered with a half-domed phospholipid membrane (Figure 7.1, middle panel); this trapped an aqueous solution in which cytosolic chemistry could evolve. Once the interior of that semi-cellular system evolved sufficiently to contain nucleic acids that served as templates for mRNA and proteins, the first cell budded off of the surface and began to self-replicate (Figure 7.1, right panel). The first cell would have been a thermophilic anaerobic obligate chemolithoautotroph whose closest current relatives would include the three methanogenic archaea mentioned earlier. The main point is that ACS and the organisms that contain this enzyme are evolutionarily ancient – catalyzing perhaps the oldest process associated with the emergence of life [3].

7.4 Reactions catalyzed by ACS

Modern ACS enzymes are far more sophisticated than their ancestral counterparts (see [12–15] for recent reviews). Extant enzymes catalyze the synthesis of acetyl-CoA from CO, CoASH and a methyl group donated by the so-called corrinoid–iron–sulfur protein (88 kDa, abbreviated CoFeSP). Heterodimeric CoFeSP contains a cobalamin and an $[Fe_4S_4]$ cluster. The reaction catalyzed by ACS is

$$CH_3^- \rightarrow Co^{III}FeSP + CO + CoASH \xrightleftharpoons{ACS\ (\alpha)} CH_3{-}C(=O){-}SCoA + Co^{I}FeSP + H^+ \qquad (7.1)$$

In reaction (7.1), the methyl group is initially coordinated to the cobalt of CoFeSP. This group is viewed as a Lewis base (with a –1 formal charge on the carbon) that donates an electron pair to form a coordinate bond with a Co^{III} ion of the cobalamin (a Lewis acid). Reaction (7.1) involves the transfer of a methyl group *cation* (+1 formal charge, six valence electrons) such that the electron pair used in forming the coordinate bond with the cobalt (designated by an arrow from the Lewis base to the Lewis acid in (7.1)) becomes property of the cobalt once the methyl group leaves. The two-electron reduction of Co^{III} yields Co^{I} as a product.

In ACS assays, CO can be added directly or it can be generated from CO_2 in a reaction catalyzed by CODH as follows:

$$CO_2 + 2e^- + 2H^+ \xrightleftharpoons{CODH\ (\beta)} CO + H_2O \qquad (7.2)$$

CODH and ACS typically associate with a 310 kDa bifunctional enzyme called CODH/ACS. Although this chapter focuses on ACS (Chapter 3 focuses on CODH), some basic properties of CODHs must be mentioned here to understand the bifunctional enzyme.

$Co^{I}FeSP$ is methylated using CH_3-tetrahydrofolate according to reaction (7.3), which is catalyzed by a methyltransferase (MT).

$$Co^{I}FeSP + \underset{(CH_3\text{–}THF)}{{>}N{-}CH_3} + H^+ \xrightleftharpoons{MT} CH_3^- \rightarrow Co^{III}FeSP + \underset{(THF)}{{>}N{-}H} \qquad (7.3)$$

ACS catalytic activity is typically initiated by mixing enzyme (ACS or CODH/ACS), a strong reductant (e.g., Ti^{III} citrate), CH_3-Co^{III}FeSP, CO and coenzyme A. Catalytic amounts of CH_3-Co^{III}FeSP are typically used along with a regenerating system that includes excess CH_3–THF and catalytic MT. When using CODH/ACS, CO_2 rather than CO may be used as a substrate. Samples are quenched at increasing times, and injected onto an HPLC column for detection and quantification. The slope of [acetyl-CoA] versus time is the steady-state rate of catalysis.

7.5 Wood/Ljungdahl pathway

Harland Wood and Lars Ljungdahl discovered the metabolic pathway involving these reactions, and that pathway has been named in their honor [16]. The Wood/Ljungdahl pathway (WLP) is an evolutionarily ancient pathway for generating organic material from CO_2 in which CODH/ACS plays the central role (Figure 7.2). This pathway plays a huge role in the global carbon cycle. It allows certain organisms (e.g., acetogens, sulfidogens and methanogens) to grow on CO_2 as the sole source of carbon and on H_2 as the source of reducing equivalents. In one branch of the pathway, CO_2 is reduced to the level of a methyl group bound to tetrahydrofolate. In the other branch, a second molecule of CO_2 is reduced to CO by the CODH portion of CODH/ACS. The products of the two branches meet at the A-cluster of ACS to generate an acetyl group, which is finally removed as acetyl-CoA. Some acetyl-CoA is used in anabolism to build cellular constituents; other acetyl-CoA molecules are hydrolyzed in catabolism to generate ATP and acetate.

The WLP is part of an energy conservation system in which membrane-bound redox enzymes pump either H^+ or Na^+ ions across that membrane to create a chemiosmotic gradient that is used to synthesize ATP [17]. CO_2 and reducing equivalents from H_2 drive the WLP. The reducing power contained in H_2 is insufficient to reduce the low-potential ferredoxin that is used, in turn, to reduce CO_2 to CO. So how can reduction occur? The most likely answer is that an "electron bifurcating" hydrogenase couples an exergonic reduction (forming NADH) to an endergonic one (forming the reduced ferredoxin) [18].

7.6 Classes of ACS enzymes

There are two major types of ACS enzymes, one from bacteria and the other from archaea [3, 4] (Figure 7.3). Bacterial CODH/ACS enzymes are heterotetramers ($\alpha_2\beta_2$) that catalyze both reactions (7.1) and (7.2). These bifunctional enzymes use an independent CoFeSP to donate the methyl group. The 82 kDa α-subunit has ACS activity while the 70 kDa β-subunit has CODH activity. The best-studied CODH/ACS is from the acetogen *Morella thermoacetica*. In recombinant genetic constructs, the α-subunit can be expressed independently of β. After incubation with $NiCl_2$, α has ACS activity as long as CO rather than CO_2 is used as a substrate. Monomeric monofunctional ACS enzymes that catalyze only reaction (7.1) are similar to the α-subunits of CODH/ACSs. CoFeSP is a separate heterodimeric protein that must be included in assays. The best characterized example of monomeric ACSs is that from the hydrogenogenic bacterium *Carboxydothermus hydrogenoformans* [19]. This organism can grow on CO as its sole source of carbon, and monomeric ACS is expressed under this condition (probably because CODH is unnecessary).

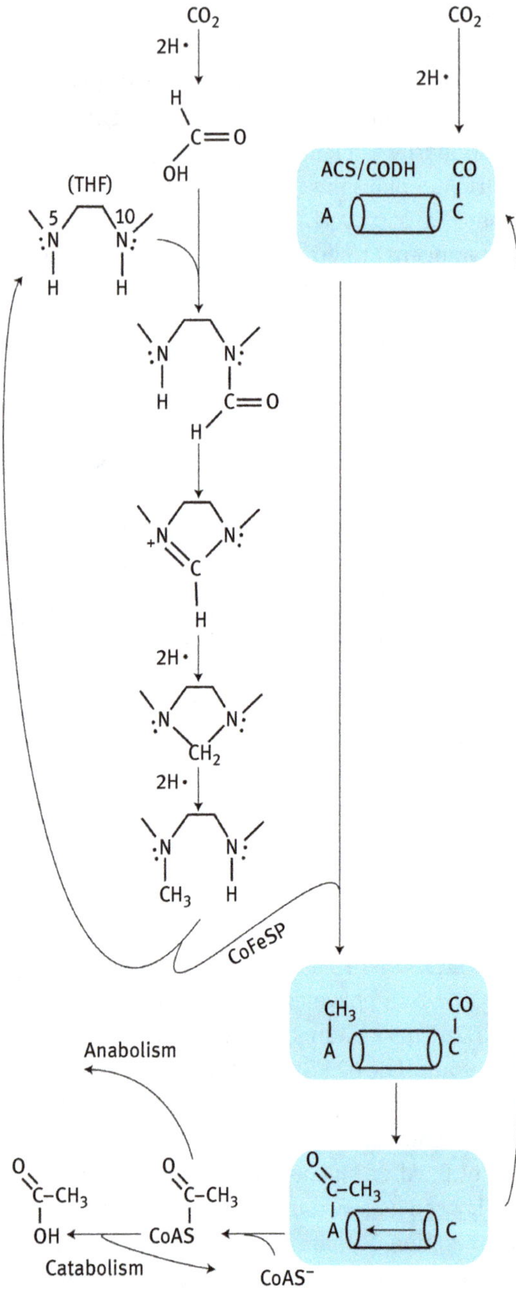

Figure 7.2: The Wood/Ljungdahl pathway. This is an autotrophic pathway in which two CO_2 molecules are reduced by eight hydrogen atom equivalents (originating as $4H_2$) to form acetyl-CoA. In acetogens, a portion of the acetyl-CoA is used in anabolism to generate all of the carbon-based cellular components. The remainder is used in catabolism to generate cellular energy (and

Archaeal enzymes are sometimes called acetyl-CoA decarbonylase/synthases (ACDSs). These bifunctional enzymes probably have an $(\alpha_2\varepsilon_2)_4\beta_8(\gamma\delta)_8$ quaternary structure [20] though some ambiguity remains. Note that the nomenclature is different; the α-subunit of ACDSs catalyzes the CODH reaction, the β-subunit catalyzes the ACS reaction and the γ- and δ-subunits of the complex are homologous to CoFeSP. The ε-subunit houses no metal centers and has no homology in bacterial ACSs or CODH/ACSs. In this chapter, the default nomenclature will be that for CODH/ACS (where ACS = the α-subunit and CODH = β-subunit). ACDSs catalyze both the CODH reaction (7.2) as well as reaction (7.4) in which acetyl-CoA is *decarbonylated* into a methyl group, CO and CoASH. The best-studied ACDS is from the methanogen *Methanosarcina thermophila*.

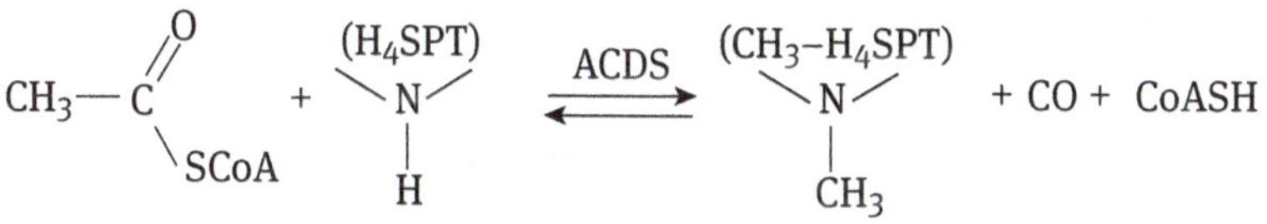

$$CH_3-C(=O)(SCoA) + (H_4SPT)N-H \xrightleftharpoons{\text{ACDS}} (CH_3-H_4SPT)N-CH_3 + CO + CoASH \quad (7.4)$$

The methyl group acceptor in ACDS-containing organisms is tetrahydrosarcinapterin (H_4SPT), which is functionally similar to THF. The methyl group of CH_3-H_4SPT is converted into methane; the overall process is called *acetoclastic methanogenesis*. This metabolic pathway allows particular methanogens to grow (heterotrophically) on acetate, producing methane as waste.

Bacteria that house CODH/ACSs can be subdivided into Terrabacteria and Gracilicutes [3]. The CODH/ACSs that have been investigated are isolated from Terrabacteria (e.g., *M. thermoacetica* and *C. hydrogenoformans*). The archaea that house these enzymes are subdivided into what I call "*hard-core*" autotrophs (formally called cluster I) such as the Methanococcales or Methanopyrales, and "*flexitarians*" (formally cluster II) such as the Methanosarcinales or Archaeoglobales, which can grow either autotrophically or heterotrophically.

Figure 7.2 (continued)

eliminate acetic acid as waste). The role of CODH/ACS is highlighted in blue. One "branch" of the pathway converts CO_2 to the methyl group of tetrahydrofolate. This is a multistep process involving numerous enzymes (see [16]). Shown here is only the portion of the TFH molecule that is being acted on. Each hydrogen equivalent is indicated as H•. The methyl group is eventually transferred to CoFeSP, and then to ACS. The other "branch" of the pathway converts CO_2 to CO, in the reaction catalyzed by the CODH portion of CODH/ACS. The two branches merge when CO migrates through the tunnel to the A-cluster, where it reacts to form an acetyl group on Ni_p. Subsequent attack by CoASH affords the product acetyl-CoA which is used in various downstream processes, some of which generate acetate.

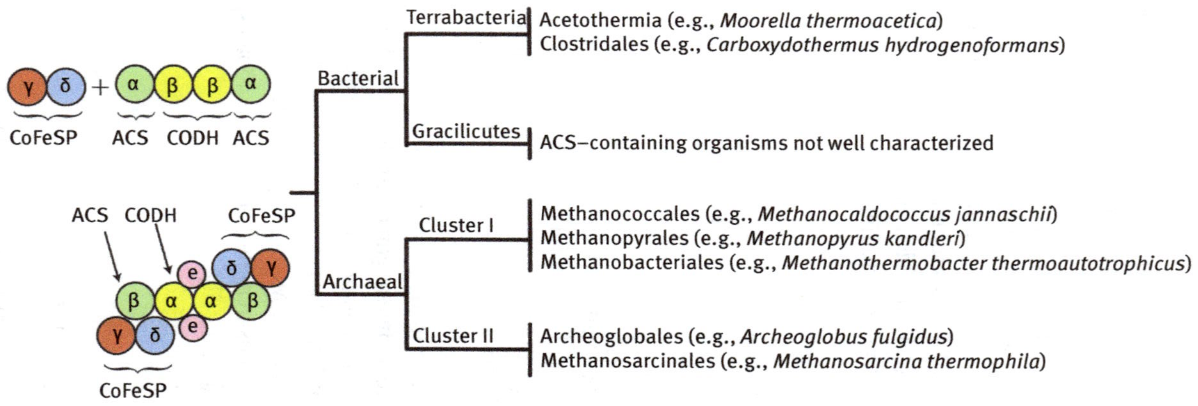

Figure 7.3: Quaternary structure of ACS enzymes and the distribution of those enzymes in living systems. The fundamental division is between bacterial- and archaeal-type enzymes (adapted from [3]). Both are bifunctional, with both CODH and ACS activities. Bacterial enzymes are $\alpha_2\beta_2$ tetramers in which the two β-subunits (catalyzing the CODH reaction) form a central core; the α-subunits, catalyzing the ACS reaction, are on the periphery of the β_2 core. The CoFeSP is an independent and autonomous heterodimeric protein. The exact quaternary structure of the archaeal ACDS enzyme is not known with certainty, and no x-ray diffraction structure has been reported. The subunit relationships suggested here are solely based on the functional connections to the better characterized bacterial enzymes. The two CODH α-subunits are presumed to form a central core, and the ACS β-subunits are presumed to bind peripherally to this core. The CoFeSP subunits are part of the complex and must be located near to the ACS subunits, since a close association between the Co and Ni_p is required for methyl group transfer. The ε-subunits have no metal centers and no known function but are associated with the CODH subunits.

7.7 Structure of the protein and the A-cluster active site

α-Subunits from bacterial ACSs and CODH/ACSs consist of three domains, including N-terminal, central and C-terminal [21, 22, 19]. The homologous β-subunits of ACDS enzymes are smaller; they consist of only the central and C-terminal domains. In bacterial $\alpha_2\beta_2$ heterotetramers, the α-subunits are bound on the periphery of a core ββ-CODH unit to create an overall linear αββα structure (Figures 7.3 and 7.4). The N-terminal domain interacts directly with the ββ core. The exact subunit arrangement of the archaeal ACDS enzymes is not known. However, it is reasonable to speculate that two CODH α-subunits, in the form of an $\alpha_2\varepsilon_2$ subcomponent also form the core [23], and that the ACS β-subunits are on the periphery, analogous

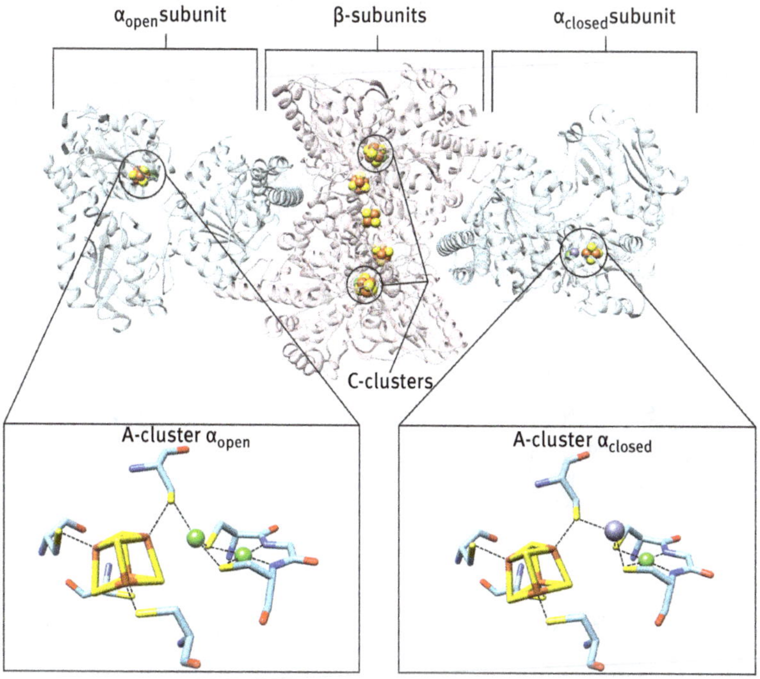

Figure 7.4: X-ray diffraction crystal structure of CODH/ACS from *M. thermoacetica*. The enzyme is approximately linear with α-subunits on the periphery and two β-subunits forming the central core. The α-subunit on the left is in the open conformation; that on the right is in the closed conformation. The structure of the A-cluster in each subunit is highlighted, with Ni atoms shown as green spheres. The A-cluster on the right has zinc (gray sphere) rather than Ni in the proximal site. The clusters in the β-subunits are also indicated, with the C-clusters circled. Other clusters in the β-subunits are standard Fe_4S_4 clusters serving redox roles. This figure and Figure 7.6 were generated as described [28].

to the bacterial enzymes. Both bacterial and archaeal enzymes show metabolic channeling of CO, and CODH and ACS activities that are coupled/synchronized [24–27], suggesting common structural features connecting the two active sites. Also, the γδ-subunits of the archaeal enzymes (corresponding to the bacterial CoFeSP) must interact with the ACS subunit in order to transfer the methyl group. These considerations suggest the overall quaternary unit structure for the archaeal enzyme shown in Figure 7.3. The probable overall $(\alpha_2\varepsilon_2)_4\beta_8(\gamma\delta)_8$ quaternary arrangement implies that four of these $(\gamma\delta)(\beta)(\alpha_2\varepsilon_2)(\beta)(\gamma\delta)$ units interact to form a huge 2 mega-Da complex.

Each α-subunit of CODH/ACS enzymes contains a single metal center called the A-cluster, which is the active site for acetyl-CoA synthase and is the only redox site in the subunit. The A-cluster consists of a dinickel unit linked to a Fe_4S_4 cubane through a bridging cysteinate ligand. The cubane is coordinated to the protein via four cysteine residues including the bridge (Figure 7.5). The nickel that is *proximal* to the cubane and bound to the bridging cysteinate, called Ni_p, is the site where CO and methyl group (and perhaps the sulfur of $CoAS^-$) bind. Besides the cysteinate that bridges to the cubane, Ni_p is also coordinated by two cysteinate sulfurs that are bridged to the *distal* nickel (Ni_d) of the A-cluster. In the two crystal structures that have Ni installed in the proximal site, there is some extra electron density indicating a small molecular species (tentatively assigned as water) coordinating to Ni_p that completes a square-planar geometry. The CoA binding site on ACS is uncertain but it must place the sulfur within a few angstroms of Ni_p.

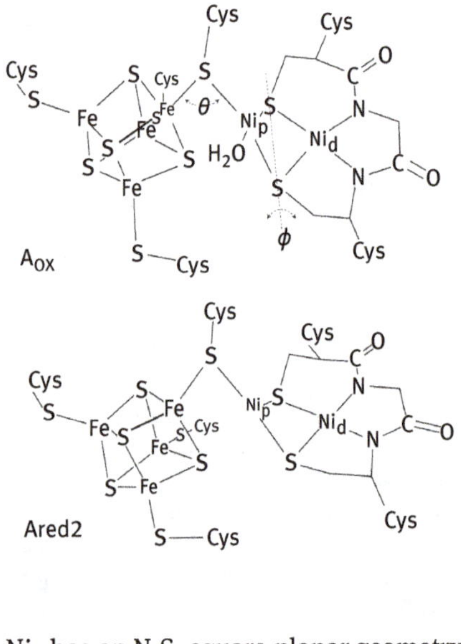

Figure 7.5: Calculated structures of the A-cluster in the A_{ox} and A_{red2} states in the open conformations. Fe-S-Ni_p and Ni_p-S_2-Ni_d angles change with redox state, illustrating the flexibility of this site. θ ranges from 70° to 100°; ϕ ranges from 120° to 130°. The cubanes can also twist relative to the dinickel unit. Adapted from [29].

Ni_d has an N_2S_2 square-planar geometry that includes the two bridging cysteinate sulfurs and two deprotonated amide nitrogens from the protein backbone. Ni_d is not labile and is "locked" into a low-spin (LS) d^8 square-planar Ni^{II} electron configuration at each step of catalysis. For this reason, we will exclude the Ni_d oxidation state from our electronic configurations for the various A-cluster states described later.

Ni_p in contrast has some unusual coordination features that endow it with novel catalytic properties. For example, all three of its endogenous cysteinate ligands are bridging to other metal centers. This partially neutralizes the negative charge on the thiolates, which in turn stabilizes Ni_p in unusually low oxidation states – at least the d^9 Nip^I state but perhaps the unprecedented d^{10} zero-valent $Ni_p{}^0$ state. (Note that $Ni_p{}^0$ is a *formal* description. If it exists, some of the electron density on the nickel would undoubtedly delocalize onto its ligands and neighboring metal centers.) The low coordination number also stabilizes Ni_p in low oxidation states. The coordinated water bound to Ni_p likely dissociates when Ni_p becomes reduced, leaving a three-coordinate site (Figure 7.5). Besides stabilizing low-valent nickel, the low coordination number of Ni_p also makes available two to three open coordinate sites for binding multiple substrates. The same property renders this metal ion labile and susceptible to chelation by the bidentate chelator 1,10-phenanthroline [30, 31]. ACS activity is abolished when Ni_p is removed, and it recovers when nickel is reinserted into that site. Copper and zinc can also bind at the proximal site, but the enzyme is inactive [32].

The mechanism of A-cluster assembly has recently been investigated [33]. The Fe_4S_4 cubane is probably installed first. Such clusters are probably assembled using "house-keeping" (ISCU-type) enzymes. AcsF then installs both Ni ions into the protein to generate active A-cluster-containing ACS. AcsF is a MinD-type ATPase in which the free energy generated from ATP hydrolysis promotes the transfer of Ni into the cubane-containing ACS. MinD is a dimeric ATPase in *Escherichia coli* that helps center the FtsZ ring used in cell division [34]. There are many members of the MinD family, including CooC, which inserts Ni into the CODH active site, and the Fe protein of nitrogenase which delivers low-potential electrons to the MoFe protein in an ATP-dependent fashion [35].

7.8 Molecular tunnel

CODH/ACS contains a molecular tunnel through which CO migrates from the C-cluster (the CODH active site in the β-subunit) to the A-cluster in the ACS α-subunit [21, 22] (Figure 7.6). Under normal physiological conditions, CO is generated at the C-cluster and migrates through the tunnel to the A-cluster where it binds Ni_p. The tunnel not only provides direct delivery of CO via metabolic channeling, it also *regulates* delivery such that CO arrives at the A-cluster at just the right step in the catalytic mechanism (called catalytic coupling). This regulated delivery of CO is among the most amazing and elegant aspects of the enzyme.

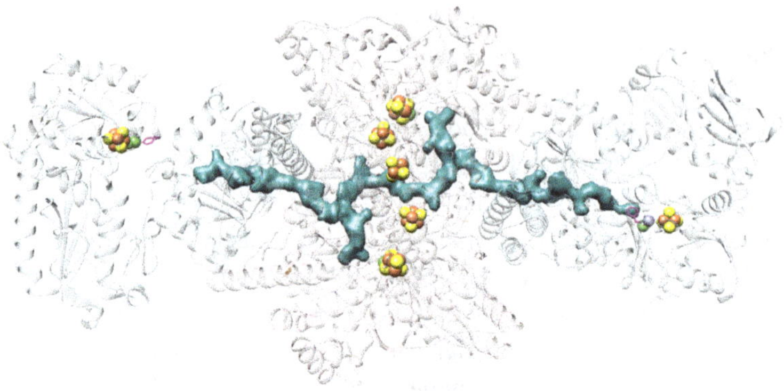

Figure 7.6: Tunnel in CODH/ACS enzymes for regulated delivery of CO to the A-cluster. The α-subunit in the closed conformation (with open tunnel) is on the right; the subunit with the open conformation (with closed tunnel) is on the left. The position of the phenylalanine that controls tunnel access to the A-cluster is highlighted as a magenta hexagon. The tunnel was generated as described [36].

The crystal structure of CODH/ACS reveals how this is done. There are two conformations of the α-subunit called open and closed (Figure 7.4; left α-subunit is open; right α-subunit is closed). The conformation affects the degree to which Ni_p is surface exposed and receptive to methyl group transfer. Ni_p is surface exposed in the open conformation and buried in the closed conformation. Ni_p *must* be surface exposed to allow methyl group transfer. The methyl group is directly handed-off from the cobalt of CH_3-Co^{III}FeSP to Ni_p. Thus, CH_3-Co^{III}FeSP must dock with ACS in the open conformation such that the {Co-CH_3 ... Ni_p} distances and angles are appropriate for an S_N2-based nucleophilic attack of the methyl group carbon by Ni_p. The X-ray crystal structure of isolated heterodimeric CoFeSP shows the cobalamin sandwiched between the two CoFeSP subunits, and unable to access Ni_p [37]. Holger Dobbek and coworkers concluded that the CoFeSP structure must be highly flexible and that its conformation changes upon docking with ACS so as to position the Co-CH_3 unit in close proximity to Ni_p. Unfortunately, an X-ray diffraction crystal structure of a CODH/ACS/CH_3-Co^{III}FeSP complex, which might show this intriguing arrangement, has not been reported.

The tunnel leading to Ni_p is blocked by a conserved phenylalanine in the open α-subunit conformation (indicated in Figure 7.6). This prevents CO from migrating to the A-cluster during the methyl group transfer step [20]. In the closed α-subunit conformation, Ni_p is buried, but the tunnel is open and the phenylalanine has moved out of the way. This allows CO to bind Ni_p exclusively after the methyl group is bound. This explanation implies an ordered mechanism in which the methyl group binds before CO. Without this regulated delivery of CO, CO could bind before the methyl group. As discussed below, the CO-bound state of the A-cluster inhibits

ACS activity. CO binding deactivates Ni_p so that the CO-bound form is insufficiently nucleophilic to accept the methyl group [20].

7.9 Heterogeneity

The A-cluster can be prepared in numerous electronic, magnetic and catalytic states. However, before introducing these states, a problem called *heterogeneity* must be mentioned. There is abundant evidence that only 30% ± 20% of CODH/ACS molecules within a population are catalytically active and the remaining molecules are inactive. This situation was discovered in ca. 1988 [38, 39] and no one has unambiguously figured out how to eliminate it despite 30 years of effort. The problem is most commonly evidenced by the low spin intensity of the $S = ½$ NiFeC EPR signal (see below); the signal historically integrates to only approximately 0.1–0.4 spins/mol α-subunit. Other properties follow this trend. Only approximately 0.3 Ni_p ions per αβ-dimer can be removed by treatment with 1,10-phenanthroline [31], only approximately 0.5 methyl groups bind per αβ-dimer [40], and only approximately 0.2 CoA molecules bind per αβ-dimer [41].

What causes heterogeneity? The problem is already present in unpurified enzyme contained in whole *M. thermoacetica* cells, that is, heterogeneity does not arise from O_2 damage during purification [42]. Heterogeneity can certainly arise from foreign metal misincorporation. For example, only half of the proximal sites in the structure of the tetrameric $\alpha_2\beta_2$ CODH/ACS complex are occupied by Ni; the other half by zinc [22]. In another CODH/ACS structure, all proximal A-cluster sites are occupied by copper [21]. Once these other metals are bound at the active site, the enzyme is inactive and the NiFeC EPR signal from the A_{red1}-CO state of the A-cluster (see below) cannot be generated. Thus, a strategy to eliminate heterogeneity has been to remove all metal ions from the proximal site (using a strong chelator) and then add back $NiCl_2$. Doing this reportedly increased the NiFeC spin intensity to 0.8 spin/α [43]. However, spin concentrations from the same lab returned to more typical values in subsequent studies (0.38–0.67 spin/α [44], 0.33 spin/α – in a sample containing 2 Ni/α [45] and 0.65 spin/α [46]). Quantifying EPR signal intensities is notoriously difficult and fraught with uncertainties. Unambiguous evidence of solving the heterogeneity problem must include Mössbauer (MB) spectra of the claimed heterogeneity-free α-subunit in the A_{red}-CO state [4]. Not only should the NiFeC signal integrate to 1 spin/α, the MB spectrum should be that of a single $[Fe_4S_4]^{2+}$ cluster possessing magnetic hyperfine interactions (due to spin coupling of an $S = ½$ Ni^I-CO moiety [38]). No one has met this challenge to date.

The problem may not be due exclusively to mismetallation. The X-ray crystal structure of monomeric ACS shows a full complement of Ni in the proximal binding site (no Zn or Cu), yet NiFeC spin intensities were only 0.14 spin/α [19]. Recombinant

M. thermoacetica α-subunit, when expressed in *E. coli* and activated with $NiCl_2$, shows the same extent of heterogeneity as complete CODH/ACS expressed in the natural host *M. thermoacetica*. Heterogeneity is observed by MB in cubane-containing recombinant α-subunit *prior* to adding $NiCl_2$ [47]. Also, low spin intensities are observed for signals from other metal sites in CODH/ACS, such as the C-cluster of CODH for which mismetallation has not been reported. These results suggest that the origin of heterogeneity is an early event at the protein level without any organism-specific factors.

Heterogeneity is an inconvenient truth, but we need to be cognizant of it when assigning electron configurations to various intermediate states of catalysis. These assignments in turn impact our understanding of the catalytic mechanism of the enzyme. If the heterogeneity problem could be solved, major advances in understanding the catalytic mechanism of the enzyme could be rapidly forthcoming. Until then, we are left with the ambiguities described below.

7.10 The A_{ox} state

This "resting" state is obtained when an enzyme is in anaerobic buffer lacking CO or a strong reductant such as dithionite or Ti^{III} citrate. It is perhaps the least controversial of any A-cluster state. The A-cluster structures determined by X-ray diffraction were probably in the A_{ox} state. Heterogeneity is not apparent in the MB spectrum of this diamagnetic state, which indicates 100% $S = 0$ $[Fe_4S_4]^{2+}$ clusters [47, 48]. The A_{ox} state exhibits no EPR signal suggesting the Ni_p^{II} oxidation state. Considered collectively, the electronic state of the A-cluster is assigned as $\{[Fe_4S_4]^{2+}\ Ni_p^{2+}\}$.

7.11 The A_{red1}-CO state

This well-studied paramagnetic state is obtained when CODH/ACS (starting in the A_{ox} state) is exposed to an atmosphere of CO. Some of the CO is oxidized to CO_2 at the C-cluster, and the electrons generated in this process are used to reduce other clusters in the enzyme, including the A-cluster. The same state is obtained when monomeric ACS (or the recombinant α subunit) is exposed to CO in the presence of a low-potential reductant such as dithionite or Ti^{III} citrate. The A_{red1}-CO state is $S = ½$ and it exhibits the NiFeC EPR signal with *g*-values of approximately 2.08, 2.07 and 2.01. This signal, discovered by Steve Ragsdale in 1982 [49], is called the *NiFeC* signal because it exhibits hyperfine broadening when the A-cluster is enriched in ^{61}Ni ($I = 3/2$), ^{57}Fe ($I = ½$) or ^{13}CO ($I = ½$). MB spectra of this state reveal that the cubane is not reduced to the $[Fe_4S_4]^{1+}$ state but rather remains in the $[Fe_4S_4]^{2+}$ state [38]. This conclusion is based on the average isomer shift of the cluster ($\delta = 0.42$ mm/s) in the A_{red1}-CO state, which is typical of standard $[Fe_4S_4]^{2+}$ clusters [38]. Thus, reducing the cubane to the $[Fe_4S_4]^{1+}$ cluster must be more difficult thermodynamically than reducing Ni_p^{II} to Ni_p^{I}. Hyperfine broadening is observed when Ni_p is exclusively enriched in ^{61}Ni; no hyperfine broadening is observed when Ni_d is enriched with the same isotope [30]. This indicates that the unpaired electron in the A-cluster is primarily associated with Ni_p^{I}. A single CO molecule binds to N_p^{I} of the A-cluster, affording Ni_p in a tetrahedral environment that includes a linear Ni–CO unit with a Ni–C distance of ~1.77 Å [46]. Considered collectively, the electronic configuration of the A_{red1}-CO state is assigned as $\{[Fe_4S_4]^{2+}\ Ni_p^{1+}\text{-CO}\}$. Starting from the A_{ox} state, attainment of the A_{red1}-CO state requires a one-electron reduction of Ni_p^{2+} and binding of CO to Ni_p^{1+}, as follows:

$$A_{ox}\left\{[Fe_4S_4]^{2+}Ni_p{}^{2+}\right\}+1e^-+CO\rightarrow A_{red1}\text{–}CO\left\{[Fe_4S_4]^{2+}Ni_p{}^{1+}-CO\right\} \qquad (7.5)$$

7.12 CO inhibition

The rate of acetyl-CoA synthesis initially increases as [CO] increases from zero, but eventually it declines, which demonstrates that CO is not only a substrate for the enzyme but also an inhibitor [20, 50, 51]. The extent of inhibition is correlated with development of the NiFeC signal. CO inhibits the activity of the wild-type (WT) A-cluster-containing enzyme from *M. thermophila* in accordance with K_i = 0.80 mM. In contrast, it inhibits the activity of the F195A mutant with K_i = 0.17 mM [20]. (F195 is the phenylalanine that blocks CO from passing through the tunnel to the A-cluster.) David Grahame and coworkers simulated their data using a mechanism in which inhibitor CO and $CH_3\rightarrow Co^{III}$FeSP compete for the same form of the enzyme (in this case from *M. thermophila*). They also monitored the NiFeC signal intensity as a function of different [CO] concentrations, and found K_d = 0.65 mM for WT and K_d = 0.12 mM for the F195A mutant. The fact that these K_ds were so similar to the K_i's demonstrates that A_{red}-CO (which gives rise to the NiFeC signal) is an inhibited state of the enzyme. For *M. thermoacetica*, CO inhibition and NiFeC signal formation develop at lower concentrations of CO: Tan et al. [50] obtained K_I ~ 5 µM and K_d ~ 3 µM for development of the NiFeC signal but the same conclusion holds. Grahame first suggested that CO was an inhibitor in 1996 [52] and my lab provided further evidence of this a year later [40]. The collective evidence is clear and unambiguous.

7.13 Reductive activation in the absence of CO

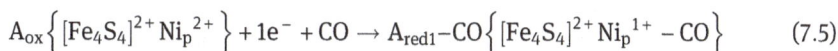

The A_{ox} state must be treated with a low-potential reductant before it can accept a methyl group from CH_3-Co^{III} FeSP. Only powerful in vitro reductants such as dithionite or Ti^{III} citrate are effective in this regard because the redox potential for A-cluster reduction is quite low, $E^{0\prime}$ ~ –0.5 V versus NHE [40, 53]. Of all the known states of the enzyme, the reductively activated state is the most controversial, with two issues swirling. The first is whether this state is generated by reducing A_{ox} by one or two electrons. We'll refer to the possible resulting reduced states as A_{red1} and A_{red2}, respectively. To examine this, the enzyme activity has been monitored as reducing equivalents are titrated into the enzyme [54]. Simulations have been fitted to the resulting plots using the Nernst equation with $n = 1$ or 2. With perfect data, the two plots could easily be distinguished because the slope of an $n = 2$ plot is steeper at the inflection point than that of an $n = 1$ plot. Using actual data, some plots fit better to $n = 1$ [53] and others to $n = 2$ [54, 47] or to two sequential $n = 1$ steps [55]. This has led to some ambiguity as to whether the fully reduced enzyme is A_{red1} ($n = 1$) or A_{red2} ($n = 2$). The reduction potential is pH dependent indicating a protonation event in association with reduction.

Another way to evaluate whether one or two electrons are used to reduce A_{ox} is by EPR. If one electron were required, then titrations with an excess of dithionite (or Ti^{III} citrate) should generate an EPR-active paramagnetic state. If two electrons were required, the final product should be diamagnetic or integer spin – either case should be EPR-silent. After treatment with a strong reductant, ACS enzymes are devoid of EPR signals that are characteristic of Ni^{I} species. However, due to heterogeneity, some ACS molecules within a population exhibit low-intensity features suggesting $\{[Fe_4S_4]^{1+}\ Ni_p^{II}\}$ clusters. Other such molecules that are EPR-silent could be reduced by two electrons, such as spin-coupled $\{[Fe_4S_4]^{1+}\ Ni_p^{I}\}$ clusters.

7.14 A_{ox}-CH_3 state

This controversial issue can be clarified by considering the reaction of the reductively activated state (A_{red1} or A_{red2}) with $CH_3 \rightarrow Co^{III}FeSP$ to form the methylated state A_{ox}-CH_3, namely

$$\{A_{red1}\text{ or }A_{red2}\} + CH_3 \rightarrow Co^{III}FeSP \rightleftarrows A_{ox}\text{-}CH_3 + Co^{I}FeSP \quad (7.6)$$

Steve Ragsdale has elegantly shown that reaction (7.6) occurs by an S_N2 nucleophilic displacement mechanism [56], resonating with an earlier result by Heinz Floss and coworkers that this step is associated with the inversion of stereochemical configuration at the methyl group [57]. Floss actually showed retention of configuration starting from CH_3–THF and ending with CH_3-C(O)-CoA. To explain this, he proposed two S_N2 methyl group displacements, each of which occurred with inversion. The first displacement involves the transfer of CH_3-THF to CoFeSP, while the second involves methyl group transfer to ACS. There is no evidence of radical chemistry.

These solid foundational results impact strongly on the mechanism of catalysis. The Co^{I} oxidation state of the $Co^{I}FeSP$ product indicates that a methyl group cation (with six valence electrons) is transferred to the A-cluster. The methylated product

A_{ox}-CH_3 is EPR-silent, and MB spectra confirm that it is diamagnetic [47]. MB spectra of A_{ox}-CH_3 reveal a quadrupole doublet with parameters typical of an oxidized $S = 0$ $[Fe_4S_4]^{2+}$ cubane. The amount of radioactive methyl that can bind the enzyme (~0.5 methyl groups per αβ) [40] matches the amount of labile Ni_p and the low spin concentration of the NiFeC signal. These data collectively support the view that the methyl cation binds to Ni_p. Moreover, there are many examples of synthetic inorganic Ni^{II}-CH_3 complexes. Thus, the electronic configuration of the A_{ox}-CH_3 state is assigned as $\{[Fe_4S_4]^{2+}\ Ni_p^{II} \leftarrow {}^{-}CH_3\}$.

Since the A-cluster product of the methyl group transfer reaction is diamagnetic and an even number of valence electrons are added to the A-cluster in the reaction, the other substrate in reaction (7.6), namely the reductively activated form of the A-cluster, must also be diamagnetic (or integer spin). This is a requirement of the conservation of spin angular momentum. This implies a two-electron reduction of A_{ox}, forming an A_{red2} state that is either diamagnetic or integer spin.

Perhaps the strongest evidence favoring the A_{red2} reductively activated state is the absence of an EPR signal associated with the methylated enzyme. If the A_{red1} state were used in methylation, the same reaction would yield a paramagnetic methylated state $S = ½$ $\{[Fe_4S_4]^{2+}\ Ni_p^{III} \leftarrow {}^{-}CH_3\}$, which is definitely not observed. Prior to ca. 1996, a one-electron $S = ½$ $\{[Fe_4S_4]^{2+}\ Ni_p^{I}\}$ A_{red1} state was presumed for the reductively activated state, because only the $S = ½$ A_{red1}-CO state had been characterized at that time. My graduate student David Barondeau and I realized that a methylation reaction with A_{red1} predicted that the methylated enzyme should be paramagnetic (e.g., $S = ½$) and EPR-active. Moreover, in this case, use of ^{13}C methyl (with $I = ½$) might display hyperfine interactions. Such a result would have demonstrated only the second organometallic bond in nature (Co–C bonds were known at the time). This realization motivated us to prepare the methylated enzyme using $^{13}CH_3$- and examine the EPR spectrum. Our discovery that the methyl intermediate was EPR silent [40] was initially disappointing, but it was also the origin of the currently best-supported catalytic mechanism (see below).

More recently, the paramagnetic A_{red1} state has been observed in frozen samples of the A_{red1}-CO state after photoirradiation in which CO dissociates; this state is evidenced by an EPR signal with $g = 2.56$, 2.10 and 2.01 [58]. In other batches, a weak EPR signal (with $g_{\parallel} = 2.10$ and $g_{\perp} = 2.03$] has been observed transiently as A_{ox} is titrated to the fully reduced EPR-silent A_{red2} state [59]. Thus, the A_{red1} state is attainable under special conditions *but is irrelevant for the catalytic mechanism.*

7.15 Electronic configuration of the A_{red2} state

The methyl group transfer from CoFeSP to ACS is the only such transfer in nature in which the methyl group transfers from one transition metal (Co) directly to another

(Ni). This adds intrigue to the reaction but it also requires that Ni_p should be a stronger nucleophile than a Co^I cobalamin (which is known to be a very strong nucleophile). Would Ni_p^I be sufficiently strong to snatch the methyl group from a Co^I cobalamin? Or would an unprecedented super-nucleophilic Ni_p^0 state be required?

We investigated this issue by analyzing MB spectra of isolated α-subunits reduced either by dithionite or Ti^{III} citrate [47]. One difficulty is that MB can only "see" the cubane component of the A-cluster, and so the redox and spin states of Ni_p must be inferred. A second difficulty is that only approximately 30% of α-subunits are in the active form (due to heterogeneity). The MB spectra of Bramlett et al. consisted of two major species; 35% of spectral intensity reflected iron in the form of $[Fe_4S_4]^{2+}$ clusters [47]. This material had an $S = 0$ system spin state. Since isolated $[Fe_4S_4]^{2+}$ clusters are diamagnetic, this result implied that Ni_p must also be diamagnetic – either Ni_p^{II} or Ni_p^0. The remaining 65% of the spectral intensity arose from an $\{S = 3/2{:}S = ½\}$ mixture of $[Fe_4S_4]^{1+}$ cubanes. Since only ~30% of the α-subunit is active (due to heterogeneity), we concluded at that time that the 35%-intensity species must represent the active form. Since A_{ox} is $\{[Fe_4S_4]^{2+}\ Ni_p^{II}]$ and A_{red2} is two electrons more reduced than A_{ox}, we assigned the A_{red2} state as $\{[Fe_4S_4]^{2+}\ Ni_p^0\}$. For a more complete argument for a zero-valent Ni_p^0 state, see [60].

We also investigated this issue using stopped-flow kinetic methods [61]. In a single experiment, we monitored both the formation of Co^IFeSP during methyl group transfer *and* the reduction of the $[Fe_4S_4]^{2+}$ cubane component of the A-cluster in the isolated α-subunit at 390 nm. Absorption at 390 nm increases with methyl group transfer (due to formation of Co^IFeSP) and it decreases with the reduction of the cubane ($[Fe_4S_4]^{2+} \rightarrow [Fe_4S_4]^{1+}$). In our experiment, a syringe contained the α-subunit in the A_{ox} state (no reductant), while the other contained the reductant Ti^{III} citrate and CH_3-Co^{III}FeSP. To accept a methyl group, A_{ox} must be reduced to A_{red2} prior to accepting the methyl group. The result was a rapid increase in absorption at 390 nm followed by a slow decline. This indicates that the A-cluster became methylated very quickly relative to the slow reduction of the cubane. This implies that the cubane remains oxidized $[Fe_4S_4]^{2+}$ when the methyl group is transferred. Since *some* component of the A-cluster must have been reduced in order for the methyl group to transfer, we conclude that the Ni_p^{II} site was reduced to the Ni_p^0 state.

Years later, we reinvestigated this issue using MB spectroscopy [59], and our results were somewhat different. In samples prepared in a slightly different way relative to in our first study, 30% of spectral intensity arose from paramagnetic $S = 3/2$ or $S = ½$ $[Fe_4S_4]^{1+}$ clusters; the remaining 70% arose from diamagnetic $S = 0$ $[Fe_4S_4]^{1+}$ clusters. This latter state caught our attention because the cubane was reduced to the 1+ state *even though the overall system spin state was diamagnetic.* This implied that Ni_p had a local spin state of $S = ½$ and that it was spin coupled to the $[Fe_4S_4]^{1+}$ cubane (i.e., it was Ni_p^I). Further support for the $\{[Fe_4S_4]^{1+}\ Ni_p^I\}$ configuration was a transient $S = ½$ EPR signal with g-values of a Ni^I species that was presumed to reflect an $\{[Fe_4S_4]^{+}\ Ni_p^I\}$ intermediate. A dissonant chord in this new perspective was that it implied that 70% of the α-subunits were in the active state. The alternative interpretation was that the active form was the 30% in the $\{[Fe_4S_4]^{1+}$ $Ni_p^{II}\}$ state prior to CO treatment. This interpretation was also difficult to reconcile because it required an electron from the cubane transfer to Ni_p upon treatment with CO (in order to form the A_{red1}-CO state $\{[Fe_4S_4]^{2+}$ Ni_p^{I}-CO$\}$. In conclusion, the current experimental landscape requires that *both* $\{[Fe_4S_4]^{2+}$ $Ni_p^{0}\}$ and $\{[Fe_4S_4]^{1+}$ $Ni_p^{I}\}$ states be considered as viable candidates for A_{red2}.

Can et al. have recently added another hypothesis into the mix [46]. Based on a recent EXAFS spectroscopic study of the A_{ox} state and the state we call A_{red2}, these authors concluded that the two states are actually isoelectronic since no spectroscopic differences were observed for either Ni or Fe X-ray absorption spectroscopy (XAS) edges. They suggested that A_{ox} is reduced by external reductants simultaneously with CO binding (forming the A_{red1}-CO state) or simultaneously with methyl group binding (forming the A_{ox}-CH_3 state) [46]. This proposal ignores previous results indicating that the enzyme, starting in the methylated state, can turnover catalytically in the absence of external reductants [47]. Ni XAS K-edge energies, which generally reflects Ni oxidation states, did not change in A_{ox}, the A_{red1}-CO state or the dithionite-reduced state. EPR of the A_{red1}-CO state clearly indicates Ni_p^{I}, whereas the diamagnetism of A_{ox} indicates Ni_p^{II}. In contrast to the results by Can et al., Schrapers et al. reported "significant nickel reduction" based on XAS spectra when samples in the A_{ox} state were treated with Ti^{III} citrate [62]; however, they interpreted this as forming Ni_p^{I} rather than Ni_p^{0}. Distinguishing Ni_p^{II} from Ni_p^{I} or Ni_p^{0} by XAS may simply not be possible – the technique may be insufficiently sensitive to Ni oxidation states in cases of delocalized Ni sites coordinated by cysteine sulfur ligands. A similar difficulty has been reported in assigning Ni oxidation states in NiFe hydrogenases [63, 64]. As with Ni_p, the Ni in these enzymes is coordinated by cysteine thiolate ligands. XAS of the so-called Ni-R state is indistinguishable from that of the two-electron-oxidized Ni-SI state [65], probably due to extensive delocalization.

7.16 Computational studies of A_{red2}

In 2003, Thomas Brunold and coworkers performed DFT (density functional theory) calculations to evaluate the electronic states of A_{red2} and A_{red1}-CO [66]. They calculated energies of the A_{red2} state alternatively assuming the electronic configurations $\{[Fe_4S_4]^{2+}$ $Ni_p^{0}\}$ and $\{[Fe_4S_4]^{1+}$ $Ni_p^{I}\}$. Their calculations favored the latter configuration, in which an $S = 1/2$ Ni_p^{I} ion is antiferromagnetic coupled to a reduced ($S = 1/2$ or $S = 3/2$) cubane. This suggests that the $[Fe_4S_4]^{2+}$ cubane is more easily reduced to the 1+ core oxidation state than Ni_p^{I} is reduced to Ni_p^{0}. However, the redox potential calculated for this reduction was <–2 V which is far lower than observed [67]. Also, positions of six carbon atoms in their structures were fixed. Webster et al.

performed DFT calculations, and concluded that $\{[Fe_4S_4]^{2+}\ Ni_p^{0}\}$ was stable [68], but they did not include the Fe_4S_4 cubane in their calculations casting doubt on their conclusions [69].

Using spin-unrestricted DTF and broken symmetry methods, Amara et al. calculated that the $\{[Fe_4S_4]^{1+}\ Ni_p^{I}\}$ configuration is more stable than $\{[Fe_4S_4]^{2+}\ Ni_p^{0}\}$ for the A_{red2} state [69]. However, they did not include the protein matrix in their study, which could affect calculated redox potentials. Moreover, they fixed the positions of the protein-based ligands to the metals during geometry optimization, which might restrict the flexibility of the A-cluster structure.

Chemielowska et al. also used broken-symmetry DFT to calculate energies of the A-cluster in different redox and protonation states [29]. The effect of the protein matrix was taken into account, and amino acid residues that could potentially form hydrogen bonds with the cluster were included. Their calculations indicated that the structure of the A-cluster is quite flexible (which may play a role in catalysis). The A_{red1} state was unstable and was concluded to be only an intermediate in forming A_{red2}. They reported significant Fe-Ni_p and Ni_p-Ni_d bonding interactions in many A-cluster states. Chemielowska et al. concluded that the A_{red2} state includes considerable $Ni_p \rightarrow Ni_d$ dative bonding interactions, which correlate with a short bond distance (2.653 Å). This stabilizes the electron-rich Ni_p in A_{red2}, as hypothesized earlier [70]. Their calculations indicated that the $\{[Fe_4S_4]^{2+}\ Ni_p^{0}\}$ configuration of A_{red2} state was 21 kcal/mol less stable than the $\{[Fe_4S_4]^{1+}\ Ni_p^{I}\}$ configuration. However, the calculated $n = 2$ redox potential of this latter configuration associated with the reduction of A_{ox} was −1.177 V which is unreasonably low. As a result, they suggested that Ni_p is protonated during this reduction, which would moderate the redox potential significantly [54]. Thus, Chemielowska et al. favored the $\{[Fe_4S_4]^{2+}\ Ni_p^{II} \leftarrow {}^{-}H\}$ configuration for A_{red2}. They realized that this would diminish the nucleophilicity of Ni_p and make the methyl group transfer from $CH_3 \rightarrow Co^{III}FeSP$ more difficult. However, they suggested that before or during methylation the proton on Ni_p was removed by a base on CoFeSP, activating it for methyl group transfer. We include this as a third viable electronic configuration for A_{red2}. The critical reaction, showing the three candidate A_{red2} configurations, is summarized in eq. (7.7). Note that the product A_{ox}-CH_3 state is the same regardless of which configuration attacks the methyl group.

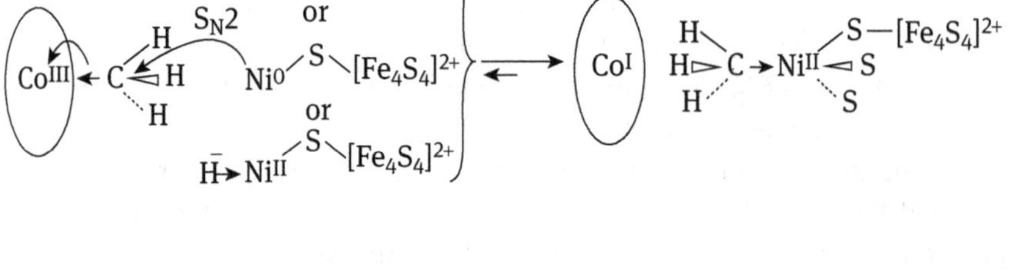

(7.7)

7.17 A_{ox}-C(O)CH_3 state

This state results when the methylated form of the enzyme is exposed to CO. However, the acetyl-bound state is difficult to study because it is also diamagnetic and EPR silent, just like the methylated enzyme. As a result, MB spectra of the acetylated enzyme are indistinguishable from that of the methylated enzyme (after adding CO to the methylated enzyme). The reaction should be

$$\left\{[Fe_4S_4]^{2+}-Ni_p^{2+}-CH_3\right\} + CO \rightleftharpoons \left\{[Fe_4S_4]^{2+}-Ni_p^{2+}-C(=O)CH_3\right\} \quad (7.8)$$

To complete the catalytic cycle, CoASH binds to the A_{ox}-acetyl state as follows:

$$\left\{[Fe_4S_4]^{2+}-Ni_p^{2+}-C(=O)CH_3\right\} + CoASH \rightleftharpoons \left\{[Fe_4S_4]^{2+}-\underset{\uparrow\,^{-}SCoA}{Ni_p^{2+}}-C(=O)CH_3\right\} + H^+ \quad (7.9)$$

Starting with the methylated intermediate, we investigated the insertion of CO and the subsequent production of acetyl-CoA by adding CoA [50]. CoASH bound to the enzyme with $K_d = 10 \pm 8\ \mu M$ [53] in one study and 52 μM in another [71]. Only 0.2 ± 0.1 mol of CoASH is bound to an αβ-dimer, consistent with other evidence of heterogeneity. CoASH does not bind to the A_{red1}-CO state but does bind to the A_{red2} state, suggesting an ordered mechanism of substrate binding. In the final step (7.10), acetyl-CoA is formed as A_{red2} is regenerated:

$$\left\{[Fe_4S_4]^{2+}-\underset{\uparrow\,^{-}SCoA}{Ni_p^{2+}}-C(=O)CH_3\right\} \rightleftharpoons A_{red2} + CoAS-C(=O)CH_3 \quad (7.10)$$

Gencic et al. generated the Ni_p^{II}-acetyl intermediate, isolated it and reacted it with CoA to produce acetyl-CoA quantitatively [26].

7.18 Inorganic model complexes

Numerous nickel-based complexes have been synthesized to mimic certain structural and/or functional properties of the A-cluster. We will not review these complexes comprehensively, but refer interested readers to recent reviews that include these aspects [12, 72]. Our objective is to illustrate important bonding and reactivity properties of the A-cluster using selected model complexes as our guides. Sufficient diversity of models is now available that general patterns have emerged.

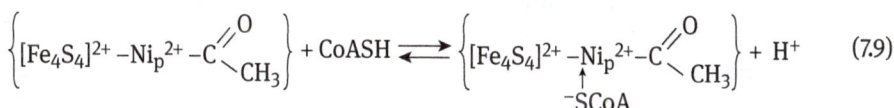

Current models can be divided into groups. *First-generation* models are mononickel complexes with Ni_p-like sites. They have ligand environments that generally include sulfur thiolate coordination but also other ligands that stabilize the nickel in low oxidation states such as Ni^I or Ni^0. Relevant spectroscopic, redox and/or reactivity properties of such complexes are often determined. These might include an EPR signal similar to the NiFeC signal, or $Ni^{II}/Ni^I/Ni^0$ redox potentials that are mild enough for low valences to be accessed. In terms of reactivity, such models are generally evaluated for their ability to bind CO and/or methyl groups.

Second-generation models are dinickel structures in which both Ni_d and Ni_p are included. In these cases, the Ni_d-like site is typically square-planar with an N_2S_2 donor set. The sulfurs are routinely thiolates that are bridging to the Ni_p-like site. The Ni_p-like site is additionally coordinated by other ligands (e.g., phosphines, CO or thiolates) that control which oxidation states can be accessed. In addition to the properties that can be compared to the A-cluster using mononuclear models, second-generation complexes can be compared in terms of Ni_d–Ni_p distance and the dihedral angle ϕ formed from the Ni_d-S_2 and Ni_p-S_2 planes (where S_2 refers to the two bridging thiolate sulfurs). For the enzyme, $\phi = 114°–138°$ [29]. This allows the possibility of Ni_d–Ni_p bonding to be considered.

There are currently no *third-generation* models, which we define as those that would include an Fe_4S_4 cubane linked to a dinickel site. Undoubtedly such models are difficult to synthesize, but they would also be very useful in resolving mechanistic issues involving spin coupling between an $S = ½$ $[Fe_4S_4]^{1+}$ cluster and an $S = ½$ $Ni_p{}^I$ site. Would a $Ni_p{}^I$ ion in such a spin-coupled arrangement react with a methylated CoFeSP-mimic using a nucleophilic displacement or radical-based mechanism? Fe–S–Ni_p angle θ could also be compared with those observed in the A-cluster. For the enzyme, $\theta \sim 70°$ [29].

The current *cutting-edge* in this field is a remarkable study by Kazuyuku Tatsumi and coworkers, which provides important insights into the mechanism of the enzyme. In 2009, Ito et al. reported the synthesis of a series of second-generation ACS models [73]. Following the approach of Tom Rauchfuss [74], they prepared an A_{red2} mimic by reacting Ni^{II}(*N*-*N′*-diethyl-3,7-diazanonane-1,9-dithiolate = $dadt^{ET}$) with Ni^0(1,5-cyclooctadiene = $cod)_2$. The resulting species called $Ni^{II}(dadt^{ET})Ni^0(cod)$ was reacted with a methylcobaloxime mimic of CH_3-Co^{III}FeSP and a bulky thiolate called $^-$SDmp. The product was a mimic of the A_{ox}-CH_3 state in which CH_3 and SDmp were the two *external* ligands of the square-planar Ni_p-like site (the two *internal* ligands were thiolates shared with the Ni_d-like site). In this state, the dihedral angle ϕ was 119°. The authors concluded that "ACS catalysis could include the Ni_d(II)-Ni_p(0) state ... reacting with methylcobalamin to afford the Ni_d(II)-Ni_p(II)-Me species." They then treated the methylated product with CO to yield $CH_3C(O)$-SDmp, a thioester of acetic acid, equivalent to acetyl-CoA. They concluded that "the formation of the acetylthioester ... probably proceeds via insertion of CO into the Ni-Me bonds (followed by) reductive elimination." They were unable to detect the acetyl intermediate because reductive elimination was faster than CO insertion. They concluded that reductive elimination occurs immediately once $CoAS^-$ coordinates to Ni_p, and that CO addition to Ni_p must be "strictly regulated" to prevent inactive carbonyl-bound states (i.e., A_{red1}-CO).

In a subsequent study, Matsumoto et al. synthesized Ni^{II}(dadtEt)Ni^{I}(SDmp)(PPh_3), a second-generation A_{red1} mimic [75]. The model complex is S = ½ and it exhibits an EPR signal with g-values of 2.62, 2.12 and 2.00. Reduction potentials were –0.73 V versus SCE for the $Ni_p^{II} + 1e^- \leftrightarrows Ni_p^{I}$ couple, and –1.68 V for the $Ni_p^{I} + 1e^- \leftrightarrows Ni_p^{0}$ couple. The dihedral angle ϕ for the A_{red1} model was 99°, smaller than the [Ni_d^{II} Ni_p^{II}-CH_3] analog and smaller than for the A-cluster. The A_{red1} model reacted with methylcobaloxime and CO, and generated the acetylthioester product, as with the Ni^0 adduct. However, in this case the yield was much lower because two equivalents of the A_{red1} model reacted with one equivalent of methylcobaloxime (CH_3-Co^{III}). The first equivalent generated the methylated Ni^{II}-CH_3 adduct and Co^{II} cobaloxime (reaction (7.11)), while the second equivalent reduced Co^{II} cobaloxime to Co^{I}:

$$Ni^{I} + {}^{-}CH_3 \rightarrow Co^{III} \rightleftharpoons Ni^{II} \leftarrow {}^{-}CH_3 + Co^{II} \quad (7.11)$$

Charlie Riordan and coworkers diagnosed a similar situation and concluded that methylation reactions involving Ni^{I} occur by a radical-based mechanism [76]. Yoo et al. synthesized a Ni^{I}-CO model compound and reacted it with alkyl iodides [77]. These reactions also proceeded by radical chemistry. Similar reactions involving Ni^0 sites proceed by S_N2 nucleophilic displacements. This discounts the possibility of a Ni_p^{I} state (e.g., as found in A_{red1}) serving as the methyl group acceptor.

7.19 Mechanism of catalysis

After walking step-by-step through the different reactions associated with ACS enzymes, and after considering results from computational studies and synthetic inorganic model complexes, we are ready to describe the mechanism of catalysis, summarized in Figure 7.7. Resting enzyme in the A_{ox} state becomes reductively activated to the A_{red2} state in assay mix containing Ti^{III} citrate. The {Ni_p^{0} $[Fe_4S_4]^{2+}$} electronic configuration of A_{red2} has been assumed (my preference) but the two other candidate configurations mentioned earlier are also possible. The intermediate A_{red1} state is presumed to form and bind CO to generate the A_{red1}-CO inhibited state. A_{red2} accepts a methyl group, forming the A_{ox}-CH_3 state. This is followed by CO binding and insertion into the Ni_p-CH_3 bond to form the Ni_p-acetyl intermediate. CoASH binds and donates a proton to a base on the enzyme, affording a $CoAS^-$ thiolate, which reacts with the acetyl group in a reductive elimination step to generate the product acetyl-CoA. Whether the sulfur of $CoAS^-$ binds directly to Ni_p is unknown. Upon releasing the product, the A_{red2} state is regenerated and the cycle is complete. The detailed kinetics of these steps have been investigated [20, 50, 51, 59].

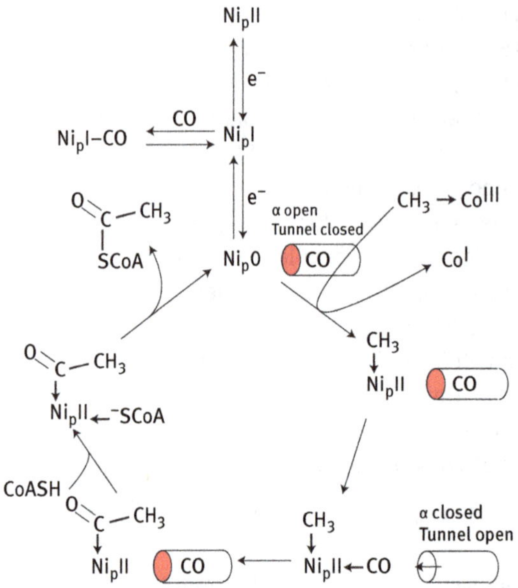

Figure 7.7: CODH/ACS mechanism of catalysis.
The enzyme in the resting A_{ox} state is reduced by two one-electron steps to the A_{red2} state, which enters the catalytic cycle. The A_{red1} state is a transient intermediate but can readily bind CO to generate an inhibitory state. In the A_{red2} state, the ACS subunit is in the open conformation but the tunnel is closed so as to prevent CO from accessing the A-cluster. Ni_p in the A_{red2} state accepts a methyl cation from methylated corrinoid iron/sulfur protein. The Ni_p^0 electronic configuration is shown but two other configurations are possible as described in the text. Once the methyl group has bound, the ACS conformation changes to the closed form in which CO can migrate through the tunnel to the A-cluster where it binds methylated Ni_p and inserts to form a Ni_p^{II}-acetyl intermediate. The step at which the ACS subunit returns to the open conformation is uncertain, as is the step at which CoASH binds. Eventually, the $CoAS^-$ thiolate attacks the acetyl group on Ni_p and the product acetyl-CoA is produced while regenerating the A-cluster to the A_{red2} state. Coordinate bonds are shown as arrows from the Lewis base to the acid. See text for other details.

When reactions are performed using CODH/ACS, and CO_2 is used as a substrate, these mechanistic events are tightly synchronized with protein conformational changes occurring in the enzyme. In the A_{red2} state, the tunnel from the C-cluster is closed, and the α-subunit is in its open surface-exposed conformation (Figure 7.7, note the red disk blocking the tunnel). This allows CH_3-Co^{III}FeSP to dock, and the methyl group to transfer all while blocking CO migration to the A-cluster. If CO would arrive before the methyl group is transferred, the A_{red1}-CO state would form and shut down catalysis. Once the methyl group transfers, the α-subunit changes to the closed conformation, the A-cluster is no longer surface exposed and the tunnel opens (the phenylalanine moves away from blocking access to the A-cluster). This allows CO to arrive at the A-cluster ready for CO insertion.

7.20 The paramagnetic mechanism

This catalytic mechanism was first proposed in 2002 [78] and it continues to be promoted even though there is little evidence for it and much evidence against it. The paramagnetic mechanism (PM) assumes that A_{ox} is reductively activated by one electron in assay solutions containing excess strong reductant. This generates the S = ½ A_{red1} state (Ni_p^I). Methyl and CO substrates bind randomly to the Ni_p^I of this state; the evidence for this is a single controversial study [45]. When CO binds first, the S = ½ A_{red1}-CO state (Ni_p^I-CO) results. This makes little sense because the A_{red1}-CO state is inhibitory. Subsequent methyl group binding affords a paramagnetic Ni_p^{III}-C(O)CH_3 state which is not observed experimentally. To rectify this problem, an electron is donated from an internal n = 1 redox site on the α-subunit to generate the observed diamagnetic Ni_p^{II}-C(O)CH_3 state. The only problem is that no internal redox site other than the A-cluster itself has been identified in three crystal structures of the α-subunit. Nevertheless, let's call this hypothetical site R, with redox states R_{red} (S = 0) and R_{ox} (S = ½). R *must* be paramagnetic in one of these states, and we arbitrarily selected R_{ox} to have half-integer spin in this example. The overall state of the enzyme in the A_{red1}-CO state should now be described as {Ni_p^I–CO (S = ½) R_{red} (S = 0)}. Immediately after methyl transfer, the state becomes {Ni_p^{III}-C(O)CH_3 (S = ½) R_{red} (S = 0}. According to the PM, R_{red} donates an electron to Ni_p^{III}, generating {Ni_p^{II}-C(O)CH_3 (S = 0) R_{ox} (S = ½)}. An internal electron transfer could certainly afford an EPR-silent A-cluster (consistent with the EPR-silence of this state) but it would also afford a signal from R_{ox}. Thus, the PM predicts the development of an EPR signal that isn't observed (because R doesn't exist). In the final step of the PM, $CoAS^-$ attacks the acetyl-bound intermediate to form acetyl-CoA in a reductive-elimination step. Typically in such reactions, Ni_p^{II} would be reduced by *two* electrons, but instead, the PM proposes that one electron goes to Ni_p^{II} (regenerating A_{red1}) and the other to R_{ox} (regenerating R_{red}). This completes the catalytic cycle.

Similar problems haunt the PM when the methyl group is presumed to bind Ni_p^I first. In this case, an S = ½ Ni_p^{III}-CH_3 state is proposed to form, whereas the observed methylated state is diamagnetic. To rectify this, the PM assumes that an electron from R_{red} quickly reduces Ni_p^{III}-CH_3, affording the observed diamagnetic Ni_p^{II}-CH_3; the overall enzyme now becomes {Ni^{II}-CH_3 (S = 0) R_{ox} (S = ½)}. An EPR signal from R_{ox} should again be observed (but isn't). CO subsequently binds, generating the diamagnetic acetyl intermediate. This is followed by the unusual electron-pair-splitting reductive elimination step to form A_{red1} and R_{red} to complete the catalytic cycle.

In summary, the only intermediate of the PM that has been observed in a stable quantitative manner is A_{red1}-CO, and this state inhibits the enzyme! All other proposed catalytic intermediate states have either not been observed experimentally, or they are only observed under transient or nonphysiological conditions. Moreover, the PM requires the participation of an imaginary n = 1 redox site on the α-subunit. Given this multitude of problems that plague the PM, I suggest that it be immediately discarded; the PM has essentially no chance of being correct.

A subtler problem is that the PM simply does not *resonate* with the chemistry of synthetic inorganic Ni complexes, which show that Ni^{I} complexes react with methyl group donors via a radical mechanism, not S_N2 nucleophilic displacements as is observed with the enzyme and as occurs with Ni^0 model complexes. Another subtle problem with the PM (or with any mechanism that assumes that CO and methyl groups bind in random order to Ni_p) involves the role of the tunnel. Ordered substrate binding is essential for understanding the role of the tunnel in the catalytic mechanism. For if CO and methyl group bind randomly to Ni_p, there would be no need for a conformationally regulated tunnel to deliver CO to Ni_p *after* the methyl group bound. Also, if binding were random, the enzyme would be inhibited whenever CO was bound first. The existence of a conformationally regulated tunnel *tells us* that the methyl and CO bind in an ordered manner.

7.21 Other lessons from nature

There are currently three viable options for the electronic configuration of the A_{red2} state, namely $\{[Fe_4S_4]^{2+}\ Ni_p{}^0\}$, $\{[Fe_4S_4]^{1+}\ Ni_p{}^I\}$ and $\{[Fe_4S_4]^{2+}\ Ni_p{}^{II}{\leftarrow}^{-}H\}$. Further studies are needed to distinguish them. If future compelling evidence favors either of the latter two configurations, this will raise new issues regarding spin coupling and the effect of protonation. That being said, it would be thrilling if $Ni_p{}^0$ was established to be utilized by the enzyme. In this case, nature might be telling us how to achieve zero-valent transition metal states generally within a biological context. The answer would look something like this:

- Use low coordination numbers (perhaps 3 is optimal).
- Use soft cysteinate sulfur donors to stabilize soft Lewis acids such as zero-valent metals.
- Use cysteinate sulfurs that are bridging to other subsidiary redox-inactive metals to neutralize their negative charge.
- Include metal→metal dative bonding interactions between the electron-rich zero-valent metal and the subsidiary metals to delocalize electron density away from the zero-valent metal.

It will be a challenge for synthetic inorganic chemists to design metal complexes with these properties, and evaluate whether the complexes can stabilize zero-valent nickel. Likewise, computational chemists will be challenged to quantify how these different factors contribute to stabilize such states. A related challenge is to accurately predict thermodynamic reduction potentials associated with the $Ni_p{}^{II/I/0}$

states and the $[Fe_4S_4]^{2+/1+}$ cubane; this might help establish which electronic configuration is stabilized in A_{red2}.

7.22 A beautiful metalloenzyme

I can't think of a more appropriate adjective to describe CODH/ACS, since so many elegant and awe-inspiring aspects converge on this one enzyme. First, what other metalloenzyme intimately engages *three* transition metals – in this case nickel, iron and cobalt – in catalysis? Cytochrome c oxidase and nitrogenase are highly complex, elegant and intriguing metalloenzymes but they only engage two transition metals – iron and copper for the former, and iron and molybdenum (or iron and vanadium) for the latter. Second, what other metalloenzyme employs a molecular tunnel that includes branches and regulatory control valves? Admittedly, carbamoyl phosphate synthetase has a longer tunnel, but one without branches or valves [79]. Third, what other metalloenzyme has a more distinguished evolutionary history, including ancestry dating back billions of years to the origin of life? Nitrogenase is undoubtedly ancient, but according to the iron–sulfur world theory, CO_2 fixation preceded N_2 fixation, and so ACS would have evolved earlier. Finally, what other metalloenzyme stabilizes (or might stabilize) the zero-valent state of a transition metal? The Ni-R state of hydrogenase is certainly a candidate but a Ni^{II} hydride seems more likely in that case. Viewed in this way, I can only conclude that CODH/ACS is a thing of beauty. Let me end by saying that no one researching this enzyme early on, including myself, realized the beauty hidden from us by our own ignorance. It was truly a joy to be a part of that dis-covering.

Acknowledgments: The author thanks all his former students who worked on this enzyme for their efforts, Seth Cory for preparing Figures 7.4 and 7.6, David A. Grahame for critically reading the manuscript prior to publication. This work was financially supported by the National Institutes of Health (GM127021) and the Robert A. Welch Foundation (A1170).

References

[1] Martin, WF., and Sousa, FL. Early microbial evolution: the age of anaerobes. Cold Spring Harbor Perspectives in Biology 2016, 8, Article Number, a018127.

[2] Weiss, MC., Sousa, FL., Mrnjavac, N., Neukirchen, S., Roettger, M., Nelson-Sathi, S., and Martin, WF. The physiology and habitat of the last universal common ancestor. Nat Microbiol 2016, 16116.

[3] Adam, PS., Borrel, G., and Gribaldo, S. Evolutionary history of carbon monoxide dehydrogenase/acetyl-CoA synthase, one of the oldest enzymatic complexes. Proc Natl Acad Sci USA 2018, 115, E1166–E1173.
[4] Lindahl, PA., and Graham, DE. Acetyl-coenzyme A synthases and Nickel-containing carbon monoxide dehydrogenases, Sigel, A., Sigel, H., and Sigel, R.K.O., Eds, Nickel and Its Surprising Impact in Nature John Wiley and Sons, West Sussex, England, Met. Ions. Life Sci., 2007, Vol. 2, 357–416.
[5] Bult, CJ. et al.. Complete genome sequence of the methonogenic archaeon, *Methanococcus jannaschii*. Science 1996, 273, 1058–1073.
[6] Lazcano, A. Historical development of origins research. Cold Spring Harbor Perspectives in Biology 2010, 2, Article Number: a002089 , DOI: 10.1101/cshperspect.a002089.
[7] Orgel, LE. Prebiotic chemistry and the origin of the RNA world. Crit Rev Biochem Mol Biol 2004, 39, 99–123.
[8] Wächtershäuser, G. Before enzymes and templates – theory of surface metabolism. Microbiol Rev 1988, 52, 452–484.
[9] Huber, C., and Wächtershäuser, G. Activated acetic acid by carbon fixation on (Fe,Ni)S under primordial conditions. Science 1997, 276, 245–247.
[10] Lindahl, PA. Stepwise evolution of nonliving to living chemical systems. Origins Life Evol Biosphere 2004, 34, 371–389.
[11] Di Giulio, M. An autotrophic origin for the coded amino acids is concordant with the coevolution theory of the genetic code. J Mol Evol 2016, 83, 93–96.
[12] Can, M., Armstrong, FA., and Ragsdale, SW. Structure, function, and mechanism of the nickel metalloenzymes, co dehydrogenase, and acetyl-CoA synthase. Chem Rev 2014, 114, 4149–4174.
[13] Boer, JL., and Mulrooney, SB. Hausinger RP nickel-dependent metalloenzymes. Arch Biochem Biophys 2014, 544, 142–152.
[14] Ferry, JG. CO in methanogenesis. Annal Microbiol 2010, 60, 1–12.
[15] Bender, G., Pierce, E., Hill, JA., Darty, JE., and Ragsdale, SW. Metal centers in the anaerobic microbial metabolism of CO and CO_2. Metallomics 2011, 3, 797–815.
[16] Ragsdale, SW., and Pierce, E. Acetogenesis and the Wood–Ljungdahl pathway of CO_2 fixation. Biochim Biophys Acta 2008, 1784, 1873–1898.
[17] Schuchmann, K., and Müller, V. Autotrophy at the thermodynamic limit of life: a model for energy conservation in acetogenic bacteria. Nat Rev 2014, 12, 809–821.
[18] Buckel, W., and Thauer, RK. Flavin-based electron bifurcation, a new mechanism of biological energy coupling. Chem Rev 118, 3862–3886.
[19] Svetlitchnyi, V., Dobbek, H., Meyer-Klaucke, W., Meins, T., Thiele, B., Romer, P., Huber, R., and Meyer, O. A functional Ni-Ni-[4Fe-4S] cluster in the monomeric acetyl-CoA synthase from Carboxydothermus hydrogenoformans. Proc Natl Acad Sci U.S.A 2004, 101, 446–451.
[20] Gencic, S., Kelly, K., Ghebreamlak, S., Duin, EC., and Grahame, DA. Different modes of carbon monoxide binding to acetyl-CoA synthase and the role of a conserved phenylalanine in the coordination environment of nickel. Biochemistry 2013, 52, 1705–1716.
[21] Doukov, TI., Iverson, TM., Seravalli, J., Ragsdale, SW., and Drennan, CL. A Ni-Fe-Cu center in a bifunctional carbon monoxide dehydrogenase/acetyl-CoA synthase. Science 2002, 298, 567–572.
[22] Darnault, C., Volbeda, A., Kim, EJ., Legrand, P., Vernede, X., Lindahl, PA., and Fontecilla-Camps, JC. Ni-Zn-[Fe4-S4] and Ni-Ni-[Fe4-S4] clusters in closed and open subunits of acetyl-CoA synthase/carbon monoxide dehydrogenase. Nat Struct Biol 2003, 10, 271–279.
[23] Gong, W., Hao, B., Wei, Z., Ferguson, DJ., Tallant, T., Krzychi, JA., and Chan, MK. Structure of the $\alpha_2\varepsilon_2$ component of the Methanosarcina barkeri acetyl-CoA decarbonylase/synthase complex. Proc Natl Acad Sci USA 2008, 105, 9558–9563.

[24] Maynard, EL., and Lindahl, PA. Evidence of a molecular tunnel connecting the active sites for CO_2 reduction and acetyl-CoA synthesis in acetyl-CoA synthase from *Clostridium thermoaceticum*. J Am Chem Soc 1999, 121, 9221–9222.
[25] Maynard, EL. Lindahl PA catalytic coupling of the active sites in Acetyl-CoA synthase, a bifunctional CO-channeling enzyme. Biochemistry 2001, 40, 13262–13267.
[26] Gencic, S., Duin, EC., and Grahame, DA. Tight coupling of partial reactions in the acetyl-CoA decarbonylase/synthase (ACDS) multienzyme complex from Methanosarcina thermophila: Acetyl C-C bond fragmentation at the A cluster promoted by protein conformational changes. J Biol Chem 2010, 285, 15450–15463.
[27] Wang, PH., Bruschi, M., De Gioia, L., and Blumberger, J. Uncovering a dynamically formed substrate access tunnel in carbon monoxide dehydrogenase/acetyl-CoA synthase. J Am Chem Soc 2013, 135, 9493–9502.
[28] Pettersen, EF., Goddard, TD., Huang, CC., Couch, GS., Greenblatt, DM., Meng, EC., and Ferrin, TE. UCSF Chimera–a visualization system for exploratory research and analysis. J Comput Chem, 2004(25), 1605–1612.
[29] Chmielowska, A., Lodowski, P., and Jaworska, M. Redox potentials and protonation of the a-cluster from acetyl-CoA Synthase. A density functional theory study. J Phys Chem A 2013, 117, 12484–12496.
[30] Shin, W., and Lindahl, PA. Discovery of a labile nickel ion required for CO acetyl-CoA exchange activity in the NiFe complex of carbon monoxide dehydrogenase from Clostridium thermoaceticum. J Am Chem Soc 1992, 114, 9718–9719.
[31] Shin, WS., Anderson, ME., and Lindahl, PA. Heterogeneous nickel environments in carbon monoxide dehydrogenase from Clostridium thermoaceticum. J Am Chem Soc 1993, 115, 5522–5526.
[32] Bramlett, MR., Tan, XS., and Lindahl, PA. Inactivation of acetyl-CoA synthase/carbon monoxide dehydrogenase by copper. J Am Chem Soc 2003, 125, 9316–9317.
[33] Gregg, CM., Goetzl, S., Jeoung, JH., and Dobbek, H. AcsF catalyzes the ATP-dependent insertion of nickel into the Ni,Ni-[4Fe4S] cluster of acetyl-CoA synthase. J Biol Chem 2016, 291, 18129–18138.
[34] Zhang, Z., Morgan, JJ., and Lindahl, PA. Mathematical model for positioning the FtsZ contractile ring in Escherichia coli. J Math Biol 2014, 68, 911–930.
[35] Jeoung, JH., Giese, T., Gruenwald, M., and Dobbek, H. CooC1 from Carboxydothermus hydrogenoformans Is a Nickel-Binding ATPase. Biochemistry 2009, 48, 11505–11513.
[36] Laskowski, RA. SURFNET: a program for visualizing molecular surfaces, cavities, and intermolecular interactions. J Mol Graph 1995, 13(323–330), 307–308.
[37] Svetlitchnaia, T., Svetlitchnyi, V., Meyer, O., and Dobbek, H. Structural insights into methyltransfer reactions of a corrinoid iron-sulfur protein involved in acetyl-CoA synthesis. Proc Natl Acad Sci USA 2006, 103, 14331–14336.
[38] Lindahl, PA., Ragsdale, SW., and Münck, E. Mössbauer studies of CO dehydrogenase from *Clostridium thermoaceticum*. J Biol Chem 1990, 265, 3880–3888.
[39] Lindahl, PA., Münck, E., and Ragsdale, SW. CO dehydrogenase from Clostridium thermoaceticum; EPR and electrochemical studies in CO_2 and argon atmospheres. J Biol Chem 1990, *265*, 3873–3879.
[40] Barondeau, DP., and Lindahl, PA. Methylation of carbon monoxide dehydrogenase from Clostridium thermoaceticum and mechanism of acetyl coenzyme A synthesis. J Am Chem Soc 1997, 119, 3959–3970.
[41] Wilson, BE., and Lindahl, PA. Equilibrium dialysis study and mechanistic implications of coenzyme A binding to acetyl-CoA synthase/carbon monoxide dehydrogenase from Clostridium thermoaceticum. J Biol Inorg Chem 1999, 4, 742–748.

[42] Shin, WS., and Lindahl, PA. Low-spin quantitation of NiFeC EPR signal from carbon-monoxide dehydrogenase is not due to damage incurred during protein-purification. Biochim Biophys Acta, Protein Struct. Mol. Enzymol. 1993, 1161, 317–322.

[43] Seravalli, J., Xiao, Y., Gu, W., Cramer, SP., Antholine, WE., Krymov, V., Gerfen, GJ., and Ragsdale, SW. Evidence that Ni-Ni acetyl-CoA synthase is active and that the Cu-Ni enzyme is not. Biochemistry 2004, 43, 3944–3955.

[44] George, SJ., Seravalli, J., and Ragsdale, SW. EPR and infrared spectroscopic evidence that a kinetically competent paramagnetic intermediate is formed when acetyl-coenzyme A synthase reacts with CO. J Am Chem Soc 2005, 127, 13500–13501.

[45] Seravalli, J., and Ragsdale, SW. Pulse-chase studies of the synthesis of acetyl-CoA by carbon monoxide dehydrogenase/acetyl-CoA synthase – Evidence for a random mechanism of methyl and carbonyl addition. J Biol Chem 2008, 283, 8384–8394.

[46] Can, M., Giles, LJ., Ragsdale, SW., and Sarangi, R. X-ray absorption spectroscopy reveals an organometallic Ni–C bond in the CO-treated form of acetyl-CoA synthase. Biochemistry 2017, 56, 1248–1260.

[47] Bramlett, MR., Stubna, A., Tan, X., Surovtsev, IV., Münck, E., and Lindahl, PA. Mössbauer and EPR study of recombinant acetyl-CoA synthase from *Moorella thermoacetica*. Biochemistry 2006, 45, 8674–8685.

[48] Xia, J., Hu, Z., Popescu, CV., Lindahl, PA., and Mössbauer, Münck E. EPR study of the Ni-activated r-subunit of carbon monoxide dehydrogenase from *Clostridium Thermoaceticum*. J Am Chem Soc 1997, 119, 8301–8312.

[49] Ragsdale, SW., Ljungdahl, LG., and DerVartanian, DV. EPR evidence for nickel-substrate interaction in carbon monoxide dehydrogenase from *Clostridium thermoaceticum*. Biochem Biophys Res Commun 1982, 108, 658–663.

[50] Tan, X., Surovtsev, IV., and Lindahl, PA. Kinetics of CO insertion and acetyl group transfer steps, and a model of the acetyl-CoA synthase catalytic mechanism. J Am Chem Soc 2006, 128, 12331–12338.

[51] Maynard, EL., Sewell, C., and Lindahl, PA. Kinetic mechanism of acetyl-CoA synthase: steady-state synthesis at variable CO/CO2 pressures. J Am Chem Soc 2001, 123, 4697–4703.

[52] Grahame, DA., Khangulov, S., and DeMoll, E. Reactivity of a paramagnetic enzyme-CO adduct in acetyl-CoA synthesis and cleavage. Biochemistry 1996, 35, 593–600.

[53] Lu, WP., and Ragsdale, SW. Reductive activation of the coenzyme A/acetyl-CoA isotopic exchange reaction catalyzed by carbon monoxide dehydrogenase from *Clostridium thermoaceticum* and its inhibition by nitrous oxide and carbon monoxide. J Biol Chem 1991, 266, 3554–3564.

[54] Bhaskar, B., DeMoll, E., and Grahame, DA. Redox dependent acetyl transfer partial reaction of the acetyl-CoA decarbonylase/synthase complex: kinetics and mechanism. Biochemistry 1998, 37, 14491–14499.

[55] Gencic, S., and Grahame, DA. Two separate one electron steps in the reductive activation of the A cluster in subunit β of the ACDS complex in *Methanosarcina thermophila*. Biochemistry 2008, 47, 5544–5555.

[56] Menon, S., and Ragsdale, SW. Role of the [4Fe-4S] cluster in reductive activation of the cobalt center of the corrinoid iron-sulfur protein from *clostridium thermoaceticum* during Acetate Biosynthesis. Biochemistry 1998, 37, 5689–5698.

[57] Lebertz, H., Simon, H., Courtney, LF., Benkovic, SJ., Zydowsky, LD., Lee, K., and Floss, HG. Stereochemistry of acetic acid formation from 5-methyltetrahydrofolate by *Clostridium thermoaceticum*. J Am Chem Soc 1987, 109, 3173–3174.

[58] Bender, G., Stich, TA., Yan, L., Britt, RD., Cramer, SP., and Ragsdale, SW. Infrared and EPR spectroscopic characterization of a Ni(I) species formed by photolysis of a catalytically competent Ni(I)-CO intermediate in the acetyl-CoA synthase reaction. Biochemistry 2010, 49, 7516–7523.
[59] Tan, X., Martinho, M., Stubna, A., Lindahl, PA., and Münck, E. Mössbauer evidence for an exchange-coupled {$[Fe_4S_4]^{1+}$ Ni_p^{1+}} A-cluster in isolated alpha subunits of acetyl-coenzyme a synthase/carbon monoxide dehydrogenase. J Am Chem Soc 2008, 130, 6712–6713.
[60] Lindahl, PA. Acetyl-coenzyme A synthase: the case for a Nip(0)-based mechanism of catalysis. JBIC, J Biol Inorg Chem 2004, 9, 516–524.
[61] Tan, XS., Sewell, C., and Lindahl, PA. Stopped-flow kinetics of methyl group transfer between the corrinoid-iron-sulfur protein and acetyl-coenzyme A synthase from *Clostridium thermoaceticum*. J Am Chem Soc 2002, 124, 6277–6284.
[62] Schrapers, P., Ilina, J., Gregg, CM., Mebs, S., Jeoung, J-H., Dau, H., Dobbek, H., and Haumann, M. Ligand binding at the A-cluster in full-length or truncated acetyl-CoA synthase studied by X-ray absorption spectroscopy. PLoS ONE 2017, 12, e0171039. DOI: 10.1371/journal.pone.0171039.
[63] Davidson, G., Choudhury, SB., Gu, Z., Bose, K., Roseboom, W., Albracht, SPJ., and Maroney, MJ. Structural examination of the nickel site in *Chromatium vinosum* hydrogenase: redox state oscillations and structural changes accompanying reductive activation and CO binding. Biochemistry 2000, 39, 7468–7479.
[64] Gu, ZJ., Dong, J., Allan, CB., Choudhury, SB., Franco, R., Moura, JJG., LeGall, J., Przybyla, AE., Roseboom, W., Albracht, SPJ., Axley, MJ., Scott, RA., and Maroney, MJ. Structure of the Ni sites in hydrogenases by X-ray absorption spectroscopy. Species variation and the effects of redox poise. J Am Chem Soc 1996, 118, 11155–11165.
[65] Maroney, MJ., and Bryngelson, PA. Spectroscopic and model studies of the Ni-Fe hydrogenase reaction mechanism. J Biol Inorg Chem 2001, 6, 453–459.
[66] Schenker, RP., and Brunold, TC. Computational studies on the A cluster of acetyl-coenzyme A synthase: geometric and electronic properties of the NiFeC species and mechanistic implications. J Am Chem Soc 2003, 125, 13962–13963.
[67] Brunold, TC. Spectroscopic and computational insights into the geometric and electronic properties of the A-cluster of acetyl-coenzyme A synthase. J Biol Inorg Chem 2004, 9, 533–541.
[68] Webster, CE., Darensbourg, MY., Lindahl, PA., and Hall, MB. Structures and energetics of models for the active site of acetyl-coenzyme A synthase: role of distal and proximal metals in catalysis. J Am Chem Soc 2003, 126, 3410–3411.
[69] Amara, P., Volbeda, A., Fontecilla-Camps, JC., and Field, MJ. A quantum chemical study of the reaction mechanism of acetyl-coenzyme a synthase. J Am Chem Soc 2005, 127, 2776–2784.
[70] Lindahl, PA. Metal-metal bonds in biology. J Inor Biochem 2012, 106, 172–178.
[71] Lu, W-P., and Ragsdale, SW. Reductive activation of the coenzyme A acetyl-CoA isotopic exchange reaction catalyzed by carbon monoxide dehydrogenase from *Clostridium thermoaceticum* and its inhibition by nitrous oxide and carbon monoxide. J Biol Chem 1991, 266, 3554–3564.
[72] Harrop, TC., Olmstead, MM., and Mascharak, PK. Synthetic analogues of the active site of the A-cluster of acetyl coenzyme A synthase/CO dehydrogenase: syntheses, structures, and reactions with CO. Inorg Chem 2006, 45, 3424–3436.
[73] Ito, M., Kotera, M., Matsumoto, T., and Tatsumi, K. Dinuclear nickel complexes modeling the structure and function of the acetyl CoA synthase active site. Proc Natl Acad Sci U.S.A 2009, 106, 11862–11866.

[74] Linck, RC., Spahn, CW., Rauchfuss, TB., and Wilson, SR. Structural analogues of the bimetallic reaction center in acetyl CoA synthase: A Ni-Ni model with bound CO. J Am Chem Soc *2003*, 125, 8700–8701.
[75] Matsumoto, T., Ito, M., Kotera, M., and Tatsumi, K. A dinuclear nickel complex modeling the Nid(II)-Nip(I) state of the active site of acetyl CoA synthase. Dalton Trans 2010, 39, 2995–2997.
[76] Ram, MS., Riordan, CG., Yap, GPA., Sands, L., Rheingold, AL., Marchaj, A., and Norton, JR. Kinetics and mechanism of alkyl transfer from organocobalt(III) to nickel(I): Implications for the synthesis of acetyl coenzyme A by CO dehydrogenase. J Am Chem Soc 1997, 119, 1648–1655.
[77] Yoo, C., Ajitha, MJ., Jung, Y., and Lee, Y. Mechanistic study on C–C bond formation of a Nickel (I) monocarbonyl species with alkyl iodides: experimental and computational investigations. Organometallics 2015, 34, 4305–4311.
[78] Seravalli, J., Kumar, M., and Ragsdale, SW. Rapid kinetic studies of acetyl-CoA synthesis: evidence supporting the catalytic intermediacy of a paramagnetic NiFeC species in the autotrophic Wood-Ljungdahl pathway. Biochemistry 2002, 41, 1807–1819.
[79] Weeks, A., Lund, L., and Raushel, FM. Tunneling of intermediates in enzyme-catalyzed reactions. Curr Opin Chem Biol 2006, 10, 465–472.

Part III: Medical Applications

Daniel Siegmund and Nils Metzler-Nolte

8 Medicinal organometallic chemistry

8.1 Introduction

Metals and their corresponding metal complexes have long been recognized as potential remedies in the treatment of various types of diseases. Starting out from ancient alchemical treatments lacking any detailed knowledge concerning the origin of diseases (and therefore actually rather mistreatments), chemotherapeutics have been developed to sophisticated, rationally based and most importantly highly efficient treatment options [1]. Along this development, metal-containing formulations were largely forgotten, replaced by purely organic molecules, and biased for extreme toxicity or severe side-effects. Essentially only irreplaceable inorganic compounds like the famous anticancer drug Cisplatin, the prime example of inorganic drugs, remained on the market. However, in recent years, it became increasingly clear that metal-containing compounds are set to make a comeback. Scientist all over the world embark on the search for novel metal-based drugs for a broad range of medicinal applications. The primary goal of many groups is the exploitation of unique properties of metal atoms that cannot easily be provided by organic molecules [2, 3]. In this way, new therapeutic approaches are created. Keeping in mind the remarkable diversity and huge successes of **organometallic compounds** in, for example, catalysis it is by no means surprising that many compounds under investigation belong to this subclass as well. By studying some of the milestone compounds in this field and emphasizing key advantages of certain metal atoms in conjunction with structural features of favorable organometallic ligands, the reader shall be equipped with a solid conceptual understanding of organometallic medicinal chemistry.

8.1.1 Organometallic drugs – some preliminary considerations

To be considered as potential drugs for usage in in vivo studies and probably later in the human body, candidate compounds need to fulfill certain obligate requirements and organometallic drugs are no exception [4, 5]. The most straightforward property that needs to be tightly controlled is the stability of the complex. By definition, organometallic compounds have at least one direct metal–carbon bond. In

Daniel Siegmund, Nils Metzler-Nolte, Inorganic Chemistry I – Bioinorganic Chemistry, Faculty of Chemistry and Biochemistry, Ruhr University Bochum, Bochum, Germany

https://doi.org/10.1515/9783110496574-008

main group organometallics, these bonds are typically polar if not ionic, and therefore highly reactive (as, e.g., in lithium alkyl reagents). On the other hand, bonds between transition metals and carbon ligands are typically less polar, and often further stabilized by a combination of σ- and π-bonding. Although many applications rely on a timely ligand exchange or in extreme cases even the partial or complete degradation of the metal complex (the so-called prodrug concept), a stable complex insensitive to air and moisture is usually needed in the first place. Special conditions at the target destination or an external stimulus may then be applied to induce a site-specific impact on the metal complex. At the very least, a candidate compound has to be sufficiently stable to ensure a considerable accumulation at the target destination, and many organometallic ligands do actually fulfill this criterion of stability [3, 6].

From the chemist's point of view, another important aspect to consider is certainly a careful tradeoff between solubility in aqueous environment and lipophilicity, both of which have decisive influence on bioavailability and biodistribution [7]. In this sense, important parameters to be controlled are the oxidation state of the complex ion and the choice of the appropriate ligand system (neutral or charged as well as lipophilic or hydrophilic). Having this in mind, it is not surprising that the organometallic ligand portfolio used for biological purposes is somewhat limited and certain structural features appear more frequently, while others remain scarce or are omitted entirely. Figure 8.1 summarizes common complex structural motifs encountered throughout medicinal organometallic chemistry [3]. Metal complexes that have π-bonded planar (hetero)aromatic rings are classified as "arene" complexes, and those with two arene ligands (e.g., the omnipresent cyclopentadienyl ligand) are called "metallocenes" (coplanar or bent, see Figure 8.1 for example). Other ligand classes are carbon monoxide (CO, leading to the famous metal carbonyl complexes) and *N*-heterocyclic carbene ligands (NHCs, bottom row of Figure 8.1).

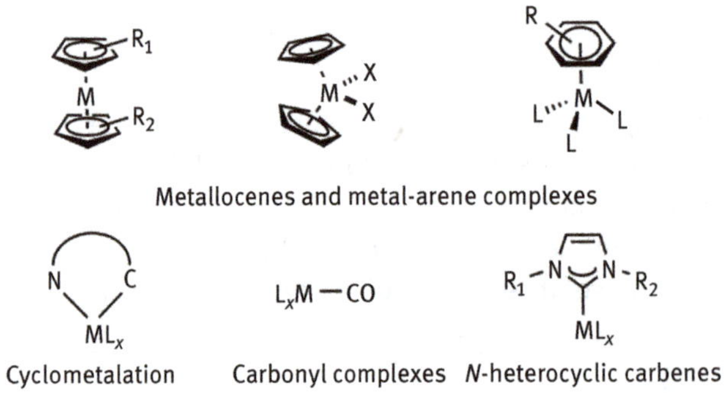

Figure 8.1: Structural motifs commonly used in medicinal organometallic chemistry.

8.1.2 Organometallic drugs – new remedies by new modes of action

Proposing organometallic complexes as drug candidates raises the obvious question for potential benefits. Clearly, there must be substantial arguments to justify the use of rare (and thus often costly) metals in medicinal applications. Moreover, certain metals are even widely regarded as poisonous. On the other hand, some of the most potent toxins are purely organic compounds, and in general, the purpose of a drug is very often to be "toxic," such as an antibiotic will be toxic for bacteria, or the function of an anticancer drug is to kill cancer cells. It is thus important to look in more detail into the particular properties and opportunities that might set metal-containing compounds apart from purely organic drugs. First, it is evident that organometallic compounds provide a huge **structural and stereochemical diversity**. In particular, their unique three-dimensional shape (e.g., square-planar, octahedral or even higher coordinated) can be hardly achieved with solely organic molecules. Further, the exchange of ligands facilitates their synthetic accessibility and allows for a simplified **rational design of structurally demanding complexes.** One of the most important aspects that sets metal complexes apart from organic molecules is the availability of a broader range of modes of action. These may be depending on well-controllable **oxidative and reductive processes**, the **catalytic activity** of metal complexes or structural alteration by the aforementioned **ligand exchange reactions.** Another important point to consider is the possibility to use **radioactive metal isotopes** as diagnostic tools or radiotherapeutic drugs. Despite the huge diversity of metals (spanning large parts of the periodic table), the lead structures and compounds presented later in this chapter can almost entirely be classified and understood by one of or by a combination of several of these key concepts [2].

8.1.3 Exploiting the "right" metal – to be spoilt by choice

As a matter of fact, around 80% of the elements in the periodic table are metals and organometallic chemistry for many representatives can be considered as quite well established nowadays. However, as probably most chemists can confirm from first-hand experience, the properties of compounds differing in just one atom can be poles apart. Thus, it is important to stress that not every metal ion will be providing the previously mentioned characteristics to the same extend. Even isostructural complexes with different metal atoms may behave quite differently in biological environment (even differences between two oxidation states of the same metal can be extremely important!). On the contrary, many conceptual aspects regarding the properties mentioned in Section 8.1.2 may be adapted and transferred to comparable complex structures. From this apparent contrasting point, it can be concluded that considering multiple metal ions may be beneficial. In a sense, there is no "correct" metal for biological systems, but there is a large periodic table of metals, some more, some less capable of delivering a certain, desired aspect. It is the complicated interplay of these aspects in a biological system that explains the huge structural diversity and large number of therapeutic approaches, which newcomers are facing in this field of research.

8.1.4 Rational design versus trial and error

Although various organometallic drugs are in principle well understandable from a conceptual point of view, it should not be concealed that many of the underlying lead structures (or rather their biological properties) were found by trial and error [8]. It is only after an initial observed activity that careful alteration of the ligand systems and the oxidation states allows for the development of detailed structure activity relationships that can be used as a basis for the rational design of improved candidate drugs. The conceptual classification can then be established by the elucidation of the mechanism of action using sophisticated techniques from molecular biology and increasingly computational biology [9, 10].

In the following sections, the presented conceptual ideas and principles are demonstrated on the basis of selected prominent examples representing milestones in medicinal organometallic chemistry. Unfortunately, the vast number of great achievements in the literature and limited space herein obviously prevent a comprehensive representation. By presenting a number of different metals in important fields of application, we nevertheless hope to inspire students and researchers to explore this field of high importance, which certainly has a great future.

8.2 Anticancer activity

Cancer, that is, the abnormal growth of cells in the human body can be considered as one of the prime health concerns of modern society. It is regarded as the leading cause of death before age 70 in many countries as estimated by the World Health Organization (WHO) [11]. Consequently, tremendous efforts are undertaken to develop potential remedies and to advance existing technologies. Still, many of the clinically relevant and successful therapies involve the application of cytotoxic drugs (chemotherapy) to target cancerous cells by exploiting key-differences to healthy cells, for example, the rapid cell division or altered metabolism. Inspired by the tremendously successful inorganic drugs of the Cisplatin family, various types of metal complexes were investigated for their potential to serve as anticancer agents. In this chapter, we outline a number of selected key concepts on how organometallic complexes may provide a valuable addition to the chemotherapeutic arsenal in the near future.

8.2.1 Enzyme inhibition

The selective inhibition of key enzymes by means of small molecules is one of the most important strategies to favorably intervene in metabolic pathways. A high percentage of commercially available drugs is in fact inhibiting one or several enzymes. This includes organic molecules as well as inorganic drugs. Thus, it is not surprising that a lot of effort is put into the development of organometallic enzyme inhibitors as well that have led to interesting discoveries confirming the unique potential of metal complexes [12–15].

8.2.1.1 Metal complexes as structural templates for noncovalent enzyme inhibitors

The most prominent examples of organometallic enzyme inhibitors were created by Meggers and coworkers who designed novel organometallic protein kinase inhibitors [16]. Protein kinases form a large family of enzymes (>500 members with distinct targets each) capable of phosphorylating other proteins to essentially turn their activity on or off depending on the exact circumstances [17]. Misregulation of these kinases upon mutation is tightly connected to tumor growth and consequently their inhibition has become an attractive target for antitumor therapy [18]. However, due to their substrate variability, it is obvious that only selected kinases play important roles in tumor growth and hence selectivity for specific subtypes has become a primary concern. Previously known (organic) substances essentially only provided rather poor selectivity to the enzymes' highly conserved ATP-binding motif (the prime target for inhibitors) [19].

Meggers et al. decided to modify the natural alkaloid and kinase inhibitor Staurosporine in a way that it can serve as a bidentate ligand for metal ions while at the same time conserving the necessary sites interacting with the enzyme's binding pocket. By replacing the glycosyl moiety with a tetrahedral or octahedral metal complex (leading to the term octasporines) with variable coordination sites, the group achieved remarkable increases in activity and selectivity for specific kinases [20, 21]. Crystal structure analyses and computational studies revealed that the increased selectivity can be attributed to interactions of metal-bound ligands with the respective enzyme. While the Staurosporine-derived bidentate ligands ensure strong binding to the kinases' ATP pocket as expected, the remaining ligands enable specific additional interactions with less conserved structural features of the proteins, which leads to an increase in selectivity [20, 22]. Of interest from an organometallic point of view are inhibitors containing CO-ligands, which, for example, were shown to play a beneficial role for the selectivity toward kinases GSK3α and Pim1 (Figure 8.2) [22]. In contrast to organic carbonyl groups, this ligand behaves rather nonpolar and, in conjunction with the unique three-dimensional geometric properties of the metal complex, enables protein interactions essentially inaccessible for purely organic drugs. The fact that several metal ions (M = Ru, Ir, Os) were successfully employed in the design of analogous kinase inhibitors emphasizes that the metal complex adopts an exclusively structural role [23]. Here, the choice of a particular metal is governed by the availability of suitable starting materials and synthetic accessibility of the intended structure. Starting from these observations, a huge family of related kinase inhibitors was already synthesized and tested, and selective inhibitors for numerous kinases were created. This includes organometallic as well as nonorganometallic complexes. It is anticipated that this concept will play an important role in the development of other enzyme inhibitors in the future.

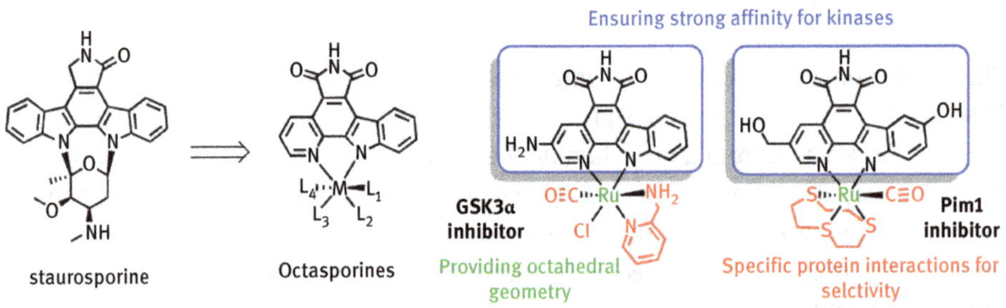

Figure 8.2: Conceptual approach of Meggers et al. for Staurosporine derivatives and two organometallic kinase inhibitors from that group.

To get an idea about the dimension of structural variability of octahedral complexes, it is worth noting that a total of up to 30 isomers are possible for a single octahedral complex bearing six different ligands (15 enantiomeric pairs), whereas a typical tetrahedral coordinated C-atom with four distinct substituents occurring in organic chemistry possesses just two isomers (one enantiomeric pair) [24].

8.2.1.2 Covalent modification of enzymes

Aside from competition for a specific substrate binding site, another prominent possibility to influence enzyme activity is the (irreversible) modification of important amino acid side chains. Although this may strictly speaking lead to an inactivation rather than an inhibition, the following examples represent well the paradigms of using organometallic complexes to target enzymes covalently.

Primary targets of such a strategy have to be amino acids that are important for catalytic reactions or essential to maintain the enzyme's structural integrity. On the other hand, they should ideally be rather seldom incorporated in enzymes to suppress undesired side-reactions. Among the natural amino acids, cysteine and selenocysteine (SelenoCys) best meet these requirements. Assuming a direct interaction between the metal and amino acid residues, attractive candidate metals for targeting Cys and SelenoCys with an organometallic complex are certainly the relatively soft gold(I) and Pt(II), both providing high affinity to organic sulfur or selenium-containing molecules according to the HSAB concept. Inspired by the findings on heavily studied Au-based antiarthritic drugs such as Auranofin, which influence a broad range of proteins and cellular processes associated with tumor growth [25, 26], novel Au(I) containing lead structures for anticancer therapy were created (Figure 8.3). For a direct covalent metalation of target structures, the complexes need to have an appropriate leaving group while at the same time being stable enough to reach the target destination (the question of stability). This careful trade-off can be achieved by incorporating halides as leaving groups and NHC as stable spectator ligands that enable the tuneability of the complexes' physicochemical properties by choosing appropriate N-substituents. In fact, NHC complexes of various transition metals emerged as interesting drug candidates and underline the huge potential and impact of this ligand class [27].

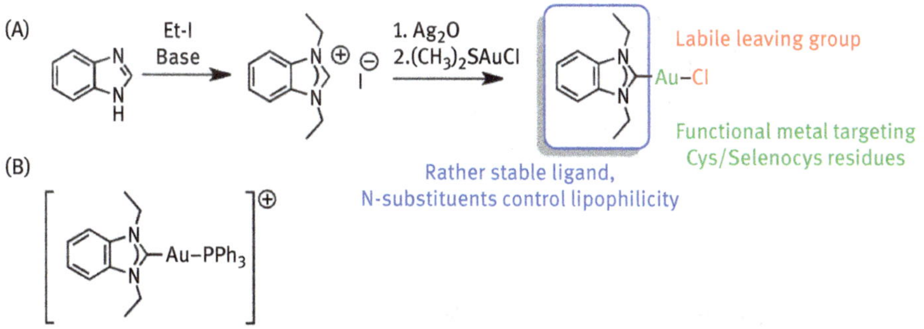

Figure 8.3: Schematic synthesis of a Au(I)-NHC anticancer lead structure (A) and cationic derivative incorporating a neutral phosphine ligand (B).

As expected, the complexes metallate the important family of selenium-containing thioredoxin reductases (TrxR), which was considered a prime target of these gold complexes [28–30]. Moreover, they were found to be antimitochondrial agents (the prime location of TxrR2), where they effectively initiate mitochondrial apoptosis pathways. However, as it is often the case for novel organometallic drug candidates, the spectrum of activity seems even more diverse. It was also shown that other selenium- or cysteine-containing enzymes could be targeted and increased values of reactive oxygen species (ROS) were observed, potentially decreasing the drug's specificity [29]. By alteration of the ligand system, for example, introducing a neutral ligand, a second carbene ligand or by increasing the lipophilicity, important factors controlling impact, selectivity and mode of action of the complexes were identified. Today, it appears that targeting selenium in key enzymes is an important, but certainly not the only, mechanism in action. Although the problem of selectivity was not entirely solved so far, interesting approaches to increase the specificity include the conjugation of the metal complex to short peptide sequences specifically targeting the desired cell type or organelle. This bioconjugation strategy is in principle suitable to alter the targeting properties of any metallodrug compatible with peptide chemistry.

In case of organometallic Au drugs, mitochondria can be considered a valuable target. Thus, a mitochondrial targeting peptide was covalently attached to a gold complex (Figure 8.4). It was later shown that a remarkable selectivity for TxrR enzymes was achieved, underlining the usefulness of this approach [31].

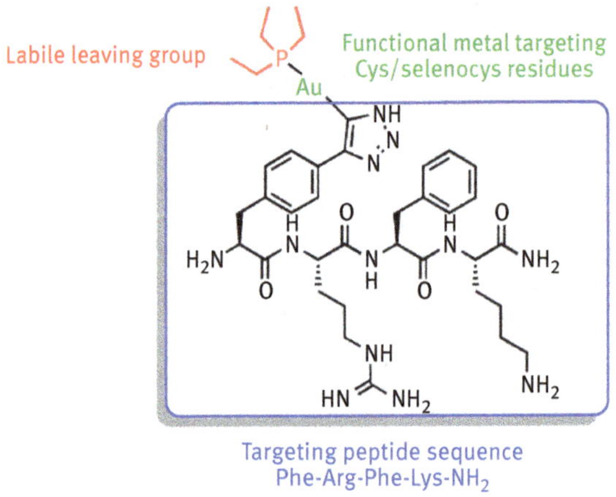

Figure 8.4: A mitochondria-targeting organometallic Au–peptide conjugate.

8.2.2 Ruthenium–arene complexes – achieving multiple biological interactions

Ruthenium complexes in the oxidation states +II and +III have gained considerable attention as potential replacements of anticancer drugs of the tremendously successful Cisplatin family, which are impaired by strong side-effects and developing tumor resistances during treatment. Famous examples of Ruthenium-based drugs are KP1019/(N)KP-1339 (from the group of Keppler) and NAMI-A (from Mestroni and Sava groups, both Figure 8.5), which entered clinical trials [32, 33]. Depending on the exact structure, the potential targets of Ru drugs are diverse and include DNA, important cellular organelles, or proteins. Furthermore, selected complexes may be exploited as photosensitizers in photodynamic therapy (PDT) [34].

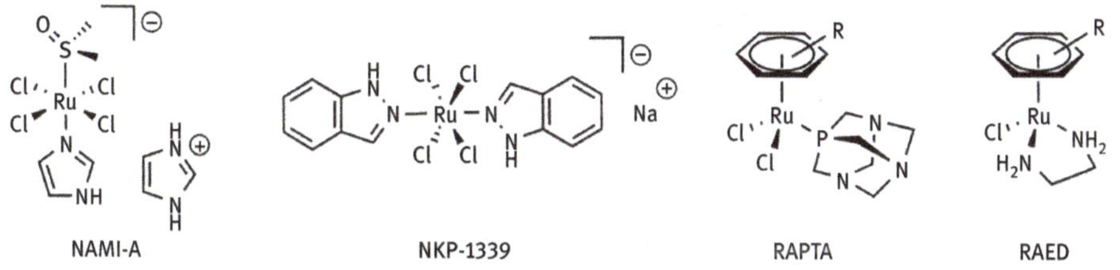

Figure 8.5: Structural formulae of intensely studied Ru anticancer drugs.

Organometallic Ru anticancer drugs are mainly based on a pseudo-octahedral Ru (II) core structure decorated with η^6-arene ligands like benzene, cymene, dihydroanthracene or related derivatives. The remaining three coordination sites are usually occupied by one or two rather labile ligands (often halides) and another monodentate or bidentate ligand. Among the most studied representatives are complexes containing the pta ligand (1,3,5-triaza-7-phosphatricyclo[3.3.1.1]decane) (RAPTA-type, Dyson et al.) or bidentate *N,N*-donor ligands like ethylenediamine (e.g., RAED by Sadler et al., Figure 8.5) [35, 36].

RAPTA-C (Figure 8.6) is probably one of the best understood examples. The compound provides antiangiogenic effects and excellent in vivo activity against metastatic tumor cells, while only minor cytotoxic effects to primary tumors and an efficient clearance from organs and the bloodstream were observed. Remarkably, RAPTA-C behaves rather nontoxic in in vitro experiments, suggesting the possibility of less pronounced side-effects in patients [34, 37].

Many Ru drugs are considered to target DNA in a manner related to Cisplatin. However, initial studies on RAPTA-C suggested a complex interplay of different intra- and extracellular events that together form the mode of action. This conclusion is likely to be true for the majority of Ru-arene drugs. Although the compound was shown to interact with DNA in a pH-dependent manner, these results failed to explain its entire cytotoxic profile. Later, it was established that interactions with proteins like histones, glutathione transferase, lysozyme and TrxR are likely also involved and probably even play a dominant role. Even so, the overall mode of action cannot be considered entirely understood in all its details so far [34].

Regarding RAPTA-C from a conceptual point of view (Figure 8.6), the halide ligands are prone to hydrolysis in intracellular environment due to low chloride concentrations. This hydrolysis will potentially accelerate the drug's impact. The pta ligand is designed to improve the solubility of the complex. Further, it may be protonated in an acidic environment that is likely present in hypoxic cancer cells. As these provide a lower pH than healthy cells, a more active complex is obtained and thus cancerous tissues are recognized. The arene ligand predominantly stabilizes the Ru (II) core structure and allows for adjustment of the lipophilic properties of the complex.

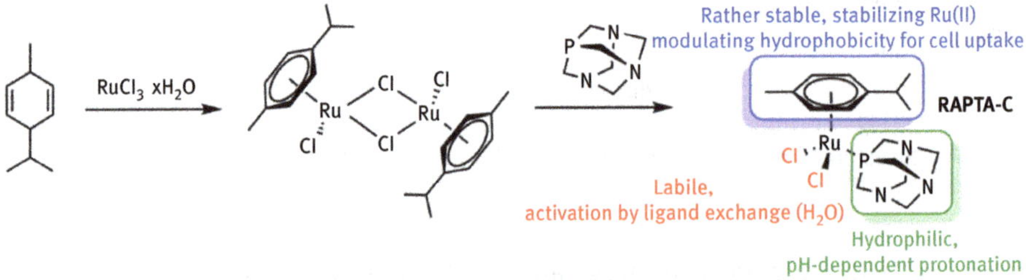

Figure 8.6: Schematic synthesis of RAPTA-C and conceptual assessment of its structure.

It is worth pointing out at this point that multiple modes of action as demonstrated for Ru-containing drugs are not a rare exception in organometallic medicinal chemistry, despite all design efforts. Although this behavior might appear chaotic or undesirable at first glance, it can in fact be a key advantage because developing resistances against these drugs will thus be much harder for cancerous cells (or bacteria and other microbes in the case of antimicrobial compounds, see Section 8.3). This also implies that the functional classification of certain drugs is by no means fixed. Depending on the context, different properties of an organometallic drug may be emphasized and conceptual similarities such as enzyme interactions are not limited to one application only, for example, anticancer therapy.

8.2.3 Titanocenes – from structural analogies to new challenges

Soon after the cis-dihalide arrangement was recognized to be a crucially important structure–activity relationship (SAR) in platinum anticancer drugs, metal complexes with a similar structural motif were screened for their antiproliferative activity. As one promising group of compounds, pseudo-tetrahedral metallocene dihalides were identified and, indeed, titanocene dichloride showed remarkable anticancer activity in mice (Figure 8.7) [38]. Titanocene dichloride has even entered clinical trials.

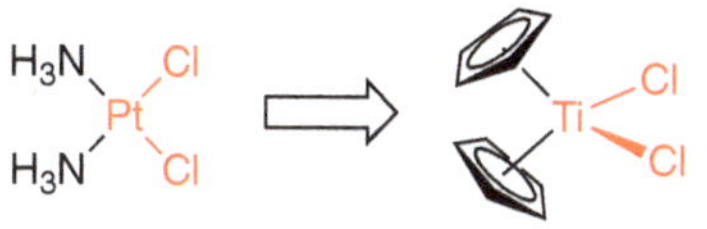

Figure 8.7: Structural analogy between Cisplatin (left) and titanocene dichloride (right), the cis-metal-dihalide arrangement is highlighted in red.

However, because of the compound's low solubility in water, problems with formulation and its problems with decomposition those trials were terminated a few years ago during phase II. The rate of hydrolysis of the halido ligands in titanocene dichloride is in the order of minutes, as compared to hours in the case of Cisplatin, and even the cyclopentadienyl rings are completely lost by hydrolysis within a few hours.

8.2.3.1 Second-generation titanocenes – chemical answers to biological challenges

To increase both the aqueous solubility and the stability of titanocenes, a number of modifications to the lead structure can be applied. For example, *ansa* derivatives have been developed where the two cyclopentadienyl rings are covalently linked, increasing stability by virtue of the chelate effect. On the other hand, amino-substituted cyclopentadienyl rings will enhance water solubility. In a screen against 36 different tumor cell lines, the *p*-methoxybenzyl-substituted derivative titanocene Y (Figure 8.8) was found to be highly active against renal cell cancer and pleura mesothelioma cell lines, in particular, for which no other effective chemotherapy is currently available. This compound also displays promising activity in vivo, and mechanistic studies revealed a desirable combination of biological effects, including antiangiogenesis, activation of the immune system, induction of apoptosis via caspases 3 and 7, but no myelosuppression. However, its hydrolytic stability still remains problematic. Therefore, inspired by the success of second-generation platinum drugs such as carboplatin, which features dicarboxylate ligands, the two chloride ligands on titanocene Y were replaced with carboxylate groups to yield equally active compounds with improved pharmacokinetics [39]. This is also in line with the harder Lewis acid character of Ti(IV), as compared to Pt(II), again according to the HSAB concept.

It should be mentioned that other early transition metals will provide structurally similar compounds, and indeed bis-cyclopentadienyl-dihalide derivatives of Zr, Hf, V, Nb, Mo and W were also investigated for their antiproliferative activity [40, 41]. While Mo in particular showed interesting activity and much improved hydrolytic stability, research into this metal has not been as intense as for the titanium analogues described above.

Despite the notion of structural similarity, which inspired the investigation of metallocene dihalide derivatives of the early transition metals as anticancer drug candidates in the first place, it is highly unlikely that they should have a mode of action similar to Cisplatin. Not only are the metal ions in higher oxidation states

(+IV in titanocenes vs. +II in Cisplatin), but also at least the first and second row transition metal ions in compounds with appreciable activity (such as Ti, Zr, V, Nb and Mo) will have much smaller radii than the third-row metal Pt. This will not only critically influence rates of ligand substitution (see the above discussion of hydrolysis kinetics), but will also result in high oxophilicity, and thus will likely alter their affinity to biological targets completely. Indeed, DNA binding was never shown to be a relevant mechanism of action in the biological experiments on any of the metallocene derivatives described in this chapter. Along the same line of thought, it should be remembered that molybdocene dihalides have two electrons more than the analogous titanocenes, and are thus Lewis bases rather than Lewis acids. This again should result in different reactivity also in a biological context, despite superficially – and deceivingly – similar structures.

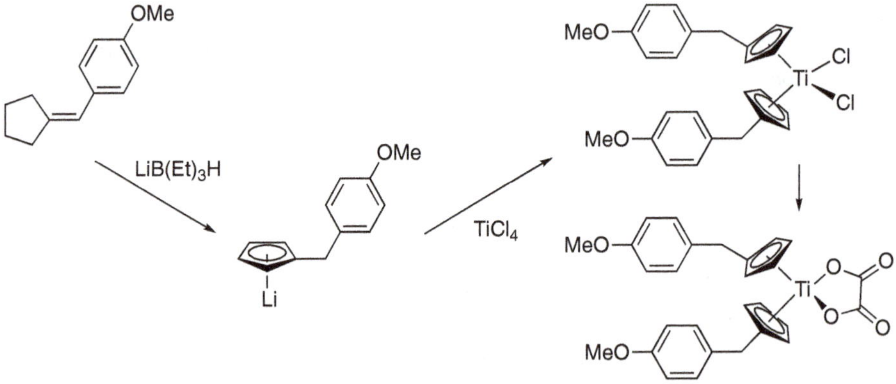

Figure 8.8: Schematic synthesis of the second-generation titanocene anticancer drug candidate (titanocene Y) via the fulvene route, and a carboxylate derivative of titanocene Y.

8.2.4 Ferrocifen – redox activation of a metallodrug

As noted in the introduction, reversible redox activity (i.e., uptake or release of electrons) is a feature that is often seen in metal complexes, including many organometallic ones. The prototype organometallic compound ferrocene (bis(cyclopentadienyl) iron) is an excellent example, with iron changing reversibly between oxidation states +II and +III without major geometrical changes (see Figure 8.9, left). This redox activity is linked to many of its biological applications [42], for example, in electrochemical biosensors, but also in drug candidates like ferrocifen (Figure 8.9, right) [43].

Jaouen et al. used the well-established anticancer drug tamoxifen as a blueprint for organometallic derivatization [44]. Phenyl rings were substituted with ferrocenyl groups, and the activity of the resulting organometallic tamoxifen derivatives (termed *ferrocifens*) was tested against different breast cancer cell lines [45]. In general, breast cancers differ in the expression of estrogen receptors. Those that express the estrogen receptor (ER+) are more common, and also more readily treated by tamoxifen-related so-called antiestrogens. Unfortunately, about one-third of all breast cancers do not overexpress the estrogen receptor (ER–), or stop its expression during treatment, and consequently also do not respond (any more) to treatment by tamoxifen. Most notably, the ferrocene derivatives were active against both, ER+ and ER– cell lines. This is a remarkable finding, which has been attributed to the redox activity of ferrocifen. Because it is still structurally similar to tamoxifen, ferrocifen will act by binding to the estrogen receptor in ER+ cells, and thus exhibits an activity very similar to tamoxifen. In ER– cells, however, a different mechanism of action is in operation, where the ferrocene core serves as a "redox antenna," leading to the formation of a quinone methide after oxidation of the phenol ring and deprotonation (Figure 8.10A). This quinone methide reacts rapidly with biomolecules, and probably induces cell death by a mechanism termed "senescence." While a similar mechanism is also possible for tamoxifen itself, the chemistry is greatly accelerated in ferrocifen and becomes the major pathway [46, 47].

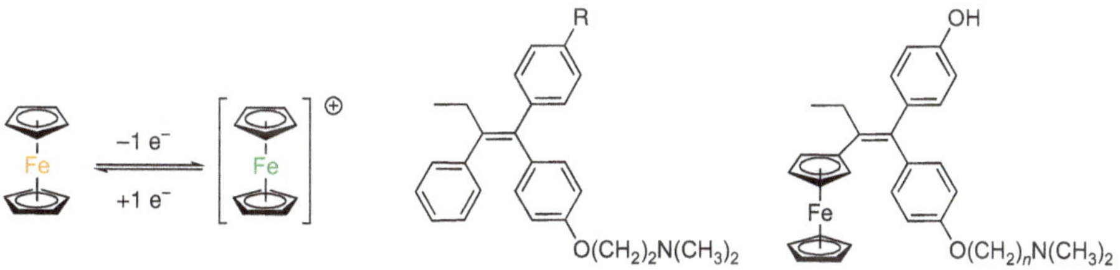

Tamoxifen (R = H)
Hydroxy-tamoxifen (R = OH)

Ferrocifens (n = 2, 3, 5, 8)

Figure 8.9: Reversible one-electron redox chemistry of ferrocene (left), and the structures of tamoxifen and ferrocifen.

Support of this unusual redox activation comes from the fact that a related ruthenocene derivative (which has very similar structural parameters as ferrocene, but does not undergo the same reversible one-electron redox chemistry) of tamoxifen was found to be inactive when evaluated against the ER– cell lines. Also interestingly, if the assumption of this "redox antenna" mechanism is correct, then the structure can be simplified and, indeed, ferrocenyl diphenols have excellent antiproliferative activity and show the same electrochemistry as the original ferrocifen lead structure (Figure 8.10B) [48]. Conversely, if the conjugation between the ferrocene ring system and the phenol is interrupted, then the antiproliferative activity is largely lost.

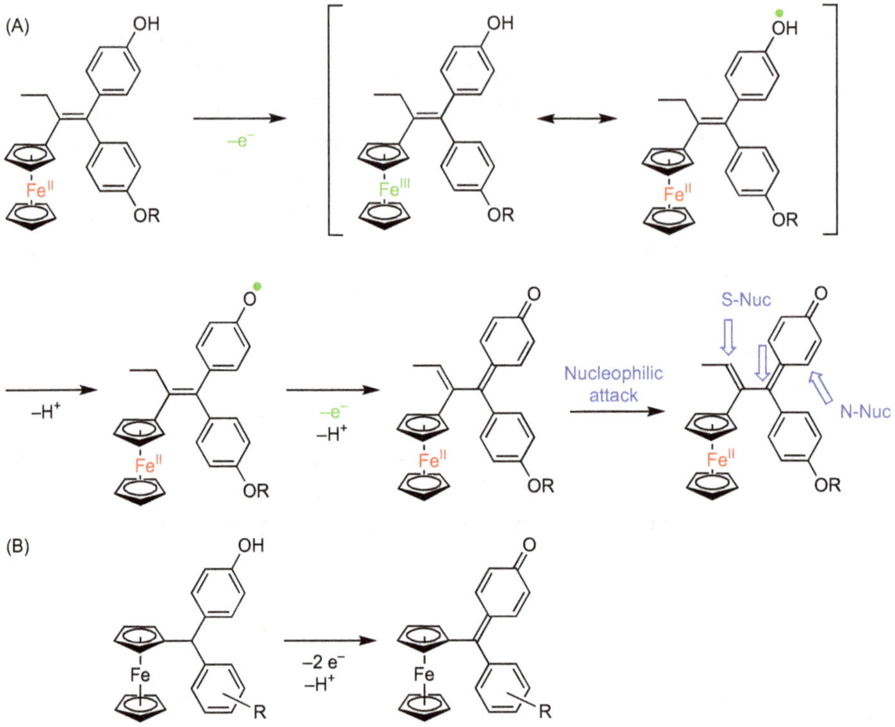

Figure 8.10: (A) "Redox antenna" mechanism of ferrocifen and (B) a simplified ferrocenyl diphenol with similar activation mechanism to form a highly reactive quinone methide.

In the case of ferrocifen and its derivatives, the synthesis is relatively straightforward, by combining two ketones using the Ti-mediated McMurry coupling (Figure 8.11). The relative stability and ease of chemical synthesis, as exemplified by ferrocifen and its derivatives, has certainly positively contributed to the success of organometallic drugs and drug candidates.

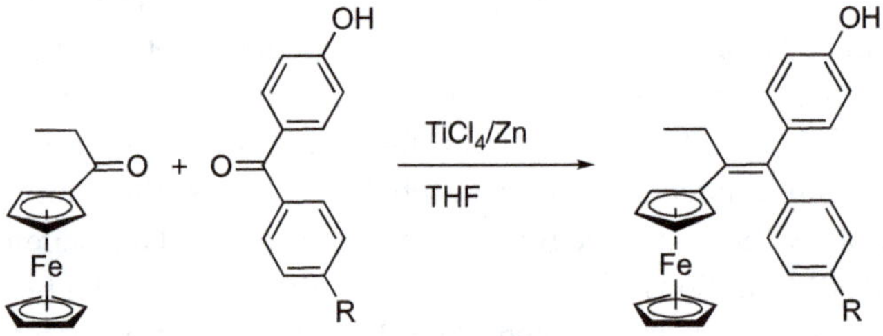

Figure 8.11: Synthesis of ferrocifen derivatives via McMurry coupling.

Finally, it is interesting to note that ferrocene has been incorporated into several other drugs in a manner that is similar to the tamoxifen -> ferrocifen transformation, that is, by substituting aromatic phenyl rings, or as an addition to other drugs and drug candidates in order to add redox activity to the molecule. While this idea seems simple yet beautiful, it has been blessed only with limited success. In most cases, an additional redox activity could either not be observed under physiological conditions, or if present it did not serve to enhance the activity of the parent drug or drug candidate molecule.

8.3 Antimicrobial activity

The discovery of effective antibiotic compounds, for example, Alexander Fleming's famous penicillin, revolutionized medicine in the 20th century and allowed deadly infective diseases to be cured routinely. However, the broad usage of antibiotics led to high evolutionary pressure on bacteria resulting in the formation of bacterial strains essentially resistant against many if not all of the well-established drugs. Nowadays, conceptually new lead structures for drugs are urgently needed to counteract this development. In this regard, metal complexes may provide new modes of action and impede the development of bacterial resistances [49].

8.3.1 Silver *N*-heterocyclic carbene (NHC) complexes

Even without detailed knowledge about the nature of infections or chemical sciences in general, silver formulations were used as antimicrobials since the 17th century. It was only much later established that the active species in these formulations are in fact Ag^+-ions. In an aqueous solution of $AgNO_3$, the metal was used for the treatment of burns, before Ag^+ with more complex organic anions was established. One member of this class of compounds is still in clinical use: Silvadene®.

In an attempt to overcome common drawbacks of these compounds, silver-NHC complexes were established by Youngs and coworkers [50]. In contrast to previously used ionic structures, the rather covalent Ag–NHC bond allows for the continuous liberation of active Ag^+ ions over time via a slow decomposition in aqueous media. By careful modification of the NHC ligand, the stability of the Ag–C bond and thus the rate of silver release can be controlled. In a similar manner, the solubility of the complexes can be optimized. As a second ligand in the linear Ag(I) complexes, the acetate anion was found to be a convenient choice to increase the solubility of the compound and to reduce its general toxicity [27].

To create a highly soluble complex with reduced toxic side effects to mammalian cells, Ag–NHC complexes derived from naturally occurring xanthines (e.g., theobromine) were created (Figure 8.12). These compounds provide excellent activity against drug-resistant pathogenic bacterial strains [51]. Also, using ligands derived from biologically active natural substances is a recurring concept in bioorganometallic chemistry. From a synthetic point of view, Ag–NHC complexes are accessible via direct metalation of suitable (benz)imidazolium salts. The latter often have to be prepared by quarternization of amine precursors. Many Ag–NHC complexes are further used as transmetalating agents to transfer the carbene ligand to other metals. This technique has been exploited many times to generate bioactive metal–NHC complexes.

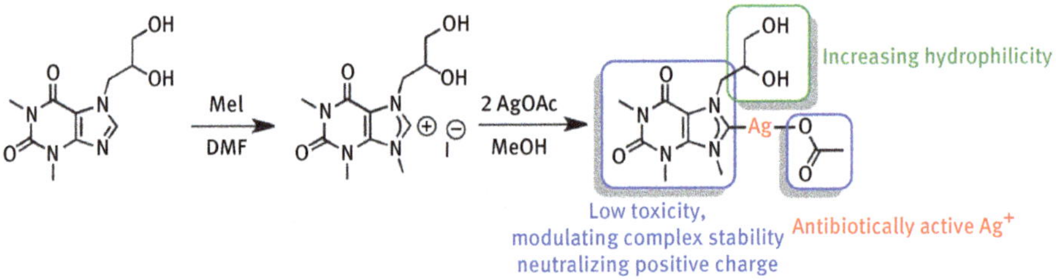

Figure 8.12: Synthesis of Ag–NHC complexes via direct metalation of suitable imidazolium derivatives.

Further progress on bioactive Ag–NHC complexes was achieved by creating cationic Ag(I) complexes with two NHC ligands, and by using asymmetrically substituted NHC ligands. Further, the incorporation of Ag–NHC complexes into nanoparticles was explored to improve systemic applications in vivo [52]. In recent years, many studies also applied Ag–NHC complexes as anticancer agents [53]. However, despite of being effective as both, antibiotic and anticancer drugs, the mechanism of action remains elusive and is subject to current research efforts.

8.3.2 Non-silver organometallic antibacterials

It is interesting to note that the first synthetic antibiotic was in fact an inorganic compound, Arsphenamine (marketed as Salvarsan®), which contains As, bonded to a substituted phenyl ring. The chemistry of Salvarsan is actually quite interesting, starting with the challenge of a correct structural description (Figure 8.13). Obviously, the formulation of Salvarsan with an As=As double bond is only a formal description. All data hint to a mixture of species in solution, and no single-crystal X-ray structure of Salvarsan could be obtained. The best available data from mass spectrometry suggest a mixture of cyclic compounds, with three- and five-membered rings being the main components [54]. While Salvarsan was highly successful as the first synthetic antibiotic on the market, research on metal-based drugs subsequently focused more on anticancer drugs (see Section 8.2), with renewed interest in organometallic antibacterials since only recently.

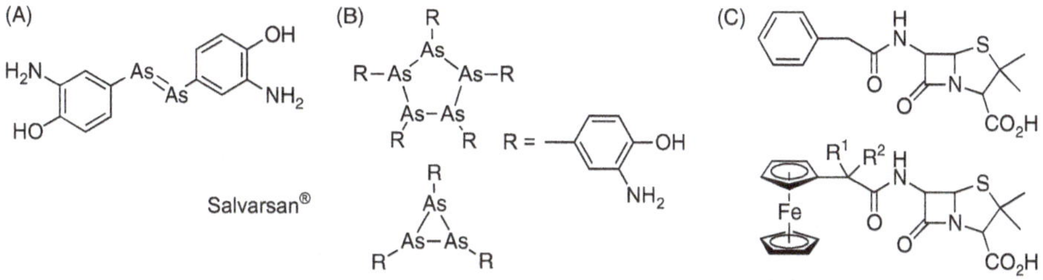

Figure 8.13: (A) Salvarsan in a formally correct description that is often used in textbooks and review articles, and (B) the three- and five-membered species of Salvarsan that were detected by mass spectrometry. (C) Penicillin G and a representative ferrocenyl analogue of penicillin.

8.3.2.1 Organometallic derivatives of established organic antibiotics

The derivatization of a well-established organic anticancer drug with a ferrocene group yielded a potent anticancer drug candidate with a novel mode of action in the case of ferrocifen (see Section 8.2.4). The same idea has also been applied to antibacterial drugs. The first examples date back to the 1970s, when penicillin derivatives with a ferrocene substitution were described (Figure 8.13) [55]. For example, the phenyl acetic acid group in penicillin G could easily be replaced by ferrocenyl acetic acid derivatives. While the compounds showed some activity, no one at the time was considering a possible alteration of the mode of action. Moreover, the pressing problem of growing resistance among bacteria to all established antibiotics was literally nonexistent 40 years ago.

Today, due to an excessive use of antibiotics and the phenomenal propensity of bacteria to adapt rapidly to hostile conditions, we are facing a situation where almost all common antibiotics have lost their power against some multiresistant bacterial strains (e.g., methicillin-resistant *Staphylococcus aureus* – MRSA, often called "superbugs" in the media). The European Union projects 10 million deaths annually worldwide due to Antimicrobial Resistance (AMR) between 2015 and 2050, and estimates in their "Action Plan" that the annual cost of AMR is in the order of 1.5 billion €, due to cost of treatment and loss of productivity. Clearly, new antibiotics are needed to combat this threat, preferably with a new mode of action, or antibiotics that combine two different modes of action so as to evade acquired resistance.

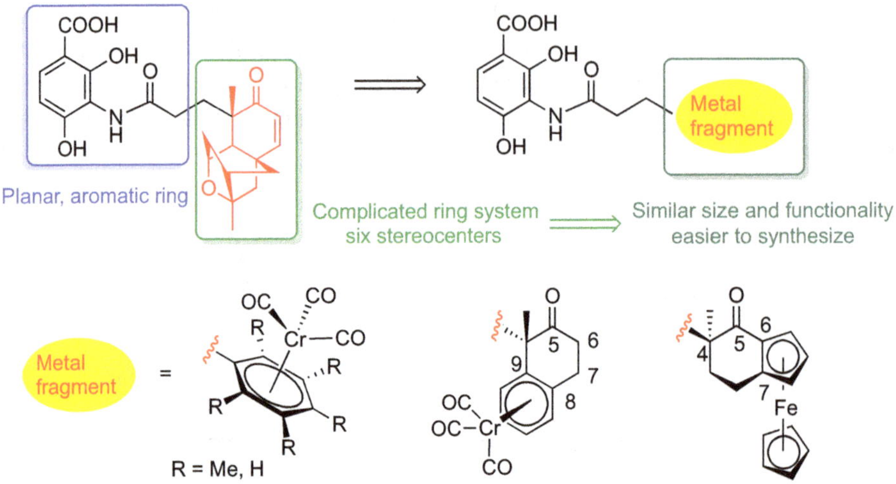

Figure 8.14: The transition from platensimycin (left) to its organometallic analogues by replacement of the tetracyclic ring. A selection of organometallic complexes that were used to replace the tetracyclic cage in platensimycin.

Following up on the idea of "organometallic replacement," the antibiotic drug candidate platensimycin was derivatized with a variety of organometallic fragments [56–60]. Platensimycin was discovered in the early 2000s as an inhibitor of the bacterial fatty acid biosynthesis, namely, the FabF enzyme [61]. It consists of an aromatic part coupled to a rather complicated tetracyclic structure with six stereocenters. While the chemical synthesis of the aromatic part of the molecule is straight-forward, the tricyclic ring system lends itself readily to replacement with organometallic fragments, such as sandwich-type ferrocene or ruthenocene derivatives, or half-sandwich cyclopentadienyl manganese or arene chromium fragments (Figure 8.14) [60]. Guided by molecular modeling, that is, computer-aided docking of the organometallic analogues into the FabF enzyme-active site pocket, numerous derivatives were indeed synthesized, which seemed to fit well into this target site. However, the overall activity of most analogues was not convincing, and later on, proteomic analysis revealed a rather unspecific toxicity as the primary mode of action for these organometallic complexes, clearly different from the intended inhibition of the fatty acid biosynthesis.

One other project in the area of organometallic antibiotics was again inspired by nature, in the area of antimicrobial peptides (AMPs). Naturally occurring AMPs typically consist of >15 amino acids, some of them positively charged, paired with lipophilic residues that in combination interact with the bacterial membranes. In one project, the activity of artificial AMPs consisting of only arginine (with a positive charge under physiological conditions) and tryptophan residues could be enhanced by addition of metallocenes to the N-terminus of the peptides [62]. Figure 8.15 shows the most active compound that was reached after optimization of the amino acid sequence and chirality (going beyond the naturally occurring L amino acids), as well as the metallocene fragment. While the starting five amino acid peptide (top left structure in Figure 8.15) had an activity >50 μM (measured as the minimum inhibitory concentration, i.e., MIC value), a more than 50-fold improvement was achieved in the final, optimized ruthenocene peptide, which had MIC values around 1 μM even against various multiresistant bacteria. Moreover, using the (nonnatural) element Ru in these peptide bioconjugates, it was possible to elucidate the mode of action of this class of AMPs in unprecedented detail [63]. Briefly speaking, these metal-substituted AMPs displace enzymes from the bacterial membrane, which stalls biosynthesis of membrane lipids and disrupts the energy supply chain of the bacteria. A key step in this investigation was localization of the metallocene–peptide conjugate in the bacterial membrane by electron microscopy and atomic absorption microscopy. Because Ru is an element with electron density much higher than all the other, lighter naturally occurring elements, it is very readily detected by electron microscopy. Moreover, its quantification by techniques like atomic absorption spectroscopy is greatly facilitated by the fact that there is no "naturally occurring" Ru – hence any Ru that is detected and quantified must originate from the peptide AMP.

8.3.3 Organometallic antimalarials

Malaria is one of the most threatening global illnesses, with more than half of the world's population at risk, mostly in countries with a low GDP and less developed health systems. According to the WHO, more than 200 million people were newly infected with malaria in the year 2017 alone, and an estimated number of 435,000 malaria-related deaths were counted in the same year [64]. Five different species are known to infect humans, among which *Plasmodium falciparum* is the most deadly one. In principle, a number of potent (and cheap) antimalarial drugs exist, such as chloroquine (CQ) and artemisinin. In terms of drug development, the biggest challenge is the development of resistance against those well-established antimalarial drugs, with more than 50% of all infections already being by resistant strains in some countries. Hence, there is an evident need for new antimalarial drugs.

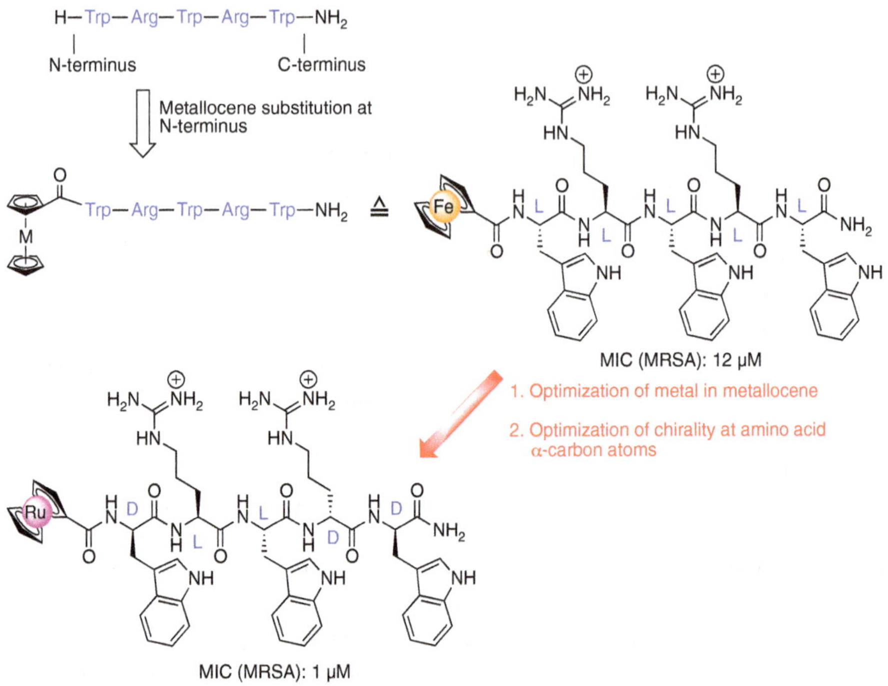

Figure 8.15: Organometallic derivatives of membrane-active antimicrobial peptides (AMPs). Systematic optimization leads to excellent activity even against methicillin-resistant *S. aureus* strains (MRSA).

8.3.3.1 Organometallic derivatives of established antimalarial drugs

In the light of the above considerations, it is not surprising that metal substitution of established antimalarial drugs was tried to enhance the activity of those existing antimalarials, or to overcome resistance. Some early examples included coordination of metal fragments to antimalarial drugs in a rather exploratory way. It was hoped that addition of a metallic moiety – possibly with a metal of known antimicrobial activity like Au – would work synergistically with the antimalarial drug. Examples of this approach are listed in Figure 8.16 as derivatives of the successful antimalarial drug chloroquine [65]. Indeed, the Au derivative of CQ shown in Figure 8.16 had ninefold higher activity in vitro than CQ against several Colombian CQ-resistant strains of *P. falciparum*. In contrast, the organometallic Rh derivative also shown in Figure 8.16 had no increased activity.

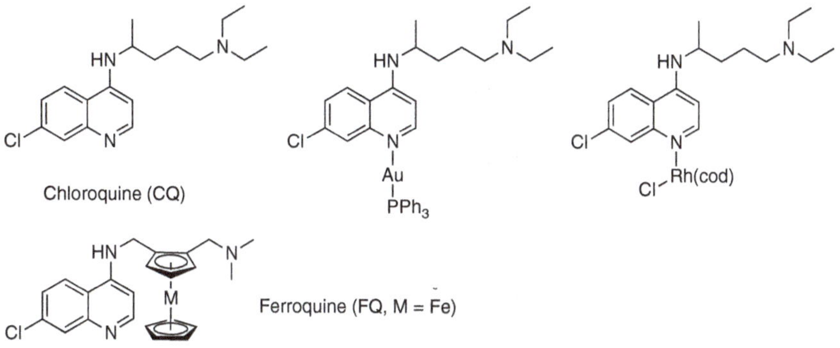

Figure 8.16: Chloroquine and organometallic derivatives (top row) and ferroquine (bottom row).

8.3.3.2 Ferroquine – an organometallic derivative of chloroquine

In 1997, Biot et al. reported a promising new lead compound, ferroquine (FQ), which is an organometallic derivative of chloroquine (Figure 8.16) [66]. The synthesis of FQ is summarized briefly in Figure 8.17 as an example how also multistep syntheses are possible on organometallic compounds [67]. While the Au and Rh derivatives discussed above had the metal coordinated to the nitrogen atom of the quinoline ring, FQ has the metallocene moiety incorporated as part of the side chain. This – seemingly small – change of the original structure of CQ has some dramatic consequences on the activity of FQ. While FQ has a similar activity as CQ on *P. falciparum*, it was also active on CQ-resistant strains. Moreover, no development of resistance was observed so far in several clinical trials [67]. The original FQ structure as well as many derivatives has been patented and further clinical development was carried out. FQ has undergone a number of trials in different countries [68]. While its use as a single antimalarial agent was stopped, it is now being explored in combination therapy with artefenomel, with the aim “single exposure, radical cure and prophylaxis” [43].

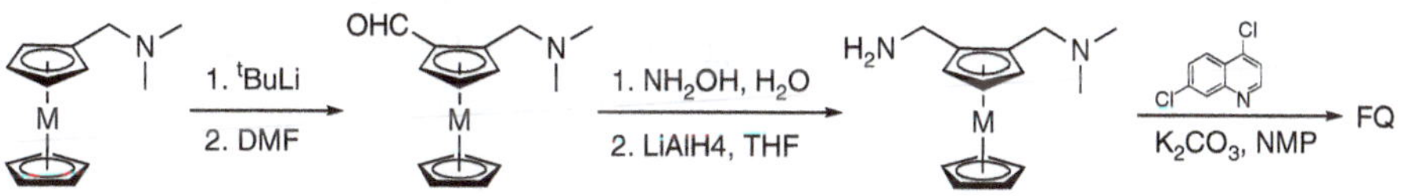

Figure 8.17: Synthesis of ferroquine.

Two major modes of action contribute to the biological activity of FQ [68, 69]. The first one is similar to that of CQ, with both CQ and FQ having a 4-aminoquinoline scaffold with basic centers that allow for accumulation in the acidic environment of the parasitic digestive vacuole. Inside the digestive vacuole, both inhibit hemozoin biomineralization, that is, the precipitation of toxic products from intracellular hemoglobin degradation. When biomineralization of these heme-based malaria pigments is inhibited, the parasite dies. The second mode of action of FQ relates to the particular redox chemistry of ferrocene, which catalytically generates hydroxyl radicals (•OH) under oxidizing conditions through a Fenton-like reaction. The presence of such hydroxyl radicals inside red blood cells that are infected with *P. falciparum* has been confirmed by fluorescence microscopy, and EPR experiments showed that such radicals were not observed for CQ or ruthenoquine (RQ, a Ru analogue of FQ, where M = Ru, Figure 8.16). This is of particular interest as FQ and RQ are isostructural and are both active in vitro against CQ-resistant *P. falciparum*. However, care has to be taken when comparing these in vitro results with in vivo studies. Obviously, the different redox chemistry of the two metallocenes (reversible one-electron oxidation for ferrocene but not for ruthenocene in aqueous solution) plays a crucial role.

Another interesting question relates to the origins of how FQ (and also RQ) overcome CQ resistance in *P. falciparum*. It has been shown that FQ and RQ do not interact with the CQ resistance transporter (pfCRT) or other CQ-exporter proteins in quinolone-resistant parasites. In this way, FQ and RQ avoid expulsion from the digestive vacuole and remain long enough inside to kill the parasite [68].

Finally, it should be emphasized that it is not only the redox-active nature of FQ that distinguishes it from CQ. For example, the presence of strong intramolecular hydrogen bonding between the secondary and tertiary amines lowers the basicity of FQ relative to CQ. Therefore, at physiological pH 7.4, FQ is predominantly in the more hydrophopbic neutral and singly protonated forms, whereas CQ is present in its doubly protonated form. Being less charged, FQ can penetrate the lipophilic digestive vacuole membrane faster than CQ. In addition, the higher lipophilicity of FQ compared to CQ also contributes to a greater accumulation of FQ in the digestive vacuole of parasites. It appears that several factors contribute to the enhanced and CQ resistance-breaking activity of FQ [43]. While most of these can be attributed to the presence of the metal complex, it would have been difficult to predict all factors. In fact, more than 100 metallocene derivatives of FQ have been prepared and tested. Out of all those, the original lead structure, FQ, still is the best drug candidate.

8.4 Others

8.4.1 Carbon monoxide releasing molecules (CORM) – release of bioactive ligands

Among the broad population, the odorless gas CO is best known for its poisonous potential and often associated with deadly accidents. Less known is the fact that low amounts of CO are actually present in the human body and act as important signaling molecules. As an endogenous messenger molecule, CO is synthesized in the human body by heme oxygenases from the breakdown of hemoglobin. In the past decades, evidence grew that associates CO with the modulation of inflammatory processes, the regulation of the cardiovascular system and neuroprotection. Beneficial effects on the human body were demonstrated by therapeutical application of low doses to patients [70–72]. However, the controlled application of gaseous CO to patients, especially by direct inhalation, is challenging and puts patients and clinic personal at a risk of CO poisoning. Moreover, achieving a target-specific delivery of CO to tissues or cells is almost impossible in this approach.

Major improvements are achieved by the application of small molecules that can release CO in a more controlled fashion (CO-releasing molecules, CORMs). Most of these molecules are in fact metal complexes carrying CO as a functional ligand that can be released either through competition with another ligand (often solvent molecules), enzymatic degradation at the target site or by applying an external trigger such as light irradiation or enzymatic activation (in a prodrug concept). Various metals have been tested for these purposes; among the most common ones are ruthenium, iron, manganese and molybdenum. In many cases, the key structural motif contains a $M(CO)_n$ core structure with the possibility to release several equivalents of CO (Figure 8.18) [71].

A heavily studied example is the water-soluble complex $[Ru(CO)_3Cl(k^2\text{-}H_2NCH_2CO_2)]$, which is one of the first examined CORMs and commonly referred to as CORM-3 (Figure 8.18). The compound was subject to detailed in vitro and in vivo investigations and provided strong evidence for the potential use of CORMs as anti-inflammatory drugs, among a number of other indications. In solution, CORM-3 undergoes complex addition and ligand-exchange reactions, leading to the loss of the first CO as CO_2 and subsequent release of CO by competition with solvent molecules or biological donor atoms if available [70, 73]. However, despite of the known CO release, the overall mode of action is likely more complex even beyond CO release and remains a matter of investigation. This once again reflects the fact that many organometallic drug candidate molecules impart their full impact via multiple modes of action. However, from a conceptual point of view, it highlights the role of the metal ion as carrier for a bioactive molecule that can be liberated upon an external trigger condition.

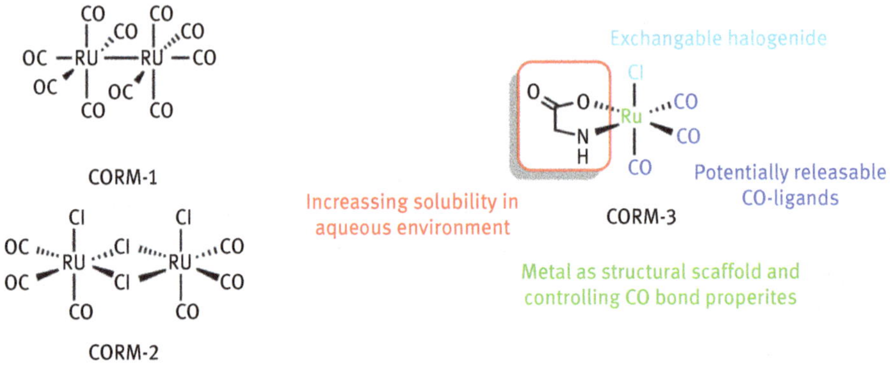

Figure 8.18: Structures of selected early CORMs and conceptual analysis of CORM-3.

Starting from the mentioned early CORMs, which still suffered from drawbacks such as insufficient stability under biological conditions, more sophisticated CORMs have been developed. The selective release of CO by irradiation with light can, for example, be achieved with photoCORMs (Figure 8.19A and B) [74]. This strategy enables both a well-timed and locally controlled application. An elegant approach is the usage of an enzyme to release CO after chemical alteration of the metal complex (enzyme-triggered CO-releasing molecules, ET-CORM) as demonstrated with a compound called **ET-CORM1**, which reacts with esterases to a complex that undergoes isomerization to a much more labile intermediate that preferentially liberates CO upon oxidation (Figure 8.19C) [75]. The key idea of this approach is to utilize the overexpression of specific enzymes in certain (abnormal) tissues that would trigger a preferential release of CO at the target site.

The potential applications for such metallodrugs are manifold and depend not only on the liberation of the CO ligands but also on the cellular interactions of the metal fragment and the coligands. This very likely causes multiple modes of action and needs to be kept in mind in related studies. Although some organometallic CORMs already entered clinical trials, the understanding of the underlying modes of action is far from complete and still demands advances such as the development of sophisticated biological assays, the application of state-of-the-art physical methods and the design of conceptually novel CO-releasing molecules.

8.4.2 Radiopharmaceuticals

An outstanding characteristic of many metal ions is the availability of radioactive isotopes that may be exploited clinically for diagnostic or therapeutic purposes [76].

Potential application fields strongly depend upon the exact radiophysical properties of the respective isotope in question. For diagnostic purposes like SPECT (single-photon emission computed tomography, which requires a γ emitter) or PET (positron emission tomography, which requires β emitters) mainly isotopes with a rather short half-life and low emitting energy are favored in order to protect the patient and the environment from extensive radioactive exposure. For therapeutic agents however, a somewhat higher radiation energy is needed to achieve an appreciable impact. Again, mainly β emitters are used for this purpose. In all cases, the radiopharmaceutical's half-life needs to be sufficiently long to enable the transport to the patient and a previous chemical modification, at least to some limited degree. Radiopharmaceuticals in general, including organometallic analogues, are a huge field, which is covered independently in books and review articles, and therefore we will only touch on one example here that is particular interesting in the (organometallic) context of this chapter.

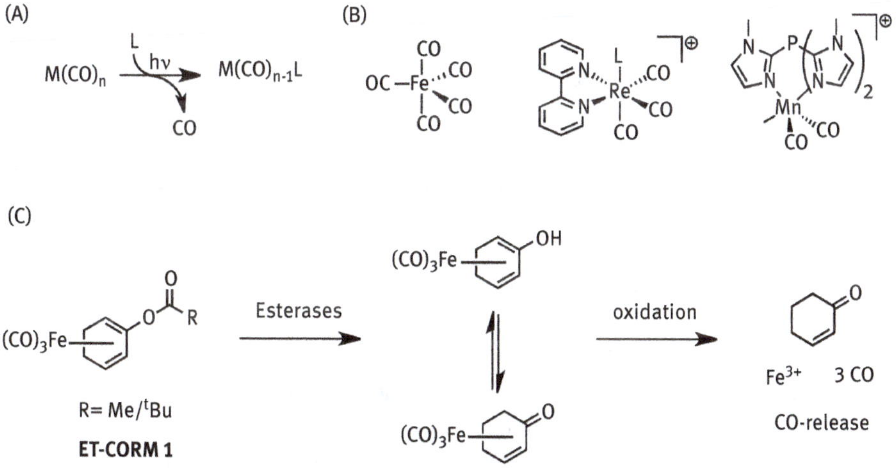

Figure 8.19: Concept (A) as well as selected examples of photoCORMs (B) and principle of ET-CORMs (C).

8.4.2.1 Radioactive $^{99m}Tc^{I}(CO)_3$ complexes and their "Cold" $Re^{I}(CO)_3$ counterparts

A particularly striking and at the same time important example for radioactive metal complexes are Tc complexes and their congeners containing Re. By far the most commonly used isotope for diagnostic purposes in SPECT is the metastable ^{99m}Tc isotope [77]. It emits γ rays with an energy of 140 keV (close to heavily used X-ray energy), which are readily detected by specialized cameras. Together with its half-life of ca. 6 h and an efficient clearance from the human body, ^{99m}Tc enables a rapid imaging data collection while keeping the patient's exposure to radiation low. A key advantage of ^{99m}Tc is the easy on-site preparation of the radioactive isotope from commercially available generators that produce a highly diluted solution of $Na^{99m}TcO_4$ (Figure 8.20A), which is commonly reacted with a reducing agent and a chelating ligand targeting the area of interest in the human body to obtain an active "hot" imaging agent. An organometallic synthon of ^{99m}Tc developed by Alberto et al. is the $[^{99m}Tc^I(CO)_3(H_2O)_3]^+$ (Figure 8.20B) complex cation [78]. Its stable d^6 low spin configuration with the readily exchangeable H_2O ligands allows for efficient chelation reactions with a broad variety of targeting multidentate organic ligands. Because of the rather short half-life of the isotope, it is important for all synthetic steps with ^{99m}Tc to ensure a high reaction efficiency, a fast reaction time and easy handling.

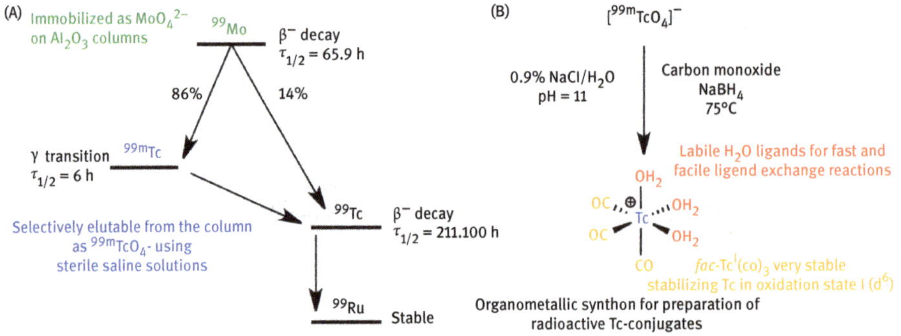

Figure 8.20: (A) Schematic representation of key processes in a commercially available Tc-generator. (B) Synthesis of an organometallic Tc-synthon.

An important advantage of $^{99m}Tc^I(CO)_3$ compounds is their similarity to the coordination chemistry of the higher homologue $Re^I(CO)_3$. The most common Re isotopes are not radioactive ("cold") and thus allow Re compounds to be used as surrogates for testing potential ligand systems and synthesis conditions under facilitated non-radioactive conditions. In the majority of cases, conclusions drawn from Re complexes can be transferred to Tc complexes. This may even include characterization data, for example, reflected in the similarity of HPLC chromatograms of Re/Tc complex pairs, which are often employed to probe a successful synthesis of ^{99m}Tc compounds.

Figure 8.21 exemplarily shows the synthesis of a radioactive Tc complex labeled with an organic targeting moiety for the prostate specific membrane antigen developed by Babich and coworkers [79]. The underlying SAR leading to the identification of the structure with the most efficient binding motif for the metal and highest affinity to the target motif was initially identified by testing the corresponding Re-complexes, before preparing the ^{99m}Tc derivative for imaging.

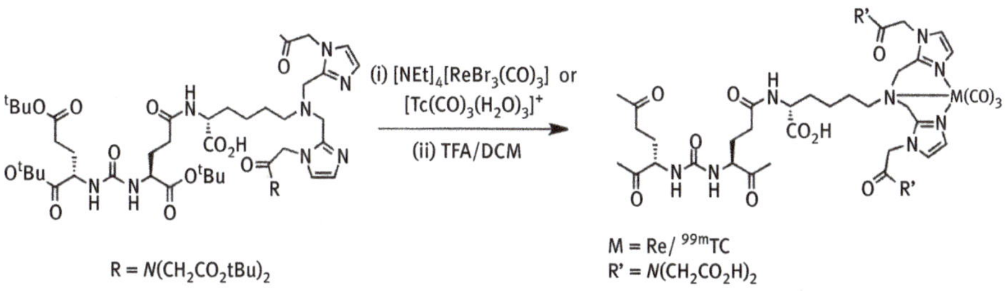

Figure 8.21: Synthesis of prostate cancer targeting ligand labeled with the organometallic Tc/Re pair.

Finally, Re itself also features two important radioactive ("hot") isotopes, ^{186}Re and ^{188}Re. The decisive difference between these Re-isotopes and Tc is the nature of emitted radiation. Both mentioned Re isotopes are "hard" ß-emitters, which render them useful for tumor-therapeutical or palliative purposes.[80] Unlike ^{99m}Tc, however, which has a relatively "soft" radiation and conveniently short half-life, the "hot" Re isotopes are difficult to handle, which limits their use to specially equipped sites and highly trained operators. Taken together, these two metals represent a remarkable potential for diagnostic and radiotherapeutic applications based on characteristics predominantly inherent to metal complexes.

8.5 Conclusions

In this chapter, we presented a survey of important concepts in medicinal organometallic chemistry by summarizing and analyzing prominent achievements in the field. It can be expected that the number of clinically relevant metal complexes will increase in the near future due to urgently needed novel therapeutic remedies. Certainly, not all metal compounds for clinical applications will need to be organometallic ones – Cisplatin, for example, is a simple coordination complex, which does not even have a single carbon atom in its structure. However, the examples above hopefully demonstrate the versatility and power both in terms of chemical synthesis and biological activity that most notably organometallic compounds possess. With this work, we hope to encourage future researchers to think outside the box of organic chemistry and to explore the potential of metal-containing molecules with all their structural and functional variability. Questions to be answered in the future are certainly related to the cellular fate of inorganic compounds in close relation to the mode of action. Establishing a detailed SAR, which enables the rational design of novel inorganic drugs, will be a tremendous challenge. Nevertheless, past results have demonstrated that such an approach is possible and the enormous potential of inorganic drugs gives us reasons to look forward full of expectation.

References

[1] Orvig, C., and Abrams, MJ. Medicinal inorganic chemistry: introduction. Chem Rev 1999, 99, 2201–2204.

[2] Gianferrara, T., Bratsos, I., and Alessio, E. A categorization of metal anticancer compounds based on their mode of action. Dalton Trans 2009, 7588–7598.

[3] Gasser, G., and Metzler-Nolte, N. The potential of organometallic complexes in medicinal chemistry. Curr Opin Chem Biol 2012, 16, 84–91.

[4] Hefti, FF. Requirements for a lead compound to become a clinical candidate. BMC Neuroscience 2008, 9, S7.

[5] Mjos, KD., and Orvig, C. Metallodrugs in medicinal inorganic chemistry. Chem Rev 2014, 114, 4540–4563.

[6] Gasser, G., Ott, I., and Metzler-Nolte, N. Organometallic anticancer compounds. J Med Chem 2011, 54, 3–25.

[7] Lipinski, CA., Lombardo, F., Dominy, BW., and Feeney, PJ. Experimental and computational approaches to estimate solubility and permeability in drug discovery and development settings. Adv Drug Delivery Rev 2001, 46, 3–26.

[8] Fricker, SP. Metal based drugs: from serendipity to design. Dalton Trans 2007, 4903–4917.

[9] Gabbiani, C., Magherini, F., Modesti, A., and Messori, L. Proteomic and metallomic strategies for understanding the mode of action of anticancer metallodrugs. Anti-Cancer Agents Med Chem 2010, 10, 324–337.

[10] Holtkamp, HU., and Hartinger, CG. Advanced metallomics methods in anticancer metallodrug mode of action studies. Trends Anal Chem 2018, 104, 110–117.

[11] Bray, F., Ferlay, J., Soerjomataram, I., Siegel, RL., Torre, LA., and Jemal, A. Global cancer statistics 2018: GLOBOCAN estimates of incidence and mortality worldwide for 36 cancers in 185 countries. Cancer J Clinicians 2018, 68, 394–424.

[12] Louie, AY., and Meade, TJ. Metal complexes as enzyme inhibitors. Chem Rev 1999, 99, 2711–2734.

[13] Anstaett, P., and Gasser, G. Organometallic complexes as enzyme inhibitors: a conceptual overview, in bioorganometallic chemistry, John Wiley & Sons, Ltd, 2014, 1–42.

[14] Kilpin, KJ., and Dyson, PJ. Enzyme inhibition by metal complexes: concepts, strategies and applications. Chem Sci 2013, 4, 1410–1419.

[15] Mulcahy, SP., and Meggers, E. Organometallics as structural scaffolds for enzyme inhibitor design, in medicinal organometallic chemistry, jaouen G and metzler-nolte N, Springer Berlin Heidelberg, Berlin, Heidelberg, 2010, 141–153.

[16] Meggers, E. Exploring biologically relevant chemical space with metal complexes. Curr Opin Chem Biol 2007, 11, 287–292.

[17] Manning, G., Whyte, DB., Martinez, R., Hunter, T., and Sudarsanam, S. The protein kinase complement of the human genome. Science 2002, 298, 1912–1934.

[18] Bhullar, KS., Lagarón, NO., McGowan, EM., Parmar, I., Jha, A., Hubbard, BP., and Rupasinghe, HPV. Kinase-targeted cancer therapies: progress, challenges and future directions. Mol Cancer 2018, 17, 48.

[19] Miljković, F., and Bajorath, J. Exploring selectivity of multikinase inhibitors across the human kinome. ACS Omega 2018, 3, 1147–1153.

[20] Feng, L., Geisselbrecht, Y., Blanck, S., Wilbuer, A., Atilla-Gokcumen, GE., Filippakopoulos, P., Kräling, K., Celik, MA., Harms, K., Maksimoska, J., Marmorstein, R., Frenking, G., Knapp, S., Essen, L-O., and Meggers, E. Structurally sophisticated octahedral metal complexes as highly selective protein kinase inhibitors. J Am Chem Soc 2011, 133, 5976–5986.

[21] Pagano, N., Maksimoska, J., Bregman, H., Williams, DS., Webster, RD., Xue, F., and Meggers, E. Ruthenium half-sandwich complexes as protein kinase inhibitors: derivatization of the pyridocarbazole pharmacophore ligand. Org Biomol Chem 2007, 5, 1218–1227.

[22] Meggers, E., Atilla-Gokcumen, GE., Bregman, H., Maksimoska, J., Mulcahy, SP., Pagano, N., and Williams, DS. Exploring chemical space with organometallics: ruthenium complexes as protein kinase inhibitors. Synlett 2007, 1177–1189.

[23] Maksimoska, J., Williams, DS., Atilla-Gokcumen, GE., Smalley, KSM., Carroll, PJ., Webster, RD., Filippakopoulos, P., Knapp, S., Herlyn, M., and Meggers, E. Similar biological activities of two isostructural ruthenium and osmium complexes. Chem Eur J 2008, 14, 4816–4822.

[24] Meggers, E. Targeting proteins with metal complexes. Chem Commun 2009, 1001–1010.

[25] Park, S-H., Lee, JH., Berek, JS., and Hu, MCT. Auranofin displays anticancer activity against ovarian cancer cells through FOXO3 activation independent of p53. Int J Oncol 2014, 45, 1691–1698.

[26] Bhabak, KP., Bhuyan, BJ., and Mugesh, G. Bioinorganic and medicinal chemistry: aspects of gold(I)-protein complexes. Dalton Trans 2011, 40, 2099–2111.

[27] Oehninger, L., Rubbiani, R., and Ott, I. N-Heterocyclic carbene metal complexes in medicinal chemistry. Dalton Trans 2013, 42, 3269–3284.

[28] Rubbiani, R., Kitanovic, I., Alborzinia, H., Can, S., Kitanovic, A., Onambele, LA., Stefanopoulou, M., Geldmacher, Y., Sheldrick, WS., Wolber, G., Prokop, A., Wölfl, S., and Ott, I. Benzimidazol-2-ylidene gold(I) complexes are thioredoxin reductase inhibitors with multiple antitumor properties. J Med Chem 2010, 53, 8608–8618.

[29] Hickey, JL., Ruhayel, RA., Barnard, PJ., Baker, MV., Berners-Price, SJ., and Filipovska, A. Mitochondria-targeted chemotherapeutics: the rational design of gold(I) N-heterocyclic carbene complexes that are selectively toxic to cancer cells and target protein selenols in preference to thiols. J Am Chem Soc 2008, 130, 12570–12571.

[30] Berners-Price, SJ., and Filipovska, A. Gold compounds as therapeutic agents for human diseases. Metallomics 2011, 3, 863–873.

[31] Köster, SD., Alborzinia, H., Can, S., Kitanovic, I., Wölfl, S., Rubbiani, R., Ott, I., Riesterer, P., Prokop, A., Merz, K., and Metzler-Nolte, N. A spontaneous gold(I)-azide alkyne cycloaddition reaction yields gold-peptide bioconjugates which overcome cisplatin resistance in a p53-mutant cancer cell line. Chem Sci 2012, 3, 2062–2072.

[32] Trondl, R., Heffeter, P., Kowol, CR., Jakupec, MA., Berger, W., and Keppler, BK. NKP-1339, the first ruthenium-based anticancer drug on the edge to clinical application. Chem Sci 2014, 5, 2925–2932.

[33] Alessio, E. Thirty years of the drug candidate NAMI-A and the Myths in the field of ruthenium anticancer compounds: a personal perspective. Eur J Inorg Chem 2017, 2017, 1549–1560.

[34] Zeng, L., Gupta, P., Chen, Y., Wang, E., Ji, L., Chao, H., and Chen, Z-S. The development of anticancer ruthenium(II) complexes: from single molecule compounds to nanomaterials. Chem Soc Rev 2017, 46, 5771–5804.

[35] Wang, F., Chen, H., Parsons, S., Oswald, IDH., Davidson, JE., and Sadler, PJ. Kinetics of aquation and anation of ruthenium(II) arene anticancer complexes, acidity and X-ray structures of aqua adducts. Chem Eur J 2003, 9, 5810–5820.

[36] Chatterjee, S., Kundu, S., Bhattacharyya, A., Hartinger, CG., and Dyson, PJ. The ruthenium (II)–arene compound RAPTA-C induces apoptosis in EAC cells through mitochondrial and p53–JNK pathways. J Biol Inorg Chem 2008, 13, 1149.

[37] Weiss, A., Berndsen, RH., Dubois, M., Müller, C., Schibli, R., Griffioen, AW., Dyson, PJ., and Nowak-Sliwinska, P. In vivo anti-tumor activity of the organometallic ruthenium(II)-arene complex [Ru(η^6-p-cymene)Cl_2(pta)] (RAPTA-C) in human ovarian and colorectal carcinomas. Chem Sci 2014, 5, 4742–4748.

[38] Köpf, H., and Köpf-Maier, P. Titanocene dichloride – the first metallocene with cancerostatic activity. Angew Chem Int Ed Engl 1979, 18, 477–478.
[39] Claffey, J., Hogan, M., Müller-Bunz, H., Pampillón, C., and Tacke, M. Oxali-titanocene Y: a potent anticancer drug. Chem Med Chem 2008, 3, 729–731.
[40] Köpf-Maier, P., and Köpf, H. Transition and main-group metal cyclopentadienyl complexes: preclinical studies on a series of antitumor agents of different structural type. Struct Bond 1988, 70, 105–185.
[41] Köpf-Maier, P., and Köpf, H. Antitumor metallocenes. Drugs Future 1986, 11, 297–320.
[42] van Staveren, DR., and Metzler-Nolte, N. Bioorganometallic chemistry of ferrocene. Chem Rev 2004, 104, 5931–5985.
[43] Patra, M., and Gasser, G. The medicinal chemistry of ferrocene and its derivatives. Nature Rev Chem 2017, 1, Article no., 0066.
[44] Jaouen, G., Top, S., Vessières, A., Leclercq, G., and McGlinchey, MJ. The first organometallic selective estrogen receptor modulators (SERMs) and their relevance to breast cancer. Curr Med Chem 2004, 11, 2505–2517.
[45] Jaouen, G., Vessieres, A., and Top, S. Ferrocifen type anti cancer drugs. Chem Soc Rev 2015, 44, 8802–8817.
[46] Schatzschneider, U., and Metzler-Nolte, N. New principles in medicinal organometallic chemistry. Angew Chem Int Ed 2006, 45, 1504–1507.
[47] Hillard, E., Vessières, A., Thouin, L., Jaouen, G., and Amatore, C. Ferrocene-mediated proton-coupled electron transfer in a series of ferrocifen-type breast-cancer drug candidates. Angew Chem Int Ed 2006, 45, 285–290.
[48] Vessieres, A., Top, S., Pigeon, P., Hillard, E., Boubeker, L., Spera, D., and Jaouen, G. Modification of the estrogenic properties of diphenols by the incorporation of ferrocene. generation of antiproliferative effects in vitro. J Med Chem 2005, 48, 3937–3940.
[49] Patra, M., Gasser, G., and Metzler-Nolte, N. Small organometallic compounds as antibacterial agents. Dalton Trans 2012, 41, 6350–6358.
[50] Hindi, KM., Panzner, MJ., Tessier, CA., Cannon, CL., and Youngs, WJ. The medicinal applications of imidazolium carbene–metal complexes. Chem Rev 2009, 109, 3859–3884.
[51] Kascatan-Nebioglu, A., Melaiye, A., Hindi, K., Durmus, S., Panzner, MJ., Hogue, LA., Mallett, RJ., Hovis, CE., Coughenour, M., Crosby, SD., Milsted, A., Ely, DL., Tessier, CA., Cannon, CL., and Youngs, WJ. Synthesis from caffeine of a mixed N-heterocyclic carbene–silver acetate complex active against resistant respiratory pathogens. J Med Chem 2006, 49, 6811–6818.
[52] Wagers, PO., Shelton, KL., Panzner, MJ., Tessier, CA., and Youngs, WJ. Synthesis and medicinal properties of silver–NHC complexes and imidazolium salts, in N-heterocyclic carbenes, John Wiley & Sons, Ltd, 2014, 151–172.
[53] Budagumpi, S., Haque, RA., Endud, S., Rehman, GU., and Salman, AW. Biologically relevant silver(I)–N-heterocyclic carbene complexes: synthesis, structure, intramolecular interactions, and applications. Eur J Inorg Chem, 2013(2013), 4367–4388.
[54] Lloyd, NC., Morgan, HW., Nicholson, BK., and Ronimus, RS. Massenspektren von Salvarsan. Angew Chem Int Ed 2005, 44, 941–944.
[55] Edwards, EI., Epton, R., and Marr, G. A new class of semi-synthetic antibiotics: ferrocenyl-penicillins and -cephalosporins. J Organomet Chem 1976, 107, 351–357.
[56] Patra, M., Merz, K., and Metzler-Nolte, N. Planar chiral (η^6-arene)$Cr(CO)_3$ containing carboxylic acid derivatives: synthesis and use in the preparation of organometallic analogues of the antibiotic platensimycin. Dalton Trans 2012, 41, 112–117.
[57] Patra, M., Gasser, G., Wenzel, M., Merz, K., Bandow, JE., and Metzler-Nolte, N. Sandwich and Half-sandwich derivatives of platensimycin: synthesis and biological evaluation. Organometallics 2012, 31, 5760–5771.

[58] Patra, M., Gasser, G., Wenzel, M., Merz, K., Bandow, JE., and Metzler-Nolte, N. Synthesis of optically active ferrocene-containing platensimycin derivatives with a C6–C7 substitution pattern. Eur J Inorg Chem 2011, 3295–3302.

[59] Patra, M., Gasser, G., Wenzel, M., Merz, K., Bandow, JE., and Metzler-Nolte, N. Synthesis and biological evaluation of ferrocene-containing bioorganometallics inspired by the antibiotic platensimycin lead structure. Organometallics 2010, 29, 4312–4319.

[60] Patra, M., Gasser, G., Pinto, A., Merz, K., Ott, I., Bandow, JE., and Metzler-Nolte, N. Synthesis and biological evaluation of chromium bioorganometallics based on the antibiotic platensimycin lead structure. Chem Med Chem 2009, 4, 1930–1938.

[61] Wang, J., Soisson, SM., Young, K., Shoop, W., Kodali, S., Galgoci, A., Painter, R., Parthasarathy, G., Tang, YS., Cummings, R., Ha, S., Dorso, K., Motyl, M., Jayasuriya, H., Ondeyka, J., Herath, K., Zhang, C., Hernandez, L., Allocco, J., Basilio, A., Tormo, JR., Genilloud, O., Vicente, F., Pelaez, F., Colwell, L., Lee, SH., Michael, B., Felcetto, T., Gill, C., Silver, LL., Hermes, JD., Bartizal, K., Barrett, J., Schmatz, D., Becker, JW., Cully, D., and Singh, SB. Platensimycin is a selective FabF inhibitor with potent antibiotic properties. Nature 2006, 441, 358–361.

[62] Albada, B., and Metzler-Nolte, N. Highly potent antibacterial organometallic peptide conjugates. Acc Chem Res 2017, 50, 2510–2518.

[63] Wenzel, M., Chiriac, AI., Otto, A., Zweytick, D., May, C., Schumacher, C., Gust, R., Albada, HB., Penkova, M., Krämer, U., Erdmann, R., Metzler-Nolte, N., Straus, SK., Bremer, E., Becher, D., Brötz-Oesterhelt, H., Sahl, H-G., and Bandow, JE. Small cationic antimicrobial peptides delocalize peripheral membrane proteins. Proc Natl Acad Sci USA 2014, 111, E1409–E1418.

[64] World malaria report, 2017, World Health Organization, Geneva, 2017.

[65] Biot, C., and Dive, D. Bioorganometallic chemistry and malaria, in medicinal organometallic chemistry (Vol 32 in topics in organometallic chemistry), Jaouen G and Metzler-Nolte N, Vol. Ch. 7, Springer, Heidelberg, Germany, 2010, 155–193.

[66] Biot, C., Glorian, G., Maciejewski, LA., Brocard, JS., Domarle, O., Blampain, G., Millet, P., Georges, AJ., Abessolo, H., Dive, D., and Lebibi, J. Synthesis and antimalarial activity and in vivo of a new ferrocene-chloroquine analogue. J Med Chem 1997, 40, 3715–3718.

[67] Dive, D., and Biot, C. Ferrocene conjugates of chloroquine and other antimalarials: the development of ferroquine, a new antimalarial. Chem Med Chem 2008, 3, 383–391.

[68] Dive, D., and Biot, C. Ferroquine as an oxidative shock antimalarial. Curr Topics Med Chem 2014, 14, 1684–1692.

[69] Biot, C., Taramelli, D., Forfar-Bares, I., Maciejewski, LA., Boyce, M., Nowogrocki, G., Brocard, JS., Basilico, N., Olliaro, P., and Egan, TJ. Insights into the mechanism of action of ferroquine. relationship between physicochemical properties and antiplasmodial activity. Mol Pharmaceutics 2005, 2, 185–193.

[70] Mann, BE. CO-releasing molecules: a personal view. Organometallics 2012, 31, 5728–5735.

[71] Schatzschneider, U. Novel lead structures and activation mechanisms for CO-releasing molecules (CORMs). British J Pharmacol 2015, 172, 1638–1650.

[72] Ling, K., Men, F., Wang, W-C., Zhou, Y-Q., Zhang, H-W., and Ye, D-W. Carbon monoxide and its controlled release: therapeutic application, detection, and development of carbon monoxide releasing molecules (CORMs). J Med Chem 2018, 61, 2611–2635.

[73] Foresti, R., Hammad, J., Clark, JE., Johnson, TR., Mann, BE., Friebe, A., Green, CJ., and Motterlini, R. Vasoactive properties of CORM-3, a novel water-soluble carbon monoxide-releasing molecule. British J Pharmacol 2009, 142, 453–460.

[74] Schatzschneider, U. PhotoCORMs: light-triggered release of carbon monoxide from the coordination sphere of transition metal complexes for biological applications. Inorg Chim Acta 2011, 374, 19–23.

[75] Romanski, S., Kraus, B., Schatzschneider, U., Neudörfl, J-M., Amslinger, S., and Schmalz, H-G. Acyloxybutadiene iron tricarbonyl complexes as enzyme-triggered CO-releasing molecules (ET-CORMs). Angew Chem Int Ed 2011, 50, 2392–2396.
[76] Alberto, R. Metal-based radiopharmaceuticals, in bioinorganic medicinal chemistry, John Wiley & Sons, Ltd, 2011, 253–282.
[77] Papagiannopoulou, D. Technetium-99m radiochemistry for pharmaceutical applications. J Labelld Compd Radiopharm 2017, 60, 502–520.
[78] Alberto, R., Schibli, R., Egli, A., Schubiger, AP., Abram, U., and Kaden, TA. A novel organometallic aqua complex of technetium for the labeling of biomolecules: synthesis of $[^{99m}Tc(OH_2)_3(CO)_3]^+$ from $[^{99m}TcO_4]^-$ in aqueous solution and its reaction with a bifunctional ligand. J Am Chem Soc 1998, 120, 7987–7988.
[79] Lu, G., Maresca, KP., Hillier, SM., Zimmerman, CN., Eckelman, WC., Joyal, JL., and Babich, JW. Synthesis and SAR of ^{99m}Tc/Re-labeled small molecule prostate specific membrane antigen inhibitors with novel polar chelates. Bioorg Med Chem Lett 2013, 23, 1557–1563.
[80] Jürgens, S., Herrmann, WA., and Kühn, FE. Rhenium and technetium based radiopharmaceuticals: development and recent advances. J Organomet Chem 2014, 751, 83–89.

Part IV: Spectroscopy Methods

Leland B. Gee, Hongxin Wang and Stephen P. Cramer

9 Nuclear resonance vibrational spectroscopy

9.1 Introduction

9.1.1 Mössbauer effect

In a Mössbauer experiment, the intensity directly scales with f_{LM}, such that the observed cross section, σ_N, is reduced from the total cross section according to $\sigma(0) \sim(\pi/2)\, \sigma_N f_{LM}$ [6]. However, there is a "sum rule" that states that the integrated cross section for a nuclear transition remains constant [7]. So, where does this missing intensity go? The missing intensity is in the "recoil fraction," $1 - f_{LM}$, where nuclear transitions couple to atomic vibrations, resulting in the creation or annihilation of phonons analogous to Stokes and anti-Stokes features in optical spectroscopies (Figure 9.1). It turns out that this fraction contains useful and detailed information about the vibrational properties of a sample.

Before the advent of modern synchrotron radiation (SR) sources, most experiments probed the recoil fraction indirectly, by observing the loss of Mössbauer intensity [8, 9]. Since the inception of high brightness third-generation synchrotron X-ray sources and high-resolution monochromators (HRMs), we can probe the recoil fraction directly by the technique known as *nuclear resonance vibrational spectroscopy* (NRVS). Other less common names for the technique include nuclear resonant inelastic X-ray spectroscopy, nuclear inelastic scattering (NIS) and the phonon-assisted Mössbauer effect (PAME).

NRVS has applications to chemistry, biology, geology, materials science and physics. It is an exciting tool because:
- NRVS yields a vibrational spectrum for a specific isotope of a specific element,
- the resulting vibrational spectrum is easily calculated and interpreted,
- the isotopic sensitivity allows labeling experiments, say of surfaces or of specific sites in metalloenzymes and
- NRVS also provides quantities such as the speed of sound and average kinetic energy.

Before we continue further, we present a pair of examples as an initial demonstration (Figure 9.1 and Figure 9.2). At one extreme is the spectrum of Fe metal, in which the

Leland B. Gee, Department of Chemistry, Stanford University, California, USA
Hongxin Wang, Stephen P. Cramer, Department of Chemistry, University of California, California, USA

https://doi.org/10.1515/9783110496574-009

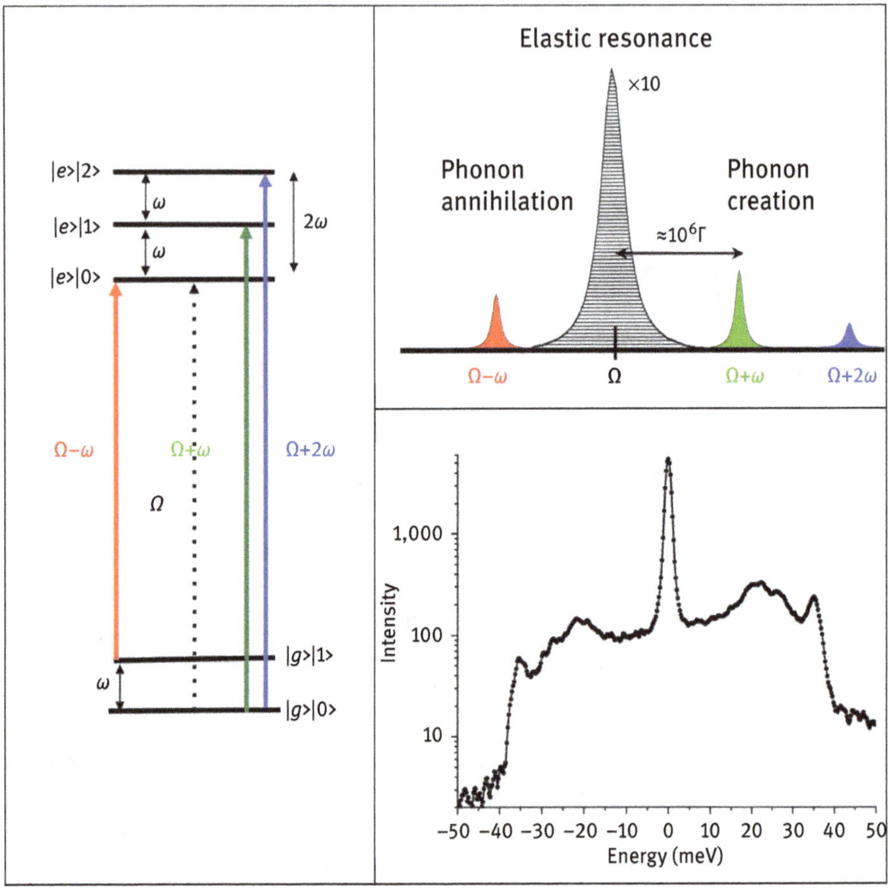

Figure 9.1: Left: State schematic of zero phonon and recoil events for a system with a single fundamental vibrational frequency. The recoil-free Mössbauer line is the |g>0> → |e>0> transition. Upper right: Comparison of energies and intensities for the same system. The phonon creation or "Stokes" region is to higher energy, with single-phonon and multiple-phonon events. The phonon annihilation or "anti-Stokes" region is at lower energy and has its intensity reduced by the Boltzmann factor (see text). Lower right: NRVS spectrum for Fe metal, typical for a system with a continuous density of states.

vibrations are described as propagating phonons with a continuous density of states. Analysis of such spectra can provide important geophysical properties such as the velocity of sound, specific heat and average vibrational force constant. At the other extreme is the spectrum of $^{57}Fe(S_2C_2H_4)(CO)_2(PMe_3)_2$, an Fe coordination complex with terminal CO ligands. The isolated optical modes from 500 to 650 cm^{-1} are dominantly Fe–CO stretching and bending in character [10].

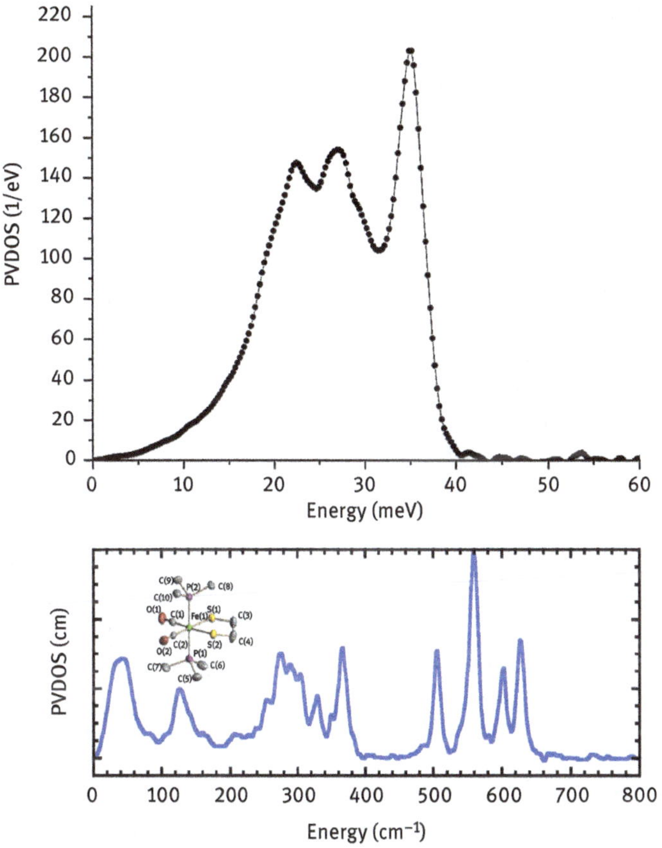

Figure 9.2: NRVS examples. top: The NRVS spectrum for ^{57}Fe Metal. bottom: Spectrum for a complex with two terminal –C≡O ligands [10].

9.1.2 History

NRVS is related to, and descended from, Mössbauer spectroscopy, which focuses on the recoil-free Mössbauer nuclear excitation to learn about the electronic structure of the atom – a recoilless excitation [1, 2, 11]. Probing vibrational dynamics through coupled nuclear transitions was explored in the early 1960s [4, 12]. Typically for Mössbauer spectroscopy, energies incident to the sample are scanned by Doppler shifting a decaying parent isotope. However, the resonances in Mössbauer spectroscopy are usually shifted on the order of neV relative to the nuclear excitation, but vibrational quanta are on the order of meV. This implies a Doppler shifted source would need to move with a velocity 6 orders of magnitude faster than a typical Mössbauer experiment to resolve vibrational modes. Although there were heroic experiments involving sources mounted onto ultracentrifuges [8, 9], such a Mössbauer spectroscopy setup converted for NRVS is considered impractical and a different type of source is required for NRVS.

The possibility of using a synchrotron as a source for nuclear resonance spectroscopies was first proposed by Ruby in 1974 [13]. The proposition was explored in the mid-1980s by Gerdau [14]. By the mid-1990s with most of the theoretical and practical groundwork laid, three teams near-simultaneously made observations of NIS caused by vibrational dynamics [15–17]. The three teams were based out of the SPring-8, APS and ESRF synchrotrons, and these remain global hubs for nuclear resonance spectroscopies to this day.

After the initial observations, there has been a boom in NRVS research. Due to its ubiquity in chemistry, materials science, biology as well as nuclear properties that are amenable to the timescale of a synchrotron, ^{57}Fe has enjoyed a particular emphasis in NRVS research. One of the first demonstrations of the efficacy of the ^{57}Fe NRVS experiment for dilute biological samples was performed on myoglobin [18, 19]. Since then many bioinorganic ^{57}Fe NRVS experiments have been performed on (not comprehensively) the NiFe site in [NiFe] hydrogenase [20–22], the diiron site in [FeFe] hydrogenase [23, 24], the P [8Fe7S]- and M [7Fe9SMo]- clusters of Mo-nitrogenase [25–28], heme systems [29, 30], nonheme systems [31, 32] and nitric-oxide sensing [4Fe4S] clusters [33]. The rate at which NRVS is answering questions about biological Fe metallocofactors shows no indication of slowing down.

9.2 NRVS intensities for discrete normal modes

How are vibrations and phonons with energies of meV coupled to nuclear transitions at the tens of keV level? The occurrence of vibrational side-bands is one way to understand the NRVS effect. A nucleus vibrating at frequency ω and illuminated by a monochromatic beam at $\omega+\Omega$ will experience a negative side-band at $(\omega+\Omega)-\omega = \Omega$ and hence be in resonance for a nuclear transition. Additionally, a nucleus that is part of a vibrational excited state will be in resonance *via* the positive side-band: $(\Omega - \omega)+\omega = \Omega$.

The overall cross section $\sigma(E)$ for nuclear resonant absorption of a photon with energy E can be factored into two terms, one of which depends on the properties of the nucleus: cross section $\sigma(E_0)$ and lifetime broadening by linewidth, Γ_0:

$$\sigma(E) = \frac{\pi}{2}\sigma(E_0)\Gamma_0 S(E - E_0) \tag{9.1}$$

where the second term, $S(E - E_0)$ is the nuclear excitation probability, which depends on the environment of the nucleus. If one assumes a harmonic lattice or molecule:

$$S(E) = f_{\mathrm{LM}}\left(\underbrace{\delta_\Gamma(E)}_{\text{Mössbauer}} + \underbrace{\sum_{n=1}^{\infty} S_n(E)}_{\text{NRVS}}\right) \tag{9.2}$$

where δΓ(E) is a Lorentzian of width Γ and the function $S_n(E)$ refers to events involving n phonons and f_{LM} is the Lamb–Mössbauer factor.

9.2.1 Stokes fundamentals

Suppose that we have a harmonic oscillator molecular system with a set of normal modes labeled α that are described by the displacements $\vec{r}_{k\alpha}$ for atom k and normal mode α. We assume a lineshape function L to account for the experimental resolution and lifetime broadening, and we convert the energy scale for our normal mode α to a frequency $\bar{\nu}_\alpha$ in wavenumbers (cm^{-1}). As a specific example, we use the $FeCl_4^-$ ion, which in T_d symmetry will have nine normal modes distributed into four bands. Accounting for degeneracy, the particular modes are labeled ν_1 – corresponding to the totally symmetric (A_1) stretch, ν_2 – corresponding to the doubly degenerate (E) bend, ν_3 – corresponding to the triply degenerate (T_2) stretch and ν_4 – corresponding to the triply degenerate (T_2) bend (Figure 9.3).

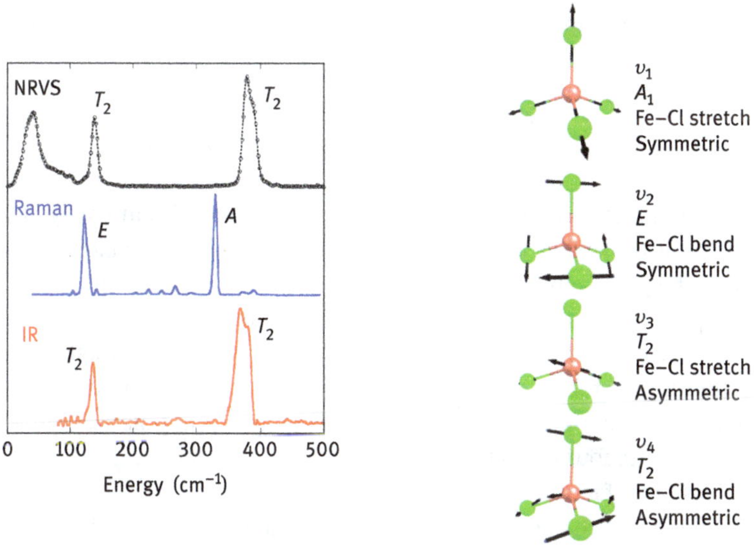

Figure 9.3: NRVS spectra for $(NEt_4)(FeCl_4)$. Left: Comparison of ^{57}Fe PVDOS with Raman and IR spectra, illustrating the dependence of mode strength on the amount of Fe motion. Right: Descriptions of atomic motion in different vibrational modes.

Combining the above, the normalized excitation probability for such a system can then be rewritten as follows:

$$S(\bar{\nu}) = f_{\mathrm{LM}}\mathcal{L}_0(\bar{\nu}) + \sum_{\Delta n_\alpha} \phi(\Delta n_\alpha)\mathcal{L}(\bar{\nu} - \sum_\alpha \Delta n_\alpha \bar{\nu}_\alpha) \tag{9.3}$$

In the above expression, $\phi(\Delta n_\alpha)$ refers to the fractional area (integrated probability) corresponding to a transition from initial population n_α to final population $n_\alpha+\Delta n_\alpha$. In the particular case of the $FeCl_4^-$ ion, α would range from 1→4, with appropriate normalization factors to account for the degeneracies.

The NRVS effect depends on the amount of active isotope nuclear motion in a particular normal mode or phonon. The critical term that captures this is the "mode composition factor," $e_{j\alpha}^2$, which is the fraction of kinetic energy associated with motion of nucleus j with mass m_j and mean square displacements $r_{j\alpha}^2$:

$$e_{j\alpha}^2 = \frac{m_j r_{j\alpha}^2}{\sum_k m_k r_{k\alpha}^2} \tag{9.4}$$

Then for a randomly oriented sample, and a nuclear transition with recoil energy $\bar{\nu}_R$ and Lamb–Mössbauer factor f_{LM}, an expression that captures the properties that govern NRVS intensity for a fundamental transition from n_α→n_α +1 is:

$$\phi_\alpha = \frac{1}{3}\left(\frac{\nu_R}{\nu_\alpha}\right)(\bar{n}_\alpha + 1) f_{\mathrm{LM}} e_{j\alpha}^2 \tag{9.5}$$

The term $\bar{n}_\alpha$ is the mean occupation number for mode α and is given by Boltzmann statistics as follows:

$$\bar{n}_\alpha = \frac{1}{\exp\left(\frac{hc\bar{\nu}_\alpha}{k_B T}\right) - 1} \tag{9.6}$$

The mode composition factor is the key term for describing the NRVS intensity of a given normal mode. It thus plays a role similar to the transition dipole moment in IR spectroscopy or the polarizability tensor in Raman spectroscopy. Due to the simplicity of the intensity mechanism, once a normal mode description is obtained from a DFT or empirical forcefield model, it is facile to calculate the normal mode composition factor for the probe atom. In contrast, IR or Raman calculations involve assumptions about molecular properties such as dipole moments or polarizabilities.

In summary, there are four terms that govern NRVS intensities:

- the Lamb–Mössbauer factor,
- a general $1/E$ dependence that reduces intensity of higher energy transitions,
- the temperature, which governs the distribution of occupied ground vibrational levels and
- most important, the mode composition factor.

Because the NRVS signal depends on motion of the nucleus of interest, some modes will be strictly forbidden by symmetry. An example is the totally symmetric stretch of a tetrahedral complex such as $FeCl_4^-$ (Figure 9.3). Modes involving light atom ligands, such as Fe–H stretches, are more difficult to observe because the light atom dominates the motion of the vibrational mode. Higher-frequency modes also suffer from the $1/E$ dependence. As will be discussed later, weak modes can be salvaged to some extent by preferentially weighting the acquisition time as done with region of interest (ROI) scans.

9.2.2 Anti-Stokes intensity

As with Raman spectroscopy, at photon energies $E_0 - \bar{\nu}_\alpha$, there are transitions involving "annihilation" of phonons. The contribution of these "anti-Stokes features' is given by:

$$\phi_\alpha = \frac{1}{3}\left(\frac{\bar{\nu}_R}{\bar{\nu}_\alpha}\right)\bar{n}_\alpha f_{\mathrm{LM}} e_{j\alpha}^2 \tag{9.7}$$

In these transitions, $n_\alpha+1 \rightarrow n_\alpha$, and the intensity is strongly temperature dependent because these transitions start from vibrational excited states. As observed in Figure 9.1 the relative strength of Stokes and anti-Stokes features depends on the temperature via the Boltzmann factor:

$$\frac{\phi_{\text{anti-Stokes}}}{\phi_{\text{Stokes}}} = \frac{\bar{n}_\alpha}{\bar{n}_\alpha + 1} = \exp\left(-\frac{hc\nu_\alpha}{k_B T}\right) \tag{9.8}$$

By rearranging this equation for T, the relative strengths of the anti-Stokes and Stokes transitions at frequencies $\pm\,\bar{\nu}_\alpha$ can be used to determine the sample temperature.

9.2.3 Multiphonon events – overtone and combination bands

Again like Raman and IR spectroscopy, an NRVS spectrum also exhibits overtone bands and combination bands involving changes of two or more phonons. How significant are they? It all depends on the Lamb–Mössbauer factor. By integrating the individual $S_n(E)$ curves, the n-phonon probabilities are obtained:

$$P_n = \int S_n(E)dE = f_{\mathrm{LM}}\frac{(-\ln f_{\mathrm{LM}})^n}{n!} \tag{9.9}$$

From this, Sturhahn gives a very simple expression for the average number of phonons excited over the nuclear excitation spectrum [34]:

$$\langle n \rangle = -\ln f_{LM} \tag{9.10}$$

As shown in Figure 9.4, this means that multiphonon contributions are less than 10% of the single phonon contribution as long as $f_{LM} > 0.83$.

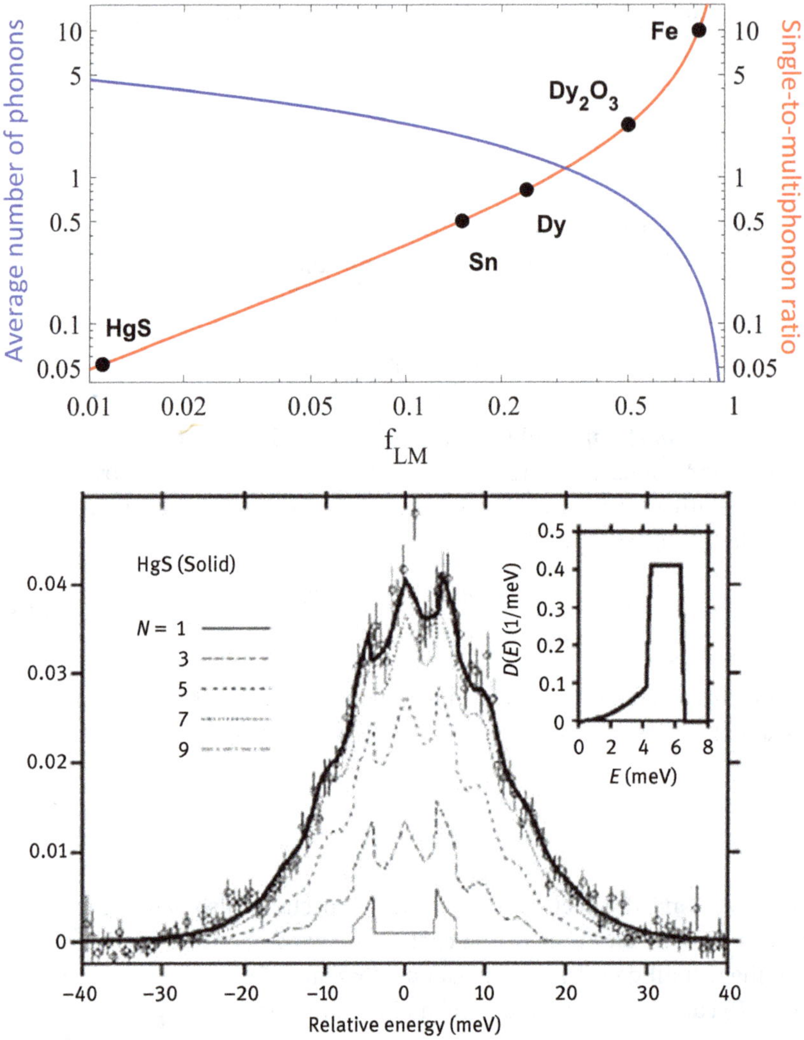

Figure 9.4: Top: The average number of phonons <n> (—) (integrated over all transition probabilities) and the ratio of single phonon to multiphonon events (—) as a function of typical elemental Lamb–Mössbauer factors f_{LM}, illustrated for a variety of typical samples and conditions: $^{161}Dy/^{161}Dy_2O_3$ at 300K [35], ^{201}HgS at 300K [36], β-^{119}Sn at 100K [37, 38], α-^{57}Fe at 300K [39]. Bottom: NRVS spectrum for HgS, an extreme case of multiphonon events (N>1) dominating the nuclear spectrum [36].

Sage and coworkers have investigated the properties of two phonon contributions in some detail [40, 41]. In the same low-temperature (high-frequency) approximation used for the single-phonon approximation, the contribution $\phi_{\alpha\alpha}$ of the overtone transition, $n_\alpha \rightarrow n_\alpha+2$, at energy $2\bar{\nu}_\alpha$ is given by eq. (9.11) [41]:

$$\phi_{\alpha\alpha} = \frac{1}{10}\left(\frac{\bar{\nu}_R}{\bar{\nu}_\alpha}\right)^2 e_{j\alpha}^4 (\bar{n}_\alpha + 1)^2 f_{\text{LM}} = \frac{9}{10}\frac{\phi_\alpha^2}{f_{\text{LM}}} \tag{9.11}$$

We can then take the ratio of eqs. (9.11) and (9.5) to determine the fraction of overtone intensity to single phonon contribution. Assuming an arbitrary value of $\phi_\alpha = 0.02$ and a favorable f_{LM}=0.85, we find that the overtone to fundamental ratio, $\frac{\phi_{\alpha\alpha}}{\phi_\alpha} \approx 2.1\%$. With the same approximations, similar expressions can be derived for the strength $\phi_{\alpha\beta}$ of the combination band, $n_\alpha \rightarrow n_\alpha+1$, $n_\beta \rightarrow n_\beta+1$, at energy $\bar{\nu}_\alpha + \bar{\nu}_\beta$; however in that case, the directions of the two different normal mode motions must also be taken into account [41].

9.2.4 Orientation dependence

Although most NRVS experiments are conducted on powder or solution samples, there is extra information to be gained when oriented or single-crystal samples are available. In the most general case, the NRVS will be different for an incident beam along three perpendicular directions in the sample, x, y and z. There will then be three distinct mode composition factors, corresponding to the projection of the nuclear motion along the three axes.

$$e_{j\alpha,x}^2 = \frac{m_j(r\cdot\hat{x})_{j\alpha}^2}{\sum_k m_k r_{k\alpha}^2},\ e_{j\alpha,y}^2 = \frac{m_j(r\cdot\hat{y})_{j\alpha}^2}{\sum_k m_k r_{k\alpha}^2},\ e_{j\alpha,z}^2 = \frac{m_j(r\cdot\hat{z})_{j\alpha}^2}{\sum_k m_k r_{k\alpha}^2} \tag{9.12}$$

As illustrated in Figure 9.5, for a crystal with isotope I and neighbor J, the intensity of the I–J stretching mode in the NRVS will vary as $\cos^2\theta$, where θ is the angle between the photon direction and the interatomic axis. This contrasts with the

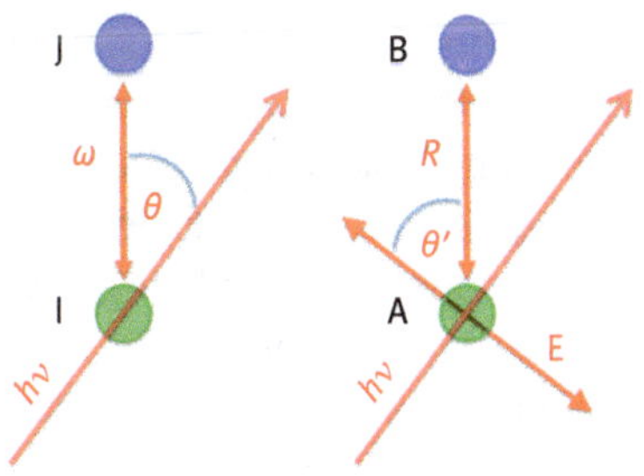

Figure 9.5: Comparison of orientation dependence for NRVS intensity of a stretching mode oriented between atoms I and J and EXAFS intensity from atom A and neighboring atom B.

$\cos^2\theta'$ angular dependence of extended X-ray absorption fine structure (EXAFS), where the θ' refers to the angle between the electric field E vector and the interatomic axis.

In one example of using this orientation dependence, the NRVS for a single crystal of FeOEP(NO) was examined, with the photon beam oriented in three orthogonal crystal axes. The observed intensities were then used to deduce the direction of Fe motion in particular normal modes, such as the Fe–NO bending motions (Figure 9.6) [42].

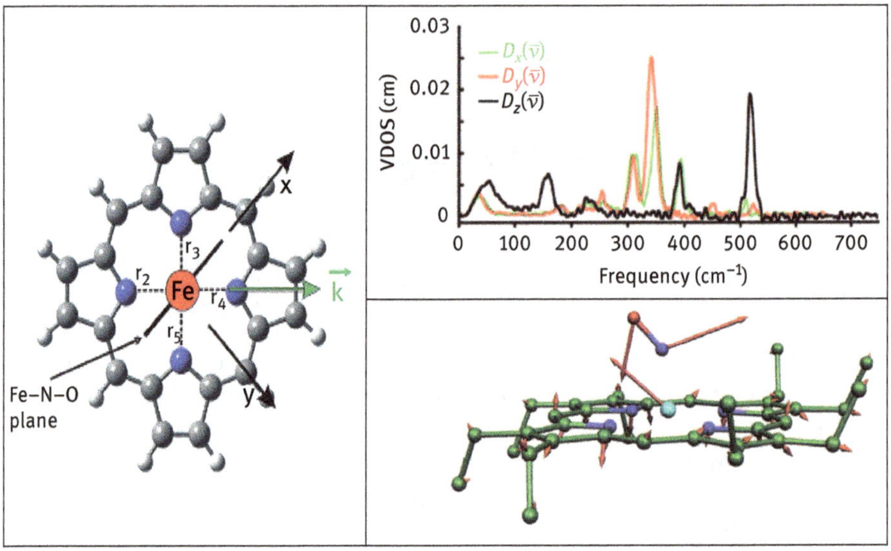

Figure 9.6: Left: Coordinate system used for NRVS on a single crystal of Fe(OEP)NO. Top right: Observed variation. Bottom right: Deduced motion for the Fe–NO bending mode.

9.2.5 Partial vibrational density of states (PVDOS) treatment

The discrete, isolated NRVS features seen in the $(NEt_4)(FeCl_4)$ spectrum are atypical. In most cases, there are many unresolved modes. In fact, apart from localized optical modes, samples will also have delocalized "acoustic" modes that involve motion of the entire unit cell and form a low-energy continuum. For $(NEt_4)(FeCl_4)$ such modes are seen below 80 cm^{-1} (Figure 9.3). For many chemical and biological applications, the low-energy modes are not of keen interest. However, the low-energy modes contain information about properties such as the speed of sound in a sample [43] that are of critical interest to geophysicists. We thus need a way to describe the NRVS for the more general case where the there is a continuous distribution of phonon energies.

These are best described by a continuous “partial vibrational density of states” or PVDOS, commonly written as $D(E)$ or $g(E)$ in the NRVS literature, and $D(E,\hat{k})$ or $g(E,\hat{k})$ if the directional dependence of the PVDOS is retained. This weighted vibrational density of states can be written in terms of photon wavevector $\hat{k}$ and mass weighted normal mode composition factors $\hat{e}_{j\alpha}$ as follows:

$$D_j\left(E,\hat{k}\right) = \sum_j \left(\hat{k}\cdot\hat{e}_{j\alpha}\right)^2 \mathcal{L}(E-E_\alpha) \tag{9.13}$$

From the equation we infer that the experimental density of states will give us a direct description of the mass-weighted normal mode composition factors. Then our goal is to derive the PVDOS from our experimental data.

In the experimental data, there will be differences in penetration depth for the elastic peak, where there is a large absorption cross section, compared to the inelastic region by the incident beam. This complicates determination of the spectral normalization factor (A) as now there is a penetration suppression factor, C, caused by saturation at the elastic peak such that the measured intensity is:

$$I(E) = A\{S(E) - C\delta(E)\} \tag{9.14}$$

Here, we take advantage of Lipkin’s sum rule [44] and recognize that the first moment of the spectrum depends only on the recoil energy of the nuclear isotope thus we obtain:

$$\int ES(E)dE = AE_r \tag{9.15}$$

Using eqs. (9.14 and 9.15) together and knowing the zeroth moment of the spectrum should be unity we can calculate the penetration suppression factor:

$$C = 1 - \frac{1}{A}\int I(E)dE \tag{9.16}$$

Next, we can further interpret the spectrum by approximating the interatomic potential V as quadratic with respect to interatomic displacements (the harmonic approximation). The total excitation probability becomes similar to eq. (9.2):

$$S(s,E) = f(s)\delta(E) + \sum S_n(s,E) \tag{9.17}$$

where S_n is the excitation probability of a phonon of order n. For the harmonic approximation, the single-phonon contribution to the probability is:

$$S_1(s,E) = \frac{E_R}{E(1-e^{-\beta E})} g(s,|E|) \tag{9.18}$$

Where $g(s,|E|)$ is our sought after partial vibrational density of states (PVDOS) and β is the thermodynamic beta ($\beta = K_B T)^{-1}$. However, our obtained spectrum includes higher-order phonons and we must isolate the single-order phonons to extract $g(s, |E|)$. Here it is critical to note that from eq. (9.18) the excitation probability for a single phonon event is limited by $E(1 - e^{-\beta E})$; this implies high-energy single-phonon features will be difficult to observe. The general excitation probability is then described with recursive phonon convolutions for $n \neq 1$:

$$S_n(s, E) = \frac{1}{nf} \int S_{n-1}(s, E') S_1(s, E - E') dE' \tag{9.19}$$

Integration of eq. (9.19) for all energy space allows for the total excitation probability as a function of phonon order:

$$P_n = \frac{f(-\ln f)^n}{n!} \tag{9.20}$$

which is identical to eq. (9.9). Here, we can see dominance in the first-order contribution to the probability as f approaches one, the $(-\ln f)$ term gets smaller. Conversely, for low f we see a rise in the higher-order phonon contribution. We can simplify the convolution in eq. (9.19) to multiplication through Fourier transformation:

$$\tilde{S}_n = \frac{f}{n!} \left(\frac{\tilde{S}_1}{f} \right)^n \tag{9.21}$$

This transformation is useful as the higher-order phonon excitation probabilities still depend recursively on (the Fourier image of) the single-phonon excitation probability. Then the Fourier image of the total energy-dependent excitation probability follows as:

$$\tilde{S} = f + \sum \tilde{S}_n = f e^{\frac{\tilde{S}_1}{f}} \tag{9.22}$$

Now we can extract the single-phonon excitation simply by solving for S_1 through inversion

$$S_1 = \mathcal{F}^{-1} \left[f \ln \left(\frac{\tilde{S}}{f} \right) \right] \tag{9.23}$$

where $\mathcal{F}^{-1}$ is the inverse Fourier-transform operator. This form of the function is more advantageous as it is dependent on Fourier image of the excitation spectrum $\tilde{S}$ and a physical property, f. Now we can relate eqs. (9.18)–(9.23) to solve for the PVDOS, $g(s, |E|)$, using accessible properties/observables E, E_R and β

$$g(s, E) = \frac{E}{E_R} \tanh \frac{\beta E}{2} \; (S_1(s, E) + S_1(s, -E)) \tag{9.24}$$

However, an exact vibrational density of states need not follow any simple expression. The Debye PVDOS increases as the square of the frequency (or energy) until it cuts off abruptly at the Debye frequency, ω_D:

$$g(\omega) = \frac{9\omega^2}{\omega_D^3} \quad or \quad D(E) = \frac{9E^2}{E_D^3} \tag{9.25}$$

If we incorporate a variety of typical numerical values for a Debye function into the above single-phonon formula (eq. (9.21)), the resulting single phonon excitation spectra are illustrated in Figure 9.7.

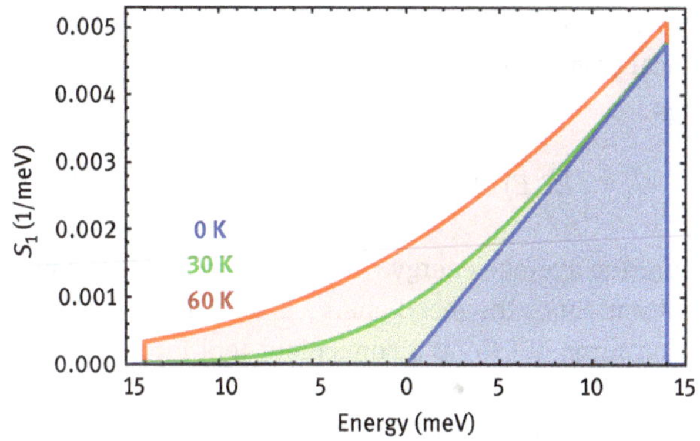

Figure 9.7: Single-phonon excitation probabilities, $S_1(E)$, for a Debye model PVDOS with a cutoff energy of 14 meV at different temperatures.

9.2.6 Other quantities from NRVS analysis – sum rules

As we mentioned earlier, there are several physical properties that can be extracted from NRVS spectra. A short list is included in Table 9.1. Many of the useful quantities are obtained from the so-called sum-rule analysis, where the sum relies on integrals yielding various moments of the excitation spectrum [7]. Specifically, the nth moment is defined as $W_n = \int E^n S(E)\, dE$ (here we use W_n instead of S_n to avoid confusion with the n-phonon absorption probability).

The first rule involving the 0th moment is trivial – the sum of probabilities over all possible events is 100% – something must happen. Thus, by definition, the zeroth moment W_0 is the integrated transition probability and hence unity:

$$W_0 = \int S(E) dE = 1 \tag{9.26}$$

Table 9.1: Summary of the properties extracted by different spectral moments.

W_0	1	By definition of probability
W_1	E_R	Recoil energy – independent of chemical environment
M_2	$4E_RK_{av}$	K_{av} is the average kinetic energy
M_3	$\frac{\hbar^2}{m}E_R\,k_{av}$	k_{av} is the average force constant

As we have seen from eq. (9.15), the first moment is useful because it turns out to be independent of the chemical environment of the nucleus under study – it just depends on the recoil energy E_R:

$$W_1 = \int ES(E)\,dE = E_r \tag{9.27}$$

This expresses a satisfying result: the average energy transfer to the lattice is equal to the recoil energy of the free atom. Since the recoil energy is already known from $E_R = E_0^2/2mc^2$, the above equation turns out to be a convenient tool for normalization of the overall spectrum.

The next two sum rules are simpler if one uses "centered moments" defined as

$$M_n = \int (E - E_r)^n S(E) dE \tag{9.28}$$

The centered second moment provides the average kinetic energy K_{av} of the nucleus under study:

$$M_2 = \int (E - E_R)^2 S(E)\,dE = 4E_R K_{av} \tag{9.29}$$

Finally, the centered third moment turns out to be proportional to the average force constant K_{av} holding an atom in its position along the average photon direction $\hat{k}$:

$$M_3 = \int (E - E_R)^3 S(E)\,dE = \frac{\hbar}{m}E_R k_{av} \tag{9.30}$$

We summarize by stating an obvious but nice fact about sum rule analysis – it requires far fewer presuppositions than building a normal mode model based on a hypothetical force field.

9.3 The NRVS experiment

Although the recoil fraction may have an overall cross section comparable to the recoil-free fraction, the intensity is spread over tens or hundreds of meV, compared to tens of neV for the Mössbauer effect. Thus, the absorption cross section at any particular energy is effectively six orders of magnitude weaker than that of the elastic resonance. So how is it ever possible to observe the NRVS effect? The key to the synchrotron experiment is to *exploit the time delay for the emission of nuclear fluorescence and internal conversion X-ray fluorescence* while ignoring unrelated faster electronic scattering events. Of course, having a high-brightness source and a HRM system is also essential [45]. A typical NRVS experimental setup is illustrated in Figure 9.8 and in this section we will discuss the role of each major component in the setup.

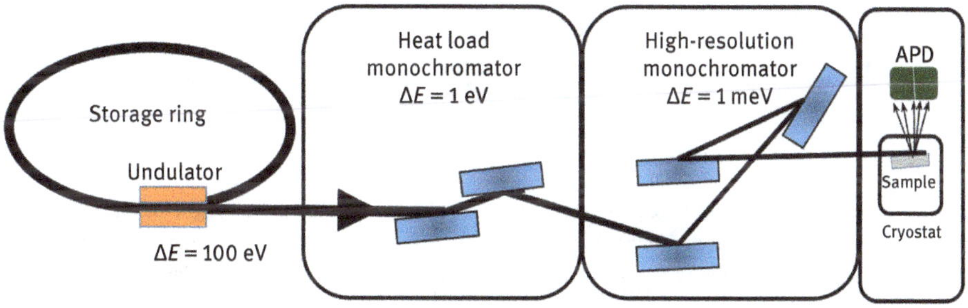

Figure 9.8: Schematic setup of an NRVS experiment.

Not every Mössbauer isotope is suitable for an NRVS experiment. First, one needs a monochromator with sufficiently narrow bandpass to resolve vibrational features, generally ~ 1 meV resolution. Currently, this limits the nuclear resonance energy to <40 keV. Second, the nuclear excited state lifetime must be long enough to allow the detector to distinguish nuclear events from "prompt" electronic events, but short enough to be accommodating to the synchrotron period and bunch frequency. The isotopes that have made the cut so far are illustrated in Figure 9.9. Currently, lifetimes from hundreds of ps to hundreds of ns are feasible. Fortunately for many materials and biological applications, one of the most accommodating isotopes for NRVS is ^{57}Fe.

9.3.1 Comparison to Inelastic Scattering Spectroscopy

NRVS is often considered a NIS spectroscopy. For inelastic spectroscopies, it is key to measure the incident E_1 and scattered energies E_2 to accurately define the energy transfer ($E_1 - E_2$), and extract the inelastically scattering signal from the significantly

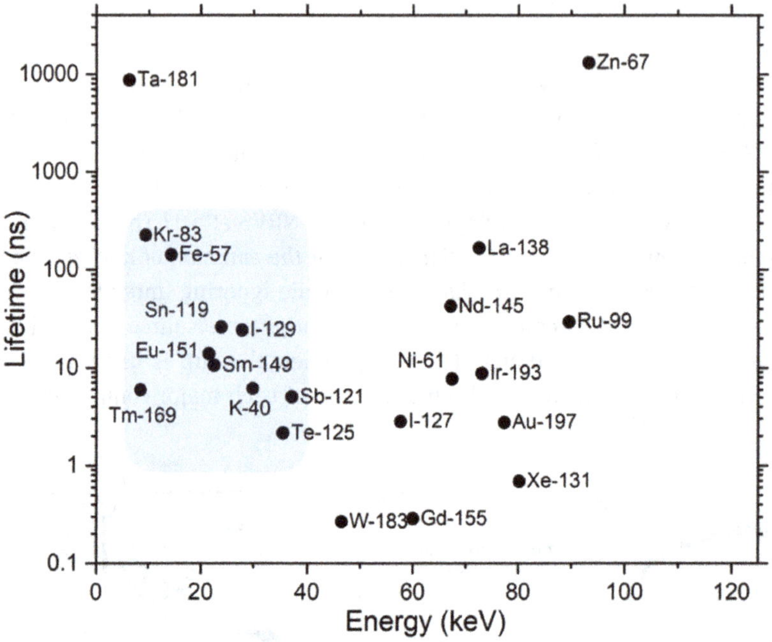

Figure 9.9: Energies and lifetimes of some nuclear isotopes with the shaded region indicating lifetimes and energies that are amenable to current synchrotron technology. Values obtained from the International Atomic Energy Agency Nuclear Data Service website (www-nds.iaea.org).

larger background. As will be discussed later, a narrow band incident energy can be provided via a HRM – where 1 meV is considered reasonable resolution for observing vibrational features. For conventional scattering spectroscopies like resonance Raman, the detection of the scattered radiation is well resolved relative to the incident beam. Also, inelastic X-ray scattering (IXS) utilizes energy-resolved detection of scattered photons that can be on the same energy scales as NRVS.

In the case of measuring NRVS, since the nuclear Mössbauer transitions often have a very narrow linewidth (~5 neV for ^{57}Fe), the nuclear transition itself provides an excellent point of reference for the E=0 in terms of vibrational energy. NRVS measures the elastic Mössbauer transition then tune the incident energy into the inelastic sidebands; thus, measurements do not utilize an energy analyzer for scattered photons, which makes this spectroscopy have a much higher photon out throughput than scattering experiments. For example, IXS uses an incident beam flux in the order of 10^{10} photons/s and produces ~1 cts/s (counts per second) for the asymmetric Fe–Cl stretch at 380 cm^{-1} in a $[FeCl_4]^-$ sample; NRVS uses a beam flux on the order of 10^9 photons/s and has ~10 cts/s of signal for the same vibrational feature. This advantage makes NRVS preferable (e.g., vs IXS) for biological measurements because it leads to a higher signal level while having a lower amount of incident radiation dose on samples.

Although the scattered energy can be well defined by the nuclear transitions themselves, the photons collected by the detector(s) include not only the signal from the nuclear scattering event but also the unwanted background counts from the electron scattering (in the order of millions of cts/s per detector). Since NRVS uses no energy analyzer, it is necessary resolve the signal from the background via another option: time. Fortunately, the nuclear scattering process has a long lifetime on the order of ns (e.g., ^{57}Fe at 14.4 keV has a $1/e$ lifetime of 141 ns), whereas X-ray-electron interactions are essentially instantaneous on this time scale, leaving the background's pulse duration the same as the SR's pulse duration (e.g., 70 ps). This makes it possible to distinguish the nuclear scattering contribution from the electron scattering background in the time domain. A practical description of this procedure will be discussed in later sections.

9.3.2 X-ray sources for NRVS

9.3.2.1 Synchrotron radiation sources

SR is the electromagnetic radiation emitted when electrons are accelerated radially while at relativistic speeds. In practice, the electrons are traveling along a circle, the acceleration is perpendicular to the velocity ($a \perp v$) and the radiation is along the tangent direction of the circle via an insertion device such as a bend magnet, wiggler or undulator. SR has many advantages, such as high flux / brilliance, high linear, elliptical or circular polarization, small angular divergence, low emittance (i.e., the product of source cross section and solid angle of emission is small), wide energy tunability and pulsed time structure. These advantages have lead to many advanced applications in areas from condensed matter physics to biology, from academic research to industrial technologies.

There are more than 100 SR rings around the world; however, most of them are suitable for NRVS measurements. The ideal narrow incident bandwidth (on the order of 1 meV) demands a specialized X-ray optics called an HRM, and the other NRVS requirements depends on the type of SR rings. For measuring ^{57}Fe nuclei ($1/e$ lifetime = 141 ns), a radiation pulse interval of ~150 ns is optimal. SR rings support multiple experimental techniques at end stations with differing requirements of flux, bunch spacing and bunch width. NRVS requires a large storage ring capable of widely spaced bunches and/or hybrid bunch modes (sections of electrons with differing bunch timings). It also requires a high energy electron beam to create sufficient flux at high X-ray energies. Currently four storage rings in the world are set up to allow NRVS experiments (Table 9.2): APS in Chicago, USA (1.1 km, 7GeV, since 1995); SPring-8 in Hyogo, Japan (1.5 km, 8GeV, since 1996); ESRF in Grenoble,

Table 9.2: Beamlines, and their approximate operating conditions, that perform NRVS spectroscopy.

Facility/ beamline	Energy (GeV) [46]	Emittance (nm-rad) [46]	Period (mm)	# of periods	Flux (photon s^{-1})/ resolution (meV) (at 14.4keV)	Possible isotopes
ESRF ID-18	6	3.0	20 [47]	60 [47]	3.6×10^9/0.84 meV [47]	^{57}Fe, ^{151}Eu, ^{149}Sm, ^{119}Sn, ^{161}Dy, ^{121}Sb, ^{125}Te, ^{129}Xe
					1.3×10^9/ 0.47 meV [47]	
APS 3-ID-D	7	3.1	27 [48]	88 [48]	5×10^9/1 meV [49]	^{57}Fe, ^{151}Eu, ^{83}Kr, ^{119}Sn, ^{161}Dy
SPring-8 BL09-XU	8	3.4	32 [50]	140 [50]	1.9×10^9/0.8 meV [51]	^{57}Fe, ^{151}Eu, ^{149}Sm, ^{119}Sn, ^{40}K, ^{125}Te, ^{121}Sb
					2.5×10^9/1.1 meV [52]	
SPring-8 BL19-LXU	8	3.4	32 [50]	780 [50]	~6×10^9/0.8 meV [53]	
PETRA III P01	6	1	32 [54]	314 [54]	12×10^9/1.0 meV [55]	^{57}Fe, ^{119}Sn, ^{121}Sb, ^{121}Te, ^{193}Ir

France (0.9 km, 6 GeV, since 1994) and Petra-III in Hamburg, Germany (2.3 km, 6 GeV, since 2010).

9.3.2.2 Undulators and monochromators

For NRVS specifically, the incident beam is invariably generated by a set of magnets called an undulator. In an undulator the radiation from different poles constructively interferes with the motion of the oscillating electrons resulting in a narrow beam energy profile. A useful metric that characterizes a wiggler/undulator is the K-factor:

$$K = \frac{eB\lambda_u}{2\pi m_e c} \tag{9.31}$$

where e is the charge of the electron, B is the magnetic field strength of the insertion device, m_e is the electron rest mass and λ_u is the period of the magnetic oscillation. The K-factor characterizes the oscillation amplitude of the particle – for K much greater than 1 the oscillation is amplitude is large and behaves like a wiggler. For K-factors much less than 1, we see the interference characteristic of an undulator. Changes in K alter the energy profile of the beam, and we can achieve the desired K by tuning the magnetic field, B. This is accomplished by modifying the "undulator gap" between the magnets above and below the electron path.

The primary advantages of using an undulator include a narrow angular divergence, a high brightness and a narrow energy distribution. The white light from an undulator at an NRVS beamline has a bandwidth on the order of 100 eV of width in energy. The gap is tuned specifically for a desired experimental energy range. As NRVS is obtained over a range of 100s of meV around a central energy on the order of keV, the undulator gap is usually not changed during the course of an experiment. The initial beam is narrowed further by a liquid nitrogen or water-cooled high heat load monochromator (HHLM), which reduces the radiation energy to about 1 eV in bandwidth, before reaching the end-station of the beamline the beam is further reduced to ~1 meV by a HRM. There are also many optics and mechanical components that shape/focus the beam. With so many requirements, experimental end-stations at NRVS beamlines tend to be 25–100 m away from the SR ring.

9.3.2.3 High-resolution monochromators

Crystals diffract X-rays with different energies at different angles characterized by Bragg's law:

$$n\lambda = 2d\ \sin\theta \tag{9.32}$$

where n is the order of diffraction, λ is the wavelength of energy being diffracted, d is the spacing between crystal planes and θ is the scattering angle. This wavelength scattering angle relationship is utilized in the design of crystal monochromators. Regular diffraction refers to the case in which the incident and the diffracted beams are symmetric with respect to the crystal's cut surface, for example a Si(1,1,1) crystal used in a heat load monochromator at an NRVS beamline. On the other hand, asymmetrical diffraction means that the cut surface of the crystal is not in parallel with the diffraction crystal plane and therefore the incident and diffracted X-rays are not symmetric with respect to the crystal's cut surface. An asymmetric diffraction can obtain a larger beam broad angular acceptance of the input beam and narrow dispersion of the output beam ($D_i \rightarrow D_o$) than a symmetric diffraction while outputting a larger beam size beam size ($S_i \rightarrow S_o$), as shown in Figure 9.10. The ~1 meV bandwidth X-ray beam used for NRVS measurements can be produced via a pair of such asymmetrically cut and high reflection index such as Si(9,7,5). To further disperse the beams with different energies, the normal angles of these two diffraction crystal planes are arranged between 90°–180°, forming a ++ (dispersive) crystal array. The pair of asymmetric crystals are referred to as the *key diffractive crystals*.

At SPring-8 BL09XU, for ^{57}Fe NRVS a three-crystal HRM is utilized, in which a pair of Si(9,7,5) crystal planes were chosen as its key diffractive crystal surfaces, and a Ge (3,3,1) crystal is used as a front mirror to adjust the direction of the input optical path.

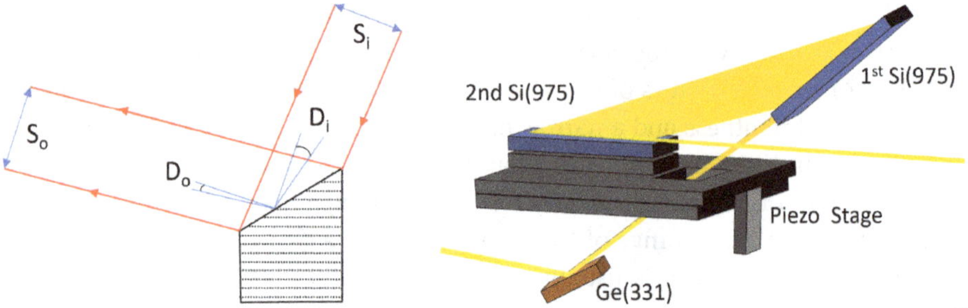

Figure 9.10: Left: A schematic of an asymmetrically cut monochromator crystal plane. Right: A diagram of the three-crystal HRM at BL-9XU at SPring-8.

APS 03ID implements a four-crystal HRM with its key crystal plane being Si(10,6,4), and the front and back manipulation crystals being Si(4,0,0); the latest Petra-III P01 nuclear scattering beamline also uses a 2Si(10,6,4)×2Si(4,0,0) HRM setup for ^{57}NRVS. The three-crystal HRM usually produces a beam with 0.1×0.6 mm^2 in cross section, while the four-crystal one produces a cross section of 0.1×0.1 mm^2. Some beamlines have Kirkpatrick–Baez focusing mirrors to produce a beam in even smaller sizes. So far, either a four- or a three-crystal HRM can provide X-rays that have ~1 meV energy resolution or better at energies (5–30keV) suitable for many NRVS experiments.

9.3.2.4 About energy calibration

Monochromator crystals can expand or contract due to fluctuations in ambient temperature. Although these environmental variables are usually controlled well at an SR facility, the precise location of the nuclear resonant peak position and the stability of the energy axis scale must be calibrated. The nuclear resonant peak can be calibrated as 0 meV with respect to vibrational quanta during data analysis. However, the energy values relative to monochromator motion still need to be calibrated with a standard sample. As shown in Figure 9.11, the $FeCl_4$ T_2 stretch peak from the same complex sample [NEt_4]($FeCl_4$) measured at different beamlines do not overlap and vary by up to 4 meV with each other. A linear correction factor (the energy axis scale) of about 0.920–1.005 can bring all of the uncalibrated spectra into very good alignment with the reported Fe–Cl T2 peak centroid (at 380 cm^{-1}).

The primary source of energy uncertainty in the monochromator comes from changes in the crystal plane *d*-spacing due to temperature. For Si, the coefficient of thermal expansion is 2.56×10^{-6} K^{-1}, leading to an energy shift of 9 cm^{-1} (or over 1 meV) per 0.03 K of temperature change. Inadvisably entering the monochromator hutch could change the crystal temperature by 0.1%. This could lead to a several meV energy drift and require hours to re-equilibrate. Therefore, beamline energies

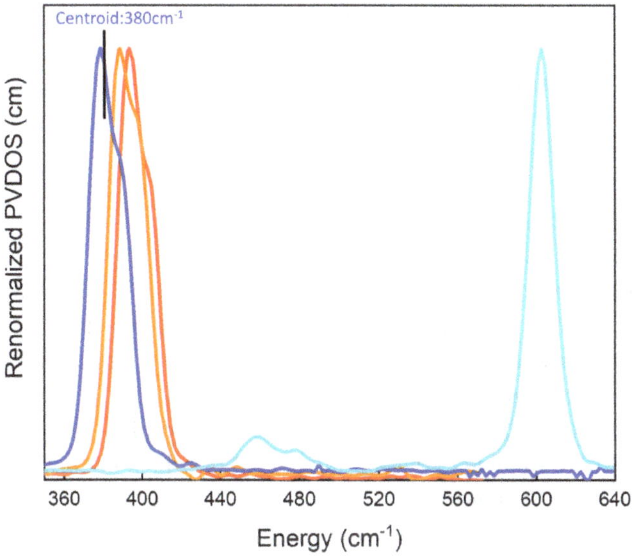

Figure 9.11: NRVS PVDOS spectra for the calibrated $(NEt_4)FeCl_4$ ion (——). An uncalibrated spectrum measured in May 2015 at BL09XU (——) and the same sample measured seven day later (——). Also, the room temperature PVDOS spectrum for $(NH_4)_2MgFe(CN)_6$ (——) as an alternative higher-energy calibrant. Major peaks renormalized from 0 to 1 for illustration.

must be frequently checked especially when the temperature of the hutch is disturbed.

However, it should be noted this temperature sensitivity can be intentionally used as a mechanism of tuning monochromator energy. For example, the sapphire backscattering HRM at Petra-III P06 is tuned by temperature scanning and it allows access to nuclear resonances above 30keV with ~1meV resolution.

The energy calibration becomes more critical for dilute biological NRVS measurements as the acquisition times can be long (>24 hours) and both the zero position and the energy scale will drift more. As sample changes take time (1 hr), the traditional energy calibration interrupting an extended biological sample measurement may not be desirable. To resolve this issue, the calibration can also be done via measuring ^{57}Fe powder at 285 cm^{-1} or $(NH_4)_2MgFe(CN)_6$ at 602 cm^{-1} at room temperature (RT) in a separate stage outside the back of the cryostat stage (similar to in situ foil calibrations used for EXAFS measurements). The calibration at this second stage can be a quickly switched to by letting the full beam bypass to the main stage (the cryostat chamber), or an *in situ* measurement that is done at the second stage while the biological sample is measured at the main stage; however, this is not recommended as it increases the mechanical deadtime of the overall measurement.

9.3.3 Detecting NRVS

As discussed previously, the most important factor for separating the NRVS signal from the prompt electronic scattering background is time resolution.

9.3.3.1 Avalanche photodiode detectors

Avalanche photodiode (APD) detectors possess properties that are suitable for measuring NRVS. The saturation level for an APD is about 6 MHz per APD element no matter how large the element area is. Therefore, using an APD array with small individual elements instead of a single APD is advantageous. Figure 9.12 (top) shows an example of 2 × 4 APD array used in SPring-8 BL09XU. The combined high-gain and high-saturation rate makes APDs suitable for measuring the NRVS, which has a weak signal but a large background at the same time.

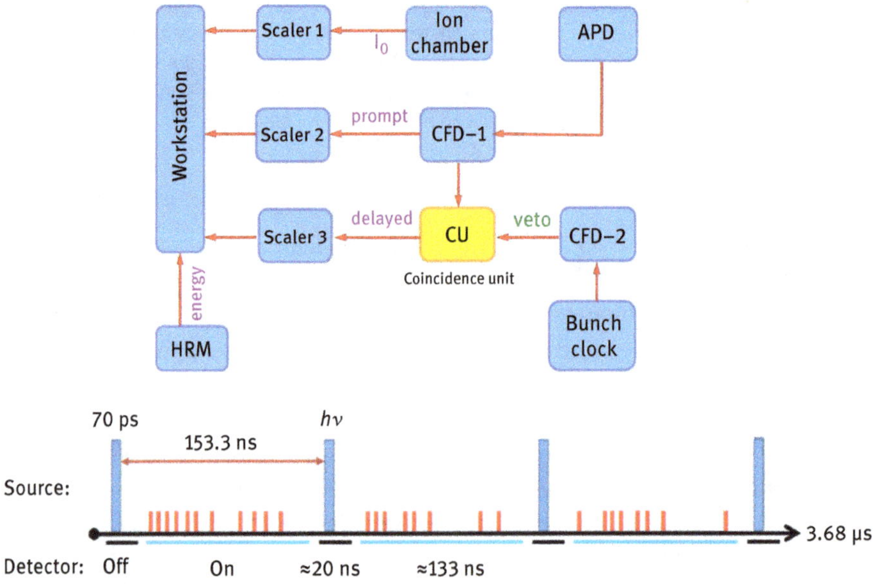

Figure 9.12: Top: Electronics used to separate delayed nuclear events from prompt -ray events. Middle: A typical timing scheme for an NRVS experiment. Detector "off" implies the counts are not being sent to "Scaler 3" in the top diagram. The specific times are typical for APS operating conditions. Bottom: Close up image of an APD array with inset being a zoomed in picture of the individual APD elements.

Modern APDs can have a time resolution of 1 ns or better, which is good for many types of NRVS experiments, including those on ^{57}Fe.

9.3.3.2 Data-acquisition electronics

APDs collect the photons but it is the data-acquisition electronics that finally separate the delayed nuclear signal from the prompt background in the time domain. A block diagram for the data-acquisition electronics is shown as in Figure 9.12 (top) to illustrate this process. One important device is the "bunch clock," which tells the electronics the initial time point of $t = 0$ when a synchrotron pulse arrives. When both the APD and bunch clock send pulses to the CU simultaneously, the latter passes the signals from the APD, which is then digitized and sent to the workstation for storage. In practice, a veto interval X is set around $t = 0$ (where X is about ±10 ns), to avoid the signal integration during this period and to allow the electronics to recover from the enormous background pulse. Then, the counts for the events that occur between $t = X$ and $150 - X$ (e.g., $t = 10$–140 ns) is passing through the CU and digitized as NRVS counts. The values without time structure, for example, I_0, prompt and energy positions are directly digitized to send to the workstation for storage.

9.4 Data analysis

9.4.1 Data processing

To compare theory and actual experiment, we need to be able to proceed in the opposite direction – extraction of the PVDOS from the experimental data. A variety of programs are available for the analysis of NRVS spectra, including DOS [57] and PHOENIX [58]. The key steps in the analysis of NRVS data are as follows:
- determination of the resolution function $R(E-E')$,
- normalization using sum rules,
- subtraction of the elastic component,
- decomposition into n-phonon contributions and
- derivation of the PVDOS.

Herein we briefly summarize the steps taken by the PHOENIX software [34]. For actual experimental data the measured intensity transforms from eq. (9.17) to:

$$I(E) = \int R(E-E')\{aS(E) - b\delta(E)\}dE' \tag{9.33}$$

We have distributed the normalization constant A into a and b and have introduced the experimental monochromator resolution function $R(E-E')$. Next, the resolution function must be precisely fit to the spectrum. The high statistics of the elastic peak give an obvious target to fit the resolution function; however, the contribution to

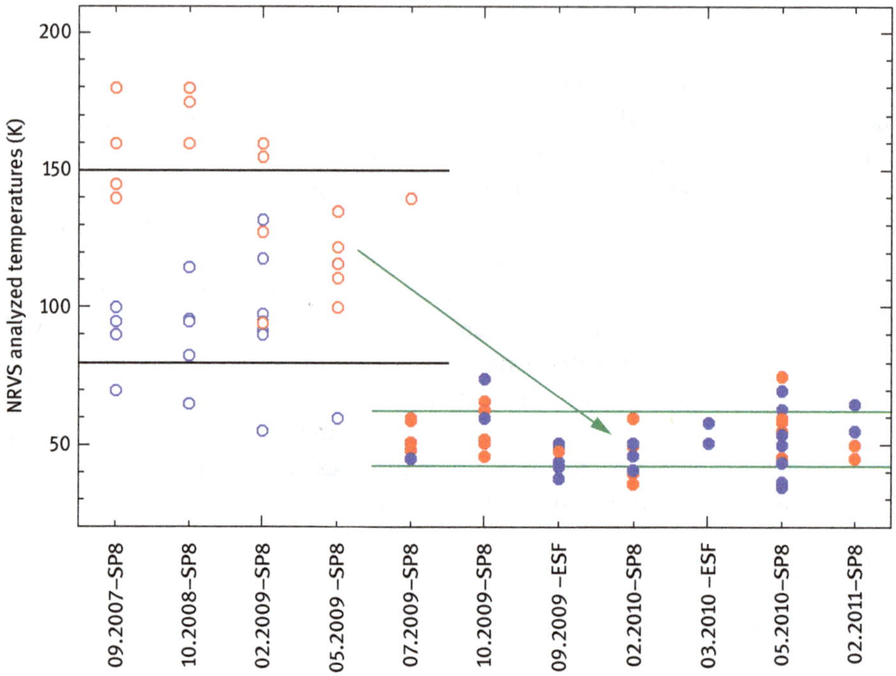

Figure 9.13: Real sample temperature during different beamtimes from 2007–2011 at SPring-8 (SP8) and ESRF. Red points are biological samples and blue points are solid samples – filled points represent samples that were attached to the cryostat by 1-propanol and open points were attached by cryogenic grease.

the area under the elastic peak also comes from the inelastic contribution – here estimated as a Debye solid – so the following function is fitted:

$$P(E) = c_1 \left\{ R(E - c_0) + \frac{c_2 \beta (E - c_0)}{\left(1 - e^{-\beta(E - c_0)}\right)} \right\} \tag{9.34}$$

All c terms are variable fitting parameters. The term c_0 allows for adjusting the position of the resonance, c_1 scales the height of the function to the elastic peak and the term c_2 adjusts the phonon contribution. The function $R(E - c_0)$ provides the resolution function as follows:

$$R(E) \left\{ = \begin{array}{ll} H\left[E\frac{2c_3}{1+c_4}\right], & E \geq 0 \\ H\left[E\frac{2c_3 c_4}{1+c_4}\right], & E < 0 \end{array} \right. \tag{9.35}$$

where $H(x)$ is a generic shape function:

$$H(x) = e^{-|x^{c_5}|} \tag{9.36}$$

All c values are fit using a least squares minimization. Next the first moment of the spectrum is normalized to the recoil energy according to eq. (9.18) determining the normalization factor a. From eq. (9.24) we see that the Lamb–Mössbauer factor must be determined for the eventual deconvolution. Utilizing the normalization factor, fitted elastic peak and resolution function, we can solve for the elastic contribution to the experimental spectrum:

$$f = 1 + \frac{c_1}{a} - \frac{1}{a}\int I(E) - c_1 R(E) dE \tag{9.37}$$

Finally, PHOENIX performs the Fourier-log deconvolution [59] a modification of eq. (9.23):

$$S_1(E) = f\mathcal{F}^{-1}\left[\tilde{R}\ln\left\{1 + \phi\frac{\tilde{I} - c_1\tilde{R}}{af\tilde{R}}\right\}\right] \tag{9.38}$$

Here $\tilde{R}$ is the Fourier image of the resolution function and $\tilde{I}$ is the Fourier image of the normalized intensity spectrum. Likewise, PHOENIX calculates higher-order phonons. The term ϕ is a mollifier function that convolutes with the subtracted spectrum to limit Fourier artifacts.

$$\phi = \min\left\{1, \left|\frac{\tilde{I}\tilde{R}(0)}{\tilde{R}\tilde{I}(0)}\right|\right\} \tag{9.39}$$

The final PVDOS is calculated identically to eq. (9.24).

From a practical standpoint, the energy axes of multiple scans are aligned based on their position of the elastic resonance, then a calibration factor is applied. As mentioned previously, this approximately linear factor can be determined by the position of three prominent peaks for $[Et_4N]FeCl_4$ (Figure 9.3). It can also be determined by fitting the Stokes and anti-Stokes regions for a sample of accurately known temperature [60].

9.4.2 Data interpretation

Once the NRVS-derived PVDOS is in hand, the interpretation depends on the problem of interest. However, among the interpretive approaches employed are as follows:

- sum rule analysis for physical observables,
- graphical analysis for speed of sound,
- isotope shifts,
- optimization of empirical force fields,

- group theory and
- interpretation via density functional theory (DFT) calculations.

As discussed previously, the sum rules provide a straightforward prescription to obtain physical observables such as the Lamb–Mössbauer factor and average force constant of the probe atom. Albeit less obvious, the speed of sound through the material can be obtained by graphical manipulation and extrapolation.

Due to the selectivity of NRVS for the probe isotope motion, many features in a spectrum can be initially assigned based on a combination of chemical intuition, literature precedents and group theory. To further refine the assignments, isotopic substitution can be employed to highlight modes involving the substituted atom motion. A normal coordinate analysis can also be employed that generates a spectrum from a set of internal coordinate force constants. The resulting force field can then be optimized to reproduce the experimental spectrum affording information about the strength and motion of vibrational modes, which, if symmetry allows, can be related back to group theory [61].

In recent years, DFT calculations have prevailed as a straightforward way to interpret vibrational spectra. For NRVS, first principle calculations of PVDOS are derived from the orthogonal matrix that diagonalizes the mass-weighted second derivative Hessian matrix [62]. The basic procedure is routinely performed by many quantum chemistry codes during the calculation of vibrational frequencies and modification to yield the PVDOS for the probe atom(s) is straightforward.

9.5 Application to model complexes

In metal complexes, metal–ligand stretching and bending frequencies range from ~50–2,000 cm^{-1}, or ~6–250 meV. NRVS is exciting, in part because it provides an isotope selective vibrational spectrum. In the case of Fe NRVS, only modes that involve displacement of the ^{57}Fe nucleus couple to the nuclear transition.

9.5.1 Coordination complexes

9.5.1.1 Fe_2O_2

Peroxo diiron complexes, abbreviated here as Fe_2O_2, are common intermediates in biological binuclear Fe oxygenases. Key to understanding the underlying biological chemistry is characterizing the binding moiety of the peroxo bridge. There are numerous possibilities for a peroxide bridge to form between two irons: a cis 1,2 bridge, a trans 1,2 bridge, two doubly bridging oxygens, a bridging plus terminally bound oxygen and a bridging oxygen with the distal oxygen not bound to either iron. NRVS spectroscopy can be utilized to understand the iron-bound peroxo species in biological intermediates, especially where Raman and/or structural data are not available. To understand how the peroxo-binding moiety affects the molecular vibrations of the diiron site, NRVS was utilized on a model complex, $[Fe_2(\mu\text{-}O_2)(N\text{-}EtHPTB)(PhCO_2)]^{2+}$ (abbreviations available in reference) [63].

As shown in Figure 9.14, the NRVS spectrum of the peroxodiiron model is dominated by an envelope of features from 150–300 cm^{-1}; these modes are dominantly Fe-(N-EtHPTB/$PhCO_2$) in nature. The region from 440–480 cm^{-1} is of special interest and the features in the natural abundance (O_2) spectrum at 446 and 458 cm^{-1} clearly shift to 446 and 458 cm^{-1} in the ^{18}O sample. Likewise, the shoulder centering at 338 cm^{-1} also shifts inward to 311 cm^{-1} in the ^{18}O isotopologue.

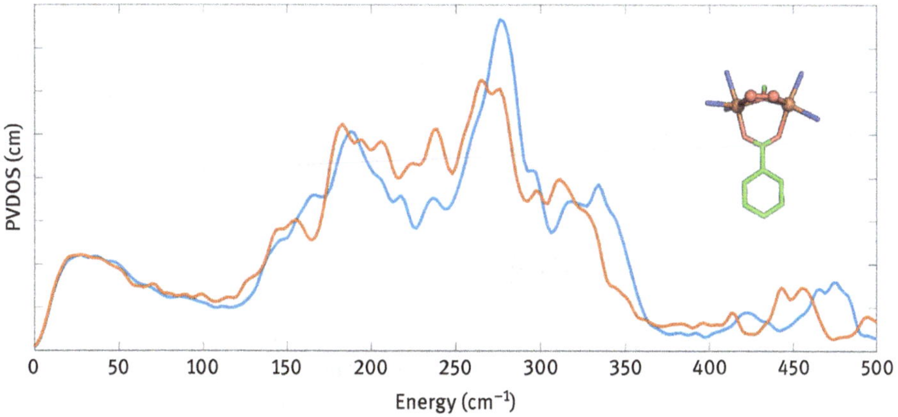

Figure 9.14: NRVS spectra for the $^{57}Fe_2(\mu\text{-}O_2)(N\text{-}EtHPTB)(PhCO_2)$ (——) and ^{18}O substituted compound (——). Inset, truncated image of the Fe_2O_2 site and neighboring atoms.

The isotope shifts, empirical force field analysis and group theory identified the higher energy features. The observed pair of highest energy peaks were characterized as a Fermi doublet [64, 65] with dominantly Fe–O–O–Fe symmetric stretching motion. The isotope-sensitive intensity centered at 325(313) cm^{-1} was not previously observed in resonance Raman. To describe this motion, the spectrum was simulated with the program Vibratz [61]. Using a set of internal coordinate force field constants it was found that the feature was a result of coupled motion of the O–O group parallel to the Fe–Fe axis and perpendicular to a pseudo-mirror plane bisecting the four atoms later referred to as a "peroxide rock."

The study did not elaborate on the modes in the 150–300 cm^{-1} region, but was comprehensively built upon in a later study [66] where DFT was utilized to describe the region as a combination of "core motions," peroxide twisting motion and a "butterfly" motion where the oxygens move perpendicular to the Fe–O–O–Fe plane in opposite motion to the irons. Collectively, the distribution of these modes in the spectrum are related to the deviation of the bridging alkoxide oxygen atom from the Fe–O–O–Fe plane created by the bridging peroxide.

These studies established correlations between NRVS spectra and structure that would be translated to the study of biological peroxo-bound diiron systems; specifically the P′ peroxo-bound intermediate of the AurF enzyme [67].

9.6 Applications to biological systems

9.6.1 Rubredoxin

The relatively simple metalloprotein, rubredoxin, contains a single Fe center coordinated to four cysteine amino acid residues. The two accessible redox states of the Fe site are formally Fe(II) and Fe(III), which facilitate the electron transfer activity of the protein. The simplicity of the Fe site and ease of handling of the protein made rubredoxin an excellent early candidate for bioinorganic ^{57}Fe NRVS [68].

The NRVS spectra for the oxidized and reduced forms of the protein are shown in and are qualitatively dominated by the three peaks. For the Fe(III) site the peaks are centered at 70, 150 and 360 cm^{-1}. Although the reduced protein spectrum is similar to its oxidized counterpart, the high-frequency envelope downshifts to 303 cm^{-1}. The original work implemented the Vibratz program for empirical force field-based modeling to derive the Fe–S force constants. The crystal structures (reduced PDB:1CAD [69], oxidized PDB:1BQ8 [70]) were used as models for the simulations. The result was a calculated 36% decrease in Fe–S bond force constant (1.24 to 0.92 mdyne/Å) upon reduction of the protein.

The peaks were assigned using the vibrational mode description from the Vibratz modeling. For the oxidized protein model, the broad region from 345 to 375 cm^{-1} is composed of three strong components at 375, 358, 350 cm^{-1} and three weaker contributions at 365, 364 and 340 cm^{-1} that are principally described as Fe-S asymmetric stretching modes. The intensity in the region from 300 to 340 cm^{-1} arises from nearly symmetric Fe–S stretch modes. It would be expected that a perfect symmetric stretch would not cause Fe motion; however, the intensity is a consequence of the deviation from ideal symmetry. The contribution in this so-called breathing region is approximately 10% that of the asymmetric stretch components. The region near 150 cm^{-1} is composed of nearly degenerate S–Fe–S bending modes. The region between 100 and 150 cm^{-1} is assigned as Fe–S–C bending modes with mixing of S–Fe–S modes. Finally, the low-energy region below 100 cm^{-1} involves significant dihedral, torsional, acoustic and delocalized vibrational modes within the greater protein.

When the same analysis is performed for the reduced Fe center, the order of assignments remains the same, but the stretching modes downshift by 18%. It is noteworthy that the symmetric stretch centroid shifts to approximately 270 cm^{-1} and start to mix with the bending modes leading to enhanced intensity throughout the 150–300 cm^{-1} region (Figure 9.15).

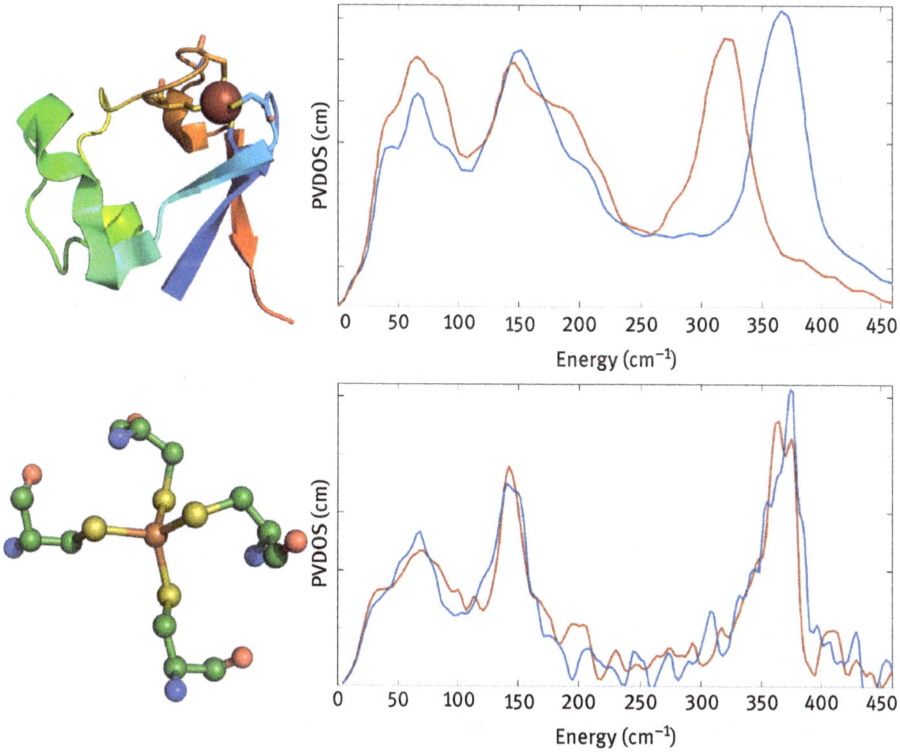

Figure 9.15: Left top: The crystal structure of rubredoxin (pdb: 1CAD). Left bottom: The Fe site in rubredoxin and proximal cysteines. Right top: The NRVS spectra of the oxidized (——) and reduced (——) rubredoxin protein. The NRVS spectra of the oxidized ^{57}Fe rubredoxin aligned along the crystallographic *a*-axis (——) and the crystallographic *c*-axis (——). Figure reproduced from previous works [68, 71].

9.6.2 Single-crystal rubredoxin

A follow-up single-crystal EXAFS and NRVS study was performed with oxidized rubredoxin – the first time such a study was done for an Fe–S protein crystal [71]. As NRVS is orientation dependent (eq. (9.12)), the vibrational modes of a single crystal of rubredoxin are enhanced when the ^{57}Fe motion is along the direction of the incident photon vector **K**. The three strong asymmetric stretch modes at 373, 358 and 352 cm^{-1} (corresponding to the modes at 375, 358 and 350 cm^{-1} in solution) were selectively enhanced by orientation of the incident photon parallel to the crystallographic *a*-axis and *c*-axis. Along the *a*-axis the average Fe–S bond length is 2.263 Å and 2.284 along the *c*-axis (determined by EXAFS). Projection of the incident beam on the *a*-axis enhanced the peak at 373 cm^{-1}, where orientation along the *c*-axis enhances the 358 cm^{-1} and completely hides the feature at 352 cm^{-1} (Figure 9.15 bottom right).

Vibrational modes in the Fe–S bending region of the rubredoxin crystal were less orientation dependent. The alignment on the *c*-axis enhanced the peak at 140 cm^{-1} and the *a*-axis orientation seemed to slightly split the peak. This implies that the ^{57}Fe bending normal mode motions are not "pure" projections onto the crystallographic axes.

The experiment was followed up by Vibratz modeling similar to the solution rubredoxin approach, but with an Fe(SCC)$_4$ model and an orientation dependent simulation. From the modeling, it is supported that the 352 cm^{-1} feature must lie perpendicular to the crystallographic *c*-axis. A key, but perhaps not surprising, observation made is that the highest frequency mode (373 cm^{-1}) enhancement is consistent with the crystallographic *a*-axis, which has the shorter average Fe–S bond length.

Collectively, the work demonstrates that single crystal NRVS coupled to EXAFS, or other structural methods, can be used to selectively probe individual vibrational modes and deconstruct vibrational envelopes. With the ultimate application of the work being to vibrationally uncouple systems containing multiple ^{57}Fe sites.

9.6.3 Non-Rieske [2Fe2S]

The logical continuation of the study of simple FeS clusters with NRVS was spectra obtaining for reduced and oxidized non-Rieske *Rhodobacter capsulatus* ferredoxin VI (Rc FdVI) [2Fe2S] cluster [72]. The [2Fe2S] cluster is characterized by two irons connected to the protein matrix by two terminal cysteinyl sulfurs (S^t) and connected to one another by two bridging sulfurs (S^b). In the oxidized form, the cluster is ferric and diamagnetic – whereas in the reduced form the cluster is a mixed ferric and ferrous antiferromagnetically coupled S=1/2 system. Unlike the single Fe counterpart in rubredoxin, the [2Fe2S] cluster in Rc FdVI has a sharper Fe–S stretching region with better-resolved components (Figure 9.16). The stretch region components in the all-ferric sample are observed at 421, 393, 352, 341 and 321 cm^{-1}. Raman spectra of [2Fe2S] clusters have a strong feature near 290 cm^{-1} that is not present in the NRVS corroborating the original assignment of the mode as a combination of two highly symmetric FeS$_4$ breathing modes with minimal Fe motion. It is useful to note that the redox sensitive Fe–S stretch modes greater

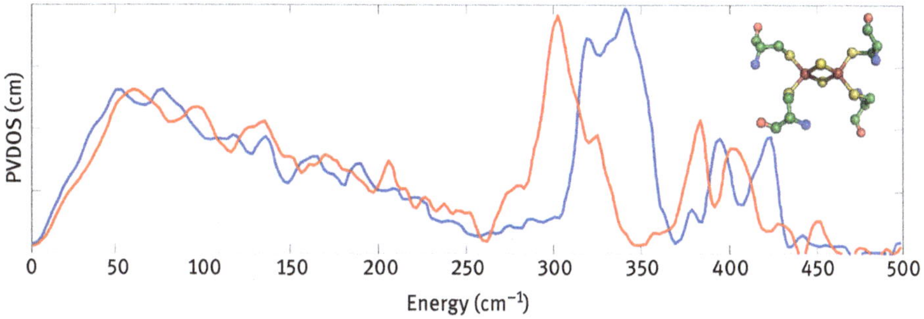

Figure 9.16: The ^{57}Fe NRVS spectrum of [2Fe2S] in the reduced (——) and oxidized (——) form. Figure reproduced from previous work [72].

than 400 cm^{-1} have become an identifier for [2Fe2S] clusters in systems with multiple clusters of varying composition [22].

The maximum of the oxidized protein NRVS spectrum occurs at 341 cm^{-1} in a region of broad unresolved components, while the highest frequency feature is found at 421 cm^{-1}. In the reduced protein, a spectral maximum occurs at 302 cm^{-1}. Likewise, the 393 cm^{-1} intensity shifts to 382 cm^{-1} in the reduced form. It is noteworthy that NRVS for the reduced protein was capable of observing a peak at 276 cm^{-1}, which was only predicted at the time to be a Raman unobserved Fe(II)–S^t stretch [73]. The spectrum for [2Fe2S] ferredoxin contains some resolvable features in the low-energy region below 250 cm^{-1}. However, since most of the intensity in this region involves numerous highly delocalized and weak vibrational motion that involve a sizable portion of the greater protein, they cannot be properly modeled using DFT or force field modeling using small model metalloclusters.

A Urey–Bradley force field simulation of the [2Fe2S] cluster spectrum was employed using Vibratz. The symmetry and models were D_{2h}:$Fe_2S_2S'_4$, C_{2h}:$Fe_2S_2(SCC)_4$ and C_1:$Fe_2S_2(Cys)_4$. The force constant treatment allowed for the development of a CHARMM-based force field that could be used for all-atom molecular dynamics simulations [74]. The vibrational analysis from all-atom molecular dynamics simulations performed using the CHARMM force field can properly account for the low-energy region of the NRVS spectrum. Extremely low energies involve completely delocalized protein skeleton motion and in-phase motion of the entire [2Fe2S] cluster moving as a rigid body. The precedent set by combining NRVS and molecular dynamics has been further developed with respect to more complex heteroclusters like the iron–molybdenum cofactor in nitrogenase [75].

9.6.4 [4Fe4S] ferredoxin

The ferredoxin of the hyperthermophillic archaeon *Pyrococcus furiosus* contains a [4Fe4S] cluster used for electron transfer. The cluster contains four bridging sulfurs (S^b) that are coordinated to the protein through three cysteinyl sulfurs and a single aspartate oxygen. The D14C mutant (PfD14C Fd) has a point mutation that ensures the [4Fe4S] cluster is coordinated to the protein by four cysteinyl sulfurs (S^t) instead of an aspartate oxygen [76]. The oxidized PfD14C Fd $[4Fe4S]^{2+}$ cluster has been shown in a crystal structure to be highly symmetric with nearly D_{2d} symmetry [77]. In the oxidized $[4Fe4S]^{1+}$ form, the four irons have a formal oxidation of 2.5 and the cluster is *S*=0. Upon reduction, two irons remain Fe(2.5), two irons become formally ferric, and the cluster becomes S=1/2.

The two oxidation states share three broadly defined vibrational regions (Figure 9.17). The intensity below 100 cm^{-1} arises from the diffuse vibrational modes coupled to protein motion as discussed previously. The region from 100 cm^{-1} to 200 cm^{-1} arises from Fe–S bending-type motion. Finally, the region from 200 cm^{-1} to slightly greater than 400 cm^{-1} involves predominantly Fe–S stretching vibrations; however, this region also has a mixture of Fe–S^t and Fe–S^b stretching vibrations at higher and lower energies, respectively. The oxidized PfD14C Fd NRVS spectrum has the most prominent feature at 148 cm^{-1} with a clear shoulder. There is also a local maximum at 84 cm^{-1}. In the canonically Fe–S stretch region, there are two prominent peaks at 354 cm^{-1} and 382 cm^{-1}. The nearly symmetric A_1 mode at 336 cm^{-1} is very weakly observed in the oxidized NRVS spectrum as it involves very little Fe motion and only S^b "breathing"-type stretch motion. In the cluster bending region, an envelope centered at 281 cm^{-1} is observed.

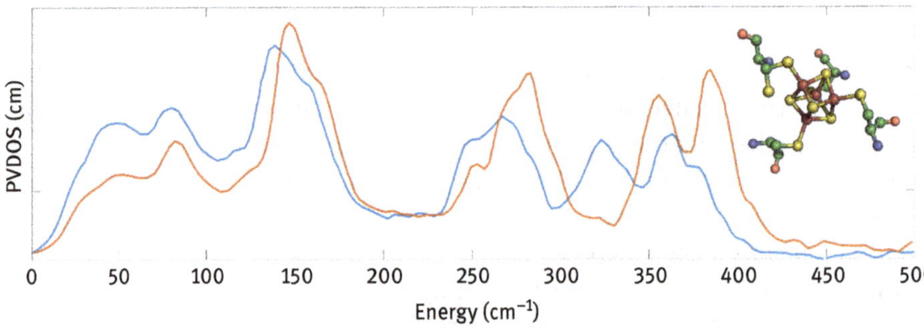

Figure 9.17: NRVS spectrum for [4Fe4S] D14C Ferredoxin in the reduced (——) and oxidized (——) form reproduced from previous work [76]. Inset is the local coordination sphere of the 4Fe4S site.

Although there are many analogies between the oxidized PfD14C Fd NRVS spectrum and existing Raman data, no such comparisons can be drawn for the reduced form as ferredoxins are notoriously difficult to probe with Raman due to fluorescence. Here, NRVS has an obvious advantage over Raman spectroscopy. After reduction, the NRVS spectrum of PfD14C Fd shows a shift of the 382cm^{-1} peak to 362 cm^{-1}, the 354 cm^{-1}, intensity shifts to 320 cm^{-1} and the middle band at 281 cm^{-1} moves to 268.5 cm^{-1}. It is notable that the low-energy region in either oxidation state, which we have defined as involving diffuse phonon-like motion in the protein, is not as dominant as it was in the [2Fe2S] and rubredoxin case.

A combined DFT, force field and molecular dynamics approach to the NRVS spectra indicates that the [4Fe4S] cluster is vibrationally isolated from the rest of the protein. Typically, the coupling between Fe–S motion and S–C–C bends of the cysteine side chains is dependent on the Fe–S–C–C dihedral angle, with a maximum coupling when all four atoms are coplanar at 0 or 180 degrees [78]. The angles in PfD14C Fd are nearer to 90 degrees. Likewise, in the case of the [4Fe4S] cluster each iron is coordinated to the protein matrix by a single sulfur, whereas in the non-Rieske [2Fe2S] and rubredoxin the irons are coordinated to the protein by two and four sulfur bonds per iron, respectively. This has implications for electron transfer – in that protein dynamics in the PfD14C Fd are largely unaffected by the redox of the cluster. In the framework of Marcus theory for electron transfer [79], the reorganization energy term is lower in PfD14C Fd when compared to non-Rieske [2Fe2S] clusters or rubredoxin.

These results help shed some light on questions about biological preferences for which type of FeS cluster is used for electron transfer [80]. Compared to [4Fe4S] clusters, [2Fe2S] clusters may act over shorter distances and/or with long range conformational changes that decouple the electron transfer protein from its redox partner. As in the case of putidaredoxin, which has a 100-fold decrease in binding affinity for its redox partner, cytochrome p450cam, after reduction of its [2Fe2S] [81].

9.6.5 Fe-containing Enzymes

^{57}Fe NRVS has been extended beyond electron-transfer proteins to enzymes with rich chemistry. In Table 9.3 we provide an extensive (but not comprehensive) list of biological NRVS work with a succinct noted observation and, in most cases, list frequencies of interesting ^{57}Fe-X modes. We note that the observations we identify do not fully convey the magnitude of each work and we encourage readers to use them as a guide to read the corresponding full references.

Table 9.3: List of biological NRVS work.

Biological System	Noted Observation	Ref.
Myoglobin Compound-II	Verification of the unprotonated Fe(IV)=O moiety with stretching mode at 805 cm^{-1}.	[82]
Myoglobin-NO	Observation of a ^{15}N-sensitive mode at 547 cm^{-1} assigned as a Fe-NO stretching mode.	[83]
Myoglobin-O_2	Assignment of Fe-O_2 stretching mode at 579 cm^{-1}.	[84]
	Fe-O_2 modes with alternatively substituted porphyrins.	[85]
Myoglobin-CO	Identified heme vibrational dynamics, including delocalization of the heme "doming" mode. Myoglobin-CO stretch at 502cm^{-1} and bend at 572 cm^{-1}.	[86]
[NiFe]-hydrogenase	Identification of the Ni-H-Fe bridging moiety with a wagging mode at 675 cm^{-1}.	[87]
	Assignment of Fe-CO vibrational region between 540–605 cm^{-1} and Fe-CN region between 450–505 cm^{-1}.	[88]
Oxygen-Tolerant [NiFe] hydrogenase	Quantitative determination of the Fe-S cluster composition, specifically identified a [2Fe2S] cluster with a stretching mode at 414 cm^{-1}.	[89]
[FeFe]-hydrogenase	First site-selective ^{57}Fe enrichment for NRVS that demonstrated the $[2Fe]_H$ cluster and $[4Fe4S]_H$ subcluster are inserted stepwise into the protein. [FeFe] site Fe-CO modes: 528, 560, 585, and 604 cm^{-1}. Fe-CN modes: 424 and 454 cm^{-1}.	[90]
	Replacement of the azadithiolate bridge of the [2Fe] site with a proton-disrupting oxodithiolate bridge allows trapping of an intermediate called H_{hyd} with Fe-H-Fe bending modes at 727 and 670 cm^{-1}.	[91]
	Vibrational characterization of the H_{hyd} intermediate, in the enzyme with an artificial azadithiolate $[2^{57}Fe]_H$ cluster. Fe-H-Fe bending modes at 744 (747) and 675 (675)cm^{-1} in CrHydA1(DdHydAB).	[92]
	Characterization of H_{hyd} in the C169S mutant (*Cr*HydA1), altering the proton transfer pathway shifting the Fe-H-Fe bending modes to 772 and 673 cm^{-1}.	[93]
	Characterization of catalytic and oxygen species in the wildtype and C169A mutant (*Cr*HydA1).	[94]
Nitrogenase Fe Protein [4Fe4S] Cluster	First NRVS comparison of a $[4Fe4S]^{2+/1+/0}$ cluster in three oxidation states identifying a linear correlation between oxidation and Fe-S stretching force constants.	[95]

Table 9.3 (continued)

Biological System	Noted Observation	Ref.
Nitrogenase P-cluster Nitrogenase M-cluster Isolated FeMo-co	NRVS spectra of the M-cluster (called "FeMo-co") [7Fe9SMo] deficient enzyme and holoenzyme allowed subtraction of the P-cluster [8Fe7S] and development of an empirical forcefield for the FeMo-co.	[96]
Nitrogenase-CO	Monitored loss of symmetry of the FeMo-co upon binding of CO through a 188 cm^{-1} "breathing mode" of the cofactor.	[97]
Benzoate 1,2-dioxygenase	Assignment of the structure of a peroxide shunt intermediate BZDOp as high-spin side-on Fe(III)-hydroperoxy species. Fe-O stretch identified at 510 cm^{-1}.	[98]
WhiD NsrR	Identification of [4Fe4S]-NO species in NO-sensing regulatory proteins.	[99]
R2lox	Characterization of an O_2 activated species in Fe/Mn and Fe/Fe R2lox indicating a terminal water bound to the Fe/Mn atom and a metal-bridging hydroxide.	[100]

9.7 Outlook

As demonstrated here, the power of NRVS is derived directly from isotopic specificity, normal mode selection and relative sensitivity. Fe is ubiquitous in biological metallocofactors either as part of electron transfer or catalytic mechanisms. Studying these cofactors with traditional vibrational spectroscopy, such as infrared (IR) or Raman spectroscopy, can be impeded by the background created by the biomolecule proper or even fluorescence of the cofactor. As such, the extreme specificity of ^{57}Fe NRVS for only ^{57}Fe motion makes the perfect tool to study vibrational dynamics of the critical Fe-containing components of the biomolecule. Without contributions from the biological sample background, NRVS enjoys very little spectroscopic noise; even modes of Fe hydrides within enzymes [21, 101] have been observed – which are typically difficult to directly identify with other techniques [102, 103]. The selection rules – or comparatively lack thereof – allows for ^{57}Fe NRVS to show both IR- and Raman-active modes that contain ^{57}Fe motion – thus providing a complete picture for Fe vibration.

Since its inception more than two decades ago, ongoing improvements in synchrotron brightness and monochromator resolution have allowed for the investigation of weaker features and more dilute biological samples. Refinement in sample preparations have allowed for high-concentration biological samples and site-of-interest-specific labeling [56]. The technique is continuing to mature, and it is solving questions where resonance Raman and IR spectroscopies cannot. The development of X-ray free electron lasers of very high flux and monochromation expands the possibilities of NRVS experiments to lower sample concentrations or coupled experiments involving pump lasers or electrochemistry.

References

[1] Mössbauer, R. Kernresonanzfluoreszenz von gammastrahlung in Ir191. Zeitschrift für Physik 1958, 151(2), 124–143.

[2] Mössbauer, R. Kernresonanzabsorption von gammastrahlung in Ir191. Naturwissenschaften 1958, 45(22), 538–539.

[3] Mössbauer Rudolf, L. Kernresonanzabsorption von γ-Strahlung in Ir191. In Zeitschrift für Naturforschung A 1959, 14(p), 211.

[4] Visscher, W. M. Study of lattice vibrations by resonance absorption of nuclear gamma rays. Annals of Physics 1960, 9(2), 194–210.

[5] Singwi, K. S., and Sjölander, A. Diffusive motions in water and cold neutron scattering. Physical Review 1960, 119(3), 863–871.

[6] Sturhahn, W. Nuclear resonant spectroscopy. J Phys Cond Matt 2004, 16, S497–S530.

[7] Lipkin, H. J. Mössbauer sum rules for use with synchrotron sources. Phys Rev B 1995, 52(14), 10073–10079.

[8] Weiss, H., and Langhoff, H. Observation of one phonon transitions in terbium by nuclear resonance fluorescence. Physics Letters A 1979, 69(6), 448–450.

[9] Weiss, H., and Langhoff, H. Observation of localized modes in TbO x using the Mößbauer effect. Zeitschrift für Physik B Condensed Matter 1979, 33(4), 365–368.

[10] Guo, Y. S., Wang, H. X., Xiao, Y. M., Vogt, S., Thauer, R. K., Shima, S., Volkers, P. I., Rauchfuss, T. B., Pelmenschikov, V., Case, D. A., Alp, E. E., Sturhahn, W., Yoda, Y., and Cramer, S. P. Characterization of the Fe site in iron-sulfur cluster-free hydrogenase (Hmd) and of a model compound via nuclear resonance vibrational spectroscopy (NRVS). Inorganic Chemistry 2008, 47(10), 3969–3977.

[11] Mössbauer, R. Kernresonanzabsorption von γ-strahlung in Ir191. In Zeitschrift für Naturforschung A 1959, 14(p), 211.

[12] Singwi, K. S., and Sjölander, A. Resonance absorption of nuclear gamma rays and the dynamics of atomic motions. Physical Review 1960, 120(4), 1093–1102.

[13] RUBY, S., L. Mössbauer experiments without conventional sources. J Phys Colloques 1974, 35 (C6), C6-209-C6-211.

[14] Gerdau, E., Ruffer, R., Winkler, H., Tolksdorf, W., Klages, C. P., and Hannon, J. P. Nuclear bragg-diffraction of synchrotron radiation in yttrium iron-garnet. Physical Review Letters 1985, 54(8), 835–838.

[15] Seto, M., Yoda, Y., Kikuta, S., Zhang, X., and Ando, M. Observation of nuclear resonant scattering accompanied by phonon excitation using synchrotron radiation. Physical Review Letters 1995, 74(19), 3828–3831.

[16] Sturhahn, W., Toellner, T., Alp, E., Zhang, X., Ando, M., Yoda, Y., Kikuta, S., Seto, M., Kimball, C., and Dabrowski, B. Phonon density of states measured by inelastic nuclear resonant scattering. Physical Review Letters 1995, 74(19), 3832–3835.

[17] Chumakov, A. I., Rüffer, R., Grünsteudel, H., Grünsteudel, H. F., Grübel, G., Metge, J., Leupold, O., and Goodwin, H. A. Energy dependence of nuclear recoil measured with incoherent nuclear scattering of synchrotron radiation. EPL (Europhysics Letters) 1995, 30(7), 427.

[18] Achterhold, K., Keppler, C., van Bürck, U., Potzel, W., Schindelmann, P., Knapp, E.-W., Melchers, B., Chumakov, A. I., Baron, A. Q. R., Rüffer, R., and Parak, F. Temperature dependent inelastic X-ray scattering of synchrotron radiation on myoglobin analyzed by the Mössbauer effect. European Biophysics Journal 1996, 25(1), 43–46.

[19] Keppler, C., Achterhold, K., Ostermann, A., van Bürck, U., Potzel, W., Chumakov, A. I., Baron, A. Q. R., Rüffer, R., and Parak, F. Determination of the phonon spectrum of iron in myoglobin using inelastic X-ray scattering of synchrotron radiation. European Biophysics Journal 1997, 25(3), 221–224.

[20] Kamali, S., Wang, H., Mitra, D., Ogata, H., Lubitz, W., Manor, B. C., Rauchfuss, T. B., Byrne, D., Bonnefoy, V., Jenney, F. E. Jr., Adams, M. W., Yoda, Y., Alp, E., Zhao, J., and Cramer, S. P. Observation of the Fe-CN and Fe-CO vibrations in the active site of [NiFe] hydrogenase by nuclear resonance vibrational spectroscopy. Angew Chem Int Ed Engl 2013, 52(2), 724–8.

[21] Ogata, H., Kramer, T., Wang, H., Schilter, D., Pelmenschikov, V., van Gastel, M., Neese, F., Rauchfuss, T. B., Gee, L. B., Scott, A. D., Yoda, Y., Tanaka, Y., Lubitz, W., and Cramer, S. P. Hydride bridge in [NiFe]-hydrogenase observed by nuclear resonance vibrational spectroscopy. Nat Commun 2015, 6, 7890.

[22] Lauterbach, L., Wang, H., Horch, M., Gee, L. B., Yoda, Y., Tanaka, Y., Zebger, I., Lenz, O., and Cramer, S. P. Nuclear resonance vibrational spectroscopy reveals the FeS cluster composition and active site vibrational properties of an O2-tolerant NAD(+)-reducing [NiFe] hydrogenase. Chem Sci 2015, 6(2), 1055–1060.

[23] Kuchenreuther, J. M., Guo, Y. S., Wang, H. X., Myers, W. K., George, S. J., Boyke, C. A., Yoda, Y., Alp, E. E., Zhao, J. Y., Britt, R. D., Swartz, J. R., and Cramer, S. P. Nuclear resonance vibrational spectroscopy and electron paramagnetic resonance spectroscopy of Fe-57-enriched [FeFe] hydrogenase indicate stepwise assembly of the H-cluster. Biochemistry 2013, 52(5), 818–826.

[24] Gilbert-Wilson, R., Siebel, J. F., Adamska-Venkatesh, A., Pham, C. C., Reijerse, E., Wang, H., Cramer, S. P., Lubitz, W., and Rauchfuss, T. B. Spectroscopic investigations of [FeFe] hydrogenase maturated with [57Fe2(adt)(CN)2(CO)4]2−. Journal of the American Chemical Society 2015, 137(28), 8998–9005.

[25] Mitra, D., George, S. J., Guo, Y. S., Kamali, S., Keable, S., Peters, J. W., Pelmenschikov, V., Case, D. A., and Cramer, S. P. Characterization of [4Fe-4S] cluster vibrations and structure in nitrogenase Fe protein at three oxidation levels via combined NRVS, EXAFS, and DFT analyses. Journal of the American Chemical Society 2013, 135(7), 2530–2543.

[26] George, S. J., Barney, B. M., Mitra, D., Igarashi, R. Y., Guo, Y. S., Dean, D. R., Cramer, S. P., and Seefeldt, L. C. EXAFS and NRVS reveal a conformational distortion of the FeMo-cofactor in the MoFe nitrogenase propargyl alcohol complex. Journal of Inorganic Biochemistry 2012, 112, 85–92.

[27] Scott, A., Pelmenschikov, V., Guo, Y., Wang, H., Yan, L., George, S., Dapper, C., Newton, W., Yoda, Y., Tanaka, Y., and Cramer, S. P. J Am Chem Soc 2014, 136, 15942.

[28] Xiao, Y., Smith, M. C., Newton, W., Case, D. A., George, S., Wang, H., Sturhahn, W., Alp, E., Zhao, J., Yoda, Y., and Cramer, S. How nitrogenase shakes – initial information about P-cluster and FeMo-cofactor normal modes from nuclear resonance vibrational spectroscopy (NRVS). Journal of the American Chemical Society 2006, 128(23), 7608–7612.

[29] Leu, B. M., Ching, T. H., Zhao, J. Y., Sturhahn, W., Alp, E. E., and Sage, J. T. Vibrational dynamics of iron in cytochrome c. Journal of Physical Chemistry B 2009, 113(7), 2193–2200.

[30] Sage, J., Durbin, S., Sturhahn, W., Wharton, D., Champion, P., Hession, P., Sutter, J., and Alp, E. Long-range reactive dynamics in myoglobin. Physical Review Letters 2001, 86(21), 4966–4969.

[31] Wong, S. D., Srnec, M., Matthews, M. L., Liu, L. V., Kwak, Y., Park, K., Bell, C. B., 3rd., Alp, E. E., Zhao, J., Yoda, Y., Kitao, S., Seto, M., Krebs, C., Bollinger, J. M. Jr., and Solomon, E. I. Elucidation of the Fe(IV)=O intermediate in the catalytic cycle of the halogenase SyrB2. Nature 2013, 499(7458), 320–3.
[32] Sutherlin, K. D., Liu, L. V., Lee, Y.-M., Kwak, Y., Yoda, Y., Saito, M., Kurokuzu, M., Kobayashi, Y., Seto, M., Que, L., Nam, W., and Solomon, E. I. Nuclear resonance vibrational spectroscopic definition of peroxy intermediates in nonheme iron sites. Journal of the American Chemical Society 2016, 138(43), 14294–14302.
[33] Serrano, P. N., Wang, H., Crack, J. C., Prior, C., Hutchings, M. I., Thomson, A. J., Kamali, S., Yoda, Y., Zhao, J., Hu, M. Y., Alp, E. E., Oganesyan, V. S., Le Brun, N. E., and Cramer, S. P. Nitrosylation of nitric-oxide-sensing regulatory proteins containing [4Fe-4S] clusters gives rise to multiple iron–nitrosyl complexes. Angewandte Chemie International Edition 2016, 55(47), 14575–14579.
[34] Sturhahn, W., CONUSS and PHOENIX: evaluation of nuclear resonant scattering data. 2000, 125, 149–172.
[35] Brown, D. E., Toellner, T. S., Sturhahn, W., Alp, E. E., Hu, M., Kruk, R., Rogacki, K., and Canfield, P. C. Partial phonon density of states of dysprosium and its compounds measured using inelastic nuclear resonance scattering. Hyperfine Interactions 2004, 153(1), 17–24.
[36] Ishikawa, D., Baron, A. Q. R., and Ishikawa, T. Nuclear resonant scattering from the subnanosecond lifetime excited state of ^{201}Hg. Phys Rev B 2005, 72, 140301(R.
[37] Barla, A., Rüffer, R., Chumakov, A. I., Metge, J., Plessel, J., and Abd-Elmeguid, M. M. Direct determination of the phonon density of states in beta-Sn. Phys Rev B 2000, 61(22), R14881–R14884.
[38] Giefers, H., Tanis, E. A., Rudin, S. P., Greeff, C., Ke, X., Chen, C., Nicol, M. F., Pravica, M., Pravica, W., Zhao, J., Alatas, A., Lerche, M., Sturhahn, W., and Alp, E. Phonon density of states of metallic Sn at high pressure. Phys Rev Lett 2007, 98, 245502.
[39] Toellner, T. S., Hu, M. Y., Sturhahn, W., Quast, K., and Alp, E. E. Inelastic nuclear resonant scattering with sub-meV energy resolution. Applied Physics Letters 1997, 71(15), 2112–2114.
[40] Sage, J. T., Paxson, C., Wyllie, G. R. A., Sturhahn, W., Durbin, S. M., Champion, P. M., Alp, E. E., and Scheidt, W. R. Nuclear resonance vibrational spectroscopy of a protein active-site mimic. J Phys Condens Matter 2001, 13, 7707–7722.
[41] Leu, B. M., Zgierski, M. Z., Wyllie, G. R. A., Ellison, Mary K., Scheidt, W. R., Sturhahn, W., Alp, E. E., Durbin, S. M., and Sage, J. T. Vibrational dynamics of biological molecules: multi-quantum contributions. J Phys Chem Solids 2005, 66, 2250–2256.
[42] Pavlik, J. W., Barabanschikov, A., Oliver, A. G., Alp, E. E., Sturhahn, W., Zhao, J., Sage, J. T., and Scheidt, W. R., Probing vibrational anisotropy with nuclear resonance vibrational spectroscopy. 2010, 49, 4400 –4404.
[43] Hu, M. Y., Sturhahn, W., Toellner, T. S., Mannheim, P. D., Brown, D. E., Zhao, J. Y., and Alp, E. E. Measuring velocity of sound with nuclear resonant inelastic x-ray scattering. Phys Rev B 2003, 67(9), 094304.
[44] Lipkin, H. J. Mössbauer sum rules for use with synchrotron sources. Hyperfine Interactions 1999, 123(1), 349–366.
[45] Toellner, T. S. Monochromatization of synchrotron radiation for nuclear resonant scattering experiments. Hyperfine Interactions 2000, 125(1), 3–28.
[46] The European synchrotron facility. http://www.esrf.eu/home/UsersAndScience/Accelerators/parameters.html.
[47] ESRF – high resolution monochromator. http://www.esrf.eu/home/UsersAndScience/Experiments/MEx/ID18/beamline_layout/optics/HRM.html (accessed 11/6/2017).

[48] Insertion devices by sector | advanced photon source. https://www1.aps.anl.gov/Magnetic-Devices/Insertion-Devices/Insertion-Devices-by-Sector.
[49] Beamlines information | advanced photon source. https://www1.aps.anl.gov/Beamlines/Directory/Details?beamline_id=6.
[50] SPring-8 – Introduction to insertion devices. http://www.spring8.or.jp/en/about_us/whats_sp8/facilities/bl/light_source_optics/sources/insertion_device/intro_insertion_devices (accessed 11/6/2017).
[51] Yoshitaka, Y., Kyoko, O., Hongxin, W., Stephen, P. C., and Makoto, S. High-resolution monochromator for iron nuclear resonance vibrational spectroscopy of biological samples. Japanese Journal of Applied Physics 2016, 55(12), 122401.
[52] Wang, H., Yoda, Y., Kamali, S., Zhou, Z. H., and Cramer, S. P. Real sample temperature: a critical issue in the experiments of nuclear resonant vibrational spectroscopy on biological samples. Journal of synchrotron radiation 2012, 19(Pt 2), 257–63.
[53] Wang, H., Yoda, Y., Ogata, H., Tanaka, Y., and Lubitz, W. A strenuous experimental journey searching for spectroscopic evidence of a bridging nickel-iron-hydride in [NiFe] hydrogenase. Journal of Synchrotron Radiation 2015, 22(6), 1334–1344.
[54] PetraIII – P01 – unified data sheet. http://photon-science.desy.de/facilities/petra_iii/beamlines/p01_dynamics/unified_data_sheet_p01/index_eng.html (accessed 11/6/2017).
[55] PetraIII – P01 – high resolution monochromators. http://photon-science.desy.de/facilities/petra_iii/beamlines/p01_dynamics/nuclear_resonant_scattering_station/high_resolution_monochromators/index_eng.html (accessed 11/6/2017).
[56] Reijerse, E. J., Pham, C. C., Pelmenschikov, V., Gilbert-Wilson, R., Adamska-Venkatesh, A., Siebel, J. F., Gee, L. B., Yoda, Y., Tamasaku, K., Lubitz, W., Rauchfuss, T. B., and Cramer, S. P. Direct observation of an iron-bound terminal hydride in [FeFe]-hydrogenase by nuclear resonance vibrational spectroscopy. J Am Chem Soc 2017, 139(12), 4306–4309.
[57] Kohn, V. G., and Chumakov, A. I. DOS: Evaluation of phonon density of states from nuclear resonant inelastic absorption. Hyperfine Interactions 2000, 125(1–4), 205–221.
[58] Sturhahn, W. CONUSS and PHOENIX: evaluation of nuclear resonant scattering data. Hyp Interac 2000, 125, 149–172.
[59] Johnson, D. W., and Spence, J. C. H. Determination of the single-scattering probability distribution from plural-scattering data. Journal of Physics D Applied 1974, 771.
[60] Zhao, J. Y., and Sturhahn, W. High-energy-resolution X-ray monochromator calibration using the detailed-balance principle. Journal of Synchrotron Radiation 2012, 19, 602–608.
[61] Dowty, E. Fully automated microcomputer calculation of vibrational spectra. Physics and Chemistry of Minerals 1987, 14(1), 67–79.
[62] Petrenko, T., Sturhahn, W., and Neese, F. First-principles calculation of nuclear resonance vibrational spectra. Hyperfine Interactions 2008, 175(1–3), 165–174.
[63] Do, L. H., Wang, H., Tinberg, C. E., Dowty, E., Yoda, Y., Cramer, S. P., and Lippard, S. J. Characterization of a synthetic peroxodiiron(III) protein model complex by nuclear resonance vibrational spectroscopy. Chemical communications (Cambridge, England) 2011, 47(39), 10945–7.
[64] Do, L. H., Hayashi, T., Moënne-Loccoz, P., and Lippard, S. J. Carboxylate as the protonation site in (Peroxo)diiron(III) model complexes of soluble methane monooxygenase and related diiron proteins. Journal of the American Chemical Society 2010, 132(4), 1273–1275.
[65] Dong, Y., Menage, S., Brennan, B. A., Elgren, T. E., Jang, H. G., Pearce, L. L., and Que, L. Dioxygen binding to diferrous centers. models for diiron-oxo proteins. Journal of the American Chemical Society 1993, 115(5), 1851–1859.
[66] Park, K., Tsugawa, T., Furutachi, H., Kwak, Y., Liu, L. V., Wong, S. D., Yoda, Y., Kobayashi, Y., Saito, M., Kurokuzu, M., Seto, M., Suzuki, M., and Solomon, E. I. Nuclear resonance

vibrational spectroscopy and DFT study of peroxo-bridged biferric complexes: structural insight into peroxo intermediates of binuclear non-heme iron enzymes. Angew Chem Int Ed Engl 2013, 52(4), 1294–8.

[67] Park, K., Li, N., Kwak, Y., Srnec, M., Bell, C. B., Liu, L. V., Wong, S. D., Yoda, Y., Kitao, S., Seto, M., Hu, M., Zhao, J., Krebs, C., Bollinger, J. M., and Solomon, E. I. Peroxide activation for electrophilic reactivity by the binuclear non-heme iron enzyme aurF. Journal of the American Chemical Society 2017, 139(20), 7062–7070.

[68] Xiao, Y., Wang, H., George, S. J., Smith, M. C., Adams, M. W., Jenney, F. E. Jr., Sturhahn, W., Alp, E. E., Zhao, J., Yoda, Y., Dey, A., Solomon, E. I., and Cramer, S. P. Normal mode analysis of Pyrococcus furiosus rubredoxin via nuclear resonance vibrational spectroscopy (NRVS) and resonance Raman spectroscopy. J Am Chem Soc 2005, 127(42), 14596–606.

[69] Day, M. W., Hsu, B. T., Joshua-Tor, L., Park, J. B., Zhou, Z. H., Adams, M. W., and Rees, D. C. X-ray crystal structures of the oxidized and reduced forms of the rubredoxin from the marine hyperthermophilic archaebacterium Pyrococcus furiosus. Protein Sci 1992, 1(11), 1494–507.

[70] Bau, R., Rees, D. C., Kurtz Jr., D. M., Scott, R. A., Huang, H., Adams, M. W. W., and Eidsness, M. K. Crystal structure of rubredoxin from Pyrococcus furiosus at 0.95 Å resolution, and the structures of N-terminal methionine and formylmethionine variants of Pf Rd. contributions of N-terminal interactions to thermostability. JBIC Journal of Biological Inorganic Chemistry 1998, 3(5), 484–493.

[71] Guo, Y., Brecht, E., Aznavour, K., Nix, J. C., Xiao, Y., Wang, H., George, S. J., Bau, R., Keable, S., Peters, J. W., Adams, M. W. W., Jenney, F. E., Sturhahn, W., Alp, E. E., Zhao, J., Yoda, Y., and Cramer, S. P. Nuclear resonance vibrational spectroscopy (NRVS) of rubredoxin and MoFe protein crystals. Hyperfine Interactions 2012, 222(S2), 77–90.

[72] Xiao, Y., Tan, M. L., Ichiye, T., Wang, H., Guo, Y., Smith, M. C., Meyer, J., Sturhahn, W., Alp, E. E., Zhao, J., Yoda, Y., and Cramer, S. P. Dynamics of rhodobacter capsulatus [2FE-2S] ferredoxin VI and aquifex aeolicus ferredoxin 5 via nuclear resonance vibrational spectroscopy (NRVS) and resonance Raman spectroscopy. Biochemistry 2008, 47(25), 6612–27.

[73] Fu, W., Drozdzewski, P. M., Davies, M. D., Sligar, S. G., and Johnson, M. K. Resonance Raman and magnetic circular dichroism studies of reduced [2Fe-2S] proteins. J Biol Chem 1992, 267 (22), 15502–10.

[74] Brooks, B. R., Brooks, C. L. 3rd, Mackerell, A. D. Jr., Nilsson, L., Petrella, R. J., Roux, B., Won, Y., Archontis, G., Bartels, C., Boresch, S., Caflisch, A., Caves, L., Cui, Q., Dinner, A. R., Feig, M., Fischer, S., Gao, J., Hodoscek, M., Im, W., Kuczera, K., Lazaridis, T., Ma, J., Ovchinnikov, V., Paci, E., Pastor, R. W., Post, C. B., Pu, J. Z., Schaefer, M., Tidor, B., Venable, R. M., Woodcock, H. L., Wu, X., Yang, W., York, D. M., and Karplus, M. CHARMM: the biomolecular simulation program. J Comput Chem 2009, 30(10), 1545–614.

[75] Gee, L. B., Leontyev, I., Stuchebrukhov, A., Scott, A. D., Pelmenschikov, V., and Cramer, S. P. Docking and migration of carbon monoxide in nitrogenase: the case for gated pockets from infrared spectroscopy and molecular dynamics. Biochemistry 2015, 54(21), 3314–9.

[76] Mitra, D., Pelmenschikov, V., Guo, Y., Case, D. A., Wang, H., Dong, W., Tan, M.-L., Ichiye, T., Jenney, F. E., Adams, M. W. W., Yoda, Y., Zhao, J., and Cramer, S. P. Dynamics of the [4Fe-4S] cluster in Pyrococcus furiosus D14C ferredoxin via nuclear resonance vibrational and resonance Raman spectroscopies, force field simulations, and density functional theory calculations. Biochemistry 2011, 50(23), 5220–35.

[77] Lovgreen, M. N., Martic, M., Windahl, M. S., Christensen, H. E., and Harris, P. Crystal structures of the all-cysteinyl-coordinated D14C variant of Pyrococcus furiosus ferredoxin: [4Fe-4S] <–> [3Fe-4S] cluster conversion. J Biol Inorg Chem 2011, 16(5), 763–75.

[78] Czernuszewicz, R. S., Kilpatrick, L. K., Koch, S. A., and Spiro, T. G. Resonance Raman spectroscopy of Iron(III) tetrathiolate complexes: implications for the conformation and force field of rubredoxin. Journal of the American Chemical Society 1994, 116(16), 7134–7141.
[79] Marcus, R. A. On the theory of oxidation—reduction reactions involving electron transfer. V. Comparison and properties of electrochemical and chemical rate constants1. The Journal of Physical Chemistry 1963, 67(4), 853–857.
[80] Meyer, J. Iron–sulfur protein folds, iron–sulfur chemistry, and evolution. JBIC Journal of Biological Inorganic Chemistry 2008, 13(2), 157–170.
[81] Pochapsky, T. C., Kostic, M., Jain, N., and Pejchal, R. Redox-dependent conformational selection in a Cys4Fe2S2 ferredoxin. Biochemistry 2001, 40(19), 5602–5614.
[82] Zeng, W., Barabanschikov, A., Zhang, Y., Zhao, J., Sturhahn, W., Alp, E. E., and Sage, J. T. Journal of the American Chemical Society, 2008, 130, 1816–1817.
[83] Zeng, W., Silvernail, N. J., Wharton, D. C., Georgiev, G. Y., Leu, B. M., Scheidt, W. R., Zhao, J., Sturhahn, W., Alp, J. T., and Sage, E. E. Journal of the American Chemical Society, 2005, 127, 11200–11201.
[84] Zeng, W. Q., Barabanschikov, A., Wang, N. Y., Lu, Y., Zhao, J. Y., Sturhahn, W., Alp, E. E., and Sage, J. T., Chemical Communications, 2012, 48, 6340–6342.
[85] Ohta, T., Shibata, T., Kobayashi, Y., Yoda, Y., Ogura, T., Neya, S., Suzuki, A., Seto, M., and Yamamoto, Y., Biochemistry, 2018, 57, 6649–6652.
[86] Sage, J. T., Durbin, S. M., Sturhahn, W., Wharton, D. C., Champion, P. M., Hession, P., Sutter, J., and Alp, E. E., Physical Review Letters, 2001, 86, 4966–4969.
[87] Ogata, H., Kramer, T., Wang, H., Schilter, D., Pelmenschikov, V., van Gastel, M., Neese, F., Rauchfuss, T. B., Gee, L. B., Scott, A. D., Yoda, Y., Tanaka, Y., Lubitz, W., and Cramer, S. P., Nat Commun, 2015, 6, 7890.
[88] Kamali, S., Wang, H., Mitra, D., Ogata, H., Lubitz, W., Manor, B. C., Rauchfuss, T. B., Byrne, D., Bonnefoy, V., Jenney, F. E. Jr., Adams, M. W., Yoda, Y., Alp, E., Zhao, J., and Cramer, S. P., Angew Chem Int Ed Engl, 2013, 52, 724–728.
[89] Lauterbach, L., Wang, H., Horch, M., Gee, L. B., Yoda, Y., Tanaka, Y., Zebger, I., Lenz, O., and Cramer, S. P., Chem Sci, 2015, 6, 1055–1060.
[90] Mitra, D., Pelmenschikov, V., Guo, Y., Case, D. A., Wang, H., Dong, W., Tan, M.-L., Ichiye, T., Jenney, F. E., Adams, M. W. W., Yoda, Y., Zhao, J., and Cramer, S. P., Biochemistry, 2011, 50, 5220–5235.
[91] Reijerse, E. J., Pham, C. C., Pelmenschikov, V., Gilbert-Wilson, R., Adamska-Venkatesh, A., Siebel, J. F., Gee, L. B., Yoda, Y., Tamasaku, K., Lubitz, W., Rauchfuss, T. B., and Cramer, S. P., J Am Chem Soc, 2017, 139, 4306–4309.
[92] Pelmenschikov, V., Birrell, J. A., Pham, C. C., Mishra, N., Wang, H., Sommer, C., Reijerse, E., Richers, C. P., Tamasaku, K., Yoda, Y., Rauchfuss, T. B., Lubitz, W., and Cramer, S. P., J Am Chem Soc, 2017, 139, 16894–16902.
[93] Pham, C. C., Mulder, D. W., Pelmenschikov, V., King, P. W., Ratzloff, M. W., Wang, H., Mishra, N., Alp, E. E., Zhao, J., Hu, M. Y., Tamasaku, K., Yoda, Y., and Cramer, S. P., Angew Chem Int Ed Engl, 2018, 57, 10605–10609.
[94] Mebs, S., Kositzki, R., Duan, J., Kertess, L., Senger, M., Wittkamp, F., Apfel, U.-P., Happe, T., Stripp, S. T., Winkler, M., and Haumann, M., Biochimica et Biophysica Acta (BBA) - Bioenergetics, 2018, 1859, 28–41.
[95] Mitra, D., George, S. J., Guo, Y. S., Kamali, S., Keable, S., Peters, J. W., Pelmenschikov, V., Case, D. A., and Cramer, S. P., Journal of the American Chemical Society, 2013, 135, 2530–2543.

[96] Xiao, Y., Wang, H., George, S. J., Smith, M. C., Adams, M. W., Jenney, F. E., Jr., Sturhahn, W., Alp, E. E., Zhao, J., Yoda, Y., Dey, A., Solomon, E. I., and Cramer, S. P., J Am Chem Soc, 2005, 127, 14596–14606.
[97] Scott, A. D., Pelmenschikov, V., Guo, Y., Yan, L., Wang, H., George, S. J., Dapper, C. H., Newton, W. E., Yoda, Y., Tanaka, Y., and Cramer, S. P., Journal of the American Chemical Society, 2014, 136, 15942–15954.
[98] Sutherlin, K. D., Rivard, B. S., Böttger, L. H., Liu, L. V., Rogers, M. S., Srnec, M., Park, K., Yoda, Y., Kitao, S., Kobayashi, Y., Saito, M., Seto, M., Hu, M., Zhao, J., Lipscomb, J. D., and Solomon, E. I., Journal of the American Chemical Society, 2018, 140, 5544–5559.
[99] Serrano, P. N., Wang, H., Crack, J. C., Prior, C., Hutchings, M. I., Thomson, A. J., Kamali, S., Yoda, Y., Zhao, J., Hu, M. Y., Alp, E. E., Oganesyan, V. S., Le Brun, N. E., and Cramer, S. P., Angewandte Chemie International Edition, 2016, 55, 14575–14579.
[100] Mebs, S., Srinivas, V., Kositzki, R., Griese, J. J., Högbom, M., and Haumann, M., Biochimica et Biophysica Acta (BBA) - Bioenergetics, 2019, 1860, 148060.
[101] Reijerse, E. J., Pelmenschikov, V., Gilbert-Wilson, R., Adamska-Venkatesh, A., Siebela, J.F., Gee, L. B., Yoda, Y., Tamasaku, K., Lubitz, W., Rauchfuss, T. B., and Cramer, S. P. Direct observation of an iron-bound terminal hydride in [FeFe]-hydrogenase by nuclear resonance vibrational spectroscopy. Submitted 2016.
[102] Rosenberg, E. Kinetic deuterium isotope effects in transition metal hydride clusters. Polyhedron 1989, 8(4), 383–405.
[103] Kaesz, H. D., and Saillant, R. B. Hydride complexes of the transition metals. Chemical Reviews 1972, 72(3), 231–281.

Maurice van Gastel

10 EPR spectroscopy

10.1 Introduction

Of all spectroscopies in the toolbox of a bio-organometallic chemist, electron paramagnetic resonance (EPR) spectroscopy arguably represents one of the most mystique-surrounded techniques. EPR spectra at first instance often look complicated; many times they are broad and without structure, and they are typically represented as first-derivative spectra. Upon asking the friendly EPR-expert as to why this is, one often obtains a short and correct but somewhat uninformative reply "for technical reasons." Moreover, asking what the wiggly line on the screen of the spectrometer's data acquisition system actually means for the system under investigation may lead to similar answers with varying degree of generality. The truth is that EPR spectra can indeed be very complicated to interpret, even for scientists with many years of expertise. EPR spectroscopy in that respect starkly contrasts with, for example, X-ray crystallography; nowadays crystallography analysis software allow for a relatively direct and, to the user, friendly and comfortable refinement of the diffraction pattern into a molecular structure. Such structure is in the vast majority of cases intuitive and easy to grasp: it provides us with an image of where the atoms are located in the crystal. Such image is oftentimes represented in the form of an ORTEP plot where the atoms are represented by small spheres.

Why there are no ORTEP plots for EPR and why are EPR spectra so a difficult to interpret? Well, says the teacher to the student, the answer is: we measure EPR spectra, because if we do manage to interpret the spectra, of which success is certainly not guaranteed, we exclusively obtain information about the redox-active electrons without background of all other electrons, and this information is unique to the EPR technique! Well, ok, the student replies, that sounds good, but then how does EPR work?

For the correct mindset of understanding the information hidden in the EPR spectrum, one needs to briefly flash back to undergraduate training in physical chemistry and remember that thinking of electrons in terms of little balls or point particles moving around the atoms is bad – very bad. Rather, quantum mechanics tells us that the electrons, say N in total, of the system are associated with an N-electron wave function and the absolute square of said wave function represents a probability distribution in $3N$ dimensional space. And electrons have spin, half-integer, associated with the spin quantum number $S = 1/2$; they are

Maurice van Gastel, Max Planck Institute for Chemical Energy Conversion, Germany

https://doi.org/10.1515/9783110496574-010

intrinsically magnetic particles. Electrons furthermore are fermions for which the Pauli principle holds, which generally states that none of the electrons are associated with the same set of quantum numbers. These considerations lead to the well-known Aufbau principle, which is probably the best place to start in this chapter's quest to understand the EPR spectrum and to lift the veil about what kind of information hides within the wiggly line of the EPR spectrum.

The Aufbau principle provides a representation of the *N*-electron wave function (within certain approximations, most importantly the complete neglect of electron correlation, whose treatment is beyond the scope of this chapter). An arrow up depicts and electron with "spin-up," $M_S = ½$ or α, an arrow down depicts a "spin-down" electron with quantum number $M_S = -½$ or β. Let us take the nitric oxide molecule, NO, as an example (Figure 10.1). NO has 11 valence electrons. When writing down the resonance structure, NO would feature a double bond and an unpaired electron at nitrogen, and the oxygen atom has two lone-pairs and nitrogen one lone-pair. Thus, according to the Aufbau principle, NO is a radical whose unpaired electron is located at nitrogen and all but one electron are paired and associated with either bonding orbitals or lone-pair orbitals! This is by far the most important realization one needs to make: all paired electrons associated with or "in" the doubly occupied orbitals do not give rise to magnetism, because the magnetic moments of the spin-up and spin-down electrons cancel! This means only the unpaired electron contributes to the EPR signal!

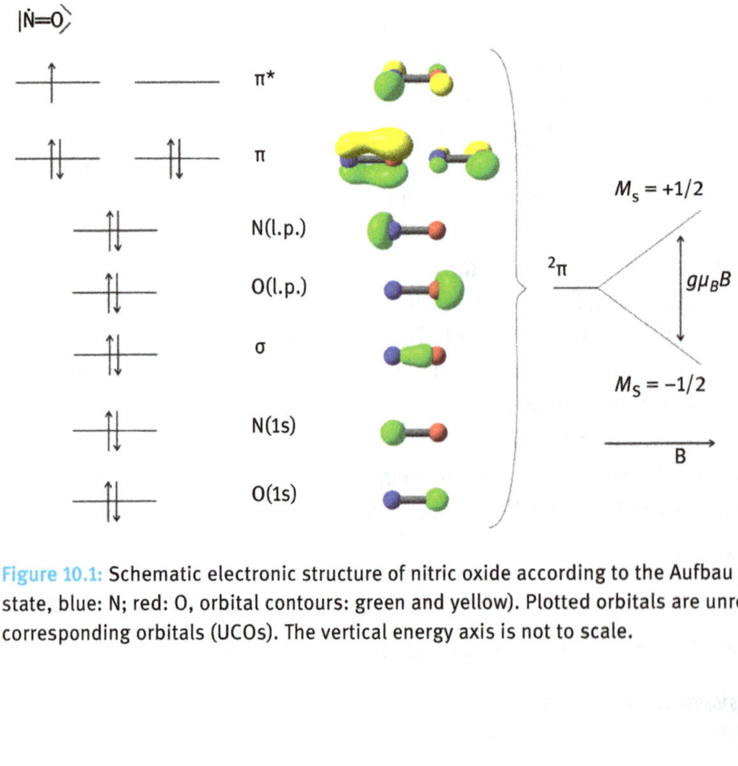

Figure 10.1: Schematic electronic structure of nitric oxide according to the Aufbau principle ($^2\Pi$ state, blue: N; red: O, orbital contours: green and yellow). Plotted orbitals are unrestricted corresponding orbitals (UCOs). The vertical energy axis is not to scale.

The unpaired electron is represented by a "dot" in the resonance structure. This electron is solely responsible for the (electron) magnetism. In Aufbau terms, the electronic structure is included in Figure 10.1(left). For the electronic structure, it is important to discriminate between orbitals and orbital energies, on the one hand, as depicted on the left and a state and the state ($^2\Pi$) energy, on the other hand, as depicted on the right: the orbitals as well as their contour plots tell us about the electron distribution. Each orbital can have an occupancy of 0, 1 or 2 and the entire configuration of orbitals and occupancies and is called a *state*; in the case of NO, the ground state would be called a "doublet-pi," $^2\Pi$ state, which, in the absence of a magnetic field, is doubly degenerate with respect to the M_S quantum number. Once a magnetic field is applied, the magnetic sublevels of this state $M_S = +\frac{1}{2}$ and $M_S = -\frac{1}{2}$ become nondegenerate. The splitting induced by the magnetic field is the well-known *Zeeman splitting*. It is proportional to the applied magnetic field, and additionally contains the Bohr magneton as a measure of the magnitude of the magnetic moment of the unpaired electron and it also contains the letter g, which is called the *g*-value. The *g*-value is one of the main experimental observables that contains chemical information of the molecule under investigation. We will return to the *g*-value and the information hidden within in the subsequent subsection, but for now suffice by stating that the *g*-value is a dimensionless proportionality constant that occurs in the Zeeman splitting, somewhat resembling a chemical shift in NMR spectroscopy.

Now the orbital structure on the left of Figure 10.1 can also be used to explain why it is so useful to be able to exclusively measure a signal related to the unpaired electron. Suppose NO is involved in redox chemistry when it becomes oxidized to NO^+, then the orbital from which it is easiest to remove an electron is the singly occupied orbital! This is because, of all occupied orbitals, the singly occupied orbital has the highest orbital energy and therefore of all electrons, the unpaired electron is bound weakest to the molecule. Likewise, if NO would become reduced to NO^-, it would be easiest to add the new electron to the singly occupied orbital (or in the particular case of NO, owing to the degeneracy, rather into the second, empty π^* orbital; according to Hund's rule this would lead to a configuration with two unpaired electrons and their spins aligned parallel, i.e., a triplet state, $S = 1$ state). One could therefore say in a somewhat overgeneralized way that if the molecule of interest displays one-electron redox chemistry, then EPR spectroscopy uniquely allows to obtain information about the orbital associated with the unpaired electron. This orbital is the "redox active" orbital; its density determines where the reactive parts of the molecule are. Information about whether redox chemistry is, for example, metal centered or ligand centered is in general extremely important to know; first for understanding how the redox reaction or catalysis may work, but also if one wants to optimize a catalyst, it is important to know which part of the catalyst actually does the chemistry.

Even with our NO example, careful comparison of the resonance structure in the upper left of Figure 10.1 with the orbital structure (which for the electrons is the equivalent of an ORTEP plot in X-ray crystallography) gives already some surprises.

In the resonance structure, one would draw the dot at nitrogen, a double bond between the atoms and two lone-pairs at O and one at N. This resonance structure is not quite in agreement with the orbital structure! Let us briefly go through this. Besides the two core 1s electron pairs, of all valence electrons, the pair associated with lowest orbital energy represents a doubly occupied σ bond. N and O then each have only one lone pair (of σ symmetry with respect to the NO bond), and not, as the resonance structure suggests, two in case of O! Both π orbitals are doubly occupied, which, together with the σ bond, would formally classify as a triple bond instead of a double bond. Lastly, the antibonding π* orbital would indeed somewhat weaken the triple bond again, but it is quite clear that the π* orbital is not located at N and rather evenly delocalized over N and O. This information about the singly occupied orbital is exactly what hides within the wiggly lines of an EPR spectrum! EPR spectroscopy is therefore an extremely useful technique to have available when one is interested in (bio)catalysis, though to get the information out of the spectrum will require analysis that may not be feasible on the fly during the actual measurement! In our example, one could thus perhaps in a slightly overgeneralized manner say that both N and O atoms in NO for redox chemistry purposes would be equally reactive.

How to extract information from an EPR spectrum will be detailed at a basic level in the subsequent sections. More advanced reading material can be found in the somewhat older but very compact and didactically written book of Carrington and McLachlan [1]. For detailed information about modern EPR methodology including pulsed EPR methodology, which we do not discuss here, the book of Schweiger and Jeschke [2] is recommended. The free software package "Easyspin" [3] (working under MatLab) is very useful for the purpose for simulating EPR spectra. For those interested in a deeper understanding of the fundamental interactions relevant for EPR on the level of Dirac theory, the monograph by Harriman [4] is recommended as reading material.

10.2 Paramagnetism: *g*-values

EPR contains the word "paramagnetic." Paramagnetism describes a system where the individual building blocks, for example, molecules in a solution, are magnetic, but the solution is sufficiently dilute that the intermolecular magnetic dipole–dipole interaction can be neglected. In this case, each molecule can be considered as a system in its own. When the concentration becomes too high, typically at concentrations above 10 mM, or in a solid, then the spins of the individual building blocks interact too strongly. This can lead to line-broadening and even to ferromagnetism in the case that all spins align and a macroscopic magnetic moment of the material is present, for example, in a steel magnet. While ferromagnetic resonance phenomena certainly

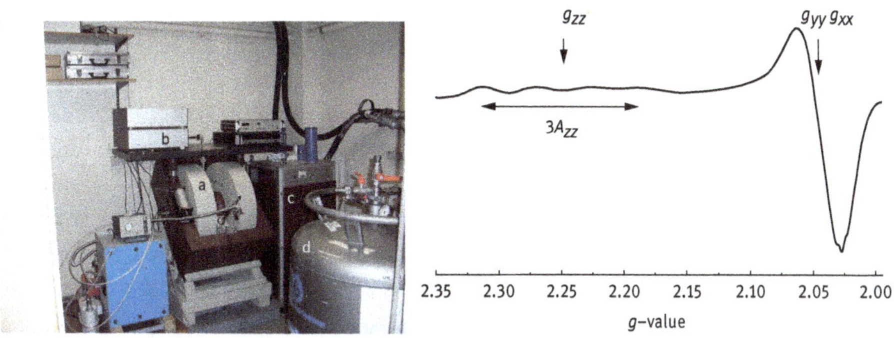

Figure 10.2: (left) Overview of an EPR spectrometer. (a) magnet; (b) microwave bridge; (c) power supply; (d) liquid helium vessel; the data acquisition system is not depicted. (right) EPR spectrum measured at a microwave frequency of 9 GHz of the blue copper site in azurin.

exist, it is not the topic of EPR spectroscopy, which is limited to paramagnetic systems. Typically, good concentrations for EPR spectroscopy amount to 1–10 mM.

In EPR spectroscopy, one needs a magnet for the purpose of generating the Zeeman splitting (cf. Figure 10.2). The magnetic sublevels are generally Boltzmann populated. At a typical magnetic field of about 0.34 T, the energy splitting amounts to 0.3 cm^{-1} or 9–10 GHz, which is so small that the low population difference (as well as fast relaxation times) at room temperature oftentimes do not allow detection of a signal. In order to have favorable Boltzmann population differences, EPR spectroscopy is typically performed at cryogenic temperatures. Moreover, in order to increase sensitivity, a resonator of fixed dimension is used, which is positioned at the center of the magnet. Next, microwaves are irradiated onto the cold sample positioned at the center of the resonator. The microwaves are generated by a source (e.g. a klystron or nowadays a Gunn diode) located in the microwave bridge. Because of the fixed dimension of the resonator (i.e. a technical reason!), the microwaves are irradiated with a fixed frequency corresponding to the Eigenfrequency of the sample-filled resonator, ν, and the magnetic field is swept. Only if the microwave energy, $h\nu$, equals the Zeeman splitting, microwaves can be absorbed by the paramagnetic system, leading to the resonance condition

$$h\nu = g\mu_B B \tag{10.1}$$

With the microwave frequency fixed, the Planck constant h and the Bohr magneton μ_B being constants, and the magnetic field being swept, the magnetic field at which a resonance occurs thus allows determination of the dimensionless g-value. Interpretation of a g-value is by no means an easy task. In fact, the system in many cases has to be described as comprised of three g-values! Before diving into the theory of g-values, the mindset for interpretation is somewhat related to that of a chemical shift in ^{1}H NMR spectroscopy: if all protons would feel the same chemical environment, the NMR spectrum would comprise of one and only one big signal at 0 ppm, which would not contain any chemical information. In NMR, the differences in chemical environment of each proton cause slight changes in the shielding and therefore the nuclear Zeeman frequency, leading to shifts in the parts-per-million (ppm) range. Similar considerations hold for the g-value: if the chemical environment is excluded, all unpaired electrons would be characterized by one-and-the-same g-value called the free-electron g-value, g_e, which numerically equals to 2.0023.

In the case of an organic radical like NO, the observed microwave absorption indeed occurs at a g-value very close but not equal to the free electron g-value. The deviation comes from the fact that the electron spin is not the only magnetic moment we have to consider. If the unpaired electron is associated with an orbital, in our case the π^* orbital in Figure 10.1, two additional interactions, that is, the orbital Zeeman interaction and spin–orbit coupling need to be considered! Both interactions involve the orbital momentum associated with the frontier orbitals (technical note: in molecules, these usually concern the first-order momenta $p_i|l_j|p_k$; the zero-order matrix elements $p_i|l_j|p_i$ are zero for all $i,j,k=x,y,z$). By second-order perturbation theory to include the orbital Zeeman interaction and spin–orbit coupling, it is possible to derive a formula [1,5], not for the g-value, but for a 3 × 3 g-tensor (or more accurately a g matrix), which can be expressed as a so-called spin Hamiltonian as follows:

$$H_{\text{Zeeman}} = g_e \begin{pmatrix} 1 & 0 & 0 \\ 0 & 1 & 0 \\ 0 & 0 & 1 \end{pmatrix} \mu_B \vec{B} \cdot \vec{s} - 2 \sum_{n \neq 0} \frac{\psi_0 | \sum_A \zeta_A \vec{l}_A \cdot \vec{s} | \psi_n \psi_n | \mu_B \vec{B} \cdot \vec{l} | \psi_0}{E_n - E_0} \tag{10.2}$$

A plethora of notes is in order with respect to this formula: (1) the denominator in the second term contains the energy difference of the nth excited state (not orbital energy differences!) and the ground state; (2) for systems with an orbitally degenerate ground state, a derivation in terms of a spin Hamiltonian formalism may not be possible; (3) the matrix elements in the summation have already been reduced to one-electron orbitals ψ_0 and ψ_n that concern those associated with the unpaired electron in the ground state and the nth excited state, respectively; we neglect electron correlation at this point so that an "Aufbau" description in terms of a single Slater determinant and uncorrelated orbitals applies; (4) we only consider excited states with the same number of unpaired electrons as the ground state (i.e. 1 unpaired electron for NO); (5) the summation over A runs over all nuclei in the molecule; the relevant orbital angular momentum $\vec{l}_A$ has its origin at the nucleus A and ζ_A is the spin–orbit coupling constant [1] of the Ath nucleus; (6) the origin of the angular momentum $\vec{l}$ in the orbital Zeeman part refers to a global origin of the coordinate system that we choose; (7) higher-order corrections to the *g* tensor are sometimes necessary for near-degenerate systems.

So, with eq. (10.2), things have become quite complicated, or "technical," in a hurry. The second part of eq. (10.2) may indeed be quite mystifying, but let us look at the formula more closely. The reason why this term, a cross term between spin–orbit coupling and the orbital Zeeman interaction, is there in second-order perturbation theory can actually be easily seen. Careful inspection of the second term reveals that it does have a linear dependence of the spin $\vec{s}$ of the unpaired electron and the magnetic field $\vec{B}$, that is, for all intents and purposes, it behaves like a Zeeman interaction, just like the first term in eq. (10.2)! The expression for the *g*-shifts (there are generally three *g*-shifts), that is, the deviation from g_e becomes

$$\Delta\vec{g} = -2\sum_{n\neq 0}\frac{\psi_0|\sum_A \zeta_A \vec{l}_A|\psi_n\psi_n|\vec{l}|\psi_0}{E_n - E_0} \tag{10.3}$$

In practice, the summation over excited states can be truncated; excited states with sufficiently large energy differences $E_n - E_0$ will not contribute to the *g*-shift. As such, the three *g*-shifts corresponding to the eigenvalues of the matrix in eq. (10.3) contain information about the orbital angular momenta associated with the orbitals of the unpaired electron in the ground state and in the low-lying excited states. The *g*-shifts are determined by and therefore contain information about the orbital structure itself.

For organic, C-, N- or O-centered radicals, the spin–orbit coupling constant [1] for these atoms is typically much smaller than the energy difference with already the first excited state, so that $\zeta/\Delta E$ is small. Organic radicals usually have very small *g*-shifts of a few parts per thousand.

We now proceed with an example of how to qualitatively use the formula in practice, using the EPR spectrum of the blue copper site in *P. aeruginosa* azurin. As is seen in Figure 10.2, the spectrum, depicted in first derivative mode has its magnetic field *x*-axis already converted to a *g*-value axis in order to allow a direct read of the *g*-values from the spectrum. Formally it is better to simulate the spectrum, for example, with Easyspin [3], or in the case of azurin, to perform EPR spectroscopy on protein crystals [6], since the latter allows accurate determination of the *g*-values. In this case, they amount to 2.039, 2.056 and about 2.25. The latter *g*-value differs considerably from the free electron *g*-value. The spin–orbit coupling parameter at copper amounts to $-829\ cm^{-1}$ (the value itself is actually positive [1]; however, for more than half-filled shells, as is the case for the $3d^9$ Cu^{2+}, it typically suffices to take the negative value, thus representing the hole in the 3d shell; a formally correct derivation would go too far into theory and the interested reader is referred to the afore mentioned books), whereas that of the next heaviest element, sulfur, amounts to $-191\ cm^{-1}$. As such, a measured large *g*-shift (>0.05) generally means that the singly occupied orbital is largely metal centered; small *g*-shifts mean that the unpaired electron is located on light atoms, typical for an organic radical. Organic radicals can have large g-shifts, but only when they would have to have a nearly degenerate ground state so that the energy difference in the denominator of eq. (10.3) becomes very small. For our purposes, it is thus reasonable to restrict the summation over nuclei (A) to the copper nucleus. We obtain

$$\Delta\vec{g} = 2\sum_{n\neq 0} \frac{829}{E_n - E_0} \psi_0|\vec{l}_{Cu}|\psi_n\psi_n|\vec{l}_{Cu}|\psi_0 \tag{10.4}$$

As a next step, we have to make an educated guess about the orbital in the ground state and the excited states. Suppose that the unpaired electron is associated with a singly occupied $d_{x^2-y^2}$ orbital with coefficient $c_{x^2-y^2}$ at copper. The excited states relevant for the g-shifts are then those where the unpaired electron is associated with the d_{xy}, d_{xz} and d_{yz} orbitals. The reason for considering these is because the first-order matrix elements $d_{x^2-y^2}|l_z|d_{xy} = -2i$, $d_{x^2-y^2}|l_y|d_{xz} = i$ and $d_{x^2-y^2}|l_x|d_{yz} = i$ [7]. The other matrix elements, $d_{x^2-y^2}|\vec{l}|d_{x^2-y^2}$ and $d_{x^2-y^2}|\vec{l}|d_{z^2}$, are zero. Insertion into eq. (10.3) gives

$$\begin{aligned}
\Delta g_{xx} &= c^2_{x^2-y^2} c^2_{yz} \frac{2\cdot 829}{E_{yz} - E_0} \\
\Delta g_{yy} &= c^2_{x^2-y^2} c^2_{xz} \frac{2\cdot 829}{E_{xz} - E_0} \\
\Delta g_{zz} &= c^2_{x^2-y^2} c^2_{xy} \frac{8\cdot 829}{E_{xy} - E_0}
\end{aligned} \tag{10.5}$$

As is seen, the expressions still formally contain many parameters (three energy differences and four wave function coefficients giving seven unknowns). Formally, this is it already; the formulae indicate that the three g-shifts are sensitive to covalency by the coefficients c and to the ligand field splitting by the energy differences. The formulae also show that based on the g-values alone, it is not possible, even if we neglect all atoms but the copper, to come to a unique determination of these seven parameters. This is why it is generally so difficult to analyze g-values and why a question like "what does this set of g-values tell me about my system?" during measurement is so difficult to answer. In many cases, the help of quantum chemical calculations has to be invoked, or, if possible, additional information about, for example, the ligand field splitting may also be obtained experimentally from UV/VIS or MCD measurements.

So is there then nothing solid that can be concluded from the g-shifts alone? Well, there is information that can be retrieved immediately. Suppose that the unpaired electron would have been associated with a d_{z^2} orbital, then evaluation of the orbital angular momentum matrix elements would have given the following expressions for the g-shift:

$$\Delta g_{xx} = c_{z^2}^2 c_{yz}^2 \frac{6 \cdot 829}{E_{yz} - E_0}$$

$$\Delta g_{yy} = c_{z^2}^2 c_{xz}^2 \frac{6 \cdot 829}{E_{xz} - E_0} \tag{10.6}$$

$$\Delta g_{zz} = 0$$

That is to say, one g-shift is zero, and two are large and almost equal. Experimentally, the g-shifts of 0.037, 0.0545 and 0.25 much more resemble those predicted by eq. (10.5) than those predicted by eq. (10.6). It thus follows directly from measurement of the g-shift that the unpaired electron is associated with a $d_{x^2-y^2}$ rather than with a d_{z^2} orbital. Even more so, if all wave function coefficients in eq. (10.5) would be set equal and if all energy differences would also be set equal in an approximation of horrendous crudity, eq. (10.5) would predict a Δg_{zz} shift to be 4 times larger than Δg_{xx} and Δg_{yy}, which would be predicted to be equal. This is in reasonable agreement with the experimentally measured g-shifts. We stress that, in practice, quantitative analysis of EPR g-shifts should only be attempted in combination with quantum chemical calculations.

As a final note for this section, if more than one unpaired electrons are present in the system under investigation, for example, in the case of high-spin Fe^{2+}, $S = 2$, an additional magnetic interaction called *zero field splitting* is present. Zero field splitting parameters can in principle also be determined from EPR spectra. These parameters, dubbed D and E, essentially derive from the magnetic dipole–dipole interaction between the unpaired electrons as well as from a, for metals usually dominant, second-order contribution from spin orbit coupling. However, if the zero field spitting, that is, all magnetic interactions of the unpaired electrons without a magnetic field, is larger than the microwave energy $h\nu$, as is prototypical for high-spin Fe^{2+}, no EPR transitions can be induced and no EPR signal will be obtained, even though the system is paramagnetic.

10.3 Nuclear hyperfine interaction

If the system under investigation contains nuclei with a nuclear spin, for example, H (I = ½), D (I = 1), N (I = 1), P (I = ½), F (I = 3/2), Cl (I = 3/2), Cu (I = 3/2) [8] additional structure may be visible in the EPR spectrum owing to the magnetic interaction between the electron spin and nuclear spin. The interaction is called the *hyperfine interaction* [1,7]. The hyperfine interaction consists of two parts: (1) the magnetic dipole–dipole interaction between electron spin and nuclear spin, which is anisotropic and its size depends on the distance and relative orientations of the spins; (2) the isotropic hyperfine interaction that results from the quantum mechanical phenomenon that electrons in s orbitals have a finite density at the nucleus.

Both contributions to the hyperfine interaction can provide valuable information about the system under investigation: if one thinks of the nuclear spins as little magnetic needles pinned at the atoms, then these needles are able to act as spies. If the probability distribution of the unpaired electrons is close to a nuclear spin, the hyperfine interaction will be large, if the unpaired electrons are effectively far away from the nuclear spin, then the hyperfine interaction will be small. The anisotropic part, originating from the dipole–dipole interaction, additionally contains orientation information, since the nuclear and electronic magnetic moments prefer to be aligned parallel along the effective connecting vector of the two spins, but they prefer to be aligned antiparallel when they are both taking orientations perpendicular to the connecting vector (i.e., exactly the same as with macroscopic bar magnets).

For example, in the EPR spectrum of azurin, Figure 10.2, a hyperfine structure of four signals is observed at the g_{zz} canonical orientation. In general, the number of signals expected for a nucleus of nuclear spin I would be $2I + 1$. If, because of symmetry, N equivalent nuclei of spin I are present, the number of signals becomes $2NI + 1$. For azurin the hyperfine structure originates from copper, whose isotopes both have nuclear spin $I = 3/2$, hence the presence of four signals. In the case of azurin, the hyperfine parameters are relatively easily retrieved from the spectra with the help of an energy-level diagram. For a nuclear spin $I = 3/2$ in interaction with an electron spin $S = ½$, the energy-level diagram is shown in Figure 10.3. The hyperfine interaction can for pencil-and-paper purposes be approximated by an effective constant, a, in the case of azurin, the A_{zz} component of the hyperfine tensor, multiplied by the product of the magnetic quantum numbers M_S and M_I.

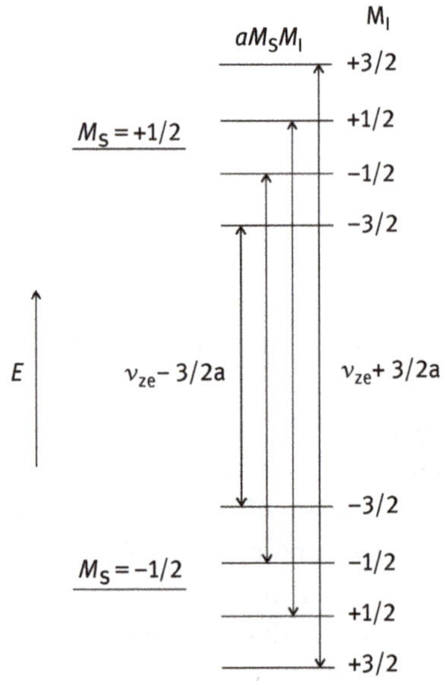

Figure 10.3: Energy level diagram for a nuclear spin $I = 3/2$ interacting with an electron spin $S = ½$. The energy axis is not to scale. The parameter ν_{Ze} represents the electron Zeeman interaction, which is typically between 9 and 10 GHz (the X-band of the microwave spectrum), whereas the hyperfine coupling constant a can vary over many orders of magnitude but usually never exceeds a value of several 100 MHz.

The EPR allowed transitions have the selection rules $|\Delta M_S| = 1$ and $|\Delta M_I| = 0$, giving rise to four allowed transitions as indicated with vertical arrows in Figure 10.3. The smallest and largest arrows differ by the amount of $3A_{zz}$. It is thus quite straightforward to estimate the A_{zz} value directly from experiment by reading the difference of the magnetic field values where the first hyperfine feature and the fourth hyperfine feature appear. Since this particular EPR spectrum is already plotted on a g-value scale for convenience of reading the g-values in the previous section, we suffice by providing the number for the hyperfine coupling A_{zz} read from the spectrum as being equal to 185 MHz.

Since hyperfine structure is not resolved at the g_{xx} and g_{yy} canonical orientations, the complete copper hyperfine tensor for azurin thus to some approximation has the form

$$\vec{A} = \begin{pmatrix} 0 & 0 & 0 \\ 0 & 0 & 0 \\ 0 & 0 & A_{zz} \end{pmatrix} = \begin{pmatrix} 0 & 0 & 0 \\ 0 & 0 & 0 \\ 0 & 0 & 185 \end{pmatrix} MHz \tag{10.7}$$

Note that the absolute sign of the hyperfine couplings cannot be determined from experiment. In principle, this number contains information about how close the unpaired electron is to the copper ion, and as such, the hyperfine interaction may even be used to estimate, for example, the parameter $c^2_{x^2-y^2}$ in eq. (10.5), although it is again stressed that it is scientifically better to perform first principles quantum chemical calculations that allow calculation of these parameters, because spin–orbit contributions to the hyperfine coupling constants are present and, therefore, the following analysis should be considered as a qualitative estimate at best! In any case, from eq. (10.7), the isotropic (a_{iso}) and anisotropic hyperfine parameter T (under axial approximation) can be determined by setting the hyperfine tensor equal to

$$\vec{A} = \begin{pmatrix} a_{iso} - T & 0 & 0 \\ 0 & a_{iso} - T & 0 \\ 0 & 0 & a_{iso} + 2T \end{pmatrix} \tag{10.8}$$

leading under the assumption from calculations that the A_{zz} component equals − 185 MHz [9] in principle to

$$a_{iso} = T = -62\ MHz \tag{10.9}$$

In particular, the anisotropic component, T, provides more information. EPR measurements for, for example, the much more ionic $Cu(NH_3)_4{}^{2+}$ give a value of $T = -314$MHz [9]. If we approximate this value as being the result of unit spin population of the unpaired electron in the $d_{x^2-y^2}$ orbital, then the ratio, −60/−314 = 20%, would simply provide an estimate that only 20% spin population of the unpaired electron is associated with the $d_{x^2-y^2}$ orbital at copper in azurin.

These considerations, albeit crude, qualitative and, to be honest, for scientific purposes not timely anymore, still have a didactic value in that they demonstrate the wealth of information that hides within the wiggly line of the EPR spectrum, but also how far one gets with pencil-and-paper analysis of the *g*-values and hyperfine parameters. That was the goal at the start of this chapter, for azurin and the wiggly line in Figure 10.2, we have deduced that the unpaired electron is associated with about approximately 20% with a $d_{x^2-y^2}$ orbital at copper. In practice, nowadays, analysis of EPR spectra in terms of electronic structure is typically combined with quantum chemical calculations [9–11], especially if quantitative agreement is sought. Such calculations also allow, for example, inclusion of electron correlation, spin–orbit contributions to the hyperfine coupling, as well as to bypass some of the approximations used in this chapter. In that respect, although density functional theory is presently used as a standard, it also has limitations [12], and exciting new developments in terms of more physically transparent methods like spin-unrestricted coupled cluster calculations are presently on the verge of making their introduction into the field of EPR spectroscopy [13].

References

[1] Carrington, A., and McLachlan, AD. Introduction to magnetic resonance, Harper & Row, New York, 1969.
[2] Schweiger, A., and Jeschke, G. Principles of pulse electron paramagnetic resonance, Oxford University Press, Oxford, 2001.
[3] Stoll, S., and Schweiger, A. EasySpin, a comprehensive software package for spectral simulation and analysis in EPR. J Magn Reson 2006, 178, 42–55.
[4] Harriman, JE. Theoretical foundations of electron spin resonance, Academic Press, New York, 1978.
[5] Stone, AJ. Gauge Invariance of the G-Tensor. Proc R Soc London, Ser A 1963, 271, 424–34.
[6] Coremans, JWA., Poluektov, OG., Groenen, EJJ., Canters, GW., Nar, H., and Messerschmidt, A. A W-band electron-paramagnetic-resonance study of single-crystal of azurin. J Am Chem Soc 1994, 116, 3097–101.
[7] Atherton, NM. Principles of electron spin resonance, Prentice Hall, New York, 1993.
[8] Bruker Biospin GmbH. Bruker EPR/ENDOR frequency table. Bruker Biospin GmbH 2013.
[9] Neese, F. Metal and ligand hyperfine couplings in transition metal complexes: the effect of spin-orbit coupling as studied by coupled perturbed Kohn-Sham theory. J Chem Phys 2003, 118, 3939–48.
[10] Neese, F. Prediction of electron paramagnetic resonance g values using coupled perturbed Hartree-Fock and Kohn-Sham theory. J Chem Phys 2001, 115, 11080–96.
[11] Neese, F. The ORCA program system. Wiley Interdiscip Rev: Comput Mol Sci 2012, 2, 73–8.
[12] Munzarova, M., and Kaupp, M. A critical validation of density functional and coupled-cluster approaches for the calculation of EPR hyperfine coupling constants in transition metal complexes. J Phys Chem A 1999, 103, 9966–83.
[13] Gauss, J., Kallay, M., and Neese, F. Calculation of electronic g-tensors using coupled cluster theory. J Phys Chem A 2009, 113, 11541–9.

Serena DeBeer

11 Introduction to X-ray spectroscopy – including X-ray absorption, X-ray emission and resonant inelastic X-ray scattering

11.1 Introduction

X-ray-based spectroscopies have played a major role in describing the geometric and electronic structure of countless metalloprotein active sites. A major advantage of X-ray spectroscopy is the element selectivity, which allows for the changes that occur at a protein active site to be probed in an element-selective way. In addition, X-ray spectroscopic methods can be applied to samples in any form (solutions, lyophilized powders or single crystals). Hence, these approaches are well suited for studying reactive intermediates and are particularly useful for proteins that are not readily crystallized.

The earliest applications of X-ray spectroscopy to biological systems focused on X-ray absorption spectroscopy (XAS), and in particular on the extended X-ray absorption fine structure (EXAFS) region of the spectrum, in order to obtain information about the identity and distance of ligands coordinated to the metal. These early EXAFS studies revealed some of the first metrical parameters for the FeS_4 active site of rubredoxin, the FeMoco cluster of nitrogenase and the Mn_4O_5Ca site of the oxygen-evolving complex of photosystem II [1–3]. At the time, the EXAFS analyses were based largely on extracting phase and amplitude parameters from small-molecule model complexes of known structure and then refining these parameters against the experimental data to obtain information about the distance, coordination number and identity of the ligands. In the years that followed, the ability to better model the EXAFS through multiple scattering-based approaches has greatly advanced the information that can be extracted from these data [4–10]. In addition, our understanding of the so-called XAS "edge" or X-ray absorption near-edge structure (XANES) has greatly advanced, enabling the low-energy "pre-edge" features to be quantitatively interpreted [11–16]. At the same time, advances in synchrotron beamlines and detectors have enabled more and more dilute systems to be measured [17, 18].

While the application of XAS to biological systems has developed over a ~40 year period, the use of X-ray emission spectroscopy (XES) in bioinorganic chemistry has occurred only during the last decade [19–28]. The recent growth in

Serena DeBeer, Department of Inorganic Spectroscopy, Max Planck Institute for Chemical Energy Conversion, Mülheim an der Ruhr, Germany

https://doi.org/10.1515/9783110496574-011

the applications of XES has been largely enabled by dedicated high-flux beamlines with moderate- to high-resolution crystal spectrometers that allow for the application of XES to dilute biological systems [29, 30]. These setups also provide a means to obtain resonant XES or so-called resonant inelastic X-ray scattering (RIXS) data [29, 30]. This is effectively a two-dimensional spectroscopy, which combines XAS and XES. The utilization of RIXS promises even greater selectivity than standard XAS or XES, and can in certain cases allow for oxidation-state, spin-state and ligand-selective XAS to be obtained [21, 31].

Herein, we provide a general introduction to XAS, XES and RIXS methods. For each spectroscopic method, we describe the basic physical process and the spectral information content. Experimental considerations and recent applications are highlighted throughout, with a focus on state-of-the-art X-ray spectroscopic studies of relevance to bioinorganic and bioorganometallic chemistry.

11.2 X-ray absorption

XAS involves the excitation of a core electron on a given photoabsorbing atom (e.g. the metal in a protein active site) first to bound states localized on the absorbing atom and then eventually to the continuum. An XAS spectrum is typically divided into two regions: (1) the so-called edge or XANES region at lower energies and (2) the EXAFS region at higher energies. The resulting XAS edges are labeled by the core level that the ionized electron originates from. For example, excitation of a 1s electron is referred to as a K-edge, excitation of a 2s or a 2p electron are L-edges, and 3s, 3p and 3d excitations comprise the M-edges. In biological applications of XAS, metal K-edges are primarily used; however, applications of ligand K-edge and metal L-edge XAS have also proved impactful [32–38].

11.2.1 Basic principles and information content

11.2.1.1 Metal K-edge

Figure 11.1 shows an iron K-edge XAS spectrum of a dilute iron-containing model complex. An XAS spectrum is the measurement of the absorption coefficient (μ) as a function of energy. When an iron atom has absorbed enough energy to excite a 1s core electron (~7.1 keV), a sharp discontinuity in the XAS spectrum results, which is known as the edge or XANES region, as noted earlier. Beyond the edge is the EXAFS region, which results after the electron has been ionized to the continuum. In the following sections, we briefly highlight the information content of each spectral region.

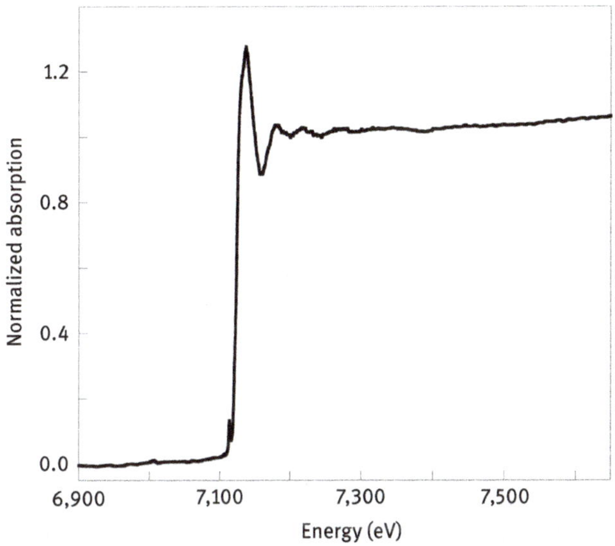

Figure 11.1: Fe K-edge XAS spectrum of a dilute (~1 mM in Fe) model complex. The low energy region corresponds to the edge or XANES region, and the high energy region (above ~7130 eV) is the EXAFS region.

11.2.1.1.1 XANES

Figure 11.2 shows an expansion of the edge region.

At lowest energy, one observes a weak 1s to 3d pre-edge feature at ~7113 eV. The pre-edge transition formally involves a change in two units of orbital angular momentum (from a 1s orbital, where $l = 0$ to a 3d orbital where $l = 2$). Hence, this transition is formally dipole forbidden, as the dipole selection rule requires that $\Delta l = \pm1$. The pre-edge transition is observed for two reasons. First, transitions with a $\Delta l = \pm2$ are quadrupole allowed. In this energy region, however, quadrupole transitions are ~100 times weaker than dipole-allowed transitions. The second mechanism by which the pre-edge gains intensity is 3d–4p orbital mixing, which imparts dipole allowed s to p character to this transition. In rigorously centrosymmetric sites, such as octahedral symmetry, 3d and 4p mixing is forbidden by group theory and the pre-edge remains weak. The observed intensity in the pre-edge in Figure 11.2 reflects a decreased site symmetry, which allows the pre-edge to gain intensity. Imagine, for example, a mononuclear iron molecule in C_{4v} symmetry. In this point group, the metal p_z orbital and the d_{z^2} orbital both transform as a_1, thus enabling symmetry-allowed p–d mixing. Similarly, the p_x and p_y orbitals and the d_{xz} and d_{yz} orbitals all transform as e, providing a second mechanism for symmetry-allowed mixing. It is for these reasons that the pre-edge provides a sensitive probe of local site symmetry [39]. A common application of the XAS pre-edge region is to use the pre-edge area to determine the coordination environment. Typically, on going from a six-coordinate (approximately O_h site

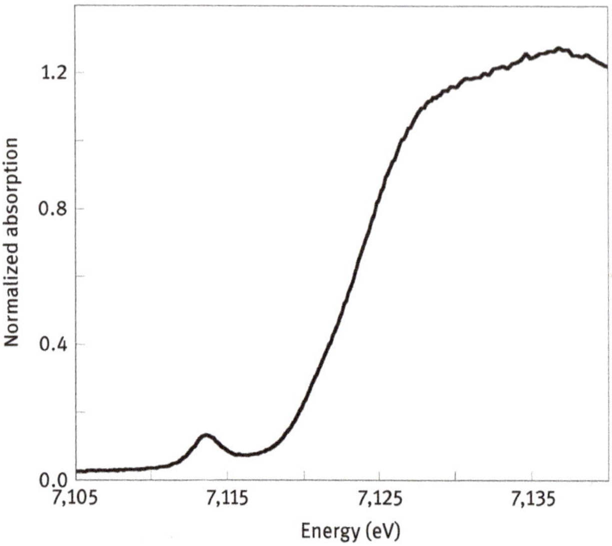

Figure 11.2: Normalized Fe K-edge XAS spectrum. The pre-edge is at ~7113 eV and the rising edge at ~7125 eV.

symmetry) to a five-coordinate (approximately C_{4V} site symmetry) iron site, the pre-edge increases in intensity, for the reasons discussed earlier. Hence, pre-edge areas have been useful for determining how the active site coordination environment changes. However, caution must be exercised, as highly distorted six-coordinate active sites can also display intense pre-edges. In addition to the pre-edge area, the pre-edge energy also contains very useful information. As the pre-edge reflects a transition to the unoccupied 3d orbitals, its energy is a reflection of both the ligand field and the oxidation state of the absorbing atom. As a general rule of thumb, the pre-edge energy increases in energy by ~1 eV for each one unit change in oxidation state [39, 40]. However, one must keep in mind that ligand field effects will also contribute to the observed energy. For example, for the same oxidation state of iron, a six-coordinate complex will have a pre-edge at ~1 eV higher in energy than a corresponding five-coordinate complex, due to the decreased ligand field destabilization in the lower coordinate complex [41]. Hence, fingerprinting of XAS pre-edges must be treated with caution. We note, however, that advances in computational approaches for interpreting and predicting the pre-edge region have greatly advanced the information that can be obtained from this spectral region [12, 14–16, 41].

To higher energy, at ~7,120 eV, one observes the dipole-allowed 1s to 4p edge transition. In general, the position of the edge reflects the effective nuclear charge on the photoabsorber. Hence, as the oxidation state of the photoabsorber increases, the amount of energy it takes to ionize a 1s electron increases, and the edge correspondingly shifts up in energy. In general, XAS reference literature indicates that the edge shifts up by ~1 eV per one unit change in oxidation state. However, this is also somewhat of an oversimplification. Figure 11.3 depicts the Fe K-edge XAS of a related series of Fe(II), Fe(III), Fe(IV) and Fe(VI) complexes [42]. Upon going from Fe(II) to Fe(III), the edge shifts up by ~2 eV in energy, while going from Fe(III) to Fe(IV) the shift is only ~1 eV. For the Fe(VI) complex, one observes that the edge is only ~1 eV higher than the Fe(IV). This is in part attributed to the fact that Z_{eff} does not increase linearly upon successive oxidations. The initial change from Fe(II) to Fe(III) is the largest, while subsequent oxidations have a smaller impact on the edge shift. Interestingly, however, the Fe(VI) complex shows a very large pre-edge area. This is due to the presence of a very short Fe(VI)-nitrido bond in this complex, which distorts the complex from centrosymmetry and gives rise to intense pre-edge transitions with dipole-allowed character.

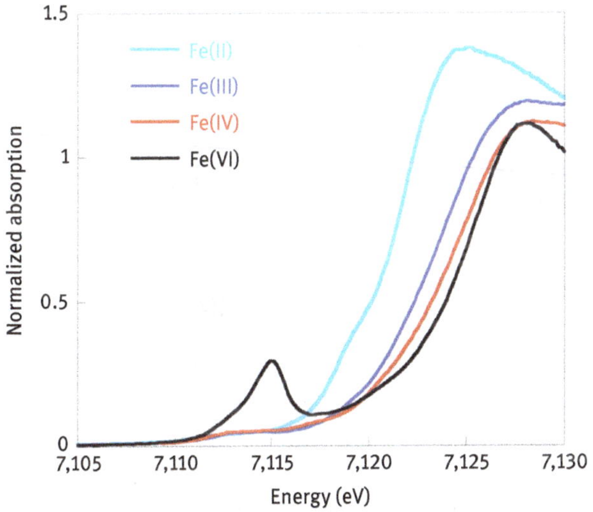

Figure 11.3: Fe K-edge XAS of a related series of Fe(II), Fe(III), Fe(IV) and Fe(VI) complexes originally reported in Reference [42].

11.2.1.1.2 EXAFS

The weak modulations that are observed just beyond the edge region in Figure 11.1 are referred to as EXAFS. In the EXAFS region, one considers that the 1s electron has already been ionized to the continuum. The ionized electron is treated as a photoelectron wave, which propagates out from the photoabsorber and may be backscattered by neighboring atoms. The backscattered waves interfere with the outgoing wave from the photoabsorber, resulting in a modulation of the absorption coefficient. The nature of the resultant interference is dependent on the nature of the backscatterers. It is for this reason that EXAFS can be used to determine information about the identity, coordination number and distance of ligands from a photoabsorbing atom [43]. As shown in Figure 11.1, the EXAFS oscillations above the edge are very weak. Hence, it is more common to choose an alternate means to display EXAFS data, namely the energy scale in eV is converted to photoelectron wavevector space (or k-space) in units of $Å^{-1}$. In order to enhance the signal at higher k-values, the data are often k^3-weighted, as depicted in Figure 11.4 (left). This conversion to k-space then allows the data to be Fourier transformed into R-space with units of Å, as shown in the right panel of Figure 11.4. In a simple picture, one can imagine that the photoabsorber is at radial position zero and that each peak in the Fourier transform represents a backscatterer at a given distance from the photoabsorber. We note, however, the Fourier transform must be phase shift corrected and hence distances cannot simply be read from the Fourier transform, but must be obtained through fitting.

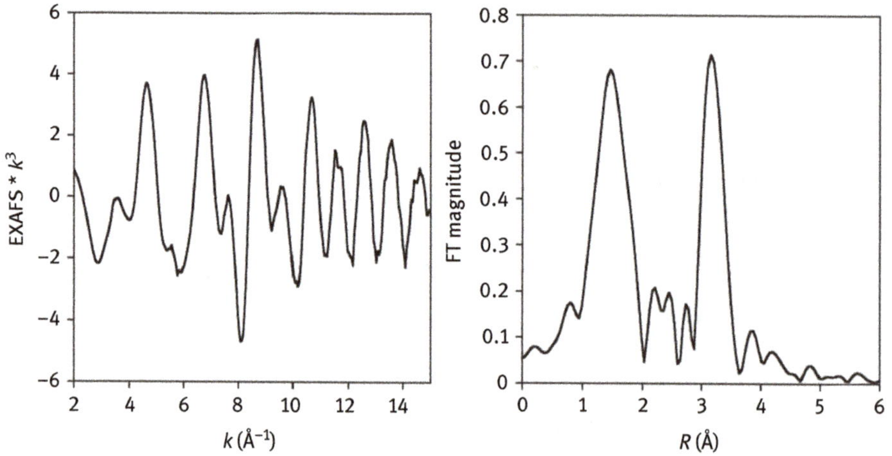

Figure 11.4: k^3-weighted EXAFS and the corresponding Fourier transform for an iron dimer. The first shell of the Fourier transform corresponds to light atom scatterers, and the outershell corresponds to Fe–Fe contributions.

Mathematically, the EXAFS region is defined in terms of $\chi(E)$, where

$$\chi(E) = \frac{\mu(E) - \mu_0(E)}{\Delta\mu_0(E_0)} \tag{11.1}$$

In eq. (11.1), $\mu(E)$ is the measured absorption coefficient and $\mu_0(E)$ corresponds to the atomic absorption coefficient of an isolated photoabsorbing atom. Typically, $\mu_0(E)$ is modeled by a smooth background function or what is referred to as a spline function. $\Delta\mu_0(E_0)$ corresponds to the measured jump in the absorption coefficient at the EXAFS onset energy E_0. As noted earlier, EXAFS are generally converted from energy space to k-space. This is achieved by defining the threshold energy E_0 and converting the energy axis using the following equation:

$$k = \sqrt{\frac{\{2m_e(E - E_0)}{\hbar^2}} \tag{11.2}$$

where m_e is the mass of an electron. We note that E_0 is often referred to as the origin of the photoelectron wavevector or more simply, the point at which the EXAFS starts. This is, however, not a rigorously defined variable. E_0 occurs at some point after the ionization of the photoelectron; hence, the energy is generally chosen slightly above the edge. However, the exact value of E_0 is typically further refined during the EXAFS fitting process. We refer the interested reader to reference [44] for a more detailed discussion of this topic.

At this point, it is useful to derive the origin of the EXAFS equation in k-space. In a simple approximation, EXAFS can be viewed as photoelectron scattering, where

$$\chi(k) = \sum_s \frac{e^{i2kR}}{kR}\left[kf(k)e^{i\delta(k)}\right]\frac{e^{i2kR}}{kR} \tag{11.3}$$

This equation can be expressed as a sine function via Euler's formula, where

$$\chi(k) = \sum_s \frac{Nf(k)}{kR^2}\sin[2kR + \delta(k)] \tag{11.4}$$

where N is the number of scatterers of a given type, $f(k)$ is the amplitude function for the backscattering atom, $\delta(k)$ is the phase shift for the absorber–backscatterer pair and R is the absorber–scatterer (i.e., metal–ligand) distance.

We note that eq. (11.4) is only for a single absorber–scatterer pair in the absence of disorder. Equation (11.5) introduces the disorder parameter σ^2 or what is often referred to as the Debye–Waller factor. This parameter accounts for contributions of both thermal and static disorders to the EXAFS signal:

$$\chi(k) = \sum_s \frac{Nf(k)}{kR^2}e^{-2k^2\sigma^2}\sin[2kR + \delta(k)] \tag{11.5}$$

In order to better understand the EXAFS equation above, it is useful to examine the effect of each parameter on both the EXAFS and the Fourier transform. Perhaps the simplest effect to understand is the impact of coordination number. As N increases, the amplitude of both the EXAFS and the Fourier transform will increase linearly. The effect of changing distance is more subtle and is best illustrated visually. Figure 11.5 shows the effect of elongating an Fe–O distance from 2 to 3 to 4 Å on both the EXAFS and the Fourier transform. The EXAFS clearly show that the overall amplitude of the EXAFS signal decreases as the distance of the oxygen from the iron atom increases. In addition, the longer the Fe–O distance, the higher the frequency of the EXAFS beat pattern. These effects are perhaps most clearly seen in the Fourier transform, which shows the peak maximum moving out to longer distance and the peak intensity decreasing with a $1/R^2$ dependence, as expected from eq. (11.5).

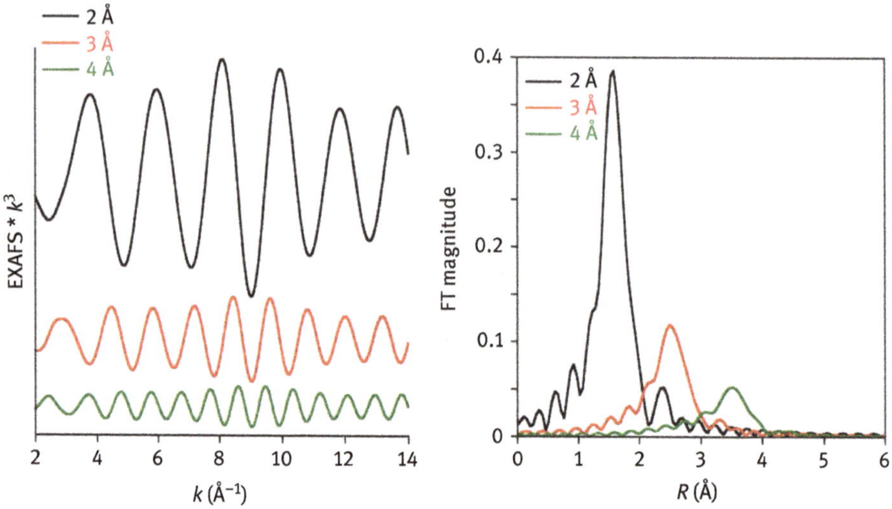

Figure 11.5: Effect of the Fe–O bond length on the calculated EXAFS (left) and Fourier transform (right).

Another important effect on the EXAFS signal is the identity of the backscatterer, which will modulate both the phase and amplitude parameters of the EXAFS signal. Figure 11.6 shows the calculated EXAFS and the corresponding Fourier transforms for Fe–O, Fe–S and Fe–Fe distances all at 2.5 Å. One clearly observes that a heavy backscatters, like Fe, gives the highest amplitude EXAFS signal and results in a modulation of the amplitude envelope, with the EXAFS signal clearly

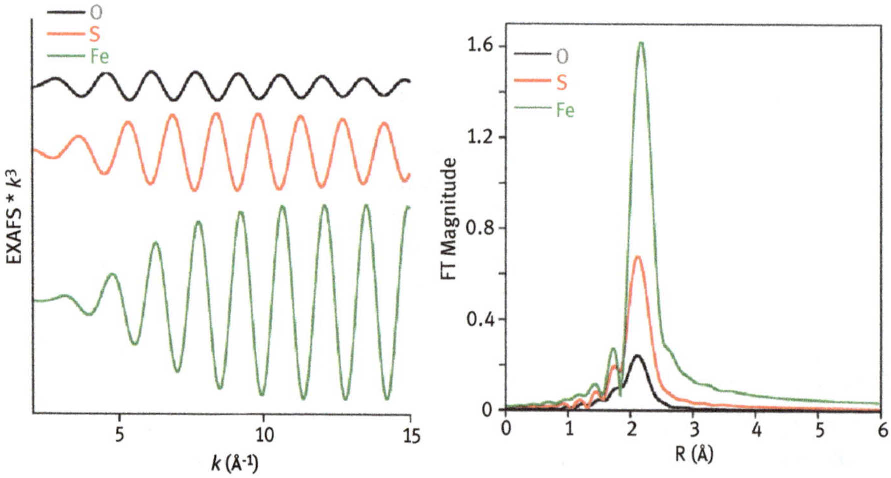

Figure 11.6: Calculated EXAFS (left) and Fourier transforms (right) for Fe–O, Fe–S and Fe–Fe contributions at 2.5 Å.

peaking at higher k. Due to the strong contributions from metal backscatters, EXAFS is particularly well suited for the determination of metal–metal distances [7]. In contrast, light atoms, such as oxygen, will result in a much weaker contribution to the EXAFS signal, particularly at a distance as long as 2.5 Å. Hence, one can see that if a measured sample has both Fe–Fe and Fe–O contributions at 2.5 Å, the Fe–Fe contribution will dominate, as the Fe–O may be challenging or impossible to fit. For Fe–S contributions, the heavier sulfur backscatterer (relative to oxygen) will result in increased contributions to the EXAFS amplitude and Fourier transform. Further, we note that similar mass atoms, such as C, N and O, cannot be distinguished by EXAFS due to their similar scattering properties.

The goal of fitting EXAFS is to deconvolute the EXAFS signal into the sum of all the sine waves that comprise it. It is important to note that the resolution of the EXAFS data is limited by the k-range of the data. The further out in k-space the data are collected, the better one can separate contributions at similar distances. As a general rule of thumb, the resolution of EXAFS data is given by the following equation:

$$\Delta R = \frac{\Pi}{2\Delta k} \tag{11.6}$$

Hence for data with a range of 2–15 $Å^{-1}$ in k-space, the resolution is 0.12 Å. This means that Fe–O distance of 1.90 and 2.00 Å would not be separable, but instead an average distance of 1.95 Å would be fit. Separating these two components would require data to at least $k = 17$ $Å^{-1}$.

Finally, we note that a major limitation in fitting EXAFS is the large number of correlated parameters. In particular, the amplitude of the EXAFS signal will depend on the coordination number, the identity of the backscatterer and the extent of disorder. As a result, the errors in coordination number derived from EXAFS fits are often quite high (on the order of 20–30%). For this reason, a close correlation of the EXAFS data to the pre-edge data is often helpful in constraining possible models.

In bioorganometallic chemistry, notable applications of EXAFS have included the earliest proposed structures of nitrogenase,[1] detailed insights into nitrogenase FeMo cofactor biosynthesis [45], insights into substrate interaction with nitrogenase cofactor [46], and most recently structural characterization of the E1 state [44]. EXAFS studies also played an important role in the early characterization of coenzyme B_{12} intermediates [47–49].

11.2.1.2 Ligand K-edges

The majority of XAS applications in bioinorganic and bioorganometallic chemistry are performed at the metal K-edge. However, ligand K-edge XAS measurements are also possible [32]. In these measurements, one directly measures the 1s excitations from carbon, nitrogen, oxygen or sulfur. However, due to the ubiquity of C, N and O in proteins, the utility of ligand K-edge XAS for these elements is rather limited in biology. Sulfur K-edge XAS, on the other hand, has had important applications in bioinorganic and bioorganometallic chemistry [32, 33, 35, 50]. It is of greatest utility when the sulfur atoms are directly coordinated to the metal active site of interest, and when there are not too many sulfur-containing amino acids outside of the active site.

Figure 11.7 depicts S K-edge XAS spectra for two iron–sulfur model complexes, with the general formula $[L_2Fe(III)_2S_2]^{2-}$ and $[L_2Fe(II)S_2]^{4-}$, corresponding to diferric and diferrous iron dimers, respectively [51]. As was the case for the metal K-edge, the ligand K-edge is discussed in two parts, the pre-edge feature at ~2,470 eV and the rising edge at ~2,475 eV. When sulfide (S^{2-}) is bound to a closed-shell metal such as Zn, in ZnS, no pre-edge is observed [52], and only a featureless rising edge corresponding to a sulfur 1s to 4p transition is seen. In contrast, when S^{2-} is bound to an open-shell ferrous or ferric ion, as shown in Figure 11.7, pre-edge features are clearly observed. In a molecular orbital picture, the pre-edge features arise from the covalent mixing of the iron 3d orbitals with the filled sulfur 3p orbitals. This mixing imparts sulfur 3p character to the unoccupied metal 3d-based orbitals and allows for a dipole-allowed sulfur 1s to sulfur 3p pre-edge transition. The intensity of the pre-edge feature reflects the extent of covalent mixing between the metal and the ligand. As such a ferric complex with shorter Fe–S bonds, and a greater number of 3d holes, has a larger pre-edge intensity than a ferrous complex, with longer Fe–S bond and fewer 3d holes.

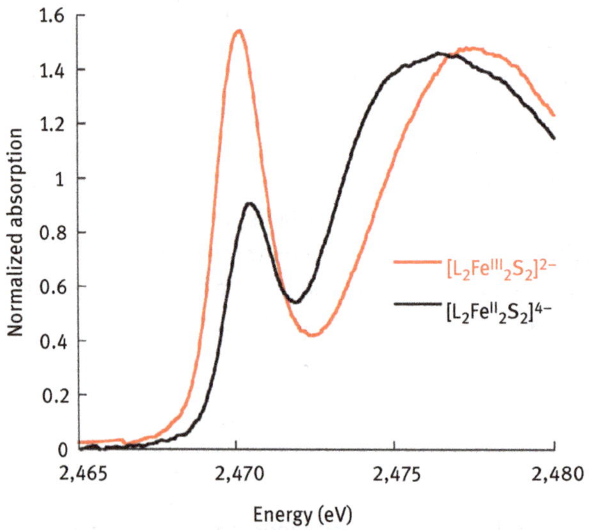

Figure 11.7: Normalized S K-edge XAS spectra of $[L_2Fe(III)_2S_2]^{2-}$ and $[L_2Fe(II)_2S_2]^{4-}$ (where L = bis (benzimidazolato). Adapted from Reference [51]. See open access article at:http://pubs.acs.org/doi/abs/10.1021/acs.inorgchem.6b00295.

Further, we note that the energy of the pre-edge relative to the rising edge serves as a reporter for the metal oxidation state. As one can see in Figure 11.7, the rising edge of the ferric complex appears to be of higher energy than that of the ferric complex. This energy shift reflects the increased charge donation of S^{2-} to Fe (III) relative to Fe(II), which shifts the sulfur 1s orbitals to deeper binding energy in the ferric case and increases the energy of the rising edge. If one considers the rising edge of a 1s to continuum-like transition, then one can align the rising edge energies and use the relative shift in the pre-edge energy to determine the metal Z_{eff}. In the present example, one sees that the pre-edge of the ferric complex is clearly shifted to much lower energy than that of the ferrous. This reflects the increased Z_{eff} in the ferric case, which stabilizes the metal 3d manifold relative to the sulfur 1s orbital. We note that this trend is opposite to what one observes at the metal K-edge, because at the metal K-edge both the metal 1s and metal 3d orbital energies contribute to the observed pre-edge energies.

In closing, we note that detailed studies by Hedman and coworkers [32] have demonstrated how ligand K-edge XAS may be used in order to extract quantitative information about metal–ligand covalency. Work by Wieghardt, Neese, DeBeer and coworkers, have shown that ligand K-edge XAS is a powerful probe of ligand radical character [53, 54]. In addition, time-dependent density functional calculations have proven very useful as a means to quantitatively interpret and predict ligand K-edge XAS spectra [11]. Notable applications in bioorganometallic chemistry include S K-edge XAS studies of nitrogenase and related model complexes [50, 55]. Most recently, Se high-resolution fluorescence detected (HERFD) XAS has also been utilized to obtain insight into the FeMoco active site [38].

11.2.1.3 Metal L-edges

Biological applications of XAS generally focus on the metal K-edge, particularly for first row transition metals. For heavier elements, where the K-edges are not accessible (such as tungsten) L-edges are often utilized. The primary advantage of using the K-edge for first row transition metals is that the 1s ionization energies are well separated from the edges of other elements, allowing for EXAFS data to be obtained to higher *k*-ranges. However, even for first row transition metals, there are certain advantages to measuring a metal L-edge over a metal K-edge.

In order to illustrate this, Figure 11.8 compares the Fe K- pre-edge region and the Fe L-edge XAS of $K_3[Fe(CN)_6]$. One immediately notes two important features, the L-edge XAS spectrum is much more intense and also far better resolved. This is due to the 2p to 3d nature of the L-edge transition, which is dipole allowed. Further, the relatively long 2p core hole lifetime results in reduced spectral broadening relative to Fe K-edge XAS [56]. Here it is of interest to consider the electronic structure of $K_3[Fe(CN)_6]$ in more detail– it is a low-spin d^5 iron complex with a $(t_{2g})^5(e_g)^0$ ground state

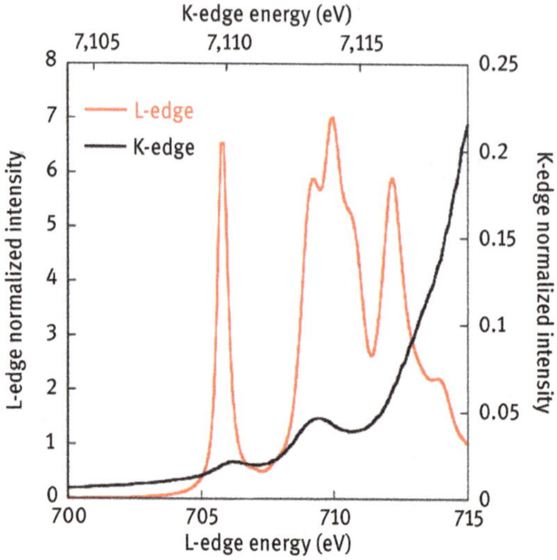

Figure 11.8: Comparison of the Fe L- and Fe K-edge of $K_3[Fe(CN)_6]$. The comparison emphasizes the difference in the selection rules and in the experimental resolution.

configuration. Hence, in a simple picture, one can think of the lowest energy feature corresponding to a transition to the t_{2g} set of orbitals and the higher energy transitions arising from transitions to the e_g set of orbitals. In the Fe L-edge an additional high-energy feature is observed, which is attributed to transitions to the empty Π* orbitals of the cyanide ligands. Hence, the Fe L-edge can serve not only as a probe of oxidation state and spin state, but can also reveal information about the extent of backbonding [57]. In the Fe K- pre-edge, one only sees two features largely due to the increased 1s core hole lifetime broadening. We note the simple molecular orbital-based interpretation presented here is, of course, a gross oversimplification, and in both cases in order to be rigorous, one must consider all possible final state multiplets. Nonetheless, this simple comparison helps highlight how L-edge XAS can provide a more detailed electronic structure picture than the metal K-edge.

11.2.2 Experimental considerations

In the previous sections, we have briefly discussed the information that can be obtained from XAS data in both the edge and EXAFS region. Here, a brief overview of the experimental setups is provided together with important sample considerations that should be taken into account. The discussion is divided into the hard (5 keV or higher), tender (2–5 keV) and soft (<2 keV) X-ray regimes.

Hard X-rays are generally defined as those having photon energies greater than 5 keV and hence having a significant path length in air. In general, metal K-edge XAS for first row transition metals all falls into the hard X-ray regime. We note that the titanium K-edge is just below 5 keV and is thus considered by many to fall into the intermediate "tender" X-ray regime. The relatively long path length of a hard X-ray in air greatly simplifies the experimental setup and the sample environment. Figure 11.9 depicts a typical setup for a hard X-ray XAS measurement. As with all XAS experiments, a tunable source of X-rays is generally the preferred way to carry out these experiments. For the majority of XAS measurements on biological samples, the X-ray source is provided by synchrotron radiation. Energy selection is achieved by utilizing a double crystal monochromator. Most often Si crystals are utilized due to the high thermal stability. Ideally, a 1 eV or better resolution of the incident X-ray beam is desirable. The flux of the incident beam is then monitored using either a diode or a gas-filled ionization chamber at I0. A similar detector can be used to monitor the beam after sample (I1). In this configuration, one performs a true absorption measurement, where the XAS signal of the sample corresponds to the log(I0/I1). We note, however, that for dilute biological samples, a true transmission measurement is rarely possible and instead one monitors the secondary fluorescent processes that occur as proportional to the absorption. For this reason the sample is often rotated at 45° relative to the incident beam in order to obtain fluorescence. The fluorescence detector may be a simple diode, but for dilute biological samples energy-resolving detectors are generally desirable. Finally an I2 detector is placed after a reference foil in order to provide an internal energy calibration. Typically, XAS edge scans span a range of just a few hundred eV, whereas EXAFS scans are on the order of 1,000 eV.

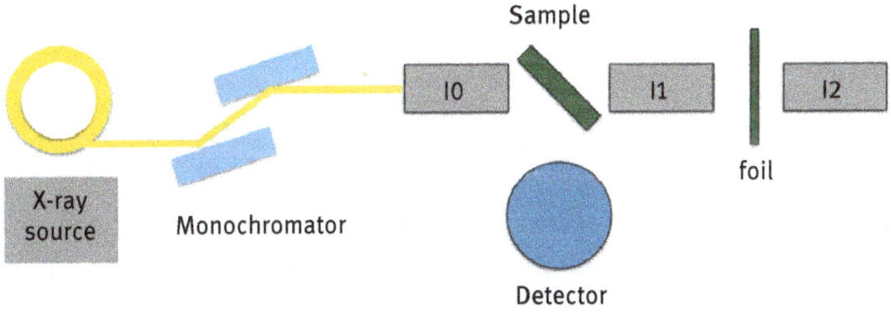

Figure 11.9: A standard XAS setup for hard X-ray measurements.

In order to minimize sample damage, biological samples are most frequently measured in liquid nitrogen or liquid helium cryostats. A further advantage of such a setup is that the thermal contributions to the Debye–Waller factor (which results in damping of the EXAFS signal at high temperatures) are minimized. We note that monitoring biological samples for beam-induced damage is an incredibly important aspect of all XAS measurements [58, 59]. Protocols must be established to determine how long a sample can be exposed to the beam before changes occur in the edge region, which are an indicator of photoreduction. For intense insertion device beamlines, the dwell time per spot may be only a matter of seconds for highly oxidized samples, thus increasing the amount of required sample. We note that in general, XAS measurements require millimolar concentrations of the photoabsorber of interest. Lower concentration can be measured, but at the cost of additional measurement time and perhaps poorer signal to noise.

In the tender X-ray regime (2–5 keV), the X-rays no longer have a significant path length in air, and the entire experiment must be enclosed in either Helium gas or in vacuum. Below 2 keV, vacuum is required. In recent years, advances have been made using differential pumping, which allow for near-ambient pressures to be utilized at low energies. Nevertheless, operating at lower X-ray energies presents significant experimental design challenges. Further, we note that the shorter path length of the X-rays results in greater absorption of X-rays by the sample, which can lead to more rapid damage rates. Hence for biological samples, measurements at low energies present even greater difficulties.

11.3 X-ray emission spectroscopy

The previous section focused on XAS, which has seen great use within the bioinorganic community over the last 40+ years. XES, in contrast, has only seen increased use in the last decade or so [19–22, 25, 27, 30, 60–63]. The increased utilization of XES may largely be attributed to the availability of high-resolution XES crystal spectrometers at synchrotron beamlines that have allowed for more routine applications of XES to be realized.

11.3.1 Basic principles and information content

An XES spectrum is the result of fluorescence decay that occurs after the ionization of a core electron [29, 30]. In this section, we focus only on the processes that occur after the ionization of a 1s electron. Further, we note that all the XES data in Section 11.3 are so-called nonresonant XES data in which the incident beam is tuned well above the ionization energy of the 1s electron. Figure 11.10 depicts the primary XES spectral regions (top) and the transitions, which give rise to each of the emission lines.

The most likely event to occur after the ionization of a 1s electron is that a 2p electron refills the core hole and produces a fluorescent photon. This is a so-called Kα line and is split into two features, the $K\alpha_1$ and $K\alpha_2$ emission lines, due to 2p core hole spin–orbit coupling in the final state. The Kα lines are relatively insensitive to changes in metal spin and oxidation state and as such are not frequently utilized to abstract chemical information. They can, however, be used to obtain higher resolution XAS spectra in resonant XES (or RIXS) measurements, as discussed in Section 11.4.

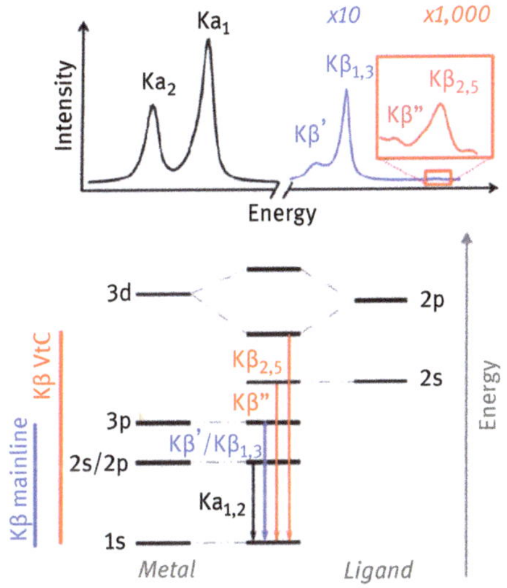

Figure 11.10: X-ray emission spectral regions (top) and the corresponding transitions (bottom). Adapted from Reference [29]. Copyright 2016 John Wiley and Sons.

At higher energy, one finds the Kβ emission lines, which are almost an order of magnitude weaker than the Kα lines. The Kβ emission lines arise from 3p to 1s transitions. Due to 3p–3d exchange in the final state, the Kβ emission lines split into the $K\beta_{1,3}$ and Kβ′ lines. The energetic splitting between the $K\beta_{1,3}$ and Kβ′ is proportional to the number of unpaired d-electrons. Hence, high-spin complexes with a large number of unpaired electrons have a large 3p–3d splitting, while low-spin complexes with fewer or no unpaired electrons will have a much smaller splitting, and no well-resolved Kβ′ feature [29, 30, 64]. To illustrate this point, Figure 11.11 compares the Kβ mainlines for high-spin and low-spin ferric model complexes. The high-spin $S = 5/2$ complex clearly has a larger (~15 eV) splitting relative to the low-spin $S = 1/2$ complex, where the Kβ′ feature is effectively absent. This example highlights the ability to use Kβ emission as a spin state marker. We note, however, that covalent delocalization of metal d character onto the ligands can greatly modify this picture [65]. Hence, using the splitting as an isolated marker of spin state can be in certain cases precarious.

At highest energy are the weak valence-to-core (VtC) XES features (Figure 11.10) [27, 30]. Despite their weak intensity, this spectral region is very useful due to its pronounced sensitivity to the metal ligation sphere. VtC XES features arise from electrons in filled ligand ns and np orbitals refilling the 1s core hole on the metal. These transitions give rise to the so-called Kβ″ and $K\beta_{2,5}$ features, respectively. The observed intensity derives from a small amount of metal np character mixing into the ligand orbitals, which imparts dipole-allowed character [66]. As such, the observed

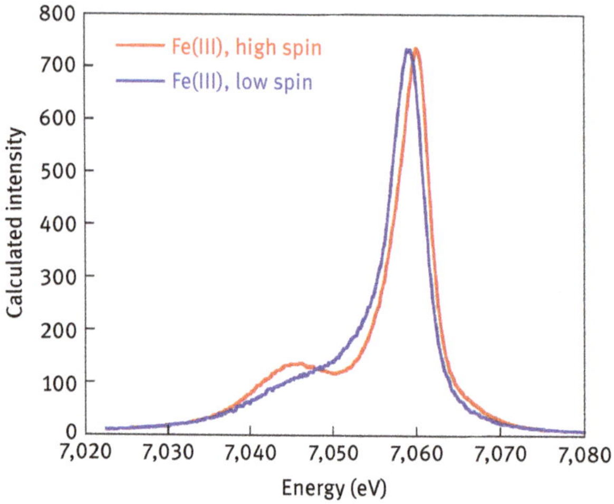

Figure 11.11: Iron K" mainline spectra of a high-spin vs a low-spin iron model complex.

intensity is dependent on metal–ligand bond lengths, with shorter bonds giving rise to greater intensities in the VtC region. The energy of the VtC features is governed largely by the ligand ionization energies. Since the 2s ionization energies of C, N, O and F vary by nearly 9 eV, the Kβ″ feature is particularly sensitive to ligand identity [27]. This thus serves as a very useful complement to EXAFS data, where similar scatterers such as C, N and O cannot be distinguished. A key bioinorganic application of VtC XES was its use to identify the central carbon atom in the FeMo cofactor of nitrogenase [20]. Figure 11.12 shows the VtC region of FeMoco compared to iron oxide and iron nitride references. The high energy of the Kβ″ feature in the protein was key to establishing that the central atom is a carbon. Subsequently, a carbide has also been identified in the FeV cofactor of the vanadium-dependent enzyme [24]. This is something that is not readily possible from EXAFS or protein crystallography [67]. The sensitivity of VtC XES to ligand ionization energies also may allow for the protonation state of the ligand to be determined. A notable example includes the protonation state of oxygen bridges in manganese dimers [60].

11.3.2 Experimental considerations

In order to measure nonresonant Kα, Kβ or VtC XES in the hard X-ray regime, one requires a high-resolution crystal analyzer setup [29, 30]. Such setups are now available at many synchrotron beamlines and there are now also an increasing number of in-house XES setups, enabling measurements in the home laboratory [68–70]. In general, these setups employ either energy-scanning crystal spectrometers or dispersive crystal spectrometers, as pictured in Figure 11.13. Scanning

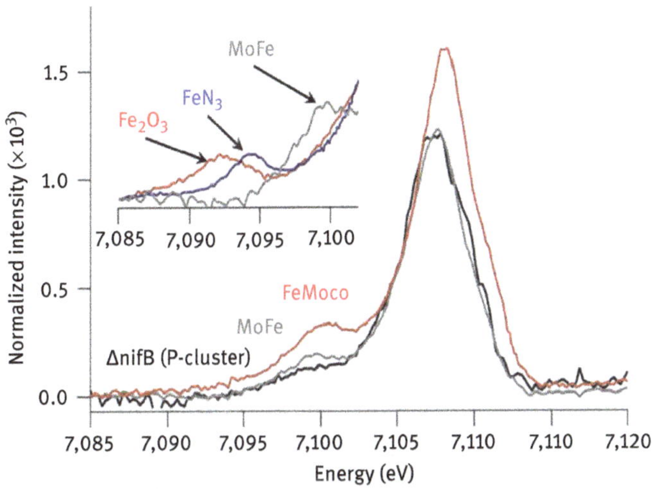

Figure 11.12: VtC XES spectra of FeMoco, the eight iron P-cluster, and MoFe protein (containing both FeMoco and the P-cluster). The inset shows a comparison of the Kβ" region of MoFe protein relative to iron oxide and iron nitride references. Adapted from Reference [20].

spectrometers utilize spherically bent crystal analyzers aligned in a Rowland geometry in order to select for a single emission energy, which is focused onto a point detector. In order to collect an entire emission spectrum, the position of both the analyzer crystal and the detector must be scanned. In contrast, a dispersive spectrometer utilizes a cylindrically bent crystal in a Von Hamos geometry to spatially separate different emission energies that are measured using a two-dimensional detector. We refer the reader to previous review articles for a more detailed explanation of the spectrometer setups [29, 71, 72]. We note that the sample requirements and beam damage issues discussed in Section 11.2.2 are also of major concern for XES studies. As the crystal analyzers require focused beams, photoreduction of biological samples can become an even greater problem.

11.4 RIXS

RIXS is an experiment in which XAS and XES are combined [29, 30, 71, 72]. Rather than performing XES in a "nonresonant" mode as was described in Section 11.3, in a RIXS experiment the incident energy is tuned to various energies through the absorption edge and the resonant emission is measured. For this reason, RIXS are also sometimes referred to a resonant XES or RXES. In effect, one can think of this experiment as a two-dimensional experiment, where XAS is measured while at the same time measuring all XES events. At the K-edge one can thus measure XAS together

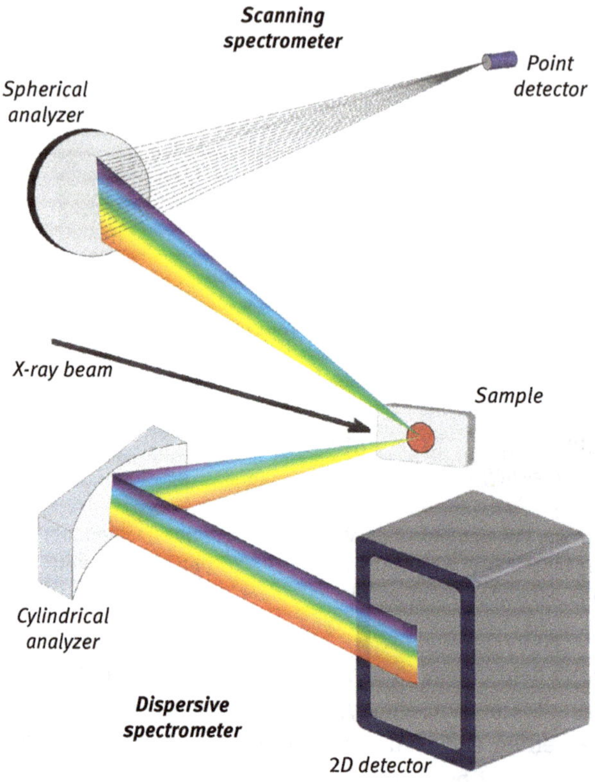

Figure 11.13: Comaprison of a scanning Rowland geometry crystal spectrometer (top) to a dispersive Von Hamos spectrometer (bottom). Adapted from Reference [29]. Copyright 2016 John Wiley and Sons.

with Kα, Kβ or VtC XES. These measurements are then referred to as 1s2p, 1s3p and 1sVtC RIXS, respectively. Alternatively, one can examine emission events after a 2p core hole ionization and follow the dipole-allowed 3d to 2p emission. These experiments are known as 2p3d RIXS. Herein, we briefly summarize the information content of each RIXS spectral region.

11.4.1 1s2p RIXS

In a 1s2p RIXS experiment, the metal K-edge XAS is measured, while simultaneously monitoring the Kα 2p1s emission. Hence, if one has a metal complex with a $1s^2 2p^6 3d^n$ ground state, then upon an XAS excitation in the pre-edge region a $1s^1 2p^6 3d^{n+1}$ intermediate state is reached. Following 2p1s emission, the final state is $1s^2 2p^5 3d^{n+1}$ or the equivalent of an L-edge-like final state (Figure 11.14). This is

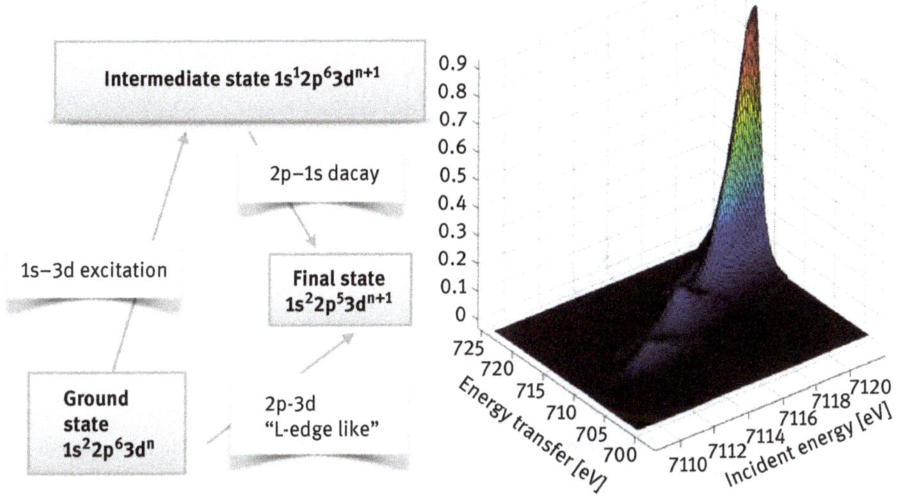

Figure. 11.14: Schematic for the states involved in 1s2p RIXS (left). A 1s2p RIXS plane (right).

of particular interest for biological applications as it allows L-edge-like information to be obtained with a hard X-ray probe. Normally, L-edge experiments require ultra-high vacuum conditions and hence can have more limited utility for proteins. Hard X-rays are also less damaging, providing further motivation for 1s2p RIXS as a means to obtain more detailed electronic structural information, particularly on radiation-sensitive biological samples. Notable applications of this approach include recent 1s2p RIXS studies on the oxygen-evolving complex in photosystem II [73] and cytochrome c, [37] using Mn and Fe 1s2p RIXS, respectively.

Measuring full 1s2p RIXS planes on dilute biological samples, however, can take a considerable amount of time and also may require large amounts of protein, due to damage rates. An alternative use of 1s2p RIXS is to measure the XAS only at the Kα maximum. This approach has been shown to overcome 1s core hole lifetime broadening, as the spectral broadening is now dominated by the 2p core hole [74]. Such measurements can result in spectra that have an approximately two- to fourfold higher resolution than standard XAS spectra. This detection method is also known as high-energy-resolution fluorescence-detected (HERFD) XAS. Figure 11.15 depicts a comparison of the Fe K pre-edge reason of standard XAS as compared to HERFD XAS [75]. The far better resolved pre-edge features are clearly evident in the HERFD spectrum. Recent example of HERFD XAS includes its use to identify a Mo(III) active site in the FeMo cofactor of nitrogenase [26] and to identify key metrical parameters in intermediate *Q* of soluble methane monooxygenase [75, 76].

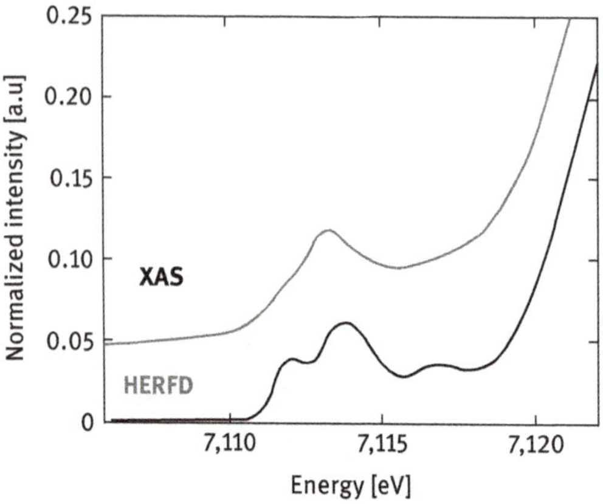

Figure 11.15: Comparison of standard total fluorescence XAS to HERFD detection in the pre-edge region of an Fe K-edge. Adapted from Reference [75]. Copyright 2017 American Chemical Society.

11.4.2 1s3p RIXS

In a 1s3p RIXS experiment, one measures XAS while monitoring the K-Beta emission, thus arriving at a $1s^2 3p^5 3d^{n+1}$ final state or the equivalent of an M-edge. To our knowledge, the information content of the M-edge energy transfer axis in 1s3p RIXS has been relatively unexplored and has not yet been applied to bioinorganic systems. To date, the primary application of 1s3p RIXS has been to obtain oxidation state and/or spin-selective XAS or EXAFS. In such an experiment, one takes advantage of the sensitivity of the K-Beta mainline to 3p–3d exchange, allowing for the XAS of specific final states to be probed. A nice example is the utilization of 1s3p RIXS to separate out the high-spin iron sulfur sites in hydrogenase from the low-spin carbonyl/cyanide-coordinated two iron clusters [19, 77]. Recently, 1s3p RIXS has also been utilized to examine spin-selective excitation channels in both molybdenum and vanadium nitrogenases [23].

11.4.3 VtC RIXS

Another application of hard X-ray RIXS is to measure the weak VtC emission features in combination with XAS. Due to the strong sensitivity of the VtC features to ligand ionization energies, such spectra have recently been shown to enable ligand-selective XAS [31]. Further, polarization information can in certain case be extracted from powder of solution samples via resonant enhancement of the VtC XES features. The theoretical principles behind these effects have recently been described [78].

11.4.4 2p3d RIXS

In the preceding sections, we have focused on hard X-ray RIXS. However, RIXS experiments are also possible in the soft X-ray regime. They are, however, considerably more challenging due to the decreased probability of radiative decay processes occurring following 2p core excitation. Nonetheless, 2p3d RIXS on first row transition metals is of great interest, as it can provide access to d–d transition energies. In a 2p3d RIXS experiment, one probes a dipole-allowed 2p to 3d excitation followed by a dipole-allowed 3d to 2p emission event [79, 80]. When one considers a transition metal with a $2p^6 3d^n$ initial state, the resulting $2p^6 3d^{n'}$ final state can either be the same as the initial state or it can represent a d–d excited state. Importantly, however, this formally dipole and parity-forbidden transition has been arrived at via two dipole-allowed processes. Further, due to 2p spin–orbit coupling in the intermediate state, spin-forbidden transitions also can gain intensity. Hence, 2p3d RIXS has great promise as a means to map out the low-lying excited states in transition metal complexes and metalloprotein active sites. At this stage, applications of 2p3d RIXS to proteins have not yet been reported, but there are a few recent applications to molecular systems [79–82].

11.5 Summary and outlook

In this chapter, an overview of XAS and XES methods has been provided. XAS has had long-standing application in bioinorganic chemistry, while applications of XES have only begun to grow in the last decade. This has largely been enabled by an increasing number of dedicated beamlines for XES measurements, and it is expected that the use of XES (at both the nonresonant and resonant limits) will continue to grow. An exciting forefront is time-resolved XES applications at both synchrotron and free electron lasers [83, 84]. Such studies have great promise to reveal important mechanistic insights into metalloenzyme mechanism.

Acknowledgments: The author gratefully acknowledges all of her collaborators and coworkers, who contributed to the work referenced within this chapter. This work was funded by the Max Planck Society.

References

[1] Cramer, SP., Gillum, WO., Hodgson, KO. et al. Molybdenum site of nitrogenase .2. comparative-study of Mo-Fe proteins and Iron-molybdenum cofactor by X-Ray absorption spectroscopy. J Am Chem Soc 1978, 100, 3814–9.

[2] Shulman, RG., Eisenberger, P., and Kincaid, BM. X-Ray absorption spectroscopy of biological molecules. Annu Rev Biophys Bio 1978, 7, 559–78.
[3] Goodin, DB., Yachandra, VK., Britt, RD., Sauer, K., and Klein, MP. The state of manganese in the photosynthetic apparatus .3. light-induced-changes in X-ray absorption (K-Edge) energies of manganese in photosynthetic membranes. Biochim Biophys Acta 1984, 767, 209–16.
[4] Newville, M., Ravel, B., Haskel, D., Rehr, JJ., Stern, EA., and Yacoby, Y. Analysis of multiple-scattering XAFS data using theoretical standards. Physica B 1995, 208, 154–6.
[5] Newville, MG., Carroll, SA., Rehr, JJ., Ravel, B., and Ankudinov, AL. Calculations of XANES spectra for transition metal oxides and sulfides. Abstr Pap Am Chem S 1997, 214, 63–GEOC.
[6] Ravel, B., Newville, M., Kas, JJ., and Rehr, JJ. The effect of self-consistent potentials on EXAFS analysis. J Synchrot Radiat 2017, 24, 1173–9.
[7] Beckwith, MA., Ames, W., Vila, FD. et al. How accurately can extended X-ray absorption spectra be predicted from first principles? implications for modeling the oxygen-evolving complex in photosystem II. J Am Chem Soc 2015, 137, 12815–34.
[8] Westre, TE., Dicicco, A., Filipponi, A. et al. Determination of the Fe-N-O Angle in (Feno)(7) complexes using multiple-scattering EXAFS analysis by GnXAS. J Am Chem Soc 1994, 116, 6757–68.
[9] Westre, TE., Dicicco, A., Filipponi, A. et al. Using GnXAS, a multiple-scattering EXAFS analysis, for determination of the Fe-N-O Angle in (Feno)(7) complexes. Physica B 1995, 208, 137–9.
[10] Westre, TE., Dicicco, A., Filipponi, A. et al. GnXAS, a multiple-scattering approach to EXAFS analysis – methodology and applications to iron complexes. J Am Chem Soc 1995, 117, 1566–83.
[11] George, SD., and Neese, F. Calibration of scalar relativistic density functional theory for the calculation of sulfur K-edge X-ray absorption spectra. Inorg Chem 2010, 49, 1849–53.
[12] George, SD., Petrenko, T., and Neese, F. Prediction of iron K-edge absorption spectra using time-dependent density functional theory. J Phys Chem A 2008, 112, 12936–43.
[13] George, SD., Petrenko, T., and Neese, F. Time-dependent density functional calculations of ligand K-edge X-ray absorption spectra. Inorg Chim Acta 2008, 361, 965–72.
[14] Roemelt, M., Beckwith, MA., Duboc, C., Collomb, MN., Neese, F., and DeBeer, S. Manganese K-edge X-ray absorption spectroscopy as a probe of the metal-ligand interactions in coordination compounds. Inorg Chem 2012, 51, 680–7.
[15] Chantzis, A., Kowalska, J. K., Maganas, D., DeBeer, S., and Neese, F. (2018) Ab Initio Wave Function-Based Determination of Element Specific Shifts for the Efficient Calculation of X-ray Absorption Spectra of Main Group Elements and First Row Transition Metals, *J Chem Theory Comput 14*, 3686–3702.
[16] Maganas, D., Kowalska, J. K., Nooijen, M., DeBeer, S., and Neese, F. (2019) Comparison of multireference ab initio wavefunction methodologies for X- ray absorption edges: A case study on [Fe(II/III)Cl-4](2-/1-) molecules, *J Chem Phys 150*.
[17] Gauthier, C., Sole, VA., Signorato, R., Goulon, J., and Moguiline, E. The ESRF beamline ID26: X-ray absorption on ultra dilute sample. J Synchrot Radiat 1999, 6, 164–6.
[18] Latimer, MJ., Ito, K., McPhillips, SE., and Hedman, B. Integrated instrumentation for combined polarized single-crystal XAS and diffraction data acquisition for biological applications. J Synchrot Radiat 2005, 12, 23–7.
[19] Lancaster, KM., Hu, YL., Bergmann, U., Ribbe, MW., and DeBeer, S. X-ray spectroscopic observation of an interstitial carbide in NifEN-bound FeMoco precursor. J Am Chem Soc 2013, 135, 610–2.
[20] Lancaster, KM., Roemelt, M., Ettenhuber, P. et al. X-ray emission spectroscopy evidences a central carbon in the nitrogenase iron-molybdenum cofactor. Science 2011, 334, 974–7.

[21] Leidel, N., Chernev, P., Havelius, KGV., Schwartz, L., Ott, S., and Haumann, M. Electronic structure of an [FeFe] hydrogenase model complex in solution revealed by X-ray absorption spectroscopy using narrow-band emission detection. J Am Chem Soc 2012, 134, 14142–57.
[22] Pushkar, Y., Long, X., Glatzel, P. et al. Direct detection of oxygen ligation to the Mn4Ca cluster of photosystem II by XES. Angew Chem Int Ed 2010, 49, 800–3.
[23] Rees, JA., Bjornsson, R., Kowalska, JK. et al. Comparative electronic structures of nitrogenase FeMoco and FeVco. Dalton T 2017, 46, 2445–55.
[24] Rees, JA., Bjornsson, R., Schlesier, J., Sippel, D., Einsle, O., and DeBeer, S. The Fe-V cofactor of vanadium nitrogenase contains an interstitial carbon atom. Angew Chem Int Edit 2015, 54, 13249–52.
[25] Martin-Diaconescu, V., Chacon, KN., Delgado-Jaime, MU. et al. K beta valence to core X-ray emission studies of Cu(I) binding proteins with mixed methionine – histidine coordination. relevance to the reactivity of the M- and H-sites of peptidylglycine monooxygenase. Inorg Chem 2016, 55, 3431–9.
[26] Bjornsson, R., Lima, FA., Spatzal, T. et al. Identification of a spin-coupled Mo(III) in the nitrogenase iron-molybdenum cofactor. Chem Sci 2014, 5, 3096–103.
[27] Pollock, CJ., and DeBeer, S. Insights into the geometric and electronic structure of transition metal centers from valence-to-core X-ray emission spectroscopy. Accounts Chem Res 2015, 48, 2967–75.
[28] Mathe, Z., Pantazis, D.A., Lee, H.B., Gnewkow, R., Van Kuiken, B., Agapie, T., DeBeer, S. (2019). Calcium Valence-to-Core X-ray Emission Spectroscopy: A Sensitive Probe of Oxo Protonation in Structural Models of the Oxygen-Evolving Complex *Inorganic Chemistry* https://doi.org/10.1021/acs.inorgchem.9b02866
[29] Kowalska, JK., Lima, FA., Pollock, CJ., Rees, JA., and DeBeer, S. A practical guide to high-resolution X-ray spectroscopic measurements and their applications in bioinorganic chemistry. Isr J Chem 2016, 56, 803–15.
[30] Glatzel, P., and Bergmann, U. High resolution 1s core hole X-ray spectroscopy in 3d transition metal complexes – electronic and structural information. Coordin Chem Rev 2005, 249, 65–95.
[31] Hall, ER., Pollock, CJ., Bendix, J., Collins, TJ., Glatzel, P., and DeBeer, S. Valence-to-core-detected X-ray absorption spectroscopy: targeting ligand selectivity. J Am Chem Soc 2014, 136, 10076–84.
[32] Glaser, T., Hedman, B., Hodgson, KO., and Solomon, EI. Ligand K-edge X-ray absorption spectroscopy: A direct probe of ligand-metal covalency. Accounts Chem Res 2000, 33, 859–68.
[33] Dey, A., Francis, EJ., Adams, MWW. et al. Solvent tuning of electrochemical potentials in the active sites of HiPIP versus ferredoxin. Science 2007, 318, 1464–8.
[34] Bjornsson, R., Delgado-Jaime, MU., Lima, FA. et al. Molybdenum L-edge XAS spectra of MoFe nitrogenase. Z Anorg Allg Chem 2015, 641, 65–71.
[35] George, SD., Metz, M., Szilagyi, RK. et al. A quantitative description of the ground-state wave function of Cu-A by X-ray absorption spectroscopy: comparison to plastocyanin and relevance to electron transfer. J Am Chem Soc 2001, 123, 5757–67.
[36] Glaser, T., Rose, K., Anxolabehere-Mallart, E. et al. The electronic structures of iron-sulfur clusters in models and proteins studied by SK-edge XAS. J Inorg Biochem 1999, 74, 142–.
[37] Kroll, T., Hadt, RG., Wilson, SA. et al. Resonant inelastic X-ray scattering on ferrous and ferric bis-imidazole porphyrin and cytochrome c: nature and role of the axial methionine-Fe bond. J Am Chem Soc 2014, 136, 18087–99.

[38] Henthorn, J. T., Arias, R. J., Koroidov, S., Kroll, T., Sokaras, D., Bergmann, U., Rees, D. C., and DeBeer, S. (2019) Localized Electronic Structure of Nitrogenase FeMoco Revealed by Selenium K-Edge High Resolution X-ray Absorption Spectroscopy, *J Am Chem Soc 141*, 13676–13688.

[39] Westre, TE., Kennepohl, P., DeWitt, JG., Hedman, B., Hodgson, KO., and Solomon, EI. A multiplet analysis of Fe K-edge 1s->3d pre-edge features of iron complexes. J Am Chem Soc 1997, 119, 6297–314.

[40] DuBois, JL., Mukherjee, P., Stack, TDP., Hedman, B., Solomon, EI., and Hodgson, KO. A systematic K-edge X-ray absorption spectroscopic study of Cu(III) sites. J Am Chem Soc 2000, 122, 5775–87.

[41] Chandrasekaran, P., Stieber, SCE., Collins, TJ., Que, L., Neese, F., and DeBeer, S. Prediction of high-valent iron K-edge absorption spectra by time-dependent density functional theory. Dalton T 2011, 40, 11070–9.

[42] Berry, JF., Bill, E., Bothe, E. et al. An octahedral coordination complex of iron(VI). Science 2006, 312, 1937–41.

[43] Penner-Hahn, JE. X-ray absorption spectroscopy in coordination chemistry. Coordin Chem Rev 1999, 190, 1101–23.

[44] Van Stappen, C., Thorhallsson, A. T., Decamps, L., Bjornsson, R., and DeBeer, S. (2019) Resolving the structure of the E-1 state of Mo nitrogenase through Mo and Fe K-edge EXAFS and QM/MM calculations, *Chem Sci 10*, 9807–9821.

[45] Corbett, MC., Hu, YL., Fay, AW., Ribbe, MW., Hedman, B., and Hodgson, KO. Structural insights into a protein-bound iron-molybdenum cofactor precursor. P Natl Acad Sci USA 2006, 103, 1238–43.

[46] George, SJ., Barney, BM., Mitra, D. et al. EXAFS and NRVS reveal a conformational distortion of the FeMo-cofactor in the MoFe nitrogenase propargyl alcohol complex. J Inorg Biochem 2012, 112, 85–92.

[47] Sagi, I., Wirt, M., Chen, E., Frisbie, S., and Chance, M. Structures of intermediates of coenzyme-B12 catalysis. Biophys J 1990, 57, A48–A.

[48] Sagi, I., Wirt, MD., Chen, EF., Frisbie, S., and Chance, MR. Structure of an intermediate of coenzyme-B12 catalysis by EXAFS – cobalt(Ii)-B12. J Am Chem Soc 1990, 112, 8639–44.

[49] Wirt, M., Sagi, I., Chen, E., Frisbie, S., and Chance, M. X-Ray edge spectroscopy of cobalt (I, Ii, Iii) B12. Biophys J 1990, 57, A49–A.

[50] Hedman, B., Frank, P., Gheller, SF., Roe, AL., Newton, WE., and Hodgson, KO. New structural insights into the iron molybdenum cofactor from Azotobacter-vinelandii nitrogenase through sulfur-K and molybdenum-L X-ray absorption-edge studies. J Am Chem Soc 1988, 110, 3798–805.

[51] Kowalska, JK., Hahn, AW., Albers, A. et al. X-ray absorption and emission spectroscopic studies of [L2Fe2S2](n) model complexes: implications for the experimental evaluation of redox states in iron-sulfur clusters. Inorg Chem 2016, 55, 4485–97.

[52] Gilbert, B., Frazer, BH., Zhang, H. et al. X-ray absorption spectroscopy of the cubic and hexagonal polytypes of zinc sulfide. Phys Rev B 2002, 66.

[53] Kapre, R., Ray, K., Sylvestre, I. et al. Molecular and electronic structures of oxo-bis(benzene-1,2-dithiolato)chromate(V) monoanions. a combined experimental and density functional study. Inorg Chem 2006, 45, 3499–509.

[54] Ray, K., George, SD., Solomon, EI., Wieghardt, K., and Neese, F. Description of the ground-state covalencies of the bis(dithiolato) transition-metal complexes from X-ray absorption spectroscopy and time-dependent density-functional calculations. Chem-Eur J 2007, 13, 2783–97.

[55] Pollock, CJ., Tan, LL., Zhang, W., Lancaster, KM., Lee, SC., and DeBeer, S. Light-atom influences on the electronic structures of iron sulfur clusters. Inorg Chem 2014, 53, 2591–7.

[56] Krause, MO., and Oliver, JH. Natural widths of atomic K-levels and L-levels, K-alpha X-ray-lines and several Kll auger lines. J Phys Chem Ref Data 1979, 8, 329–38.
[57] Hocking, R. K., Wasinger, E. C., de Groot, F. M. F., Hodgson, K. O., Hedman, B., and Solomon, E. I. (2006) Fe L-edge XAS studies of K-4[Fe(CN)(6)] and K-3[Fe(CN)(6)]: A direct probe of back-bonding, *J Am Chem Soc 128*, 10442–10451.
[58] van Schooneveld, MM., and DeBeer, S. A close look at dose: toward L-edge XAS spectral uniformity, dose quantification and prediction of metal ion photoreduction. J Electron Spectrosc Relat Phenom 2015, 198, 31–56.
[59] Yano, J., Kern, J., Irrgang, KD. et al. X-ray damage to the Mn4Ca complex in single crystals of photosystem II: A case study for metalloprotein crystallography. P Natl Acad Sci USA 2005, 102, 12047–52.
[60] Lassalle-Kaiser, B., Boron, TT., Krewald, V. et al. Experimental and computational X-ray emission spectroscopy as a direct probe of protonation states in oxo-bridged MnIV dimers relevant to redox-active metalloproteins. Inorg Chem 2013, 52, 12915–22.
[61] Messinger, J., Robblee, JH., Bergmann, U. et al. Absence of Mn-centered oxidation in the S-2 -> S-3 Transition: Implications for the mechanism of photosynthetic water oxidation. J Am Chem Soc 2001, 123, 7804–20.
[62] Leidel, N., Chernev, P., Havelius, KGV., Ezzaher, S., Ott, S., and Haumann, M. Site-selective X-ray spectroscopy on an asymmetric model complex of the [FeFe] hydrogenase active site. Inorg Chem 2012, 51, 4546–59.
[63] Schuth, N., Mebs, S., Gehring, H. et al. Biomimetic mono- and dinuclear Ni(I) and Ni(II) complexes studied by X-ray absorption and emission spectroscopy and quantum chemical calculations. J Phys Conf Ser 2016, 712.
[64] Lee, N., Petrenko, T., Bergmann, U., Neese, F., and DeBeer, S. Probing valence orbital composition with iron K beta X-ray emission spectroscopy. J Am Chem Soc 2010, 132, 9715–27.
[65] Pollock, CJ., Delgado-Jaime, MU., Atanasov, M., Neese, F., and DeBeer, S. K beta mainline X-ray emission spectroscopy as an experimental probe of metal-ligand covalency. J Am Chem Soc 2014, 136, 9453–63.
[66] Pollock, CJ., and DeBeer, S. Valence-to-Core X-ray emission spectroscopy: a sensitive probe of the nature of a bound ligand. J Am Chem Soc 2011, 133, 5594–601.
[67] Delgado-Jaime, MU., Dible, BR., Chiang, KP. et al. Identification of a single light atom within a multinuclear metal cluster using valence-to-core X-ray emission spectroscopy. Inorg Chem 2011, 50, 10709–17.
[68] Anklamm, L., Schlesiger, C., Malzer, W., Grotzsch, D., Neitzel, M., and Kanngiesser, B. A novel von Hamos spectrometer for efficient X-ray emission spectroscopy in the laboratory. Rev Sci Instrum 2014, 85.
[69] Mortensen, DR., Seidler, GT., Ditter, AS., and Glatzel, P. Benchtop nonresonant X-ray emission spectroscopy: coming soon to laboratories and XAS beamlines near you?. J Phys Conf Ser 2016, 712.
[70] DeBeer, S., and Bergmann, U. X-ray emission spectroscopic techniques in bioinorganic applications, Scott RA, ed., Encyclopedia of inorganic and bioinorganic chemistry, John Wiley, Chichester, 2016.
[71] Malzer, W., Grotzsch, D., Gnewkow, R., Schlesiger, C., Kowalewski, F., Van Kuiken, B., DeBeer, S., and Kanngiesser, B. (2018) A laboratory spectrometer for high throughput X-ray emission spectroscopy in catalysis research, *Rev Sci Instrum 89*.
[72] Szlachetko, J., Nachtegaal, M., de Boni, E. et al. A von Hamos x-ray spectrometer based on a segmented-type diffraction crystal for single-shot x-ray emission spectroscopy and time-resolved resonant inelastic x-ray scattering studies. Rev Sci Instrum 2012, 83.

[73] Glatzel, P., Yano, J., Bergmann, U. et al. Resonant inelastic X-ray scattering (RIXS) spectroscopy at the MnK absorption pre-edge – a direct probe of the 3d orbitals. J Phys Chem Solids 2005, 66, 2163–7.

[74] Hamalainen, K., Siddons, DP., Hastings, JB., and Berman, LE. Elimination of the inner-shell lifetime broadening in X-ray-absorption spectroscopy. Phys Rev Lett 1991, 67, 2850–3.

[75] Castillo, RG., Banerjee, R., Allpress, CJ. et al. High-energy-resolution fluorescence-detected X-ray absorption of the Q intermediate of soluble methane monooxygenase. J Am Chem Soc 2017, 139, 18024–33.

[76] Cutsail, G. E., Banerjee, R., Zhou, A., Que, L., Lipscomb, J. D., and DeBeer, S. (2018) High-Resolution Extended X-ray Absorption Fine Structure Analysis Provides Evidence for a Longer Fe center dot center dot center dot Fe Distance in the Q Intermediate of Methane Monooxygenase, *J Am Chem Soc 140*, 16807–16820.

[77] Lambertz, C., Chernev, P., Klingan, K., Leidel, N., Sigfridsson, K. G. V., Happe, T., and Haumann, M. (2014) Electronic and molecular structures of the active-site H-cluster in [FeFe]-hydrogenase determined by site-selective X-ray spectroscopy and quantum chemical calculations, *Chem Sci 5*, 1187–1203.

[78] Maganas, D., DeBeer, S., and Neese, F. A restricted open configuration interaction with singles method to calculate valence-to-core resonant X-ray emission spectra: a case study. Inorg Chem 2017, 56, 11819–36.

[79] Van Kuiken, BE., Hahn, AW., Maganas, D., and DeBeer, S. Measuring Spin-Allowed and Spin-forbidden d-d excitations in vanadium complexes with 2p3d resonant inelastic X-ray scattering. Inorg Chem 2016, 55, 11497–501.

[80] Hahn, AW., Van Kuiken, BE., Al Samarai, M. et al. Measurement of the ligand field spectra of ferrous and ferric iron chlorides using 2p3d RIXS. Inorg Chem 2017, 56, 8203–11.

[81] Van Kuiken, B. E., Hahn, A. W., Nayyar, B., Schiewer, C. E., Lee, S. C., Meyer, F., Weyhermuller, T., Nicolaou, A., Cui, Y. T., Miyawaki, J., Harada, Y., and DeBeer, S. (2018) Electronic Spectra of Iron-Sulfur Complexes Measured by 2p3d RIXS Spectroscopy, *Inorg Chem 57*, 7355–7361.

[82] Hahn, A. W., Van Kuiken, B. E., Chilkuri, V. G., Levin, N., Bill, E., Weyhermuller, T., Nicolaou, A., Miyawaki, J., Harada, Y., and DeBeer, S. (2018) Probing the Valence Electronic Structure of Low-Spin Ferrous and Ferric Complexes Using 2p3d Resonant Inelastic X-ray Scattering (RIXS), *Inorg Chem 57*, 9515–9530.

[83] Chen, LX., Zhang, X., and Shelby, ML. Recent advances on ultrafast X-ray spectroscopy in the chemical sciences. Chem Sci 2014, 5, 4136–52.

[84] Chergui, M. Time-resolved X-ray spectroscopies of chemical systems: New perspectives. Struct Dynam-US 2016;3.

Index

A-cluster 286
Acetyl-coenzyme A synthase (ACS) 6, 285
adenosylcobamide-dependent enzymes 256
Ammonia 159, 196, 255
antimalarial drugs 337, 338
Antivitamins B_{12} 264, 270
ATP 48, 75, 78, 118, 159, 172, 270, 289, 295
Auranofin 325
Autotrophic 147, 286, 290
Avalanche photodiode (APD) 374
2-aza-propane-1,3-ditholate bridge (adt) 16

B_{12} derivatives 6, 243
B_{12}-aptamers 265
B_{12}-binding oligonucleotides 265
B_{12}-riboswitches 265
back-bonding 66, 137, 150
back-donating 66
Base-off 247
Base-on 247
Bis(μ-oxo)diiron(IV) 228

Carbide 4, 164, 168, 183, 422
Carbon monoxide dehydrogenase (CODH) 1, 54, 87, 113
Carbon monoxide releasing molecules (CORMs) 341
Chemiosmotic potential 48
Chemolithoautotrophy 286
Chlamydomonas reinhardtii (Cr) 17, 20
chloroquine 337, 338, 339
Cisplatin 320, 327, 329, 345
Clostridium pasteurianum (Cp) 3, 13, 22, 177
CO inhibition 68, 299
Cobalt corrin 243, 250
Corrinoids 243
Cu,Mo-containing CODHs 140
Cyanocobalamin 243
cyclam 154
Cytosolic bidirectional [NiFe] hydrogenases 50
Cytosolic H_2-uptake [NiFe] hydrogenases 50

Dehalogenase 262
density function theory (DFT) calculations 29, 116, 215, 303, 358, 378
Desulfovibrio desulfuricans (Dd) 3, 13
Diamagnetic 151, 171, 186, 212, 253, 298, 309
Diironperoxo complexes 228
Distal 14, 30, 34, 64, 143, 195, 294

electron paramagnetic resonance spectroscopy (EPR) 26, 55, 57, 58, 70, 151, 164, 170, 174, 187, 191, 212, 229, 297, 340, 384, 395
ENDOR 22, 165, 171, 193
Escherichia coli (E. coli) 20, 79, 113, 127, 182, 197, 258, 295
EXAFS 212, 223, 303, 361, 381, 411

Fe-nitrogenase 161, 167, 197
Ferredoxin 51, 143, 210, 289, 382, 384
Ferrocifens 331
Ferroquine 339
flavin adenine dinucleotide (FAD) 142
frustrated Lewis pair 16, 35, 62
FTIR spectroscopy 26

GTP 17, 78, 119
guanylylpyridinol 106, 107, 110

H-cluster 2, 7, 14, 18, 25, 29, 125
Haber–Bosch reaction 159
Heterobimetallic 54, 74, 81, 86
heterocubanes 153
Heteroleptic 83
Heterotrophic 286
Hieber base reaction 3
High-spin 58, 212, 230, 387, 403, 421
Homoleptic 83
HydA, HydE, HydF, HydG 17
Hydride 2, 25, 28, 33, 35, 61, 66, 84, 106, 138, 194
Hydrogen (H_2) conversion 1, 22, 46, 52, 59
Hydrogen (H_2) heterolysis 61
Hydrogen (H_2) homolysis 61
Hydrogen formation 35
Hydrogen oxidation 2, 24, 29
Hydrogenases 1, 13, 45, 106, 141, 303
hydrothermal vents 74
[Fe] hydrogenase 50, 106
[FeFe] hydrogenases 7, 13, 53, 64, 66

https://doi.org/10.1515/9783110496574-012

[NiFe] hydrogenases 3, 45
Hyp 78

iron-guanylylpyridinol (FeGP) cofactor 107
Iron-sulfur cluster 17, 108
Iron-Sulfur World 286, 287, 311

K-edge XAS 186, 190, 191, 408
Kinase 51, 68, 323, 324
Knallgas reaction 47

L-cluster 160, 182, 387
Lamb–Mössbauer factors 360
Low-spin 55, 62, 112, 250, 295, 417, 421
Lowe–Thorneley reaction model 193
LUCA 74

M-cluster 167, 387
maturase 17, 22
maturation 18, 76, 160, 183, 189
membrane-bound H_2-uptake [NiFe] hydrogenases 50
metal–dihydrogen σ-bond complex 62, 65
Metallocenes 320, 337, 340
Metalloenzymes 1, 64, 106, 112, 126, 219, 311, 353
Methane monooxygenases (MMO) 5, 207
methenyl-tetrahydromethanopterin 106
methicillin-resistant Staphylococcus aureus (MRSA) 335
Methylcobalamin 1, 243, 244, 268, 306
methylene-tetrahydromethanopterin 107
methyltransferases 258
Mo-nitrogenase 4, 5, 160
Monsanto acetic acid process 6
Mössbauer effect 165, 353, 367
Mössbauer spectroscopy 7, 22, 112, 151, 165, 175, 212, 224, 353
Multiphonon event 359, 360

N-heterocyclic carbene (NHC) 320, 333
NADH 51, 70, 75, 208, 210, 289
NADPH 74, 269
Ni,Fe-containing CODHs 147
nif-gene 160
NifDK 4, 160, 167
NifH 4, 160
nitrogen fixation 1, 50, 159
nitrogenase 160
Nuclear hyperfine interaction 403
nuclear resonance vibrational spectroscopy (NRVS) 7, 25, 28, 30, 353

O_2-tolerant [NiFe] hydrogenases 48, 67, 70, 77, 87

P-cluster 160, 387, 423
Paramagnetic 55, 151, 164, 245, 253, 298, 309, 395
Partial vibrational density of states (PVDOS) 362
photodynamic therapy 327
photosynthetic 87, 257
Photosystem 80, 407, 425
Positron emission tomography (PET) 343
prebiotic catalysis 74
primordial soup 286
proton-coupled electron transfer (PCET) 27, 29, 155
Proximal 14, 53, 63, 66, 70, 72, 143, 293, 381

Radical trap 250
Radioactive complexes 6, 301, 321, 343
Raman 213, 224, 357, 382
RAPTA 327, 328
RIXS 408, 420, 423
RNA world 286
Rotated state 7, 16, 22
Rubredoxin 62, 81, 82, 85, 380, 407

Salvarsan 334, 335
SAM-dependent radical methyl transferases 260
Selenium 24, 325
Selenocysteine 54, 325, 326
Single-photon emission computed tomography (SPECT) 343
structure–activity relationship (SAR) 328
Sulfoxygenation 73
Surface metabolist 287

Tetrahydrofolate 77, 140, 148, 258, 288
Thiocyanate 76, 145
thioredoxin reductases (TrxR) 325

V-nitrogenase 161, 167, 168, 170, 197
Vitamin B_{12} 1, 243

Water gas shift reaction 4, 87, 138
water splitting 87
Wood– Ljungdahl pathway 140

X-ray absorption fine structure (EXAFS) 212, 223, 229, 303, 361, 373, 381, 407
X-ray absorption near-edge structure (XANES) 408, 409
X-ray absorption spectroscopy (XAS) 145, 163, 164, 173, 186, 190, 303, 407
X-ray crystallography 24, 55, 117, 119, 145, 162, 395
X-ray emission spectroscopy 164, 407, 420
X-ray-based spectroscopies 7, 407
XANES 409

π-basicity 137

www.ingramcontent.com/pod-product-compliance
Lightning Source LLC
Chambersburg PA
CBHW080136220326
41598CB00032B/5082
9783110496505